CONTEMPORARY MATHEMATICS

708

An Alpine Bouquet of Algebraic Topology

Alpine Algebraic and
Applied Topology Conference
August 15–21, 2016
Saas-Almagell, Switzerland

Christian Ausoni
Kathryn Hess
Brenda Johnson
Ieke Moerdijk
Jérôme Scherer
Editors

AMS AMERICAN MATHEMATICAL SOCIETY

2010 *Mathematics Subject Classification.* Primary 11S40, 18Axx, 18Gxx, 19Dxx, 55Nxx, 55Pxx, 55Sxx, 55Uxx, 57Nxx.

Library of Congress Cataloging-in-Publication Data

Names: Conference on Alpine Algebraic and Applied Topology (2016: Saas-Almagell, Switzerland) | Ausoni, Christian, 1968– editor.

Title: An alpine bouquet of algebraic topology: Conference on Alpine Algebraic and Applied Topology, August 15–21, 2016, Saas-Almagell, Switzerland / Christian Ausoni [and four others], editors.

Description: Providence, Rhode Island: American Mathematical Society, [2018] | Series: Contemporary mathematics; volume 708 | Includes bibliographical references.

Identifiers: LCCN 2017055549 | ISBN 9781470429119 (alk. paper)

Subjects: LCSH: Algebraic topology–Congresses. | AMS: Number theory – Algebraic number theory: local and p-adic fields – Zeta functions and L-functions. msc | Category theory; homological algebra – General theory of categories and functors – General theory of categories and functors. msc | Category theory; homological algebra – Homological algebra – Homological algebra. msc | K-theory – Higher algebraic K-theory – Higher algebraic K-theory. msc | Algebraic topology – Homology and cohomology theories – Homology and cohomology theories. msc | Algebraic topology – Homotopy theory – Homotopy theory. msc | Algebraic topology – Operations and obstructions – Operations and obstructions. msc | Algebraic topology – Applied homological algebra and category theory – Applied homological algebra and category theory. msc | Manifolds and cell complexes – Topological manifolds – Topological manifolds. msc

Classification: LCC QA612 .C5927 2018 | DDC 514/.2–dc23

LC record available at https://lccn.loc.gov/2017055549

DOI: http://dx.doi.org/10.1090/conm/708

∞ The paper used in this book is acid-free and falls within the guidelines
established to ensure permanence and durability.
Visit the AMS home page at http://www.ams.org/

10 9 8 7 6 5 4 3 2 1   23 22 21 20 19 18

Contents

Preface

After four conferences held in Arolla, it was time to search for new landscapes and summits. This 2016 edition of the Alpine conferences in algebraic topology took place farther east in the Saas Valley, in the village of Saas-Almagell, which offered many new challenging paths to explore. This led us to new perspectives on rational equivariant homotopy theory, topological Hochschild homology and K-theory, stable and unstable homotopy theory, functor calculus, and an incursion into applied algebraic topology.

Our sincerest thanks go to all of the speakers and participants, whose talks and interest made this conference happen. We are very grateful for the authors, whose results are published here, and thank them for their contributions. Many thanks also to the referees for their anonymous work and to Chris Thivierge at the AMS for her efficient help and guidance.

We would like to thank the following institutions and organizations for their financial support and sponsorship of the 2016 Alpine Algebraic and Applied Topology Conference at Saas:

- U.S. National Science Foundation,
- Foundation Compositio Mathematica,
- Faculté des Sciences de la Vie, École Polytechnique Fédérale de Lausanne,
- Conférence Universitaire de Suisse Occidentale,
- Centre National de la Recherche Scientifique, and
- Université Sorbonne Paris Cité.

This Saas edition of the Alpine topology conferences was hosted by the Hotel Kristall Saphir in Saas-Almagell. We express our gratitude to Mrs. Rosine Breuer and her family and staff for not sparing any effort in making 70 exuberant topologists feel at home. It was a pleasure organizing this conference with their help.

Ch. Ausoni, K. Hess, B. Johnson, I. Moerdijk, and J. Scherer

Fluchthorn summit, August 15, 2016

List of Conference Talks

Monday, August 15, 2016

Cristina Costoya	*Kahn's realizability problem*

Tuesday, August 16, 2016

John Greenlees	*Rational equivariant cohomology theories and affine formal covers of the sphere*
David Barnes	*Rational S^1-equivariant ring spectra*
Christian Wimmer	*Rational global homotopy theory*
Adeel Khan	*A^1-homotopy invariance in spectral algebraic geometry*
Vesna Stojanoska	*Galois extensions of motivic ring spectra*
Nitu Kitchloo	*Landweber flat real pairs and $ER(n)$-cohomology*
Tse Leung So	*Homotopy types of gauge groups over 4-manifolds*

Wednesday, August 17, 2016

Vincent Franjou	*Lannes' T vs Harish-Chandra restriction*
Anssi Lathinen	*Modular characteristic classes for representations over finite fields*
Marc Hoyois	*On the vanishing of negative equivariant K-theory*
Steffen Sagave	*Logarithmic topological Hochschild homology*
Lars Hesselholt	*Topological Hochschild homology and the Hasse-Weil zeta function*

Thursday, August 18, 2016

Agnès Beaudry	*The $K(2)$-local Picard group at $p = 2$*
Tilman Bauer	*Chromatic unstable homotopy, plethories, and the Dieudonné correspondence*
David Chataur	*Intersection homotopy theory*

Friday, August 19, 2016

Vidit Nanda	*The discrete flow category*
Wojciech Chachólski	*Multi-persistence*
Ran Levi	*Topological analysis of neural networks*
Mark Grant	*The Poincaré-Hopf theorem for line fields (revisited)*
Wolfgang Pitsch	*Relative homological algebra via truncations*

Saturday, August 20, 2016

Pedro Boavida de Brito	*Configuration categories and embedding spaces*
Emanuele Dotto	*Equivariant calculus and the tower of the identity on pointed G-spaces*
Kristine Bauer	*A combinatorial model for certain Taylor towers*
Marc Stephan	*Categorical suspension and stable Postnikov data*
Sune Precht Reeh	*Saturated fusion systems as stable retracts of groups*

List of Participants

Arlettaz, Dominique
University of Lausanne

Ausoni, Christian
University Paris 13

Baker, Andrew
University of Glasgow

Bandklayder, Lauren
Northwestern University

Barnes, David
Queen's University Belfast

Bauer, Kristine
University of Calgary

Bauer, Tilman
KTH

Beaudry, Agnès
University of Chicago

Boavida de Brito, Pedro
University of Lisbon

Bobkova, Irina
University of Rochester

Chachólski, Wojciech
KTH

Chataur, David
UPJV

Ching, Michael
Amherst College

Costoya, Cristina
Universidade da Coruña

Dłotko, Paweł
Inria, Saclay

Dotto, Emanuele
MIT

Dundas, Bjørn Ian
University of Bergen

Elmendorf, Tony
Purdue University Calumet

Filakovsky, Marek
IST Austria

Franjou, Vincent
University of Nantes

Fresse, Benoit
University of Lille

Grant, Mark
University of Aberdeen

Greenlees, John
University of Sheffield

Harper, John E.
Ohio State University

Hess, Kathryn
EPFL

Hesselholt, Lars
Nagoya University/University of Copenhagen

Hiroshi, Naruse
University of Yamanashi

Hoyois, Marc
MIT

Jeitziner, Rachel
EPFL

Joachim, Michael
University of Muenster

Johnson, Brenda
Union College

Kędziorek, Magdalena
EPFL

Khan, Adeel Yusufzai
University Duisburg-Essen

Kitchloo, Nitu
Johns Hopkins University

Lahtinen, Anssi
University of Hamburg

Lesh, Kathryn
Union College

Levi, Ran
University of Aberdeen

Merrill, Leanne
University of Oregon

Mischaikow, Konstantin
Rutgers University

Moerdijk, Ieke
Utrecht University

Mycroft, William
University of Sheffield

Nakade, Apurva
Johns Hopkins University

Nanda, Vidit
University of Pennsylvania

Oliver, Robert
University Paris 13

Pitsch, Wolfgang
Universidad Autonoma de Barcelona

Recio Mitter, David
University of Aberdeen

Reeh, Sune Precht
MIT

Richter, Birgit
University of Hamburg

Rovelli, Martina
EPFL

Sagave, Steffen
Radboud University Nijmegen

Scheinmann, Nathan
EPFL

Scherer, Jérôme
EPFL

Scolamiero, Martina
EPFL

So, Tse Leung
University of Southampton

Spreemann, Gard
EPFL

Stephan, Marc
University of Chicago

Stojanoska, Vesna
University of Illinois at Urbana-Champaign

Stonek, Bruno
University Paris 13

Upmeier, Markus
University of Augsburg

VanKoughnett, Paul
Northwestern University

Verrette, Jean
EPFL

Viruel, Antonio
University of Malaga

Vokřínek, Lukáš
Masaryk University

Werndli, Kay
EPFL

Wimmer, Christian
University of Bonn

Woo, Calvin
Indiana University

Yeakel, Sarah
University of Illinois at Urbana-Champaign

Zaganidis, Dimitri
EPFL

Ziegenhagen, Stephanie
KTH

Zielenkiewicz, Magdalena
University of Warsaw

Contemporary Mathematics
Volume **708**, 2018
http://dx.doi.org/10.1090/conm/708/14265

Characteristics for $\mathcal{E}_\infty$ ring spectra

Andrew Baker

ABSTRACT. We introduce a notion of characteristic for connective p-local $\mathcal{E}_\infty$ ring spectra and study some basic properties. Apart from examples already pointed out by Markus Szymik, we investigate some examples built from Hopf invariant 1 elements in the stable homotopy groups of spheres and make a series of conjectures about spectra for which they may be characteristics; these appear to involve hard questions in stable homotopy theory.

CONTENTS

Introduction

In ordinary ring theory, the characteristic of a unital ring is really part of the structure, although often not introduced in elementary courses except in the context of fields. Less standard is a generalisation of the notion to algebras over a commutative ring and we discuss this in Section 1. The main aim of this note is introduce an appropriate notion of characteristic for derived commutative rings, at least in the topological context of commutative S-algebras (also known as $\mathcal{E}_\infty$ ring spectra). Our approach could be extended to other versions of derived commutative rings such as simplicial commutative algebras or $\mathcal{E}_\infty$-algebras over a fixed commutative ring, but we focus on the topological version.

In [**16, 17**], Markus Szymik introduced a notion of characteristic for a commutative S-algebra. We consider what properties a more general notion of characteristic

2010 *Mathematics Subject Classification.* Primary 55P43; Secondary 55P42, 55P48.

Key words and phrases. Stable homotopy theory, $\mathcal{E}_\infty$ ring spectrum, commutative S-algebra, cell algebra, power operations.

The mathematics described in this paper is partly based upon work supported by the National Science Foundation under Grant No. 0932078 000 while the author was in residence at the Mathematical Sciences Research Institute in Berkeley California, during the Spring 2014 semester.

might be expected to have in this setting, at least for connective algebras localised at a prime. We do not attempt to work in the chromatic setting since we rely on the theory of minimal atomic commutative S-algebras which does not seem to extend to such an intrinsically non-connective context.

As well as setting up a general notion of characteristic, we discuss possible candidates for characteristics of some important standard examples, and state conjectures which appear to involve non-trivial questions in stable homotopy theory. One possible approach to proving these might involve old work of Joel Cohen [**11**], however, to date we have been unable to carry out such a programme. Further related ideas are explored in [**4**].

We will assume the reader is familiar with the basic theory of $\mathcal{E}_\infty$ ring spectra in their avatar as commutative S-algebras [**12**], discussions of cellular aspects can be found in [**1, 2, 5**].

1. Motivation: characteristics in algebra

If R is a (not necessarily commutative) ring with unity $1 \neq 0$, there is a ring homomorphism $\eta_R\colon \mathbb{Z} \to R$ called the *unit* or *characteristic* homomorphism, defined by

$$\eta_R(n) = n1 = \begin{cases} \underbrace{1 + \cdots + 1}_{n} & \text{if } n > 0, \\ -(\underbrace{1 + \cdots + 1}_{-n}) & \text{if } n < 0, \\ 0 & \text{if } n = 0. \end{cases}$$

Since $1 \in R$ is non-zero, $\ker \eta_R$ is a proper ideal of $\mathbb{Z}$ and there is a quotient monomorphism $\overline{\eta}_R\colon \mathbb{Z}/\ker \eta_R \to R$ which allows us to identify the quotient ring $\mathbb{Z}/\ker \eta_R$ with image $\eta_R\mathbb{Z} \subseteq R$, the *characteristic subring* of R. Thus there is a unique non-negative integer $\operatorname{char} R \geqslant 0$ such that $\ker \eta_R = (\operatorname{char} R) \lhd \mathbb{Z}$, and this is called the *characteristic* of R.

We can generalise this to unital $\Bbbk$-algebras over a commutative ring $\Bbbk$. Here $\Bbbk$-algebra is used in its most general sense: a $\Bbbk$-*algebra* A is a unital ring equipped with a unital homomorphism $\eta_A\colon \Bbbk \to A$ whose image is central. The ideal $\ker \eta_A \lhd \Bbbk$ is not necessarily principal, and the quotient homomorphism $\overline{\eta}_A\colon \Bbbk/\ker \eta_A \to A$ defines what might reasonably be called the *characteristic subalgebra* of A. This construction is functorial with respect to $\Bbbk$-algebra homomorphisms, i.e., given an algebra homomorphism $\varphi\colon A \to B$, there is a commutative diagram

$$\begin{array}{ccc} \Bbbk/\ker \eta_A & \xrightarrow{\varphi_0} & \Bbbk/\ker \eta_B \\ {\scriptstyle \overline{\eta}_A}\big\downarrow & & \big\downarrow{\scriptstyle \overline{\eta}_B} \\ A & \xrightarrow{\varphi} & B \end{array}$$

and in particular, if φ is an isomorphism, so is φ_0.

The latter generalisation seems natural and we use it as motivation for our discussion of the analogue for commutative S-algebras. In that context there do not appear to be obvious notions of ideals or quotient objects (see [**13**] for recent work on related questions), so care is needed in making suitable definitions. There are some features of a notion of characteristic in that setting which seem desirable, in particular some kind of functoriality and homotopy invariance. We tacitly assume that a characteristic of a commutative S-algebra R should be a factorisation of its unit $\iota_R\colon S \to R$ of the form

$$S \xrightarrow{\iota_{R_0}} R_0 \xrightarrow{\bar{\iota}_R} R \qquad (\text{composite } \iota_R)$$

where $\bar{\iota}_R$ is a morphism of commutative S-algebras. Motivated by obvious functoriality properties of characteristics for $\Bbbk$-algebras, the following basic properties appear to be reasonable requirements.

Functoriality: If $f\colon R \to R'$ is a morphism of commutative S-algebras, then there is a diagram of commutative S-algebras

$$\begin{array}{ccc} R_0 & \xrightarrow{f_0} & R'_0 \\ {\scriptstyle \bar{\iota}_R}\downarrow & & \downarrow{\scriptstyle \bar{\iota}_{R'}} \\ R & \xrightarrow{f} & R' \end{array}$$

which is homotopy commutative.

Homotopy invariance: If $g\colon R \xrightarrow{\sim} R''$ is a weak equivalence of commutative S-algebras then there is a diagram of commutative S-algebras

$$\begin{array}{ccc} R_0 & \underset{\sim}{\xrightarrow{g_0}} & R''_0 \\ {\scriptstyle \bar{\iota}_R}\downarrow & & \downarrow{\scriptstyle \bar{\iota}_{R''}} \\ R & \underset{\sim}{\xrightarrow{g}} & R'' \end{array}$$

which is homotopy commutative.

Functoriality implies that characteristics of homotopy equivalent commutative S-algebras are homotopy equivalent.

We will show that our definition of characteristic does possess homotopy invariance, but does not appear to satisfy functoriality in this sense, but nevertheless it does satisfy a weaker version of this property.

2. Background material on S-modules and commutative S-algebras

We will assume the reader is familiar the framework provided by [**12**], in particular we will work with the simplicial monoidal model category of S-modules $\mathscr{M}_S$ and the associated simplicial model category of commutative S-algebras $\mathscr{C}_S$. Actually we will work with the p-local versions of these for some prime p, and later S will denote the p-local sphere spectrum but no essential differences occur in that setting. We will write $\iota_A\colon S \to A$ for the unit of a commutative S-algebra, which is taken to be part of its structure.

We choose a cofibrant replacement $S^0 \xrightarrow{\sim} S$ for S in model category $\mathscr{M}_S$ of S-modules of [**12**] (for example we could use the functorial cofibrant replacement). We may consider the slice category $S^0/\mathscr{M}_S$ of S-modules under S^0. Every commutative S-algebra A admits a canonical morphism of S-modules

$$S^0 \underset{\sim}{\longrightarrow} S \xrightarrow{\iota_A} A$$

making it an object of $S^0/\mathscr{M}_S$. This gives rise to a functor

$$\widetilde{\mathbb{U}}\colon \mathscr{C}_S \to S^0/\mathscr{M}_S$$

which has a left adjoint

$$\widetilde{\mathbb{P}}\colon S^0/\mathscr{M}_S \to \mathscr{C}_S,$$

the reduced free functor [**1**, section 5] which gives a Quillen adjunction

$$\mathscr{C}_S \underset{\widetilde{\mathbb{U}}}{\overset{\widetilde{\mathbb{P}}}{\leftrightarrows}} S^0/\mathscr{M}_S$$

In $S^0/\mathscr{M}_S$, coproducts are defined using pushouts in $\mathscr{M}_S$, and we use the symbol $\overset{S^0}{\vee}$ to indicate such coproducts. So for $X, Y \in S^0/\mathscr{M}_S$, $X \overset{S^0}{\vee} Y$ is the pushout of $X \leftarrow S^0 \to Y$ in $\mathscr{M}_S$, and since the reduced free algebra functor $\widetilde{\mathbb{P}}\colon \mathscr{M}_S \to \mathscr{C}_S$ is a left adjoint it preserves coproducts, hence

$$\widetilde{\mathbb{P}}(X \overset{S^0}{\vee} Y) \cong \widetilde{\mathbb{P}}X \wedge \widetilde{\mathbb{P}}Y.$$

In the p-local connective setting, we will use ideas on minimal atomic S-modules and commutative S-algebras which may be found in [**5**, **6**, **14**]. In particular, the notions of nuclear CW S-modules and commutative S-algebras will play a central rôle in our work. This depends in turn on the theory of cellular and CW objects [**12**] in $\mathscr{M}_S$ and $\mathscr{C}_S$. When discussing cellular constructions we will refer to multiplicatively defined cell objects built in $\mathscr{C}_S$ using the phrase $\mathcal{E}_\infty$ *cell*, and reserve *cell* for objects built in $\mathscr{M}_S$. Given a cell object X in $\mathscr{M}_S$ we will write $X^{[n]}$ for the n-skeleton, while for a cell object Y in $\mathscr{C}_S$ we will write $Y^{\langle n\rangle}$. Finally, given a morphism f out of a cell object we will write $f^{[n]}$ or $f^{\langle n\rangle}$ for the restriction to the n-skeleton.

Following the Referee's suggestion, we summarise briefly some of the key ideas about minimal atomic S-modules and commutative S-algebras developed in [**5**, **6**]. In the following we switch viewpoints and regard spectra as S-modules.

First we recall from [**6**, section 1] that p-local connective S-module X with $\pi_0(X)$ non-trivial and cyclic is *minimal atomic* if any map $f\colon Y \to X$ for which $\pi_0(f)\otimes\mathbb{F}_p$ and $\pi_n(f)$ $(n \geqslant 0)$ are monomorphisms is a weak equivalence. Then X is minimal atomic if and only if the Hurewicz homomorphism $\mathrm{h}\colon \pi_*(X) \to H_*(X;\mathbb{F}_p)$ is trivial in all positive degrees. Notice that this characterisation does not depend on a choice of CW structure on X; another useful way to characterise minimal atomic spectra is in terms of a special kind of CW structure. Following [**6**, section 2], a CW S-module Z is *nuclear* if $Z^{[0]} = S^0$ and for each $n \geqslant 1$, the attaching map of the $(n+1)$-cells

$$f_n\colon \bigvee_i S^n \to Z^{[n]}$$

satisfies

$$\ker \pi_n(f_n) \subseteq p\,\pi_n\Big(\bigvee_i S^n\Big).$$

Then X is minimal atomic if and only if there is a weak equivalence $f\colon Z \to X$ where Z is a nuclear CW S-module.

Following [**5**, section 3], a p-local connective commutative S-algebra A with $\pi_0(A)$ non-trivial and cyclic is *minimal atomic* if given any morphism of commutative S-algebras $g\colon B \to A$ for which $\pi_0(g) \otimes \mathbb{F}_p$ and $\pi_n(g)$ $(n \geqslant 0)$ are monomorphisms is a weak equivalence. Then A is minimal atomic if and only if the TAQ-Hurewicz homomorphism (induced by the universal derivation $A \to \Omega_S(A)$) $\mathrm{taq}\colon \pi_*(X) \to \mathrm{TAQ}_*(A, S; \mathbb{F}_p)$ is trivial in all positive degrees. There is also a notion of nuclear commutative S-algebra defined in terms of attaching $\mathcal{E}_\infty$ cells. A CW commutative S-algebra C is *nuclear* if $C^{\langle 0\rangle} = S$ and for each $n \geqslant 1$, the attaching map of its $(n+1)$-dimensional $\mathcal{E}_\infty$ cells is induced from a map

$$g_n\colon \bigvee_i S^n \to C^{\langle n\rangle}$$

by passing to the induced morphism of commutative S-algebras $\mathbb{P}\Big(\bigvee_i S^n\Big) \to C^{\langle n\rangle}$ where

$$\ker \pi_n(g_n) \subseteq p\,\pi_n\Big(\bigvee_i S^n\Big).$$

Then A is minimal atomic if and only if there is a weak equivalence $b\colon C \to X$ where C is a nuclear CW commutative S-algebra.

3. Characteristics of connective p-local commutative S-algebras

Let $\iota_R\colon S \to R$ be such a connective commutative S-algebra. We remark that if we started with R being non-connective then we could replace it with its connective cover in our discussion below, so we do not lose anything by assuming connectivity. The induced ring homomorphism $(\iota_R)_*\colon \pi_0(S) \to \pi_0(R)$ could have a non-trivial kernel, and also might not be surjective. If we focus on $\pi_0(-)$ it might seem reasonable to define the characteristic of R to be this kernel. However, this neglects the kernel of $(\iota_R)_*\colon \pi_*(S) \to \pi_*(R)$ in positive degrees. So another definition might be the (graded) kernel of $(\iota_R)_*$. These definitions are closely wedded to the algebra, and instead we propose a different approach which makes the characteristic a commutative S-algebra equipped with a morphism into R.

Conventions: For ease of notation and other simplifications, from now on we work with connective p-local commutative S-algebras A for some fixed rational prime $p > 0$. Thus S is to be interpreted as the p-local sphere spectrum, and we will assume that $\pi_0(A)$ is a cyclic $\mathbb{Z}_{(p)}$-module. The use of *finite-type* is always in the p-local context of p-local cells or $\mathbb{Z}_{(p)}$-modules.

Let R be a connective p-local commutative S-algebra.

DEFINITION 3.1. A *characteristic morphism* of R is a morphism of commutative S-algebras $j\colon T \to R$ where T is a finite-type CW commutative S-algebra, where the $\mathcal{E}_\infty$ skeleta of T are defined inductively using maps of the form

$$\bigvee_i S^n \longrightarrow S \xrightarrow{\iota_{T^{\langle n\rangle}}} T^{\langle n\rangle}, \qquad f^n = \text{composite}$$

factoring through the unit of $T^{\langle n\rangle}$, and which satisfy the conditions

$$\ker[f^n_*\colon \pi_n(\bigvee_i S^n) \to \pi_n(T^{\langle n\rangle})] \subseteq p\,\pi_n(\bigvee_i S^n), \tag{3.1a}$$

$$\operatorname{im} f^n_* = \operatorname{im}[(\iota_{T^{\langle n\rangle}})_*\colon \pi_n(S) \to \pi_n(T^{\langle n\rangle})] \cap \ker[j^{\langle n\rangle}_*\colon \pi_n(T^{\langle n\rangle}) \to \pi_n(R)]. \tag{3.1b}$$

The domain of any characteristic morphism is called a *characteristic* for R.

Note: Condition (3.1a) says that the CW structure on T is *nuclear*, hence T is also *minimal atomic*. Further properties of such commutative S-algebras are discussed in [**5**, section 3].

Of course this definition begs the question of whether such characteristics exist and also whether or not they are in any sense unique. Notice also that the attaching maps of $\mathcal{E}_\infty$ cells all originate as maps into the sphere spectrum S.

LEMMA 3.2. *Let R be a connective p-local commutative S-algebra.*
(a) *Characteristics for R exist.*
(b) *Suppose that $f\colon R \to R'$ is a morphism of commutative S-algebras and that $j\colon T \to R$ and $j'\colon T' \to R$ are characteristic morphisms. Then there is a morphism of commutative S-algebras $T \to T'$.*
(c) *Suppose that T_1 and T_2 are two characteristics for R. Then there is a homotopy equivalence of commutative S-algebras $T_1 \xrightarrow{\simeq} T_2$. Therefore characteristics are unique up to homotopy equivalence.*

PROOF. We will make use of the notation in Definition 3.1.
(a) We can build the skeleta of a nuclear CW commutative S-algebra inductively making sure that conditions of (3.1) are satisfied. In detail, assuming the n-skeleton $T^{\langle n\rangle}$ as been constructed, consider the epimorphism

$$\begin{array}{c}\iota_{T^{\langle n\rangle}}^{-1}\left(\operatorname{im}[(\iota_{T^{\langle n\rangle}})_*\colon \pi_n(S) \to \pi_n(T^{\langle n\rangle})] \cap \ker[j_*^{\langle n\rangle}: \pi_n(T^{\langle n\rangle}) \to \pi_n(R)]\right) \subseteq \pi_n(S)\\ \Big\downarrow \\ \operatorname{im}[(\iota_{T^{\langle n\rangle}})_*\colon \pi_n(S) \to \pi_n(T^{\langle n\rangle})] \cap \ker[j_*^{\langle n\rangle}: \pi_n(T^{\langle n\rangle}) \to \pi_n(R)] \subseteq \pi_n(T^{\langle n\rangle})\end{array}$$

and after choosing a minimal set of generators for the codomain, lift them to elements of the domain. These can be used to form a suitable composition

$$f^n\colon \bigvee_i S^n \longrightarrow S \xrightarrow{\iota_{T^{\langle n\rangle}}} T^{\langle n\rangle}$$

satisfying (3.1). The $(n+1)$-skeleton $T^{\langle n+1\rangle}$ is defined by the following pushout diagram in $\mathscr{C}_S$,

$$\begin{array}{ccc}\mathbb{P}(\bigvee_i S^n) & \longrightarrow & \mathbb{P}(\bigvee_i D^{n+1})\\ {\scriptstyle \widetilde{f^n}}\Big\downarrow & \ulcorner & \Big\downarrow\\ T^{\langle n\rangle} & \longrightarrow & T^{\langle n+1\rangle}\end{array}$$

where $\widetilde{f^n}$ is induced using the freeness of the functor $\mathbb{P} = \mathbb{P}_S\colon \mathscr{M}_S \to \mathscr{C}_S$. The existing morphism $j^{\langle n\rangle}\colon T^{\langle n\rangle} \to R$ extends to a morphism $j^{\langle n+1\rangle}\colon T^{\langle n+1\rangle} \to R$.
(b) We will inductively build compatible morphisms of commutative S-algebras $g^n\colon T^{\langle n\rangle} \to T'$.

Assume that for some $n \geqslant 0$, we have a morphism of commutative S-algebras $g^n\colon T^{\langle n\rangle} \to T'$ making the following diagram of solid arrows commute, where ι

always denotes a suitable unit.

(3.2)

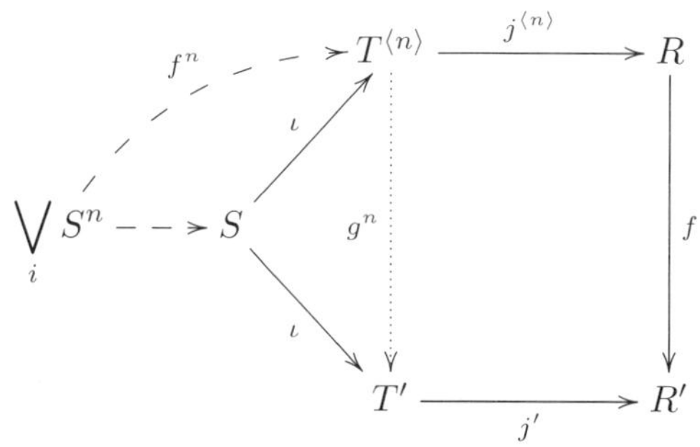

Note that we are not asserting the the right hand square with dotted edge commutes, however the adjacent triangle does. The dashed arrows represent the factorisation of the attaching map f^n of the $\mathcal{E}_\infty$ $(n+1)$-cells of T and the diagram of solid and dashed arrows commutes. By construction, $j^{\langle n\rangle} \circ f^n$ is null homotopic, hence so is the lower composition

$$\bigvee_i S^n \longrightarrow S \xrightarrow{\iota} T' \xrightarrow{j'} R'.$$

Therefore the image of the induced group homomorphism

$$\pi_n(\bigvee_i S^n) \to \pi_n(T')$$

is contained in $\ker[(j')_* \colon \pi_n(T') \to \pi_n(R)]$. It follows that there is an extension of g^n to a morphism of commutative S-algebras $g^{n+1} \colon T^{\langle n+1\rangle} \to T'$. By induction on n and passing to the colimit, we obtain a morphism $g \colon T \to T'$.
(c) We make use of the fact that nuclear complexes are (minimal) atomic; see [**6**, proposition 2.3] and [**5**, theorem 3.4] for the multiplicative case. Using this, it is enough to construct morphisms of commutative S-algebras

$$T_1 \underset{g_2}{\overset{g_1}{\rightleftarrows}} T_2$$

by applying (b) to the identity morphism $R \to R$. Since the compositions $g_2 \circ g_1$ and $g_1 \circ g_2$ are weak equivalences and therefore homotopy equivalences by Whitehead's Theorem, therefore so is each g_r. □

LEMMA 3.3. *Let $g \colon R \to R'$ be a weak equivalence of connective p-local commutative S-algebras and let $k \colon T \to R'$ be a characteristic morphism. Then there is a morphism of commutative S-algebras $j \colon T \to R$ such that $g \circ j \simeq k$ and j is a characteristic morphism for R.*

PROOF. We can make use of the simplicial structure of $\mathscr{C}_S$ to form cylinder objects; for an algebra $A \in \mathscr{C}_S$ the cylinder $A \otimes I$ is the domain for homotopies between morphisms. We also require a version of Peter May's HELP in $\mathscr{C}_S$, see [**12**]:

Given a commutative diagram of solid arrows

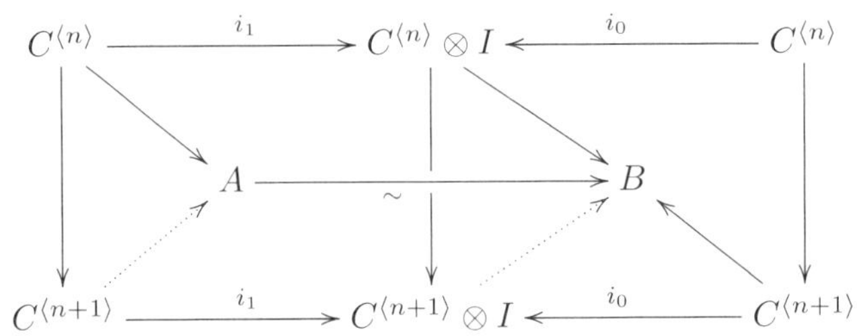

there is an extension to the larger commutative diagram with dotted arrows. Here C is a connective CW algebra and $i_0, i_1\colon X \to X\otimes I$ are the two morphisms corresponding to the ends of the cylinder on X; the vertical morphisms involve inclusions of skeleta.

We apply HELP to give the inductive step in constructing a morphism $T \to R$. Assume that we have a suitable morphism $j^{\langle n\rangle}\colon T^{\langle n\rangle} \to R$ so that $g\circ j^{\langle n\rangle} \simeq k^{\langle n\rangle}$. Given a homotopy $h_n\colon T^{\langle n\rangle}\otimes I \to R'$ with $h_n\circ i_0 = k^{\langle n\rangle}$ and $h_n\circ i_1 = g\circ j^{\langle n\rangle}$, there is a commutative diagram of solid arrows

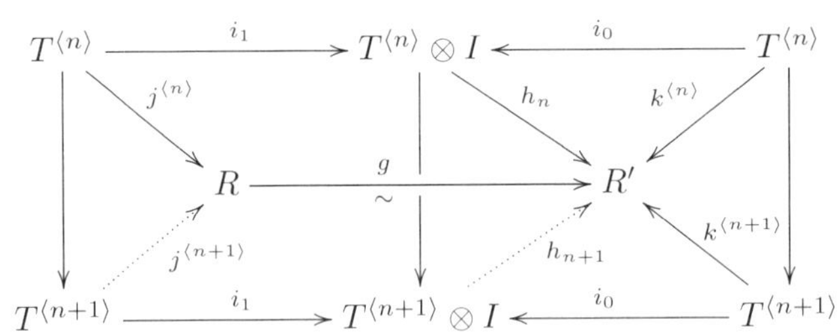

and by HELP an extension to a larger diagram exists. The induction is grounded in the case $n = 0$ where $T^{\langle 0\rangle} = S$. $\square$

Here is a summary of our results which provide substitutes for the functoriality and homotopy invariance conditions of Section 1.

THEOREM 3.4. *Let $\mathscr{C}_S$ be the category of p-local commutative S-algebras.*
(a) *Every connective object $R \in \mathscr{C}_S$ has a characteristic* $\operatorname{char} R$ *which is well-defined up to homotopy equivalence in $\mathscr{C}_S$.*
(b) *Given a morphism of commutative S-algebras $R \to R'$, there is a morphism* $\operatorname{char} R \to \operatorname{char} R'$.
(c) *Given a weak equivalence $g\colon R \to R'$, there are characteristic morphisms $j\colon T \to R$ and $k\colon T \to R'$ which fit into the following homotopy commutative diagram.*

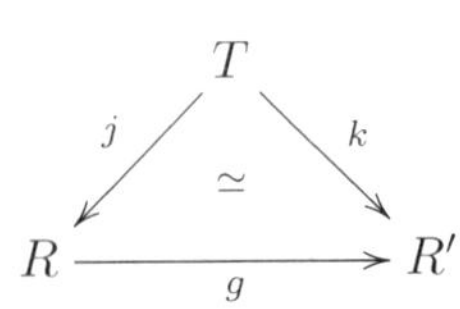

The next result shows that our notion of characteristic really only depends on the kernel of the induced algebraic unit; in particular all commutative S-algebras with torsion free homotopy have the same characteristics.

PROPOSITION 3.5. *Suppose that* $\iota_R\colon S \to R$ *and* $\iota_{R'}\colon S \to R'$ *are two commutative* S*-algebras so that* $\ker(\iota_R)_* \subseteq \ker(\iota_{R'})_* \subseteq \pi_*(S)$. *Then there is a morphism of commutative* S*-algebras* $\operatorname{char} R \to \operatorname{char} R'$.

PROOF. It is straightforward to show that for $T = \operatorname{char} R$ and $T' = \operatorname{char} R'$, given a morphism $T^{\langle n\rangle} \to T'$ on the n-skeleton, there is an extension to the $(n+1)$-skeleton, giving a commutative diagram

$$\begin{array}{ccc} & S & \\ \swarrow & & \searrow \\ T^{\langle n\rangle} & \longrightarrow & T^{\langle n+1\rangle} \\ \searrow & & \swarrow \\ & T' & \end{array}$$

and by induction on n, it follows that a morphism $T \to T'$ exists. □

COROLLARY 3.6. *Suppose that* $\iota_R\colon S \to R$ *and* $\iota_{R'}\colon S \to R'$ *are two commutative* S*-algebras for which* $\ker(\iota_R)_* = \ker(\iota_{R'})_* \subseteq \pi_*(S)$. *Then there is a homotopy equivalence of commutative* S*-algebras* $\operatorname{char} R \xrightarrow{\simeq} \operatorname{char} R'$.

PROOF. By the Proposition there are morphisms $\operatorname{char} R \to \operatorname{char} R' \to \operatorname{char} R$ whose compositions are weak equivalences since $\operatorname{char} R$ and $\operatorname{char} R'$ are both (minimal) atomic. □

We end this discussion with an observation that relates the notion of a characteristic to that of a nuclear S-module. Let R be a connective p-local commutative S-algebra. Then as described in [**6**], beginning with the 0-cell S^0 we can construct a nuclear CW complex X by attaching cells only to the bottom cell, and a map $X \to R$ which induces a monomorphism on $\pi_*(-)$ and $\mathbb{F}_p \otimes \pi_0(-)$ (i.e., a core for R). There is a unique extension to a morphism of commutative S-algebras $\widetilde{\mathbb{P}}X \to R$.

PROPOSITION 3.7. *If* $\widetilde{\mathbb{P}}X$ *is a minimal atomic commutative* S*-algebra then the morphism* $\widetilde{\mathbb{P}}X \to R$ *described above is a characteristic for* R.

PROOF. Suppose that $R_0 \to R$ is a characteristic morphism for R. Then it is straightforward to construct a map $X \to R_0$ under S^0, and this has a unique extension to a morphism of commutative S-algebras $\widetilde{\mathbb{P}}X \to R_0$. Another cellular argument constructs a morphism of commutative S-algebras $R_0 \to \widetilde{\mathbb{P}}X$. If $\widetilde{\mathbb{P}}X$ is minimal atomic then the two composite endomorphisms are weak equivalences and so each morphism is also a weak equivalence. □

The reason for the conditional statement here is that in general for a CW complex $S^0 \to Y$, while $\widetilde{\mathbb{P}}Y$ being minimal atomic implies Y is minimal atomic, the converse need not be true. An example for $p = 2$ is provided by $Y = \Sigma^{-2}\Sigma^\infty\mathbb{C}\mathrm{P}^\infty$, see [**9**]. In the examples we will consider later, this condition will in fact be satisfied.

4. Examples

4.1. Prime power characteristics. We begin with a generalisation of examples discussed in [**16**]. For a prime p and $r \geqslant 1$, we may form $S/\!/p^r$ as the pushout of the solid square in

$$\begin{array}{ccccc} & & \mathbb{P}S^0 & \rightarrowtail & \mathbb{P}D^1 \\ & \swarrow^{\widetilde{p^r}} & \downarrow & \ulcorner & \downarrow \\ S & \overset{\sim}{\twoheadleftarrow} & \widetilde{S} & \rightarrowtail & S//p^r \end{array}$$

where $\widetilde{p^r}$ is the multiplicative extension of a map $S^0 \to S$ of degree p^r and the left-hand triangle is the functorial cofibration/acyclic fibration factorisation of $\widetilde{p^r}$. There is also a homotopy equivalence of commutative S-algebras

$$\widetilde{\mathbb{P}C_{p^r}} \xrightarrow{\simeq} S//p^r$$

where C_{p^r} is the mapping cone of a map $S^0 \to S^0$ of degree p^r viewed as an object of $S^0/\mathscr{M}_S$. Notice that $\pi_0(S//p^r) = \mathbb{Z}/p^r$, so this ring has characteristic p^r. Furthermore, there is a (non-unique) morphism of commutative S-algebras $S//p^r \to H\mathbb{Z}/p^r$.

When $r = 1$, Steinberger's splitting result [**10**, theorem III.4.1] (see also [**3**, section 10] for a more recent approach) implies that the spectrum $S//p^r$ splits as a wedge of suspensions of the Eilenberg-Mac Lane spectrum $H\mathbb{F}_p = H\mathbb{Z}/p$. This means that the unit induces a ring homomorphism $\pi_*(S) \to \pi_*(S//p)$ whose kernel contains p and all positive degree elements of the domain. This shows that $S//p$ is a characteristic of $H\mathbb{F}_p$. More generally, we have the following.

LEMMA 4.1. *Let R be a p-local commutative S-algebra for which the ring $\pi_0(R)$ has characteristic p. Then $S//p$ is a characteristic of R and R is a wedge of suspensions of $H\mathbb{F}_p$.*

PROOF. The unit map $S \to R$ factors through the mapping cone C_p so there is an associated morphism of commutative S-algebras $S//p \to R$ which also factors the unit. By the above discussion, the kernel of $\pi_*(S//p) \to \pi_*(R)$ contains p and all positive degree elements. Furthermore, $\pi_*(R)$ is a graded $\mathbb{F}_p$-vector space and a standard argument shows that given a basis b_i $(i \in I)$ there is a weak equivalence

$$R \sim \bigvee_{i \in I} \Sigma^{|b_i|} H\mathbb{F}_p. \qquad \square$$

When $r > 1$, we still have $\pi_0(S//p^r) = \mathbb{Z}/p^r$, but the splitting result of [**10**, theorem III.4.2] requires further conditions and does not imply a splitting of $S//p^r$. When p is odd, the element $\alpha_1 \in \pi_{2p-3}(S)$ survives in $\pi_{2p-3}(S//p^r)$, and we can attach an $\mathcal{E}_\infty$ cell to kill its image, giving a commutative S-algebra $S//(p^r, \alpha_1)$ for which

$$\pi_{2p-2}(S//(p^r, \alpha_1)) = \mathbb{Z}_{(p)} u_1',$$

where the new cell gives an integral homology class $x'_{2p-2} \in H_{2p-2}(S//(p^r, \alpha_1))$ so that the Hurewicz image of u_1' is px'_{2p-2}. In order to obtain a commutative S-algebra satisfying the condition that 1 is in the image of $\beta\mathcal{P}^1_*$ acting on $H_{2p-1}(-)$, we need to attach an $\mathcal{E}_\infty$ cell of dimension $2p-1$ to kill u_1', giving $S//(p^r, \alpha_1, u_1')$ which does split as a wedge of suspensions of the Eilenberg-Mac Lane spectra $H\mathbb{Z}/p^s$ for $0 \leqslant s \leqslant r$. It is tempting to state the following.

CONJECTURE 4.2. For an odd prime p and $r > 1$, $S//(p^r, \alpha_1)$ is a characteristic for $H\mathbb{Z}/p^r$.

When $p = 2$, we have a similar situation with η in place of α_1. Then $1 \in H_0(S//(2^r, \eta, u_1'))$ is in the image of Sq^3 acting on $H_3(S//(2^r, \eta, u_1'))$, so $S//(2^r, \eta, u_1')$ splits as a wedge of suspensions of the Eilenberg-Mac Lane spectra $H\mathbb{Z}/2^s$ for $0 \leqslant s \leqslant r$. Again we are led to make a conjecture.

CONJECTURE 4.3. For $r > 1$, $S//(2^r, \eta)$ is a characteristic for $H\mathbb{Z}/2^r$.

The reader is warned that we have no direct evidence for this and the discussion of Section 4.3 suggests that it may be far too optimistic.

4.2. Odd-primary examples associated with Hopf invariant one elements. Let p be an odd prime. Then we can kill the element α_1 to form $S//\alpha_1$ which has an $\mathcal{E}_\infty$ morphism $S//\alpha_1 \to \ell$ to the connective cover of the Adams summand of $KU_{(p)}$ which is known to possess an essentially unique $\mathcal{E}_\infty$ structure by results of [**7**, **8**]. Then

$$\begin{aligned} \pi_{2p-2}(S//\alpha_1) &= \mathbb{Z}_{(p)} u_1, \\ H_{2p-2}(S//\alpha_1; \mathbb{Z}_{(p)}) &= \mathbb{Z}_{(p)} x_{2p-2}, \end{aligned}$$

where the Hurewicz image of u_1 is $p x_{2p-2}$.

As in the discussion for $S//p^r, \alpha_1, u_1'$, we can apply Steiberger's splitting results to show that $S//\alpha_1, u_1$ splits as a wedge of $H\mathbb{Z}_{(p)}$ and suspensions of Eilenberg-Mac Lane spectra $H\mathbb{Z}_{p^r}$ for various $r \geqslant 1$.

CONJECTURE 4.4. For an odd prime p and $r > 1$, $S//\alpha_1$ is a characteristic for $H\mathbb{Z}_{(p)}$.

4.3. 2-primary examples associated with Hopf invariant one elements. In this section we take $p = 2$ and consider the four elements $2 \in \pi_0(S)$, $\eta \in \pi_1(S)$, $\nu \in \pi_3(S)$ and $\sigma \in \pi_3(S)$ of Hopf invariant 1. We also set $H = H\mathbb{F}_2$ and denote the mod 2 dual Steenrod algebra by $\mathcal{A}_* = \mathcal{A}(2)_*$ and the Steenrod algebra by $\mathcal{A}^* = \mathcal{A}(2)^*$. By results of [**2**], the mod 2 homology of $S//\eta$, $S//\nu$ and $S//\sigma$ are all polynomial on admissible Dyer-Lashof monomials on generators $x_1 \in H_1(S//2)$, $x_2 \in H_2(S//\eta)$, $x_4 \in H_4(S//\nu)$ and $x_8 \in H_4(S//\sigma)$:

$$\begin{aligned} H_*(S//2) &= \mathbb{F}_2[\mathrm{Q}^I x_1 : \mathrm{exc}(I) > 1], & H_*(S//\eta) &= \mathbb{F}_2[\mathrm{Q}^I x_2 : \mathrm{exc}(I) > 2], \\ H_*(S//\nu) &= \mathbb{F}_2[\mathrm{Q}^I x_4 : \mathrm{exc}(I) > 4], & H_*(S//\sigma) &= \mathbb{F}_2[\mathrm{Q}^I x_8 : \mathrm{exc}(I) > 8]. \end{aligned}$$

The $\mathcal{A}_*$-coaction on the generator x_{2^d} is given by

$$\psi(x_{2^d}) = \zeta_1^{2^d} \otimes 1 + 1 \otimes x_{2^d},$$

and the coaction on a generator $\mathrm{Q}^I x_2$ can be found using formulae in [**3**], at least in principal. Dually, the Steenrod action satisfies

$$\mathrm{Sq}_*^{2^d}(x_{2^d}) = 1,$$

and we also have

$$\mathrm{Sq}_*^{2^{d+k}}(x_{2^d}^{2^k}) = 1.$$

This suggests that ν and σ might map to zero in $\pi_*(S//\eta)$, at least modulo Adams filtration 2. We will examine this in detail later.

There are $\mathcal{E}_\infty$ morphisms

$$\begin{array}{ccccccc} S/\!/2 & & S/\!/\eta & & S/\!/\nu & & S/\!/\sigma \\ & \searrow & \searrow & & \swarrow & \swarrow & \\ & & & H & & & \end{array}$$

which induce ring homomorphisms

$$\begin{array}{ccccccc} H_*(S/\!/2) & & H_*(S/\!/\eta) & & H_*(S/\!/\nu) & & H_*(S/\!/\sigma) \\ & \searrow & \searrow & & \swarrow & \swarrow & \\ & & & H_*(H) & & & \\ & & & \| & & & \\ & & & \mathcal{A}_* & & & \end{array}$$

which allow us to identify their images as $\mathcal{A}_*$-subcomodule algebras of the dual Steenrod algebra $\mathcal{A}_*$.

LEMMA 4.5. *The above homomorphisms give epimorphisms of $\mathcal{A}_*$-comodule algebras*

$$H_*(S/\!/2) \xrightarrow{\cong} \mathbb{F}_2[\zeta_s : s \geqslant 1] \qquad H_*(S/\!/\eta) \twoheadrightarrow \mathbb{F}_2[\zeta_s^2 : s \geqslant 1]$$

$$H_*(S/\!/\nu) \twoheadrightarrow \mathbb{F}_2[\zeta_s^4 : s \geqslant 1] \qquad H_*(S/\!/\sigma) \twoheadrightarrow \mathbb{F}_2[\zeta_s^8 : s \geqslant 1]$$

onto subalgebras of $\mathcal{A}_*$.

PROOF. It is easy to see that in each case, the generator x_{2^d} maps to $\zeta_1^{2^d}$. It follows that

$$\mathrm{Q}^{(i_1,\ldots,i_\ell)}x_{2^d} \longmapsto \begin{cases} (\mathrm{Q}^{(i_1/2^d,\ldots,i_\ell/2^d)}\zeta_1)^{2^d} & \text{if every } i_r \text{ is divisible by } 2^d, \\ 0 & \text{otherwise,} \end{cases}$$

hence the image is a subring of $\mathbb{F}_2[\zeta_s^{2^d} : s \geqslant 1]$. Making use of Steinberger's determination of the Dyer-Lashof action on $\mathcal{A}_*$ we see that the image contains all of the elements $\zeta_s^{2^d}$, therefore the image is exactly this subring of 2^d-th powers. $\square$

There is a commutative diagram of $\mathcal{E}_\infty$ morphisms as indicated by solid arrows

(4.1)

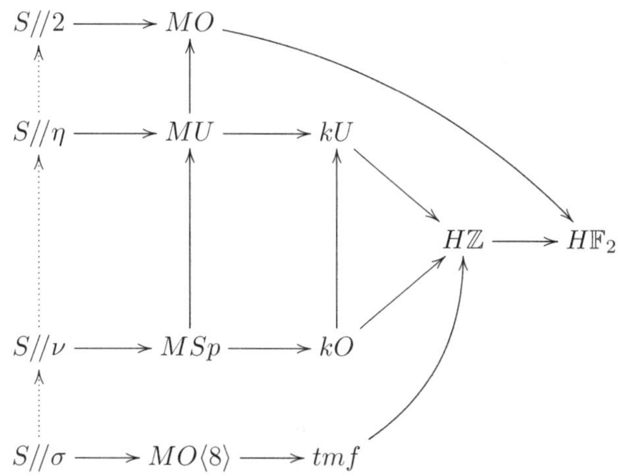

where the left-most horizontal morphisms exist because the bottom cells of the Thom spectra support non-trivial actions of Steenrod operations of the form Sq^{2^d} by the Wu formulae. We will consider the possible existence of suitable morphisms corresponding to the vertical dotted arrows.

Steinberger's work shows that there is a splitting of $S//2$ into a wedge of suspensions of $H = H\mathbb{F}_2$, hence the ring homomorphism $\pi_*(S) \to \pi_*(S//2)$ induced by the unit is trivial in positive degrees. It follows that there are $\mathcal{E}_\infty$ morphisms to $S//2$ from each of $S//\eta$, $S//\nu$ and $S//\sigma$. In each case, under the induced homomorphism in homology, $x_{2^d} \mapsto \zeta_1^{2^d}$. For the case of $S//\eta$ we can also deduce that such a map exists using the fact that the inclusion of the bottom cell induces an isomorphism

$$\pi_1(S^0) \xrightarrow{\cong} \pi_1(C_2),$$

so in $H_*(S//2)$ the 2-cell is attached to the bottom cell by η since $\mathrm{Sq}^2(x_1^2) = 1$.

LEMMA 4.6. *Under the ring homomorphism $\pi_*(S) \to \pi_*(S//\eta)$ induced by the unit $S \to S//\eta$, $\nu \mapsto 0$, hence there is an $\mathcal{E}_\infty$ morphism $S//\nu \to S//\eta$ inducing a ring homomorphism $H_*(S//\nu) \to H_*(S//\eta)$ under which $x_4 \mapsto x_2^2$.*

PROOF. Recall that there is a homotopy equivalence of $\mathcal{E}_\infty$ ring spectra

$$\widetilde{\mathbb{P}}C_\eta \simeq S//\eta.$$

The long exact sequence for the homotopy of the mapping cone C_η has an exact portion

$$\begin{array}{ccccccccc}
\pi_3(S^1) & \xrightarrow{\eta} & \pi_3(S^0) & \longrightarrow & \pi_3(C_\eta) & \longrightarrow & \pi_2(S^1) & \xrightarrow{\eta} & \pi_2(S^0) \\
\downarrow{\cong} & & \| & & & & \downarrow{\cong} & & \| \\
\pi_2(S^0) & & & & & & \pi_1(S^0) & & \\
\| & & & & & & \| & & \\
\mathbb{Z}/2\,\eta^2 & & \mathbb{Z}/8\,\nu & & & & \mathbb{Z}/2\,\eta & & \mathbb{Z}/2\,\eta^2 \\
 & & & & & & & & \\
\eta^2 & \longmapsto & \eta^3 = 4\nu & & & & \eta & \longmapsto & \eta^2
\end{array}$$

showing that

$$\pi_3(S^0)/\eta^3 = \pi_3(S^0)/4\nu \xrightarrow{\cong} \pi_3(C_\eta).$$

In $H_*(S/\!/\eta)$ we have $\mathrm{Sq}^4(x_2^2) = 1$. If we realise $S/\!/\eta$ as a minimal CW S-module, its 4-skeleton has one cell in each of the dimensions 0, 2 and 4. The attaching map of the 4-cell to the 2-skeleton is detected by Sq^4, therefore it must be one of the generators of $\pi_3(C_\eta)$ and so homotopic to a map which factors through S^0 where it agrees with $\pm\nu \bmod 4$. Thus there is a map $C_\nu \to S/\!/\eta$ which extends to a morphism of $\mathcal{E}_\infty$ ring spectra

$$\widetilde{\mathbb{P}}C_\nu \xrightarrow{\simeq} S/\!/\nu \to S/\!/\eta,$$

inducing the stated homomorphism in homology. □

The situation for $S/\!/\nu$ and $S/\!/\sigma$ is more involved.

LEMMA 4.7. *There is no morphism of $\mathcal{E}_\infty$ ring spectra $S/\!/\sigma \to S/\!/\nu$ for which the induced homology homomorphism sends x_8 to x_4^2.*

PROOF. The mapping cone of ν has an associated long exact sequence

$$\begin{array}{ccccccccc}
\pi_7(S^3) & \xrightarrow{\nu} & \pi_7(S^0) & \longrightarrow & \pi_7(C_\nu) & \longrightarrow & \pi_7(S^4) & \xrightarrow{\nu} & \pi_7(S^1) \\
\downarrow{\cong} & & \| & & & & \downarrow{\cong} & & \downarrow{\cong} \\
\pi_4(S^0) & & & & & & \pi_3(S^0) & & \pi_6(S^0) \\
\| & & & & & & \| & & \| \\
0 & & \mathbb{Z}/16\,\sigma & & & & \mathbb{Z}/8\,\nu & & \mathbb{Z}/2\,\nu^2
\end{array}$$

where the element $2\nu \in \pi_3(S^0)$ has Adams filtration 2. It follows that

$$\pi_7(C_\nu) \cong \pi_7(S^0) \oplus 2\pi_7(S^4) \cong \mathbb{Z}/16\,\sigma \oplus \mathbb{Z}/4\,\widetilde{2\nu}.$$

In $H_*(S/\!/\nu)$ we have $\mathrm{Sq}^8(x_4^2) = 1$, but it does not follow that the attaching map of the 8-cell to the 4-skeleton C_ν can be taken to be σ. In fact, the attaching map can be seen to be $\sigma + \widetilde{2\nu}$ since the natural symmetrisation map

$$C_\nu \wedge C_\nu \to E\Sigma_2 \ltimes_{\Sigma_2} (C_\nu \wedge C_\nu)$$

induces the fold map on the 4-cells and a careful analysis in integral homology shows that the attaching map is as claimed (I learnt this argument from Peter Eccles; see Figure 4.1). The upshot is that there is no morphism of $\mathcal{E}_\infty$ ring spectra $S//\sigma \to S//\nu$ since there can be no map $C_\sigma \to S//\nu$ extending the unit. $\square$

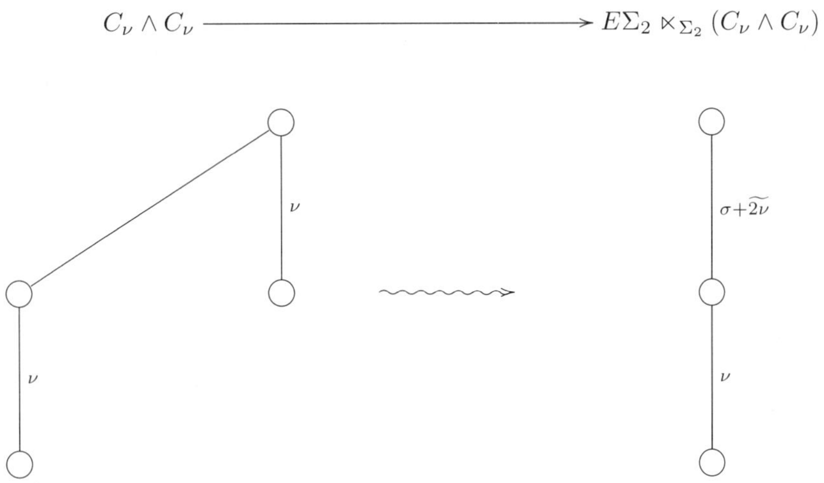

FIGURE 4.1

It follows that the mapping cone of the composition

$$S^3 \vee S^7 \xrightarrow[\nu\vee\sigma]{} S^0 \vee S^0 \xrightarrow[\text{fold}]{} S^0 \qquad (\text{composite } \nu+\sigma)$$

gives rise to the minimal atomic commutative S-algebra

$$S//(\nu,\sigma) = \widetilde{\mathbb{P}}C_{\nu+\sigma}.$$

Then there are $\mathcal{E}_\infty$ morphisms

$$S//\nu \to S//(\nu,\sigma) \to MSp \to kO$$

and in fact $S//(\nu,\sigma) \to MSp$ is an 8-equivalence in the classical sense. By [**5**, proposition 5.1], all of these are minimal atomic spectra.

For $S//\sigma$, there is a morphism of commutative S-algebras $S//\sigma \to tmf$ and we expect this to be a characteristic. There is a morphism $S//\sigma \to H\mathbb{F}_2$ inducing a ring homomorphism

$$H_*(S//\sigma) \to H_*(H\mathbb{F}_2) = \mathcal{A}_*$$

whose image consists of the eighth powers.

The following result is even more challenging to verify.

LEMMA 4.8. *There is no morphism of $\mathcal{E}_\infty$ ring spectra $S//\sigma \to S//\eta$ for which the induced homology homomorphism sends x_8 to x_2^4.*

PROOF. The point is that the image of σ in $\pi_*(S//\eta)$ is non-zero. This was shown in an unpublished calculation by John Rognes (private communication 2013). This image has order 4 (instead of 16), and in the Adams spectral sequence, it has filtration 2 rather than 1. $\square$

CONJECTURE 4.9. Each of the following is a characteristic, where the maps are compositions of those mentioned above with standard ones.

$$\begin{array}{ccccc} & & S//(\eta,\sigma) & & \\ & \swarrow & \downarrow & \searrow & \\ MU & \longrightarrow & kU & \longrightarrow & H\mathbb{Z}_{(2)} \end{array}$$

$$\begin{array}{ccc} & S//(\nu,\sigma) & \\ \swarrow & & \searrow \\ MSp & \longrightarrow & kO \end{array}$$

$$S//\sigma \to tmf$$

Let us consider what the statements in this conjecture really amount to. The first is equivalent to the unit homomorphism $\pi_*(S) \to \pi_*(S//(\eta,\sigma))$ being trivial in positive degrees. The second statement is equivalent to the equalities

$$\ker[\pi_*(S) \to \pi_*(S//\nu,\sigma)] = \ker[\pi_*(S) \to \pi_*(kO)] = \ker[\pi_*(S) \to \pi_*(MSp)],$$

where the second equality was proved by Stan Kochman [**15**, part I], but the first would also imply it. The third statement is equivalent to the equality

$$\ker[\pi_*(S) \to \pi_*(S//\sigma)] = \ker[\pi_*(S) \to \pi_*(tmf)].$$

To end, we mention some further results on these spectra.

PROPOSITION 4.10. *The 2-local morphisms $S//\eta \to kU$ and $S//\nu \to kO$ induce epimorphisms on $\pi_*(-)$.*

PROOF. For $S//\eta \to kU$, it suffices to show that it induces an epimorphism on $\pi_2(-)$. This is clear since the Toda bracket $\langle 2, \eta, 1_{S//\eta}\rangle \subseteq \pi_2(S//\eta)$ is defined, where $S//\eta$ is viewed as an S-module, and the bracket is to be taken as a function on the set

$$\pi_0(S) \times \pi_1(S) \times \pi_1(S//\eta).$$

By naturality, this maps to the Toda bracket $\langle 2, \eta, 1_{kU}\rangle \subseteq \pi_2(kU)$ which contains a generator of $\pi_2(kU)$ and has indeterminacy $2\pi_2(kU)$ as pointed out in [**18**, page 64].

For $S//\nu \to kO$, it is sufficient to show that the generators a, b of the groups $\pi_4(kO), \pi_4(kO)$ come from $\pi_4(S//\nu), \pi_8(S//\nu)$. We can appeal to [**18**, page 64] (see also [**6**, lemma 7.3]), where it is shown that the Toda brackets $\langle 8, \nu, 1_{kO}\rangle$ and $\langle 8, \nu, a\rangle$ contain a, b respectively. Analogues of these brackets can be defined in $\pi_*(S//\nu)$ and are preimages of the kO versions. Zhouli Xu has pointed out that the Toda brackets $\langle \eta^2, \eta, 2\rangle$ and $\langle \eta, \eta^2, \eta, \eta^3\rangle$ in $\pi_*(kO)$ also contain a, b, furthermore they make sense in $\pi_*(S//\nu)$; this time we use the traditional Toda brackets defined in the homotopy of a ring spectrum. □

There are factorisations of these $\mathcal{E}_\infty$ morphisms

$$S//\eta \to T_{kU} \to kU, \quad S//\nu \to T_{kO} \to kO,$$

where each second factor is a characteristic morphism. Hence these characteristic morphisms induce epimorphisms on $\pi_*(-)$. Motivated by these examples, we are led to make a conjecture on a characteristic morphism for tmf.

CONJECTURE 4.11. A 2-local characteristic morphism $T_{tmf} \to tmf$ induces an epimorphism on $\pi_*(-)$.

Acknowledgments

The author would like to thank the following for helpful comments: Peter Eccles, Paul Goerss, Rolf Hoyer, Niko Naumann, Nigel Ray, Birgit Richter, Vesna Stojanoska, John Rognes, Markus Szymik, Grant Walker and Zhouli Xu.

References

[1] A. Baker, *Calculating with topological André-Quillen theory, I: Homotopical properties of universal derivations and free commutative S-algebras* (2012), available at `arXiv:1208.1868(v8+)`.

[2] A. Baker, *BP: close encounters of the E_∞ kind*, J. Homotopy Relat. Struct. **9** (2014), no. 2, 553–578. MR3258694

[3] A. Baker, *Power operations and coactions in highly commutative homology theories*, Publ. Res. Inst. Math. Sci. **51** (2015), no. 2, 237–272. MR3348109

[4] A. Baker, *$\mathcal{E}_\infty$ ring spectra and elements of Hopf invariant 1*, Bol. Soc. Mat. Mex. (3) **23** (2017), no. 1, 195–231. MR3633133

[5] A. Baker, H. Gilmour, and P. Reinhard, *Topological André-Quillen homology for cellular commutative S-algebras*, Abh. Math. Semin. Univ. Hambg. **78** (2008), no. 1, 27–50. MR2501527

[6] A. J. Baker and J. P. May, *Minimal atomic complexes*, Topology **43** (2004), no. 3, 645–665. MR2041635

[7] A. Baker and B. Richter, *On the Γ-cohomology of rings of numerical polynomials and E_∞ structures on K-theory*, Comment. Math. Helv. **80** (2005), no. 4, 691–723. MR2182697

[8] A. Baker and B. Richter, *Uniqueness of E_∞ structures for connective covers*, Proc. Amer. Math. Soc. **136** (2008), no. 2, 707–714. MR2358512

[9] A. Baker and B. Richter, *Some properties of the Thom spectrum over loop suspension of complex projective space*, An alpine expedition through algebraic topology, Contemp. Math., vol. 617, Amer. Math. Soc., Providence, RI, 2014, pp. 1–12. MR3243390

[10] R. R. Bruner, J. P. May, J. E. McClure, and M. Steinberger, *H_∞ ring spectra and their applications*, Lecture Notes in Mathematics, vol. 1176, Springer-Verlag, Berlin, 1986. MR836132

[11] J. M. Cohen, *The decomposition of stable homotopy*, Ann. of Math. (2) **87** (1968), 305–320. MR0231377

[12] A. D. Elmendorf, I. Kriz, M. A. Mandell, and J. P. May, *Rings, modules, and algebras in stable homotopy theory*, Mathematical Surveys and Monographs, vol. 47, American Mathematical Society, Providence, RI, 1997. With an appendix by M. Cole. MR1417719

[13] M. Hovey, *Smith ideals of structured ring spectra* (2014), available at `arXiv:1401.2850`.

[14] P. Hu, I. Kriz, and J. P. May, *Cores of spaces, spectra, and E_∞ ring spectra*, Homology Homotopy Appl. **3** (2001), no. 2, 341–354. Equivariant stable homotopy theory and related areas (Stanford, CA, 2000). MR1856030

[15] S. O. Kochman, *Symplectic cobordism and the computation of stable stems*, Mem. Amer. Math. Soc. **104** (1993), no. 496, x+88. MR1147349

[16] M. Szymik, *Commutative $\mathbb{S}$-algebras of prime characteristics and applications to unoriented bordism*, Algebr. Geom. Topol. **14** (2014), no. 6, 3717–3743. MR3302977

[17] M. Szymik, *String bordism and chromatic characteristics* (2012), available at `arXiv:1211.3239`.

[18] G. W. Whitehead, *Recent advances in homotopy theory*, American Mathematical Society, Providence, R.I., 1970. Conference Board of the Mathematical Sciences Regional Conference Series in Mathematics, No. 5. MR0309097

SCHOOL OF MATHEMATICS & STATISTICS, UNIVERSITY OF GLASGOW, GLASGOW G12 8QW, SCOTLAND

Email address: `a.baker@maths.gla.ac.uk`

URL: `http://www.maths.gla.ac.uk/~ajb`

Contemporary Mathematics
Volume **708**, 2018
http://dx.doi.org/10.1090/conm/708/14271

Segal objects and the Grothendieck construction

Pedro Boavida de Brito

ABSTRACT. We discuss right fibrations in the ∞-categorical context of Segal objects in some category $\mathcal{V}$ and prove some basic results about these.

Right fibrations and Cartesian fibrations (and their dual notions) form the backbone of ∞-category theory [**13**], [**16**]. The purpose of this note is to investigate them from the point of view of Rezk's Segal spaces and, more generally, Segal objects in some enriching category $\mathcal{V}$. Due to the known comparison between (complete) Segal spaces and quasi-categories, our arguments also give an alternative route to proving the quasi-categorical analogues, a direct proof of which is often technically more involved. More importantly for us, Segal spaces provide natural models for infinity categories in geometric contexts, and it seems useful to have some theory developed specifically for them.

Let us briefly state the main definitions and results. Throughout, $\mathcal{S}$ denotes the category of spaces (*space* may be used as a synonym for *simplicial set*). Let B be a Segal space, not necessarily complete (for definitions, see section 1). We refer to B_0 as the space of objects, B_1 as the space of morphisms, d_0 is the operator *source* and d_1 is the operator *target*. A **right fibration** over B is a Segal space X together with a map $X \to B$ having the property that the square

$$\begin{array}{ccc} X_0 & \xleftarrow{\;d_1\;} & X_1 \\ \downarrow & & \downarrow \\ B_0 & \xleftarrow{\;d_1\;} & B_1 \end{array} \tag{0.1}$$

is homotopy cartesian. A map between right fibrations over B is a weak equivalence if it is a weak equivalence on 0-simplices (alias object spaces).

After defining a model structure whose homotopy theory is that of right fibrations, we prove the following strictification result.

THEOREM A. *Let $\mathcal{C}$ be a category (possibly simplicially enriched or, more generally, internal to spaces). The Grothendieck construction (see section 1) is a simplicial Quillen equivalence between the right fibration model structure on the category*

2010 *Mathematics Subject Classification.* Primary 55U40, 55U35, 18G55.
The author is supported by a FCT grant SFRH/BPD/99841/2014.

$s\mathcal{S}_{/NC}$ of simplicial spaces over the nerve of $\mathcal{C}$ and the projective model structure on $\mathsf{Fun}(\mathcal{C}^{op}, \mathcal{S})$, the category of contravariant simplicial functors from $\mathcal{C}$ to spaces.

The next step in generality are Cartesian fibrations. Still for a fixed Segal space B (which we view as a bisimplicial space with (m,n)-simplices B_m), we say that a map $X \to B$ of bisimplicial spaces is a **Cartesian fibration** if the simplicial spaces $X_{\bullet,n}$ and $X_{m,\bullet}$ are Segal spaces for all n and m, and the map

$$X_{\bullet,n} \to B_{\bullet,n}$$

is a right fibration for every n. Moreover, a map $X \to Y$ between Cartesian fibrations is a weak equivalence if is it a Dwyer-Kan equivalence on 0-simplices $X_{0,\bullet} \to Y_{0,\bullet}$.

After defining a model structure whose underlying homotopy theory is that of Cartesian fibrations, we prove the analogous strictification statement below.

THEOREM B. *The Grothendieck construction is a right Quillen equivalence between the right fibration model structure on $ss\mathcal{S}_{/NC}$ and the projective model structure on $\mathsf{Fun}(\mathcal{C}^{op}, \mathsf{Cat}_\infty)$. It is moreover an equivalence of $(\infty,2)$-categories.*

Here Cat_∞ refers to a category of $(\infty,1)$-categories, which for concreteness we take to be Rezk's model category of complete Segal spaces [**17**].

The definition of a right fibration of (complete) Segal spaces is not new nor unexpected; Michael Shulman has mentioned a similar definition in unpublished notes of Charles Rezk, and Ilan Barnea pointed out to us that a definition has appeared (independently) in [**15**]. The definition of a Cartesian fibration above has a different flavor to the construction that usual bears that name (in categories or quasi-categories). Theorem B argues that from a certain point of view both try to achieve the same (more on this in remark 3.3).

Theorems A and B have counterparts in the quasi-categorical setting that have appeared in the work of Joyal [**13**], Lurie [**16**] and Heuts-Moerdjik [**8**], [**9**]. An exception is the internal version of these theorems, as far as we know. Rezk's theory of complete Segal spaces is manifestly under-developed when compared to that of quasi-categories, and we hope this fact is enough justification for this paper. Moreover, the approach used here seems well suited to generalizations, for example to the (∞, n) setting.

Outline. In section 1, after some preliminary work, we prove theorem A. The key ingredient is a version of the Yoneda lemma (lemma 1.31). We also show that right fibrations are fibrations in the *complete* Segal space model structure, and directly relate our definition of a right fibration to the quasi-categorical one. In section 2 we define the homotopy theory underlying right fibration objects in a general category $\mathcal{V}$, and in section 3 we apply that to define the homotopy theory of Cartesian fibrations and prove theorem B. In section 4 we describe (co)limits of right fibrations. In section 5 we give an explicit formula for Kan extensions of a right fibration in spaces and then show that the homotopy theory of right fibrations is invariant under (Dwyer-Kan) equivalences of the base.

Notation. As customary, we denote the simplex category by Δ. Objects in Δ are finite totally ordered sets $[n] = \{0 < 1 < \cdots < n\}$, for $n \geq 0$, and morphisms are order-preserving maps. Often we implicitly view $[n]$ as a category $0 \to 1 \to \cdots \to n$.

We write $s\mathcal{S}$ for the category of simplicial spaces, or more generally, $s\mathcal{V}$ for the category of simplicial objects in a category $\mathcal{V}$, that is, contravariant functors from Δ to $\mathcal{V}$. We frequently and unapologetically regard a simplicial set X as a simplicial *discrete* space with n-simplices given by the constant simplicial set X_n. In particular, we keep the notation $\Delta[n]$ for the simplicial discrete space (which Rezk refers to as $F[n]$). For a fixed object C in $s\mathcal{V}$, we write $s\mathcal{V}_{/C}$ for the category of simplicial objects over B.

For a category $\mathcal{C}$, which may be enriched or internal in spaces, we write $N\mathcal{C}$ for the nerve of $\mathcal{C}$, i.e. the simplicial *space* whose space of 0-simplices is the space $\mathrm{ob}(C)$ of objects of $\mathcal{C}$, the space of 1-simplices is the space $\mathrm{mor}(\mathcal{C})$ of morphisms of $\mathcal{C}$ and whose space of n-simplices is the space of n composable morphisms

$$c_0 \leftarrow \cdots \leftarrow c_n$$

of $\mathcal{C}$ (which is expressed as an iterated pullback). Applied to such a string, the face map d_i drops c_i and the degeneracy map s_i replaces c_i by an identity map. With these conventions, we follow Bousfield-Kan [**6**], but we warn the reader that these differ from other sources (e.g. Rezk).

Acknowledgments. Thanks to Geoffroy Horel and Mike Shulman for discussions and encouragement; it was their suggestion that we upgrade the result to internal categories and include the relation with quasi-categories. The author is also grateful to the referee for a careful reading and helpful comments. An earlier iteration of this paper appeared in the author's PhD thesis.

1. Left and right fibrations

1.1. Segal spaces. The following definition is due to Charles Rezk. It is a model for the umbrella term ∞-*category*.

DEFINITION 1.1 ([**17**]). A **Segal space** is a simplicial space X such that, for each $n \geq 2$, the map

$$X_n \xrightarrow{(\alpha_0^*,\ldots,\alpha_{n-1}^*)} \mathrm{holim}(X_1 \xrightarrow{d_0} X_0 \xleftarrow{d_1} X_1 \xrightarrow{d_0} \ldots \xleftarrow{d_1} X_1) \tag{1.1}$$

is a weak equivalence of spaces, where α_i is the map $\Delta[1] \to \Delta[n]$ sending $0 < 1$ to $i < i + 1$.

Elements of X_0 are called *objects* and elements of X_1 are called *morphisms.*

EXAMPLE 1.2. The nerve of an internal category in spaces is a Segal space if either the source or target map is a fibration, using that the category of simplicial sets is right proper. The nerve of a simplicially enriched category is always a Segal space. Indeed, it satisfies the non-derived version of (1.1) and we may assume that each space X_i is fibrant (for example by applying Kan's Ex^∞ functor which commutes with finite limits). Then the source-target operators map onto a discrete space and so are fibrations and in that case the relevant pullbacks are homotopy pullbacks.

DEFINITION 1.3. Let X be a Segal space. Given two objects $x, y \in X_0$, the space of morphisms $\mathrm{mor}_X(x,y)$ is defined as the homotopy fiber, over (x,y), of

$$X_1 \xrightarrow{(d_0,d_1)} X_0 \times X_0 \; .$$

Two maps $f, g \in \mathrm{mor}_X(x,y)$ are said to be homotopic if they lie in the same component. The *homotopy category* of X is the category $\mathsf{Ho}(X)$ with objects X_0 and morphisms $\mathrm{map}_{\mathsf{Ho}(X)}(x,y) := \pi_0 \mathrm{mor}_X(x,y)$. A map $f \in \mathrm{mor}_X(x,y)$ is a *homotopy equivalence* if it admits a left and right inverse in $\mathsf{Ho}(X)$.

The space of homotopy equivalences is denoted X_1^{he}. It is necessarily a union of components of X_1 since if a map can be connected by a path to a homotopy equivalence, then it must itself be a homotopy equivalence [**17**, lemma 5.8].

DEFINITION 1.4. A simplicial map $i : C \to D$ between Segal spaces is a *Dwyer-Kan equivalence* if it is fully faithful, i.e. for each pair of objects $x, y \in C_0$, the induced map

$$\mathrm{mor}_C(x,y) \to \mathrm{mor}_D(f(x), f(y))$$

is a weak equivalence of spaces, and $\mathsf{Ho}(f) \colon \mathsf{Ho}(X) \to \mathsf{Ho}(Y)$ is essentially surjective.

EXAMPLE 1.5. Suppose $(\mathcal{C}, \mathcal{W})$ is a category $\mathcal{C}$ with a subcategory $\mathcal{W}$ containing all objects. Such a pair $(\mathcal{C}, \mathcal{W})$ goes by the name of *relative category* in [**3**]. From a relative category $(\mathcal{C}, \mathcal{W})$, Rezk defines a Segal space $N(\mathcal{C}, \mathcal{W})$, the *Rezk nerve* (Rezk calls it the *classification diagram*) which retains the homotopical information of the pair. The simplicial set $N(\mathcal{C}, \mathcal{W})_m$ of m-simplices is defined as the nerve of the category of contravariant functors $[n] \times [m] \to \mathcal{C}$ which send all the arrows in the n-direction to $\mathcal{W}$.

An older, closely related construction is the *hammock localization* of Dwyer-Kan [**7**]. This is a simplicially enriched category $L^H\mathcal{C}$ which is a simplicial enhancement of the classical localization, in the sense that the category obtained by taking π_0 of the morphism spaces in $L^H\mathcal{C}$ is the classical localization of $\mathcal{C}$ with respect to $\mathcal{W}$. If $\mathcal{C}$ happens to be a simplicially enriched model category, then the morphism spaces in $L^H\mathcal{C}$ are identified with the derived mapping spaces computed relative to the model structure on $\mathcal{C}$.

In [**2**], Barwick and Kan show that the Rezk nerve and the usual nerve of the hammock localization agree. More precisely, they show that $N(L^H\mathcal{C})$, the nerve of $L^H\mathcal{C}$, is naturally weakly equivalent (in the complete Segal space sense of [**17**]) to $N(\mathcal{C}, \mathcal{W})$[1]. A map between Segal spaces is a complete Segal weak equivalence if and only if it is a Dwyer-Kan equivalence. So if $N(\mathcal{C}, \mathcal{W})$ is a Segal space (e.g. if $\mathcal{C}$ has a model category structure with weak equivalences given by $\mathcal{W}$) then $N(\mathcal{C}, \mathcal{W})$ is Dwyer-Kan equivalent to $N(L^H\mathcal{C})$.

[1] In more detail, Barwick-Kan show that the Rezk nerve of $(\mathcal{C}, \mathcal{W})$ is weakly equivalent (in the model structure for complete Segal spaces) to the Rezk nerve of the relativization of $L^H\mathcal{C}$ [**2**, proposition 3.1], and in [**2**, proposition 1.11] that the Rezk nerve of the relativization of a simplicially enriched category A is naturally weakly equivalent to $N(A)$, the (usual) nerve of A. (Warning: Barwick-Kan use the letter N for the Rezk nerve, and the letter Z for what we refer to as the nerve.)

1.2. Right fibrations. Let F be a *contravariant* functor from $\mathcal{C}$ to spaces (possibly enriched). The *Grothendieck construction* of F is a simplicial space $\mathcal{G}(F)$ over $N\mathcal{C}$ defined as

$$\mathcal{G}(F)_n := \coprod_{c_0,\dots,c_n} \mathrm{map}_{\mathcal{C}}(c_n, c_{n-1}) \times \cdots \times \mathrm{map}_{\mathcal{C}}(c_1, c_0) \times F(c_0)$$

Degeneracies select the identity maps. Face maps encode functoriality. For example, the maps d_0 and d_1

$$\coprod_{c_0,c_1} \mathrm{map}_{\mathcal{C}}(c_1, c_0) \times F(c_0) \to \coprod_{c_0} F(c_0)$$

are composition and projection, respectively (remember F is contravariant). As defined, $\mathcal{G}(F)$ is the nerve of an internal category in spaces (i.e. a category object in spaces, which we may assume are all fibrant as in Example 1.2) and the target map is a projection, hence is a fibration. So $\mathcal{G}(F)$ is a Segal space. There is reference functor from $\mathcal{G}(F)$ to the nerve of $\mathcal{C}$ and it is easy to see that $\mathcal{G}$ is fully faithful as a functor $\mathsf{Fun}(\mathcal{C}^{op}, \mathcal{S}) \to s\mathcal{S}_{/N\mathcal{C}}$. Clearly, the Grothendieck construction $\mathcal{G}(F) \to N\mathcal{C}$ of a *contravariant* functor F is a right fibration.

REMARK 1.6. The dual notion to a right fibration is a *left fibration*. A map $p : X \to B$ between Segal spaces is a left fibration if $X^{op} \to B^{op}$ is a right fibration[2]. Alternatively, p is a left fibration if the variant of the square (0.1) with d_1 replaced by d_0, is homotopy cartesian. The Grothendieck construction of a *covariant* functor is a *left* fibration.

PROPOSITION 1.7. *Suppose that the square* (0.1) *is homotopy cartesian for a map of simplicial spaces* $X \to B$ *and* B *is a Segal space. The following statements are equivalent.*

(1) *X is a Segal space (i.e. $X \to B$ is a right fibration)*
(2) *For each $n \geq 1$, the diagram*

$$\begin{array}{ccc} X_0 & \xleftarrow{a_n^*} & X_n \\ \downarrow & & \downarrow \\ B_0 & \xleftarrow{a_n^*} & B_n \end{array}$$

is homotopy cartesian, where $a_n : [0] \to [n]$ is the map sending 0 to 0.

(3) *For each $n \geq 1$, the diagram*

$$\begin{array}{ccc} X_{n-1} & \xleftarrow{d_n} & X_n \\ \downarrow & & \downarrow \\ B_{n-1} & \xleftarrow{d_n} & B_n \end{array}$$

is homotopy cartesian, where d_n is induced by the face map $d^n : [n-1] \to [n]$ which misses n.

[2]By definition, the opposite of a Segal space Y is the Segal space Y^{op} obtained by reversing the order of the simplicial maps, i.e. $Y_n^{op} := Y_n$, $d_i : Y_n^{op} \to Y_{n-1}^{op}$ is defined as $d_{n-i} : Y_n \to Y_{n-1}$ and similarly for s_i.

PROOF. This is rather formal and it works for any model category which, like $\mathcal{S}$, is right proper. Suppose (1) holds. We show (2) holds by induction on n. The case $n = 1$ holds by assumption. Suppose, for the induction hypothesis, that (2) holds for $n - 1$. Factor the square in (2) as in the diagram below.

$$\begin{array}{ccccc} X_0 & \longleftarrow & X_{n-1}\times^h_{X_0}X_1 & \longleftarrow & X_n \\ \downarrow & & \downarrow & & \downarrow \\ B_0 & \longleftarrow & B_{n-1}\times^h_{B_0}B_1 & \longleftarrow & B_n \end{array}$$

(rows induced by $\Delta[0] \to \Delta[n-1] \amalg_{\Delta[0]} \Delta[1] \to \Delta[n]$; the first map is given by the inclusion $\alpha_1 = d^1 : \Delta[0] \to \Delta[1]$, the second is the Segal map)

The right-hand square is homotopy cartesian since the horizontal maps are weak equivalences (using the assumption that X and B are Segal spaces). The left-hand square is also homotopy cartesian; this follows by commuting homotopy pullbacks, the induction hypothesis and the assumption that $X \to B$ is a right fibration.

The equivalence between (2) and (3) is straightforward.

It remains to show that (3) implies (1). We will verify the Segal condition by showing that the commutative square

$$\begin{array}{ccc} X_n & \xrightarrow{d_0} & X_{n-1} \\ {\scriptstyle d_n}\downarrow & & \downarrow{\scriptstyle d_{n-1}} \\ X_{n-1} & \xrightarrow{d_0} & X_{n-2} \end{array} \tag{1.2}$$

is homotopy cartesian, for each $n \geq 2$. It is not hard to deduce that this is an equivalent formulation of Segal's condition. Now prolong the above square to the diagram

$$\begin{array}{ccccc} X_n & \xrightarrow{d_0} & X_{n-1} & \longrightarrow & B_{n-1} \\ {\scriptstyle d_n}\downarrow & & \downarrow{\scriptstyle d_{n-1}} & & \downarrow{\scriptstyle d_{n-1}} \\ X_{n-1} & \xrightarrow{d_0} & X_{n-2} & \longrightarrow & B_{n-2} \end{array}$$

Here, the right-hand square is homotopy cartesian by assumption. So it remains to show that the outer rectangle is homotopy cartesian. That outer rectangle is identified with the outer rectangle in the diagram:

$$\begin{array}{ccccc} X_n & \longrightarrow & B_n & \xrightarrow{d_0} & B_{n-1} \\ {\scriptstyle d_n}\downarrow & & \downarrow{\scriptstyle d_n} & & \downarrow{\scriptstyle d_{n-1}} \\ X_{n-1} & \longrightarrow & B_{n-1} & \xrightarrow{d_0} & B_{n-2} \end{array}$$

But both squares in this diagram are homotopy cartesian, the left-hand square by assumption (2) and the right-hand square since B is a Segal space. $\square$

1.3. Homotopy theory of right fibrations. We begin by recalling the Segal space model structure on the category of simplicial spaces. There is, throughout the paper, an underlying model structure on the category of simplicial spaces with *levelwise* weak equivalences from which all the other model structures are obtained by localization. There are at least two choices for this model structure, depending on whether we ask the cofibrations or the fibrations to be given levelwise. It is well known that the identity map gives a Quillen equivalence between the two model structures (usually called injective and projective model structure, respectively). Because of this, we refer to either generically as the *levelwise* model structure. So, whenever we say that a map is a weak equivalence, fibration or cofibration of *simplicial spaces* or *in the levelwise model structure*, we mean that in either of the two senses above (but the meaning should of course be fixed, at least locally).

PROPOSITION 1.8. *(Rezk) There is a left proper, simplicial model structure structure on $s\mathcal{S}$ – called the* ***Segal space model structure*** *– uniquely characterized by the following properties:*

- *An object X is fibrant if it is a Segal space (and is fibrant as a simplicial space).*
- *A morphism $X \to Y$ between (fibrant) Segal spaces is a weak equivalence if $X_n \to Y_n$ is a weak equivalence for each $n \geq 0$.*
- *A morphism is a cofibration if it is a cofibration of simplicial spaces.*

More generally, a morphism $f : X \to Y$ between any two simplicial spaces is a weak equivalence if the induced map on derived mapping spaces

$$\mathbb{R}\mathrm{map}(f, Z) : \mathbb{R}\mathrm{map}(Y, Z) \to \mathbb{R}\mathrm{map}(X, Z)$$

(computed with respect to the levelwise model structure) is a weak equivalence of spaces for every Segal space Z.

REMARK 1.9. (Alternative characterization of the Segal condition) For each $n \geq 2$, the map

$$(\alpha_0, \ldots, \alpha_{n-1}) : \Delta[1] \amalg_{\Delta[0]} \cdots \amalg_{\Delta[0]} \Delta[1] \hookrightarrow \Delta[n]$$

is a weak equivalence by definition of the Segal space model structure. Here is a folklore statement that resonates with the definition of a quasi-category. For $n \geq 2$, let $\iota : \Lambda^k[n] \hookrightarrow \Delta[n]$ be an inner horn inclusion, i.e. $0 < k < n$. Then ι is a weak equivalence in the Segal space model structure. One can show this inductively on n. The case $n = 2$ is already part of the Segal condition. For higher n, factor the map $(\alpha_0, \ldots, \alpha_{n-1})$ as

$$\Delta[1] \amalg_{\Delta[0]} \cdots \amalg_{\Delta[0]} \Delta[1] \hookrightarrow \Lambda^k[n] \to \Delta[n]$$

The left-hand map can be given as a composition of inner horn extensions, i.e. pushouts along inner horn inclusions. These inner horn inclusions are all of *lower* dimension and so are weak equivalences by induction. It follows that each extension is a pushout along a weak equivalence (which is also a cofibration in the injective structure) and so a weak equivalence. By two-out-of-three, we conclude that the right-hand map in the display is a weak equivalence.

PROPOSITION 1.10. *Fix a Segal space B. There is a left proper, simplicial model structure structure on $s\mathcal{S}_{/B}$ – called the* **right fibration model structure** *– uniquely characterized by the following properties:*

- *An object $X \to B$ is fibrant if it is a right fibration and a fibration of simplicial spaces.*
- *A morphism $X \to Y$ between right fibrations over B is a weak equivalence if $X_0 \to Y_0$ is a weak equivalence, i.e. if the maps of homotopy fibers $X_b \to Y_b$ are weak equivalences for each $b \in B_0$.*
- *A morphism is a cofibration if it is a cofibration of simplicial spaces.*

More generally, a morphism $f : X \to Y$ between any two simplicial spaces over B is a weak equivalence if the induced map on derived mapping spaces

$$\mathbb{R}\mathrm{map}_B(f, W) : \mathbb{R}\mathrm{map}_B(Y, W) \to \mathbb{R}\mathrm{map}_B(X, W)$$

(computed with respect to the levelwise model structure) is a weak equivalence of spaces for every fibrant object $W \to B$.

PROOF. Start with the Segal space model structure on simplicial spaces. Take the overcategory model structure. Note that a fibrant object is a fibration in the Segal space model structure (which, since B is fibrant there, is equivalent to X being Segal and the map being a levelwise fibration). Then take the left Bousfield localization with respect to the set of maps

$$\{\mathbb{L}\Delta[0] \xrightarrow{\mathbb{L}d^1} \mathbb{L}\Delta[1] \xrightarrow{f} B : f \in B_1\}$$

where $\mathbb{L}d^1$ refers to a replacement of d^1 by a cofibration in the levelwise model structure (note that no replacement is needed for the injective model structure.) □

PROPOSITION 1.11. *Right fibrations are stable under base change.*

PROOF. Suppose $X \to B$ is a right fibration and $B' \to B$ a map. Let X' denote the homotopy pullback $X \times^h_B B'$. We have a commutative diagram

$$\begin{array}{ccccc} X'_1 & \xrightarrow{d_1} & X'_0 & \longrightarrow & X_0 \\ \downarrow & & \downarrow & & \downarrow \\ B'_1 & \xrightarrow{d_1} & B'_0 & \longrightarrow & B_0 \end{array}$$

The right-hand square is homotopy cartesian by assumption. The outer rectangle is also homotopy cartesian, since it is also given as the composition of two different homotopy cartesian squares. So the left-hand square is homotopy cartesian. □

PROPOSITION 1.12. *Suppose $f : X \to Y$ and $g : Y \to Z$ are maps between Segal spaces.*

(1) *f and g are right fibrations $\Rightarrow$ gf is a right fibration*
(2) *g and gf are right fibrations $\Rightarrow$ f is a right fibration*
(3) *f and gf are right fibrations and f induces a surjection $\pi_0 X_0 \to \pi_0 Y_0 \Rightarrow$ g is a right fibration.*

PROOF. (1) and (2) are straightforward. (3) follows from [**5**, lemma B.9]. □

REMARK 1.13. Proposition 1.7 can be rephrased as follows. For every string of morphisms $b = (b_0 \leftarrow \cdots \leftarrow b_n) \in B_n$, the maps

$$\Delta[0] \xrightarrow{\alpha_n} \Delta[n] \xrightarrow{b} B \qquad \text{and} \qquad \Delta[n-1] \xrightarrow{d^n} \Delta[n] \xrightarrow{b} B$$

are weak equivalences in the right fibration model structure.

1.4. Right fibrations are fibrations in Rezk's model structure of complete Segal spaces.

DEFINITION 1.14. A simplicial space X is **complete** if either (and therefore all) of the maps d_0, d_1, s_0 induce a weak equivalence $X_1^{\text{he}} \simeq X_0$.

REMARK 1.15. The following reformulation of the completeness condition is useful. Let E denote the (nerve of the) groupoid with two objects x, y and exactly two (non-identity) isomorphisms $x \to y$ and $y \to x$. A simplicial space X is complete if and only if the map

$$\mathbb{R}\mathrm{map}(E, X) \to \mathbb{R}\mathrm{map}(\Delta[0], X)$$

induced by any of the (two) maps $\Delta[0] \to E$, is a weak equivalence. The *complete* Segal space model structure is constructed by localizing the Segal space model structure with respect to these maps. The weak equivalences between Segal spaces (not necessarily complete) become the Dwyer-Kan equivalences.

DEFINITION 1.16. A simplicial space X over B is **fiberwise complete** if the square

$$\begin{array}{ccc} X_0 & \xleftarrow{d_1} & X_1^{\text{he}} \\ \downarrow & & \downarrow \\ B_0 & \xleftarrow{d_1} & B_1^{\text{he}} \end{array}$$

is homotopy cartesian.

REMARK 1.17. A levelwise fibration $p : X \to B$ is a fiberwise complete Segal space if and only if every diagram

$$(1.3) \qquad \begin{array}{ccc} \Delta[0] & \longrightarrow & X \\ {\scriptstyle \text{target}}\downarrow & & \downarrow \\ E & \longrightarrow & B \end{array}$$

has a contractible space of lifts for each pair of horizontal maps. More precisely, for each $x \in B_1^{\text{he}}$, the map

$$(1.4) \qquad \mathbb{R}\mathrm{map}_B(f, X) : \mathbb{R}\mathrm{map}_B(E, X) \to \mathbb{R}\mathrm{map}_B(\Delta[0], X)$$

induced by $\Delta[0] \to E \xrightarrow{x} B$ is a weak equivalence of spaces.

PROPOSITION 1.18. *Let B be a Segal space. A map $p : X \to B$ is a fibration in the complete Segal space model structure if and only if p is fiberwise complete and a levelwise fibration. In particular, a map which is a right fibration and a levelwise fibration is a fibration in the complete Segal space model structure.*

PROOF. The first assertion is shown in [**5**, Appendix B]. To prove the second assertion, we suppose $X \to B$ is a right fibration (and a levelwise fibration) and will show that X is fiberwise complete over B. Right fibrations are conservative; that is, if a morphism f in X is mapped to an identity, then f is an identity. So if $g : b' \to b$ and $h : b \to b'$ are homotopy inverse in B and $x \in X_0$ mapping to b, then by the right fibration condition these lift uniquely to a pair of morphisms $\overline{g}$ and $\overline{h}$ in X, which by conservativity must be homotopy inverse. □

COROLLARY 1.19. *If $X \to B$ is a right fibration and B is a complete Segal space, then X is a complete Segal space.*

REMARK 1.20. Proposition 1.18 implies that the left Bousfield localization of the (overcategory) *complete* Segal space model structure with respect to the same maps as in proposition 1.10 defines the same model category.

PROPOSITION 1.21. *Suppose $f : X \to Y$ is a Dwyer-Kan equivalence between Segal spaces over B. Then f is a weak equivalence in the right fibration model structure. If in addition f is a right fibration then f is a levelwise weak equivalence.*

PROOF. Dwyer-Kan equivalences are the weak equivalences between Segal spaces in the complete Segal space model structure, and the right fibration model structure on B is a left Bousfield localization of the overcategory of the complete Segal space model structure, so the first claim follows.

As for the second claim, using proposition 1.18 the assumption is that f is an acyclic fibration in the complete Segal space model structure. But in any left Bousfield localization, the set of acyclic fibrations coincides with the set of acyclic fibrations of the underlying model structure (which in this case is the levelwise model structure on simplicial spaces). □

1.5. Comparison with quasi-categories. Recall that a space (simplicial set) X is viewed as a simplicial discrete space (bisimplicial set) by taking the pull-back along $p_1 : \Delta \times \Delta \to \Delta$, the projection onto the first factor. (The space (set) of n-simplices of p_1^*X is X_n, viewed as a constant simplicial set.) The functor p_1 has a left adjoint, i_1, given by $i_1([n]) = ([n], [0])$, so there is an adjunction

$$p_1^* : \mathcal{S} \leftrightarrows s\mathcal{S} : i_1^* X$$

The set of n-simplices of i_1^*X is the set of 0-simplices of X_n, the space of n-simplices of X. By [**14**], the pair (p_1^*, i_1^*) forms a Quillen equivalence between the model structure for quasi-categories (alias the *Joyal* model structure) and the model structure for complete Segal spaces (with underlying Reedy model structure on simplicial spaces). We can use that to prove that the notion of right fibration for Segal spaces agrees with the quasi-categorical one:

THEOREM 1.22. *Let B be a (Reedy fibrant) complete Segal space. Then (i_1^*, p_1^*) forms a Quillen equivalence between $s\mathcal{S}_{/B}$ and $\mathcal{S}_{/i_1^*B}$, both equipped with the right fibration model structure.*

PROOF. According to [**8**], when Y is a quasi-category, the right fibration model structure on $\mathcal{S}_{/Y}$ is obtained by left Bousfield localization of the overcategory of the Joyal model structure, with respect to right horn inclusions, i.e. maps $\Lambda^n[0] \hookrightarrow \Delta[n] \to Y$ for each $n \geq 1$, where $\Lambda^n[0]$ is the union of all the faces except d^0. (Note that Moerdijk-Heuts' use different conventions, so their definitions are the mirror of ours.) For a complete Segal space B, we have established (c.f. remark 1.20) that the right fibration model structure on $s\mathcal{S}_{/B}$ can be obtained as the left Bousfield localization of the overcategory of the complete Segal space model structure with respect to the maps $p_1^*\Delta[0] \to p_1^*\Delta[1] \to Y$ (left-hand map is d^1).

The claim is that the overcategory model structures are being localized at the same set of maps. In other words, the maps $p_1^*(\Lambda^n[n]) \hookrightarrow p_1^*(\Delta[n])$ (over $\mathbb{L}p_1^* i_1^* B \simeq B$) are weak equivalences in the complete Segal space side. For $n = 1$ there is nothing to prove. For higher n, factor d^n as

$$\Delta[n-1] \to \Lambda^n[0] \to \Delta[n]$$

(Here, and in line with the rest of the paper, we have omitted p_1^* from the notation.) The composite map is a weak equivalence by proposition 1.18. The left-hand map can be given as a composition of lower dimensional right horn extensions (that is, pushouts along right horn inclusions $\Lambda^k[0] \to \Delta[k]$ for $k < n$), each of which is an acyclic cofibration by induction. Therefore, the left-hand map in the display is a weak equivalence and so, by two-out-of-three, the right-hand map is a weak equivalence. □

REMARK 1.23. The condition that B is *complete* can be dropped. Indeed, a Segal space B which is not necessarily complete can be completed: in [**17**], Rezk constructs a Dwyer-Kan equivalence $B \to B^\sharp$ to a *complete* Segal space $B^\sharp$. The homotopy theories of right fibrations over B and $B^\sharp$ are equivalent (this will be proved in corollary 5.6). Therefore, by the theorem above, the homotopy theory of right fibrations over B is equivalent to the one of quasi-categorical right fibrations over $i_1^* B^\sharp$.

1.6. Terminal objects and the generalized Yoneda lemma.

DEFINITION 1.24. Let X and Y be simplicial spaces. The (derived) internal hom construction X^Y is the simplicial space with n-simplices given by

$$\mathbb{R}\mathrm{map}(\Delta[n] \times Y, X)$$

where $\mathbb{R}$map denotes the derived mapping space in the levelwise model structure. Henceforth, this will be the standard interpretation, unless mentioned otherwise. To make sure this defines a functor in the variable $[n]$, one should fix cofibrant/fibrant replacement functors in $s\mathcal{S}$.

REMARK 1.25. The construction above *may not be* the derived functor (in a model categorical sense) of the usual internal hom functor in simplicial spaces, e.g. if the projective model structure is taken. If one works with the injective model structure as Rezk does, then it is. That is a consequence of the fact that this model structure is *closed* (see [**17**]). We'll be cavalier about this point and simply use the definition above. The reason for that is we will often want to use the projective model structure.

DEFINITION 1.26 (Generalized overcategory). Let X be a Segal space. For a space K and map $\alpha : K \to X_0$, define X/K to be the simplicial space obtained as the homotopy pullback of

$$X^{\Delta[1]} \xrightarrow{d_1} X \xleftarrow{\alpha} \Delta[0] \times K$$

REMARK 1.27. When $K = X_0$ and α is the identity map, X/X_0 is the Segal space whose objects are morphisms $x \to y$ in X. A morphism is a string $x \to x' \to y$ (its source is the composite $x \to y$ and its target is $x' \to y$). Note that this construction is not what is often referred to as the arrow category. Rather, it is a overcategory-type construction, as exemplified by the case $K = *$ and α the map selecting an object x.

PROPOSITION 1.28. *The map $d_0 : X/K \to X$, induced by $d_0 : X^{\Delta[1]} \to X$, is a right fibration.*

PROOF. Rezk shows [**17**] that $X^{\Delta[1]}$ is a Segal space. This implies that X/K is a Segal space, so we only need to verify the fibration condition (0.1). We will do this for $Y := X/X_0$. The general case then follows, using that $X/K \simeq Y \times_{X_0} K$.

By construction, Y_0 is identified with X_1. Let us describe Y_1 in words, first. It is the space of commutative squares in X, i.e. $gf = g'f'$, with the extra condition that g' is an identity morphism. This last condition obviously forces the equality $f' = gf$. That is, Y_1 is identified with the space of two composable morphisms in X, alias X_2. More formally, use the decomposition

$$\Delta[1] \times \Delta[1] = \Delta[2] \amalg_{\Delta[1]} \Delta[2]$$

of the square in two triangles, to write Y_1 as the homotopy limit of

$$X_2 \xrightarrow{d_1} X_1 \xleftarrow{d_1} X_2 \xrightarrow{d_2} X_1 \xleftarrow{s_0} X_0$$

Since X is a Segal space, we have that $X_2 \times_{X_1} X_0 \simeq X_1$ and hence $Y_1 \simeq X_2$, as claimed. □

DEFINITION 1.29. An object $x \in X$ is *terminal* if $d_0 : X/x \to X$ is a weak equivalence of simplicial spaces.

This is equivalent to the condition that the homotopy fiber of the target map $d_1 : X_1 \to X_0$ over x be weakly equivalent to X_0.

The crucial observation is

LEMMA 1.30. *Let $p : X \to B$ be a simplicial space over B and suppose X has a terminal object x. Then for every right fibration $q : Y \to B$ the map*

$$x^* : \mathbb{R}\mathrm{map}_B(X, Y) \to \mathbb{R}\mathrm{map}_B(\Delta[0], Y) \tag{1.5}$$

given by precomposition with $x : \Delta[0] \to X$ is a weak equivalence. That is to say, the inclusion $\Delta[0] \xrightarrow{x} X$ (over B) is a weak equivalence in the right fibration model structure.

The following immediate corollary may be regarded as an $(\infty,1)$-categorical form of the Yoneda lemma.

LEMMA 1.31 (Yoneda lemma). *The inclusion $\Delta[0] \to Z/z$ over Z which selects $id_z := s_0(z)$ is a weak equivalence in the right fibration model structure.*

We will use the following simple observation repeatedly in the proof below. Suppose B is a fibrant simplicial space. By definition, the derived mapping space $\mathbb{R}\mathrm{map}_B(X, Y)$ is weakly equivalent to $\mathrm{map}_B(X^c, Y^f)$ where $Y^f \to B$ is a fibrant replacement of $Y \to B$ (a levelwise fibration) and X^c a cofibrant replacement of X

(cofibrant as a simplicial space). The latter space is of course the pullback (alias fiber) of

$$* \to \mathrm{map}(X^c, B) \leftarrow \mathrm{map}(X^c, Y^f)$$

where the right-hand map is given by composition with $Y^f \to B$. The right hand-map is a fibration, and both mapping spaces in the display are weakly equivalent to the derived mapping spaces. Therefore, for a map $Y \to B$, $\mathbb{R}\mathrm{map}_B(X,Y)$ is naturally weakly equivalent to the homotopy pullback of

$$* \to \mathbb{R}\mathrm{map}(X, B) \leftarrow \mathbb{R}\mathrm{map}(X, Y)\ .$$

PROOF OF LEMMA 1.30. Without loss of generality, we may assume that $Y \to B$ is fibrant (a right fibration and fibration of simplicial spaces) and X is cofibrant and B is fibrant as a simplicial space. Let $y \in Y_0$ with $q(y) = b$. Define $\mathbb{R}\mathrm{map}((X,x),(Y,y))$ as the homotopy fiber of the map

$$\mathbb{R}\mathrm{map}(X,Y) \xrightarrow{x^*} \mathbb{R}\mathrm{map}(\Delta[0], Y) = Y_0$$

over $y \in Y$, and define $\mathbb{R}\mathrm{map}((X,x),(B,b))$ similarly. By commuting homotopy limits, the homotopy fiber of (1.5) over y is identified with the homotopy fiber of

$$\mathbb{R}\mathrm{map}((X,x),(Y,y)) \to \mathbb{R}\mathrm{map}((X,x),(B,b))$$

over $p : X \to B$.

We have a commutative square

$$\begin{array}{ccc}\mathbb{R}\mathrm{map}((X,x),(Y,y)) & \longrightarrow & \mathbb{R}\mathrm{map}((X/x, id_x),(Y/y, id_y)) \\ \downarrow & & \downarrow \\ \mathbb{R}\mathrm{map}((X,x),(B,b)) & \longrightarrow & \mathbb{R}\mathrm{map}((X/x, id_x),(B/b, id_b))\end{array} \tag{1.6}$$

(Here we follow Rezk and write $id_z \in Z_1$ for the image under s_0 of $z \in Z_0$.)

The horizontal maps are weak equivalences because x is the terminal object of X. In more detail, the composition

$$\mathbb{R}\mathrm{map}((X,x),(Y,y)) \to \mathbb{R}\mathrm{map}((X/x, id_x),(Y/y, id_y)) \xrightarrow{d_0} \mathbb{R}\mathrm{map}((X/x, id_x),(Y,y))$$

agrees with precomposition with $d_0 : X/x \to X$ and so it is a weak equivalence (because x is the terminal object of X). This shows that the top horizontal map in (1.6) has a homotopy left inverse.

Now let X/id_x denote the over category $(X/x)/id_x$ whose objects are 2-morphisms in X with source id_x. The composition

$$\begin{array}{ccc}\mathbb{R}\mathrm{map}((X/x, id_x),(Y/y, id_y)) & \longrightarrow & \mathbb{R}\mathrm{map}((X/x, id_x),(Y,y)) \\ & & \downarrow{\scriptstyle d_0} \\ & & \mathbb{R}\mathrm{map}((X/id_x, id_{id_x}),(Y/y, id_y))\end{array}$$

agrees with precomposition with $d_0 : X/id_x \to X/x$ and so is a weak equivalence (because id_x is the terminal object of X/x). This shows that the top horizontal map in (1.6) also has a homotopy right inverse, and so is a weak homotopy equivalence. The same arguments show that the lower horizontal map is also a weak equivalence.

It remains to show that the right-hand map in (1.6) is a weak equivalence. Since $Y \to B$ is a right fibration, the induced map between horizontal homotopy fibers (over $y \in Y_0$ and $b \in B_0$) in the diagram

$$\begin{array}{ccc} Y_0 & \xleftarrow{d_1} & Y_1 \\ \downarrow & & \downarrow \\ B_0 & \xleftarrow{d_1} & B_1 \end{array}$$

is a weak equivalence. That is, $Y/y \xrightarrow{\simeq} B/b$. This finishes the proof. □

REMARK 1.32. There are parametrized versions of lemmas 1.30 and 1.31. Let X be a Segal space over B. If $d_0 : X/K \to X$ is a weak equivalence (i.e. K is a terminal subspace of the space objects of X) then

$$\alpha : K \times \Delta[0] \to X$$

(over B) is a weak equivalence in the right fibration model structure on $s\mathcal{S}_{/B}$. Therefore, the map $K \times \Delta[0] \to X/K$ is a weak equivalence in the right fibration model structure on $s\mathcal{S}_{/X}$.

1.7. Proof of theorem *A***.** We begin with

PROPOSITION 1.33. *The Grothendieck construction $\mathcal{G}$ has a simplicial left adjoint*

$$L : s\mathcal{S}_{/N\mathcal{C}} \leftrightarrows \mathsf{Fun}(\mathcal{C}^{op}, \mathcal{S}) : \mathcal{G}$$

This adjunction is Quillen when both sides are equipped with the projective model structure.

PROOF. Suppose X is a simplicial space over $N\mathcal{C}$. Then X can be presented as the colimit (alias coequalizer) in $s\mathcal{S}$ of

$$\coprod_{[n_0]} \Delta[n_0] \times X_{n_0} \leftleftarrows \coprod_{[n_0]\to[n_1]\in\Delta} \Delta[n_0] \times X_{n_1} \tag{1.7}$$

This is a well known consequence of the Yoneda lemma. Note that the map from this colimit to X is an isomorphism over $N\mathcal{C}$.

Left adjoints preserve colimits, and so L is uniquely determined (up to isomorphism) by its value on objects of the form $\alpha : \Delta[n] \times K \to N\mathcal{C}$ for some space K. Suppose given one such object corresponding via the Yoneda lemma to a map $\alpha : K \to N\mathcal{C}_n$ from K to the space of n composable morphisms in $\mathcal{C}$. Denote by α' the map $K \to N\mathcal{C}_n \to N\mathcal{C}_0$ given by applying α and then $d_1 d_2 d_3 \dots d_n$, the operation selecting the ultimate target object.

Define

$$L(\Delta[n] \times K \xrightarrow{\alpha} N\mathcal{C}) := \coprod_{c \in im(\alpha')} K_c \times \mathrm{map}_{\mathcal{C}}(-, c)$$

where K_c is the fiber of $\alpha' : K \to N\mathcal{C}_0$ over the object c in the image of α'. Note that when $K = *$, $L(\alpha : \Delta[n] \to N\mathcal{C})$ is the representable functor $\mathrm{map}_{\mathcal{C}}(-, c_0)$ where c_0 is the ultimate target object of $\alpha = (c_0 \leftarrow \cdots \leftarrow c_n)$.

It is a mechanical verification using the Yoneda lemma that

$$\mathrm{Hom}(L(\alpha), F) \cong \mathrm{map}_{N\mathcal{C}}(K \times \Delta[n], \mathcal{G}(F))$$

for every functor F. This settles the claim that $(L, \mathcal{G})$ form a simplicial adjunction. Clearly, $\mathcal{G}$ preserves fibrations and weak equivalences, and so the pair is Quillen. □

REMARK 1.34. In fact, $\mathcal{G}$ also has a right adjoint. Let c be an object of $\mathcal{C}$ and $\mathcal{C}/c$ denote the overcategory (this is an internal category, with *space* of objects consisting of maps with target c). Let $N(\mathcal{C}/c) \to N(\mathcal{C})$ be the map induced by the forgetful functor $\mathcal{C}/c \to \mathcal{C}$. The functor $R : s\mathcal{S}_{/N\mathcal{C}} \to \mathsf{PSh}(\mathcal{C})$ which to $X \in s\mathcal{S}_{/N\mathcal{C}}$ associates the presheaf $R(X)$ defined by

$$R(X)(c) := \mathrm{map}_{N\mathcal{C}}(N(\mathcal{C}/c), X)$$

is the (simplicial) right adjoint to $\mathcal{G}$.

The substance of theorem A is encompassed in the following step.

LEMMA 1.35. *For each object $X \to N\mathcal{C}$, the derived unit map $X \to \mathbb{R}\mathcal{G}\mathbb{L}L(X)$ is a weak equivalence in the right fibration model structure on the category of simplicial spaces over $N\mathcal{C}$.*

PROOF OF LEMMA 1.35. Since $\mathcal{G}$ preserves weak equivalences, it is enough to show that the (non-derived) unit map $X \to \mathcal{G}L(X)$ is a weak equivalence for each cofibrant simplicial space X. That is, we are required to show that for each fibrant (i.e. right fibration) Z, the induced map

$$\mathbb{R}\mathrm{map}_{N\mathcal{C}}(\mathcal{G}L(X), Z) \to \mathbb{R}\mathrm{map}_{N\mathcal{C}}(X, Z) \tag{1.8}$$

is a weak equivalence of spaces.

Let $Y_\bullet$ be the canonical resolution of X by representables. Namely, Y is a simplicial object in $s\mathcal{S}_{/N\mathcal{C}}$ with simplicial space Y_k of k-simplices given by

$$\coprod_{[n_0]\to[n_1]\to\cdots\to[n_k]} X_{n_k} \times \Delta[n_0]$$

The Yoneda lemma provides a map $Y_\bullet \to X$, where X is viewed as a constant simplicial object in $s\mathcal{S}_{/N\mathcal{C}}$. The induced map $|Y_\bullet| \to X$ is well known to be a levelwise weak equivalence in $s\mathcal{S}_{/N\mathcal{C}}$ (this can be proved by exhibiting a simplicial contracting homotopy).

We can thus reduce the problem to the case of representables. By Ken Brown's lemma [**10**, 7.7.2], the left Quillen functor L preserves weak equivalences between cofibrant objects. So $L(X) \xrightarrow{\simeq} L(|Y_\bullet|)$ since X is cofibrant by hypothesis and $|Y_\bullet|$ is the geometric realization of a Reedy cofibrant object, hence cofibrant. Therefore,

$$\mathbb{R}\mathrm{map}_{N\mathcal{C}}(\mathcal{G}L(X), Z) \simeq \mathbb{R}\mathrm{map}_{N\mathcal{C}}(|\mathcal{G}L(Y_\bullet)|, Z) \simeq \mathrm{Tot}\,\mathrm{map}_{N\mathcal{C}}(\mathcal{G}L(Y_\bullet), Z)$$

The first equivalence is a consequence of the fact that both L and $\mathcal{G}$ commute with geometric realization (and colimits in general, since both are left adjoints). In the second equivalence, we have used that $|\mathcal{G}L(Y_\bullet)|$ is cofibrant since it is the geometric realization of a Reedy cofibrant object. Similarly, $\mathbb{R}\mathrm{map}_{N\mathcal{C}}(X, Z) \simeq \mathrm{Tot}\,\mathrm{map}_{N\mathcal{C}}(Y_\bullet, Z)$.

The upshot is that it suffices to check that (1.8) is a weak equivalence for X of the form $f : \Delta[n] \to N\mathcal{C}$ corresponding to some string of n morphisms. Because the map $\Delta[0] \to \Delta[n]$ (selecting the first vertex 0) is a weak equivalence in the right fibration model structure, it suffices to prove the claim for $X = \Delta[0] \to N\mathcal{C}$ corresponding to an object $b \in N\mathcal{C}$. This verification occupies the remainder of the proof.

When $X = (\Delta[0] \xrightarrow{b} N\mathcal{C})$, the target of (1.8) is $\mathbb{R}\mathrm{map}_{N\mathcal{C}}(\Delta[0], Z) \simeq Z_b$. We now investigate the source. Recall that $L(f) = \mathrm{map}(-, b)$. Therefore, $\mathcal{G}L(f) = N\mathcal{C}/b$. The inclusion

$$\Delta[0] \xrightarrow{id_b} N\mathcal{C}/b$$

over $\{b\} \in N\mathcal{C}_0$, induces a map

$$\mathbb{R}\mathrm{map}_{N\mathcal{C}}(N\mathcal{C}/b, Z) \to \mathbb{R}\mathrm{map}_{N\mathcal{C}}(\{b\}, Z) = Z_b \tag{1.9}$$

which agrees with (1.8). Clearly $\{id_b\}$ is a terminal object of $N\mathcal{C}/b$. This implies, by proposition 1.30, that (1.9) is a weak equivalence. □

Moreover, if X is a right fibration over $N\mathcal{C}$, then we immediately obtain the strictification result that X is levelwise equivalent to an actual (*strict*) presheaf:

COROLLARY 1.36. *Suppose X is a right fibration over $N\mathcal{C}$ and X is cofibrant as a simplicial space. Then the unit map $X \to \mathcal{G}L(X)$ is* levelwise *weak equivalence.*

PROOF. Weak equivalences between fibrant objects in $s\mathcal{S}_{/N\mathcal{C}}$ are determined levelwise, by proposition 1.10. □

PROOF OF THEOREM A. We have already established that $(L, \mathcal{G})$ forms a Quillen pair. The unit of the induced adjunction on homotopy categories

$$\mathsf{Ho}L : \mathsf{Ho}(s\mathcal{S}_{/N\mathcal{C}}) \leftrightarrows \mathsf{Ho}(\mathsf{Fun}(\mathcal{C}^{\mathrm{op}}, \mathcal{S})) : \mathsf{Ho}\mathcal{G} \tag{1.10}$$

is an isomorphism by lemma 1.35. Moreover, $\mathcal{G}$ reflects weak equivalences between fibrant objects. It follows that the counit is also an isomorphism. □

1.8. Explicit description of the left adjoint.

PROPOSITION 1.37. *There is a weak equivalence*

$$\operatorname*{hocolim}_{\Delta} N(c/\mathcal{C}) \times_{N\mathcal{C}} X \to L(X)(c)$$

natural in $c \in \mathcal{C}$. Here $c/\mathcal{C}$ is the internal (under)category with space of objects consisting of maps in $\mathcal{C}$ with source c. The left-hand side is clearly functorial in $\mathcal{C}$ since a map $c \to d$ in $\mathcal{C}$ induces a map of nerves $N(d/\mathcal{C}) \to N(c/\mathcal{C})$.

PROOF. It suffices to consider the case when $X = \Delta[n] \to N\mathcal{C}$. For each $c \in \mathcal{C}$, the simplicial space $N(c/\mathcal{C}) \times_{N\mathcal{C}} \Delta[n]$ is equivalent to one with k-simplices given by

$$\coprod_{0 \leq i_0 \leq i_1 \leq \cdots \leq i_k \leq n} \mathrm{map}_{\mathcal{C}}(c, b_{i_0})$$

where the face maps are given by composition and the degeneracy are obvious. This has a simplicial contraction onto $\mathrm{map}_{\mathcal{C}}(c, b_n)$, using the maps $b_{i_j} \to b_n$. □

1.9. Internal version. There exists a variant of theorem A for category objects in $\mathcal{S}$ (alias internal categories in spaces).

DEFINITION 1.38. Let $\mathcal{C}$ be an internal category in spaces. An internal (contravariant) functor $F : \mathcal{C}^{op} \to \mathcal{S}$ (also called *discrete fibration*) consists of an internal category $\mathcal{F}$ together with a functor $\mathcal{F} \to \mathcal{C}$ of internal categories (i.e., a simplicial map between nerves) with the property that the square

$$\begin{array}{ccc} \mathcal{F}_1 & \xrightarrow{\text{target}} & \mathcal{F}_0 \\ \downarrow & & \downarrow \\ \mathcal{C}_1 & \xrightarrow{\text{target}} & \mathcal{C}_0 \end{array}$$

is cartesian (alias a pullback). The subscript 1 stands for space of morphisms and the subscript 0 stands for space of objects.

We adopt the same notation as before and denote by $\mathsf{Fun}(\mathcal{C}^{op}, \mathcal{S})$ the (internal) category of internal functors $\mathcal{C}^{op} \to \mathcal{S}$. Horel has proved in [**11**] – under the relatively harmless condition that the target map $\mathcal{C}_1 \to \mathcal{C}_0$ is a fibration in $\mathcal{S}$ – that the category $\mathsf{Fun}(\mathcal{C}^{op}, \mathcal{S})$ has a projective model structure in which a natural transformation $\mathcal{F} \to \mathcal{F}'$ over $\mathcal{C}$ is a weak equivalence (resp. fibration) if the map $\mathcal{F}_0 \to \mathcal{F}'_0$ is a weak equivalence (resp. fibration) in $\mathcal{S}$.

The promised mild extension of theorem A is

THEOREM 1.39. *Let $\mathcal{C}$ be an internal category in $\mathcal{S}$ such that the target map $\mathcal{C}_1 \to \mathcal{C}_0$ is a fibration. Then*

$$L : s\mathcal{S}_{/N\mathcal{C}} \leftrightarrows \mathsf{Fun}(\mathcal{C}^{op}, \mathcal{S}) : \mathcal{G}$$

is a simplicial Quillen equivalence, where the left-hand side is equipped with the right fibration model structure. Here $\mathcal{G}$ assigns to an internal functor $\mathcal{F} \to \mathcal{C}$ the map of simplicial spaces $N\mathcal{F} \to N\mathcal{C}$ given by the nerve.

PROOF. We need to modify slightly the construction of the left adjoint L from the proof of proposition 1.33. For an object $\alpha : \Delta[n] \times K \to N\mathcal{C}$, declare $L(\alpha)$ to be the category with object space $L(\alpha)_0$:

$$L(\alpha)_0 := \text{pullback}(\mathcal{C}_1 \xrightarrow{\text{target}} \mathcal{C}_0 \leftarrow K)$$

where the right-hand map is the composition $K \to \mathcal{C}_n \to \mathcal{C}_0$ of α with the ultimate target operator. The morphism space $L(\alpha)_1$ is defined as $\mathcal{C}_1 \times_{\mathcal{C}_0} L(\alpha)_0$. The source map

$$L(\alpha)_1 \to L(\alpha)_0$$

is induced by the composition map $\mathcal{C}_1 \times_{\mathcal{C}_0} \mathcal{C}_1 \to \mathcal{C}_1$ in $\mathcal{C}$. The target map is the projection. The forgetful map $L(\alpha) \to N\mathcal{C}$ gives an internal functor $\mathcal{C}^{op} \to \mathcal{S}$. We leave to the reader the verification that L as prescribed above (and extended to all simplicial spaces by colimits) is the left adjoint to $\mathcal{G}$.

The condition that the target map $\mathcal{C}_1 \to \mathcal{C}_0$ is a fibration guarantees that $\mathcal{G}$ preserves fibrant objects and, more generally, fibrations. It is equally clear that $\mathcal{G}$ preserves weak equivalences. So the pair $(L, \mathcal{G})$ is Quillen.

The argument for the enriched case goes through mostly without substantial changes to show that $(L, \mathcal{G})$ is a Quillen equivalence. The key step is to show that, for an object $\alpha : \Delta[0] \times K \to N\mathcal{C}$ with K a cofibrant space, the unit map

$$\alpha \to \mathcal{G}(L(\alpha))$$

is a weak equivalence in the right fibration model structure on $s\mathcal{S}_{/N\mathcal{C}}$. In other words, that for every right fibration $Z \to N\mathcal{C}$, the induced map

$$\mathbb{R}\mathrm{map}_{N\mathcal{C}}(\mathcal{G}(L(\alpha)), Z) \to \mathbb{R}\mathrm{map}_{N\mathcal{C}}(\alpha, Z)$$

is a weak equivalence. This can be proved as before; but now both sides are weakly equivalent to $\mathbb{R}\mathrm{map}_{\mathcal{C}_0}(K, Z_0)$. For the left-hand side, this follows from lemma 1.31 and remark 1.32. □

2. Segal objects in a $\mathcal{V}$-category

In this section, we discuss a generalization of the previous constructions, by considering Segal objects in a category $\mathcal{V}$.

To conform with model categorical setups, we need to impose a set of technical conditions on $\mathcal{V}$. We assume that $\mathcal{V}$ is a monoidal category with respect to the cartesian product and that it is a cartesian closed model category (see e.g. [**12**, Ch. 4], [**17**, 2.5]). We denote the unit (i.e. the terminal object) by $*$, and we assume it is a cofibrant object. Finally, we assume that $\mathcal{V}$ is left proper and combinatorial.

The category $s\mathcal{V}$ of simplicial objects in $\mathcal{V}$ is tensored and cotensored over $\mathcal{V}$. For $X \in s\mathcal{V}$ and $K \in \mathcal{V}$, the tensor $X \times K$ is given by $[n] \mapsto X_n \times K$, and the cotensor X^K is given by $[n] \mapsto \mathrm{map}(K, X_n)$, where map here refers to the internal mapping object in $\mathcal{V}$. It follows that $s\mathcal{V}$ becomes a $\mathcal{V}$-model category (see [**12**, 4.2.18] for a definition) for the Reedy model structure, and is left proper and combinatorial (from the assumptions we made on $\mathcal{V}$).

The obvious definition works:

DEFINITION 2.1. A *Segal object* in $\mathcal{V}$ is a functor $X : \Delta^{op} \to \mathcal{V}$ such that

$$X_n \to X_1 \times^h_{X_0} \cdots \times^h_{X_0} X_1$$

is a weak equivalence in $\mathcal{V}$.

DEFINITION 2.2. Let Z be a Segal object in $\mathcal{V}$. A map $X \to Z$ in $s\mathcal{V}$ is said to be a **right fibration** if X is a Segal object in $\mathcal{V}$ and the square analogous to (0.1) is homotopy cartesian.

The technical assumptions on $\mathcal{V}$ made above guarantee the existence of enriched left Bousfield localizations (see [**1**]). So we can define a model category structure on $s\mathcal{V}_{/B}$, the category of simplicial objects in $\mathcal{V}$ over B, whose homotopy theory is that of right fibrations over B.

PROPOSITION 2.3. *Fix a Segal object B. There is a left proper, $\mathcal{V}$-model structure structure on $s\mathcal{V}_{/B}$ – called the* ***right fibration model structure*** *– uniquely characterized by the following properties:*

- *An object $X \to B$ is fibrant if it is a right fibration and a fibration in $s\mathcal{V}$*
- *A morphism $X \to Y$ between right fibrations over B is a weak equivalence if $X_0 \to Y_0$ is a weak equivalence*
- *A morphism is a cofibration if it is a cofibration in $s\mathcal{V}$.*

More generally, a morphism $f : X \to Y$ between any two simplicial objects over B is a weak equivalence if the induced map on derived mapping objects

$$\mathbb{R}\mathrm{map}_B(f, W) : \mathbb{R}\mathrm{map}_B(Y, W) \to \mathbb{R}\mathrm{map}_B(X, W)$$

(computed with respect to the levelwise model structure) is a weak equivalence in $\mathcal{V}$ for every fibrant object $W \to B$.

We note that, by the assumptions made on $\mathcal{V}$, the internal mapping object $\mathrm{map}(-,-)$ is a Quillen bifunctor, and so it admits a derived functor $\mathbb{R}\mathrm{map}$.

3. Cartesian fibrations

In this section, we take $\mathcal{V}$ to be Rezk's category of complete Segal spaces Cat_∞. A Segal object in Cat_∞ is what one might call a double Segal space, that is, a bisimplicial space $X_{m,n}$ such that $X_{\bullet,n}$ is a Segal space (for each fixed $n \geq 0$) and $X_{m,\bullet}$ is a Segal space (for each fixed $m \geq 0$). This is reminiscent of a double category (a category object in categories) not of a 2-category.

A Segal space Z is an example of a double Segal space: set $X_{m,n} = Z_m$ for $m, n \geq 0$.

REMARK 3.1. A double Segal space as defined is not a fibrant object in $s\mathcal{V}$, the missing condition being that $X_{\bullet,n}$ is complete as a Segal space.

For the rest of this section, B denotes a given double Segal space.

DEFINITION 3.2. A map $X \to B$ of bisimplicial spaces is a **Cartesian fibration** if X is a double Segal space and the map

$$X_{\bullet,n} \to B_{\bullet,n}$$

is a right fibration (of simplicial spaces) for every n.

REMARK 3.3. A word of warning about terminology: the definition above is not the literal analogue of the classical definition of a Cartesian fibration (or its quasi-categorical version), so it may be argued that it's not fully deserving of that name. For example, definition 3.2 is one categorical level higher when compared to the standard definition. We decided to keep the potentially conflicting terminology, in part due to theorem B (both perspectives give weak models for functors with values in Cat_∞) and because it is plausible that a direct comparison should exist.

3.1. Homotopy theory of Cartesian fibrations.

Because of its importance, we restate proposition 2.3 for the special case $\mathcal{V} = \mathsf{Cat}_\infty$ below.

PROPOSITION 3.4. *There is a left proper, Cat_∞-enriched model structure on the category of bisimplicial spaces over B uniquely characterized by the following data.*

- *An object $X \to B$ is fibrant if it is a Cartesian fibration and $X_{m,\bullet}$ is a complete Segal space for each m.*
- *A map $f : X \to Y$ over B is a cofibration if it is so in the underlying model structure on bisimplicial spaces.*

A map $f : X \to Y$ over B is a weak equivalence if $f_{m,\bullet}$ is a weak equivalence in the complete Segal space model structure for all m, and $f_{\bullet,n}$ is a weak equivalence in the right fibration model structure for all n.

REMARK 3.5. Suppose B is a Segal space (viewed as a double Segal space). Let X and Y be over B, but $X_{n,\bullet}$ and $Y_{n,\bullet}$ not necessarily complete. A map $f : X \to Y$ over B is a weak equivalence if and only if the map

$$f_{0,\bullet} : X_{0,\bullet} \to Y_{0,\bullet}$$

is a Dwyer-Kan equivalence or, equivalently, the map

$$f_{b,\bullet} : X_{b,\bullet} \to Y_{b,\bullet}$$

is a Dwyer-Kan equivalence for every $b \in B_0$.

3.2. Strictification. Let $F : \mathcal{C}^{op} \to s\mathcal{S}$ be an simplicially enriched functor. Consider $N\mathcal{C}$ as a bisimplicial space trivially, i.e. $N(\mathcal{C})_{m,n} = N(\mathcal{C})_m$. Form a bisimplicial space $\mathcal{G}(F)$ with space of (m,n)-simplices given by

$$\mathcal{G}(F)_{m,n} = \coprod_{(c_0,\dots,c_m)} \mathrm{map}_{\mathcal{C}}(c_m, c_{m-1}) \times \cdots \times \mathrm{map}_{\mathcal{C}}(c_1, c_0) \times F(c_0)_n$$

We call $\mathcal{G}$ the *Grothendieck construction.* (We point out that this is not the standard usage of the term when applied to functors with values in categories.) There is an adjunction

$$L : ss\mathcal{S}_{/N\mathcal{C}} \leftrightarrows \mathsf{Fun}(\mathcal{C}^{op}, s\mathcal{S}) : \mathcal{G}$$

It is easy to check that

$$L(\Delta[m,n] \xrightarrow{f} N\mathcal{C}) = \Delta[n] \times \mathrm{map}_{\mathcal{C}}(-, c_0)$$

where f corresponds to a string of morphisms $c_0 \leftarrow \cdots \leftarrow c_m$ in $\mathcal{C}$.

If $F(c)$ is a Segal space for each $c \in \mathcal{C}$, then $\mathcal{G}(F) \to N\mathcal{C}$ is a Cartesian fibration. The pair $(L, \mathcal{G})$ lifts to a Quillen pair between the Cartesian model structure on bisimplicial spaces over $N\mathcal{C}$ and the projective model structure on $\mathsf{Fun}(\mathcal{C}^{op}, \mathsf{Cat}_\infty)$.

LEMMA 3.6. *The right adjoint $\mathcal{G}$ preserves weak equivalences.*

PROOF. This follows rather directly from the definitions of a weak equivalence in $\mathsf{Fun}(\mathcal{C}^{op}, \mathsf{Cat}_\infty)$ and the proposition above. □

When $X = \Delta[m,n]$, the following is essentially the Yoneda lemma.

LEMMA 3.7. *For X a cofibrant bisimplicial space and a map $X \to N\mathcal{C}$, the non-derived unit map $X \to \mathcal{G}L(X)$ (over $N\mathcal{C}$) is a weak equivalence in the Cartesian fibration model structure.*

PROOF. Let X be a representable $f : \Delta[m,n] \to N\mathcal{C}$ corresponding to a string $c = c_0 \leftarrow \cdots \leftarrow c_m$ of morphisms in $\mathcal{C}$. Then $L(f) = \Delta[n] \times \mathrm{map}_{\mathcal{C}}(-, c)$ and $\mathcal{G}L(f)$ is the *perpendicular* (or box) product $N(\mathcal{C}/c) \square \Delta[n]$ of $N(\mathcal{C}/c)$ and $\Delta[n]$, i.e. the bisimplicial space with space $N(\mathcal{C}/c)_i \times \Delta[k]_j$ of (i,j)-simplices.

The map $\Delta[0,n] \to \Delta[m,n]$ (over $N\mathcal{C}$) that selects the object c is a weak equivalence by definition. The map $\Delta[0,n] \to N(\mathcal{C}/c) \square \Delta[n]$ over $N\mathcal{C}$ is a weak equivalence in the Cartesian fibration model structure as we now explain. Suppose $W \to N\mathcal{C}$ is a Cartesian fibration. By adjunction, we have

$$\mathbb{R}\mathrm{map}_{N\mathcal{C}}(N(\mathcal{C}/c) \square \Delta[n], W) \simeq \mathbb{R}\mathrm{map}_{N\mathcal{C}}(N(\mathcal{C}/c), W_{\bullet,n})$$

But the right-hand mapping space is weakly equivalent to $\mathbb{R}\mathrm{map}_{N\mathcal{C}}(\Delta[0,n], W)$ since $W_{\bullet,n}$ is a right fibration over $N\mathcal{C}$ for every n by assumption. This proves the claim for representable objects. For an arbitrary cofibrant X, resolve by representables. □

PROOF OF THEOREM B. Since $\mathcal{G}$ preserves weak equivalences, the derived unit is weakly equivalent to the the non-derived unit for cofibrant objects, which is a weak equivalence by lemma 3.7. Moreover, the Grothendieck construction $\mathcal{G}$ reflects weak equivalences between fibrant objects. Both $ss\mathcal{S}_{/N\mathcal{C}}$ and $\mathsf{Fun}(\mathcal{C}^{op}, \mathsf{Cat}_\infty)$ are enriched model categories over Cat_∞ and the pair $(L, \mathcal{G})$ is a Cat_∞-enriched adjunction. Hence, the derived adjunction $(\mathbb{L}L, \mathbb{R}\mathcal{G})$ is an equivalence of $(\infty, 2)$-categories. □

REMARK 3.8. There is a variant of theorem B when $\mathcal{C}$ is a bicategory.

4. (Co)limits for a general $\mathcal{V}$

4.1. Colimits. Let C be a Segal object in $\mathcal{V}$ and $p : X \to C$ a right fibration. We define the *homotopy colimit* of p as the homotopy colimit of $X : \Delta^{op} \to \mathcal{V}$. This is justified by the following.

PROPOSITION 4.1. *The homotopy colimit of p is the homotopy left adjoint to the constant diagram functor*

$$s\mathcal{V} \to s\mathcal{V}_{/C}$$

which sends Y to $Y \times C \to C$. Here both categories are given the right fibration model structure (with underlying projective model structure on $\mathcal{V}$).

PROOF. Write $t : C \to *$ for the terminal map and t^* for the constant diagram functor. Then t^* has a left adjoint $t_!$ given by $t_!(X \to C) = X$ and $(t_!, t^*)$ form a Quillen pair for the projective model structure on $s\mathcal{V}$. Because t^* preserves right fibrations, this pair is also Quillen with respect to the right fibration model structures. Note that, for the right fibration model structure on $s\mathcal{V} = s\mathcal{V}_{/*}$, fibrant objects are the homotopically constant simplicial objects (i.e. all maps are weak equivalences).

Write $c : \mathcal{V} \to s\mathcal{V}$ for the constant diagram functor. It remains to show that $t_!(p) = X$ is weakly equivalent to $c(\mathrm{hocolim} X)$ in the right fibration model structure on $s\mathcal{V}$. For a fibrant (i.e. homotopically constant) simplicial object Z, we have that $c(\mathrm{hocolim} Z) \simeq Z$. So the result follows from the chain of weak equivalences:

$$\begin{aligned} \mathbb{R}\mathrm{map}(X, Z) &\simeq \mathbb{R}\mathrm{map}(X, c(\mathrm{hocolim} Z)) \\ &\simeq \mathbb{R}\mathrm{map}(\mathrm{hocolim} X, \mathrm{hocolim} Z) \\ &\simeq \mathbb{R}\mathrm{map}(c(\mathrm{hocolim} X), c(\mathrm{hocolim} Z)) \end{aligned}$$

where the second equivalence comes from the adjunction between hocolim and c. □

4.2. Limits. If $\mathcal{V}$ is a cartesian closed model category, then $s\mathcal{V}$ is a cartesian closed model category with the injective model structure. It follows from the definition that, for every cofibrant object B, the adjunction

$$- \times B : s\mathcal{V} \leftrightarrows s\mathcal{V} : (-)^B$$

forms a Quillen pair. This implies that (and is in fact equivalent to) the following pair is also Quillen:

$$t^* : s\mathcal{V} \leftrightarrows s\mathcal{V}_{/B} : t_*$$

where $t^*(X) = X \times B \to B$ and $t_*(Y \to B) = \mathrm{map}_A(A, Y)$, viewed as a constant simplicial object. We then have

PROPOSITION 4.2. *Let B be a (cofibrant) Segal object in $\mathcal{V}$. The constant diagram functor t^* has a right adjoint*

$$t^* : s\mathcal{V} \leftrightarrows s\mathcal{V}_{/B} : t_*$$

and the pair is Quillen with respect to the right fibration model structure (with underlying injective model structure on $s\mathcal{V}$). Moreover, $\mathbb{R}t_*$ evaluated at a right fibration $p : Y \to B$ is identified with the derived sections object $\mathbb{R}\mathrm{map}_B(B, Y)$, i.e. the pullback of

$$* \to \mathbb{R}\mathrm{map}(B, B) \leftarrow \mathbb{R}\mathrm{map}(B, Y)$$

computed in $\mathcal{V}$. We call $\mathbb{R}t_*(p)$ the (homotopy) limit of p.

5. Kan extensions

We now deal with Kan extensions of right fibrations with values in spaces. Suppose $i : C \to D$ is a map between Segal spaces. Base change along i induces a functor

$$i^* : s\mathcal{S}_{/D} \to s\mathcal{S}_{/C}$$

by setting $i^*(Y \to D) = Y \times_D C$.

PROPOSITION 5.1 (Left Kan extensions exist). *Base change along i is the right adjoint in a simplicial Quillen pair*

$$i_! : s\mathcal{S}_{/C} \leftrightarrows s\mathcal{S}_{/D} : i^*$$

where both categories are equipped with the right fibration model structure. The derived left adjoint $\mathbb{L}i_!$ is called the homotopy left Kan extension along i.

PROOF. The left adjoint $i_!$ is given by post-composition

$$i_!(Z \xrightarrow{f} C) = Z \xrightarrow{i \circ f} D$$

When $Z = \Delta[n]$, we have

$$\mathrm{map}_D(i_!(f), X) \xrightarrow{\cong} \mathrm{map}_C(f, i^*X)$$

This map is induced, via the Yoneda lemma, by $X_{i(f)} \xrightarrow{\cong} (i^*X)_f$. The general case follows.

By construction, $i_!$ preserves cofibrations since it preserves generating cofibrations. We also know that right fibrations are stable under base change (proposition 1.11) which implies that i^* preserves fibrant objects (i.e. those right fibrations which are also levelwise fibrations). □

The next topic is how to characterize of Kan extensions. Let $i : C \to D$ be a map of Segal spaces. The key construction is a bisimplicial space D/i which in bidegree (m, n) consists of strings of composable morphisms in $\mathcal{D}$ of the form

$$i(c_0) \leftarrow \cdots \leftarrow i(c_n) = d_0 \leftarrow \cdots \leftarrow d_m \leftarrow d$$

(the last n morphisms come from morphisms in C). Functoriality in each simplicial direction is given by composing, or forgetting, morphisms (face maps) or adding identities (degeneracy maps).

More precisely, let us define

$$(D/i)_{m,n} := D_{m+n+1} \times_{D_n} C_n$$

where $D_{m+n+1} \to D_n$ is the map induced by the inclusion $\gamma := d_{n+1} \ldots d_{m+n+1} : [n] \to [m + n + 1]$, i.e. $\gamma(i) = i$.

On morphisms, D/i is constructed as follows. Fix m. The degeneracy maps in the n-direction

$$s_i : (D/i)_{m,n} \to (D/i)_{m,n+1}$$

(for $0 \leq i \leq n$) are given by (s_{i+m+1}, s_i). The face maps are defined as (d_{i+m+1}, d_i) (inspection shows that it induces a map of pullbacks). Now fix n. Given a map $j : [m] \to [m']$ in Δ, extend by the identity to a map $J : [m+n+1] \to [m'+n+1]$. That is,

$$J(i) = \begin{cases} i & \text{if } i \leq n \\ j(i-n) & \text{if } n \leq i \leq m+n \\ i-m-n & \text{if } i > m+n \end{cases}$$

Then declare

$$(D/i)_{m',n} \to (D/i)_{m,n}$$

to be the map given by (J^*, id_{C_n}). We leave to the reader the verification that $(D/i)_{m,n}$ as defined is indeed a bisimplicial space.

There is an alternative definition of D/i which highlights functoriality. Namely, for $n \geq 0$, let $(D/i)_m$ be the simplicial space defined as the homotopy limit of the diagram

$$D_m \times \Delta[0] \to D^{\Delta[m]} \xleftarrow{d_0} D^{\Delta[m+1]} \xrightarrow{d_1 \dots d_{m-1}} D \leftarrow C$$

(recall that $d^1 \dots d^{m-1} : [0] \to [n]$ induces the ultimate target map). This is clearly functorial in $[m]$. Under the assumption that C and D are Segal spaces, it is not hard to see that the two definitions of D/i are equivalent.

Note that, for each $n \geq 0$, there is a map of simplicial spaces

$$(D/i)_{\bullet,n} \to D \tag{5.1}$$

induced by $\beta^* : D_{m+n+1} \to D_m$, where $\beta : [m] \to [m+n+1]$ sends i to $n+1+i$. With the second description of D/i, this map is simply the projection onto the $D_m \times \Delta[0]$ factor.

LEMMA 5.2. *For each $n \geq 0$, (5.1) is a right fibration.*

PROOF. Since D is a Segal space, we have that $D_m \times_{D_0} \times D_{n+1} \simeq D_{m+n+1}$, and so it follows that $D_m \times_{D_0} (D/i)_{0,n} \simeq (D/i)_{m,n}$. □

PROPOSITION 5.3. *Let X be a right fibration over C. The simplicial space given by*

$$[m] \mapsto \underset{[n]\in\Delta}{\operatorname{hocolim}}\, (D/i)_{m,n} \times^h_{C_n} X_n$$

is a right fibration over D and computes $\mathbb{L}i_!(X)$, the homotopy left Kan extension of X along i.

PROOF. For the duration of this proof, we will sometimes write $\mathcal{L}$ for the simplicial space over D in the statement of the proposition. Because in $\mathcal{S}$ homotopy colimits are stable under homotopy base change, we have that

$$\left(\underset{\Delta}{\operatorname{hocolim}}\, (D/i)_0 \times^h_C X\right) \times^h_{D_0} D_m \simeq \underset{\Delta}{\operatorname{hocolim}} \left((D/i)_0 \times^h_{D_0} D_m \times^h_C X\right)$$

where we have suppressed n from the notation. By lemma 5.2, the target is weakly equivalent to

$$\underset{\Delta}{\operatorname{hocolim}}\, (D/i)_m \times^h_C X = \mathcal{L}_m$$

and so, by proposition 1.7, we conclude that $\mathcal{L} \to D$ is a right fibration.

The second claim is that there is a natural zigzag of weak equivalences relating $\mathbb{L}i_!(X)$ and $\mathcal{L}$ in the right fibration model structure.

We may reduce statement to representables, i.e. X of the form $C/K \to C$ for $K \subset C_0$. We take $K = C_0$ (the general case is analogous). In that case, $\mathbb{L}i_!(X)$ is identified with $i_0 : \Delta[0] \times C_0 \to D$ – first, using the Yoneda lemma, we replace X with $\Delta[0] \times C_0 \to C$, and then we apply (non-derived) $i_!$.

We are left to identify $i_0 : \Delta[0] \times C_0 \to D$ and $\mathcal{L}$. Applying the Yoneda lemma again, we identify $i_0 : \Delta[0] \times C_0 \to D$ with

$$D/C_0 \to D$$

Note that $(D/C_0)_m = (D/i)_{m,0}$. In the remainder of the proof, we will construct a weak equivalence[3]

$$\operatorname*{hocolim}_{[n]\in\Delta} (D/i)_{m,n} \times_{C_n} C_{n+1} \simeq (D/i)_{m,0}$$

The claimed weak equivalence will be in fact given as a zigzag of weak equivalences.

As a first step, we can replace the hocolim in the display above with

$$\operatorname*{hocolim}_{[n]\in\Delta} (D/i)_{m,n+1} \tag{5.2}$$

Indeed, there is a map of simplicial objects

$$(D/i)_{m,n+1} \to (D/i)_{m,n} \times_{C_n} C_{n+1} \tag{5.3}$$

induced by the map between homotopy pullbacks of the rows in the diagram

$$\begin{array}{ccccc}
D_{m+n+2} & \xrightarrow{\gamma^*} & D_{n+1} & \xleftarrow{p} & C_{n+1} \\
\downarrow{\scriptstyle d_0} & & \downarrow{\scriptstyle d_0} & & \downarrow{\scriptstyle id} \\
D_{m+n+1} & \xrightarrow{\gamma^*} & D_n & \xleftarrow{d_0 p} & C_{n+1}
\end{array}$$

(recall γ is the map $\gamma(i) = i$, wherever defined). But the left-hand square in the diagram is homotopy cartesian since D is a Segal space, and it follows that the induced map on homotopy pullbacks is a weak equivalence. Thus, (5.3) is a levelwise weak equivalence of bisimplicial spaces.

Now, the face operator $d_1 : D_{m+2} \to D_{m+1}$ (which corresponds to the composition of the last two morphisms) yields a map

$$(D/i)_{m,1} \to (D/i)_{m,0}$$

which is an augmentation for $[n] \mapsto (D/i)_{m,n+1}$, and thus gives rise to a map from (5.2) to $(D/i)_{m,0}$. It remains to see that this map is a levelwise weak equivalence of simplicial spaces (in the variable m). Since both simplicial spaces are right fibrations over D, it is enough to show that it is a weak equivalence on 0-simplices, i.e. for $m = 0$. This can be done by exhibiting a contracting homotopy using the extra degeneracy maps provided by $s_0 : D_k \to D_{k+1}$. □

Setting $D = *$ in proposition 5.3, we recover proposition 4.1 with $\mathcal{V} = \mathcal{S}$.

Remark 5.4. When i is the nerve of a functor of $\mathcal{S}$-categories and F is a presheaf on C, we can set $X = \mathcal{G}(F)$ in proposition 5.3 to recover the classical formulae computing the homotopy left Kan extension of F along i. In particular, when D is a point, $\mathbb{L}i_!\mathcal{G}(F)$ computes the usual homotopy colimit of F as in [**6**].

5.1. Invariance under ∞-categorical equivalences.

Proposition 5.5. *If $i : C \to D$ is a Dwyer-Kan equivalence between Segal spaces, then*

$$i_! : s\mathcal{S}_{/C} \leftrightarrows s\mathcal{S}_{/D} : i^*$$

is a simplicial Quillen equivalence, where both sides have the right fibration model structure.

[3] This can be regarded as an analogue of the Yoneda lemma in the following formulation:

$$\operatorname{map}_{\mathcal{D}}(-, i(-)) \otimes_{\mathcal{C}} \operatorname{map}_{\mathcal{C}}(-, c) \cong \operatorname{map}_{\mathcal{D}}(-, i(c)).$$

PROOF. We show that the derived unit is a weak equivalence and that $\mathbb{R}i^*$ is homotopy conservative. Let X be a simplicial space over C. The derived unit is the morphism (over C)

$$X \to \mathbb{R}i^*\mathbb{L}i_!X \tag{5.4}$$

We begin with the case $X = \{c\} : \Delta[0] \xrightarrow{c} C$. By the Yoneda lemma (lemma 1.31), the vertical maps in the diagram below are weak equivalences in the right fibration model structure

$$\begin{array}{ccc} \{c\} & \longrightarrow & \mathbb{R}i^*\mathbb{L}i_!(\{c\}) \\ \simeq\downarrow & & \downarrow\simeq \\ C/c & \longrightarrow & \mathbb{R}i^*\mathbb{L}i_!(C/c) \end{array}$$

But $i_!(\{c\}) = D/i(c)$ and $\{c\}$ is cofibrant, and so $\mathbb{R}i^*\mathbb{L}i_!(\{c\}) \simeq \mathbb{R}i^*(D/i(c))$. Moreover, $D/i(c) \to D$ is a right fibration, so $\mathbb{R}i^*(D/i(c)) \simeq i^*(D/i(c))$. This shows that the top right-hand space in the square is weakly equivalent to $i^*(D/i(c))$.

Now, by hypothesis i is fully faithful. That is to say, the map $C/c \to i^*(D/i(c))$ is a weak equivalence. This shows that the horizontal maps in the diagram above are weak equivalences.

When X is of the form $f : \Delta[n] \to C$ corresponding to a string of maps $c_0 \leftarrow \cdots \leftarrow c_n$, the claim follows from the previous case by proposition 1.7. The general case can be proved by reduction to the previous two cases, in the same spirit as the proofs of the previous propositions. Given a simplicial space X over C, resolve it by representables $Y_\bullet$. Since $i_!$ and i^* commute with colimits (the latter since it is a left adjoint, the former since colimits in spaces are stable under base change), the map (5.4) is a weak equivalence if

$$Y_k \to \mathbb{R}i^*\mathbb{L}i_!Y_k$$

is a weak equivalence for each $k \geq 0$. This concludes the proof that the derived unit is a weak equivalence.

It remains to prove that $\mathbb{R}i^*$ is homotopy conservative, i.e. that it detects weak equivalences between fibrant objects. Suppose that $f : X \to Y$ is a map of right fibrations over D and that the induced map $i^*X \to i^*Y$ is a weak equivalence of right fibrations over C. This means that for each object $c \in C_0$ the induced map of homotopy fibers

$$X_{i(c)} \to Y_{i(c)}$$

is a weak equivalence of spaces. But, since i is essentially surjective, for every $d \in D_0$ there exists some $c \in C_0$ such that $i(c) \simeq d$. Since right fibrations are fiberwise complete (proposition 1.18), the spaces $X_{i(c)}$ and X_d are weakly equivalent, and similarly for Y. It follows that f is a weak equivalence in degree 0, and so in every degree. □

COROLLARY 5.6. *Suppose B is a Segal space (not necessarily complete) and let $i : B \to B^\sharp$ be its Rezk completion, i.e. a Dwyer-Kan equivalence to a* complete *Segal space $B^\sharp$. Then i^* is a Quillen equivalence between the right fibration model structures on B and $B^\sharp$.*

REMARK 5.7. Since, by [**4**], any Segal space B may be strictified (i.e. is Dwyer-Kan equivalent) to some simplicially enriched category $\mathcal{C}$, theorem A and the invariance statement above together imply that any right fibration over B may be strictified to a space-valued presheaf on $\mathcal{C}$.

References

[1] C. Barwick, *On left and right model categories and left and right Bousfield localizations*, Homology Homotopy Appl. **12** (2010), no. 2, 245–320. MR2771591

[2] C. Barwick and D. M. Kan, *A characterization of simplicial localization functors and a discussion of DK equivalences*, Indag. Math. (N.S.) **23** (2012), no. 1-2, 69–79. MR2877402

[3] C. Barwick and D. M. Kan, *Relative categories: another model for the homotopy theory of homotopy theories*, Indag. Math. (N.S.) **23** (2012), no. 1-2, 42–68. MR2877401

[4] J. E. Bergner, *Three models for the homotopy theory of homotopy theories*, Topology **46** (2007), no. 4, 397–436. MR2321038

[5] P. Boavida de Brito and M. Weiss, *Spaces of smooth embeddings and configuration categories*, J. Topol. **11** (2018), no. 1, 65–143.

[6] A. K. Bousfield and D. M. Kan, *Homotopy limits, completions and localizations*, Lecture Notes in Mathematics, Vol. 304, Springer-Verlag, Berlin-New York, 1972. MR0365573

[7] W. G. Dwyer and D. M. Kan, *Calculating simplicial localizations*, J. Pure Appl. Algebra **18** (1980), no. 1, 17–35. MR578563

[8] G. Heuts and I. Moerdijk, *Left fibrations and homotopy colimits*, Math. Z. **279** (2015), no. 3-4, 723–744. MR3318247

[9] G. Heuts and I. Moerdijk, *Left fibrations and homotopy colimits II*, arXiv preprint arXiv:1602.01274, 2016.

[10] P. S. Hirschhorn, *Model categories and their localizations*, Mathematical Surveys and Monographs, vol. 99, American Mathematical Society, Providence, RI, 2003. MR1944041

[11] G. Horel, *A model structure on internal categories in simplicial sets*, Theory Appl. Categ. **30** (2015), No. 20, 704–750. MR3356341

[12] M. Hovey, *Model categories*, Mathematical Surveys and Monographs, vol. 63, American Mathematical Society, Providence, RI, 1999. MR1650134

[13] A. Joyal, *Notes on quasi-categories*, preprint, 2008.

[14] A. Joyal and M. Tierney, *Quasi-categories vs Segal spaces*, Categories in algebra, geometry and mathematical physics, Contemp. Math., vol. 431, Amer. Math. Soc., Providence, RI, 2007, pp. 277–326. MR2342834

[15] Y. Varshavskiĭ and D. Kazhdan, *The Yoneda lemma for complete Segal spaces* (Russian, with Russian summary), Funktsional. Anal. i Prilozhen. **48** (2014), no. 2, 3–38; English transl., Funct. Anal. Appl. **48** (2014), no. 2, 81–106. MR3288174

[16] J. Lurie, *Higher topos theory*, Annals of Mathematics Studies, vol. 170, Princeton University Press, Princeton, NJ, 2009. MR2522659

[17] C. Rezk, *A model for the homotopy theory of homotopy theory*, Trans. Amer. Math. Soc. **353** (2001), no. 3, 973–1007. MR1804411

DEPT. OF MATHEMATICS, INSTITUTO SUPERIOR TECNICO, UNIV. OF LISBON, AV. ROVISCO PAIS, LISBOA, PORTUGAL

Email address: pedrobbrito@tecnico.ulisboa.pt

Contemporary Mathematics
Volume **708**, 2018
http://dx.doi.org/10.1090/conm/708/14263

Blown-up intersection cohomology

David Chataur, Martintxo Saralegi-Aranguren, and Daniel Tanré

Abstract. In previous works, we have introduced the blown-up intersection cohomology and used it to extend Sullivan's minimal models theory to the framework of pseudomanifolds, and to give a positive answer to a conjecture of M. Goresky and W. Pardon on Steenrod squares in intersection homology.

In this paper, we establish the main properties of this cohomology. One of its major feature is the existence of cap and cup products for any filtered space and any commutative ring of coefficients, at the cochain level. Moreover, we show that each stratified map induces an homomorphism between the blown-up intersection cohomologies, compatible with the cup and cap products. We prove also its topological invariance in the case of a pseudomanifold with no codimension one strata. Finally, we compare it with the intersection cohomology studied by G. Friedman and J.E. McClure.

A great part of our results involves general perversities, defined independently on each stratum, and a tame intersection homology adapted to large perversities.

Contents

2010 *Mathematics Subject Classification.* Primary 55N33, 57N80, 55S05.

Key words and phrases. Intersection cohomology, cap product, cup product, topological invariance.

The third author is partially supported by the MINECO and FEDER research project MTM2016-78647-P and ANR-11-LABX-0007-01 "CEMPI".

Introduction

Intersection homology was defined by M. Goresky and R. MacPherson in the case of PL-pseudomanifolds in [**16**] and with sheaf theory in [**17**] where the authors restore Poincaré duality for pseudomanifolds: if X is a compact oriented pseudomanifold of dimension n and $\overline{p},\overline{q}$ are two complementary perversities, there is an isomorphism

$$H_r^{\overline{p}}(X;\mathbb{Q}) \cong H_{\overline{q}}^{n-r}(X;\mathbb{Q}),$$

with $H_{\overline{q}}^{n-r}(X;\mathbb{Q}) := \hom(H_{n-r}^{\overline{q}}(X;\mathbb{Q}),\mathbb{Q})$. This restoration also presents some disadvantages :

(1) If we don't restrict ourselves to perversities comprised between the zero $\overline{0}$ and the top $\overline{t}$ perversities respecting some growth condition, then we don't have Poincaré duality anymore and we need to use the more general *tame intersection cohomology* $\mathfrak{H}_{\overline{p}}^*(X;R)$ of X (see Definition 5.5),
(2) The cohomology is defined via a linear dual process and not via a geometrical one,
(3) This duality is not satisfied for any commutative ring R. Especially for a commutative ring properties of "torsion freeness" for the links of the singularities are needed.

Blown-up intersection cohomology aims to overcome some of these difficulties. The goal in this article is to develop the main properties of the Blown-up intersection cohomology $\mathscr{H}_{\overline{p}}^*(X;R)$ for filtered spaces and more precisely for CS sets. This theory was successfully used in [**4**] to extend the theory of Sullivan's minimal models for intersection cohomology and in [**8**] to answer a conjecture on Steenrod squares.

The first sections contain some reminders about the type of stratified spaces and maps used in the rest of the paper. After that come the definitions by a local system of the blown-up complex $\widetilde{N}_{\overline{p}}^*(X;R)$, first for a regular simplex, then by a limit argument for a general weighted simplicial complex in Proposition 2.8 and finally for a perverse space $(X,\overline{p})$ in Definition 3.2. The interesting features are that

(1) the perversities $\overline{p}$ used here are completely general, lifting the restriction on the perversities used for classical intersection cohomology,
(2) it is well defined for any commutative ring R as coefficients ring, regardless of its characteristic.

The existence and properties of a cup product for this cohomology are proved in Proposition 4.2 by the existence of an associative multiplication

$$-\cup-\colon \widetilde{N}_{\overline{p}}^k(X;R)\otimes\widetilde{N}_{\overline{q}}^\ell(X;R)\to\widetilde{N}_{\overline{p}+\overline{q}}^{\ell+k}(X;R),$$

inducing a graded commutative multiplication with unity

$$- \cup - : \mathscr{H}_{\overline{p}}^{k}(X;R) \otimes \mathscr{H}_{\overline{q}}^{\ell}(X;R) \to \mathscr{H}_{\overline{p}+\overline{q}}^{\ell+k}(X;R).$$

The same goes for the cap product, proving in Proposition 6.7 that the cap product allows an interaction between the blown-up intersection cohomology $\mathscr{H}_{\overline{p}}^{*}(X;R)$ and the tame intersection homology $\mathfrak{H}_{*}^{\overline{p}}(X;R)$.

The first part ends with two sections detailing the behavior of a stratified map with respect to blown-up intersection cohomology, first at the local level and then globally. The key result being here the Theorem A stating that any stratified map $f\colon X \to Y$ as defined in the Definition 1.5 induces a chain map

$$f^{*}\colon \widetilde{N}_{\overline{p}}^{k}(Y;R) \to \widetilde{N}_{\overline{p}}^{k}(X;R)$$

compatible with the cup product. This theorem also states that the induced maps f_{*} and f^{*} are compatible with the interaction between the blown-up intersection cohomology $\mathscr{H}_{\overline{p}}^{*}(X;R)$ and the tame intersection homology $\mathfrak{H}_{*}^{\overline{p}}(X;R)$ of Proposition 6.7.

The second part of the paper is devoted to the various properties of the blown-up intersection cohomology and their proofs. The first property comes from the Theorem B stating the existence of the complex of $\mathcal{U}$-small chains $\widetilde{N}_{\overline{p}}^{*,\mathcal{U}}(X,R)$ where $\mathcal{U}$ is an open cover of the perverse space $(X,\overline{p})$ and the existence of a quasi-isomorphism $\widetilde{N}_{\overline{p}}^{*}(X,R) \to \widetilde{N}_{\overline{p}}^{*,\mathcal{U}}(X,R)$. This leads then to the Theorem C where the Mayer- Vietoris exact sequence is proved. The two next sections detail results about explicit computations with the blown-up intersection cohomology, the first one is the Theorem D determining the blown-up intersection cohomology of the product of a filtered space X with the line $\mathbb{R}$, from the existence of an isomorphism

$$\mathscr{H}_{\overline{p}}^{*}(X \times \mathbb{R};R) \cong \mathscr{H}_{\overline{p}}^{*}(X;R).$$

The second result is the Theorem E giving the computation for the blown-up intersection cohomology of an open cone $\mathring{c}X$ over a compact filtered space X.

The paper finishes with the two following theorems. The Theorem F compares the blown-up intersection cohomology $\mathscr{H}_{\overline{p}}^{*}(X;R)$ to the already known tame intersection cohomology $\mathfrak{H}_{\overline{p}}^{*}(X;R)$ for $(X,\overline{p})$ a paracompact separable perverse CS set and proves that if R is a field or if the space X is a locally $(D\overline{p},R)$-torsion free pseudomanifold there is an isomorphism

$$\mathscr{H}_{\overline{p}}^{*}(X;R) \cong \mathfrak{H}_{D\overline{p}}^{*}(X;R)$$

where $D\overline{p}$ is the complementary perversity. The last Theorem G states that the blown-up intersection cohomology is a topological invariant when working with GM-perversities and separable paracompact CS sets with no codimension one strata .

We fix for the sequel a commutative ring R with unity. All (co)homologies in this work are considered with coefficients in R. For a topological space X, we denote by $\mathtt{c}X = X \times [0,1]/X \times \{0\}$ the *cone* on X and $\mathring{\mathtt{c}}X = X \times [0,1[/X \times \{0\}$ the *open cone* on X.

We thank the anonymous referee for her/his comments and suggestions which have in particular contributed to improve the writing and the organization of this introduction.

Part 1. Blown-up intersection cohomology. Stratified maps

We introduce the main notion of this work: the blown-up intersection cohomology and its associated cup and cap products.

1. Some reminders.

This section contains the basic definitions and properties of the main notions used in this work.

DEFINITION 1.1. A *filtered space* is a Hausdorff topological space endowed with a filtration by closed sub-spaces

$$\varnothing = X_{-1} \subseteq X_0 \subseteq X_1 \subseteq \ldots \subseteq X_{n-1} \subsetneq X_n = X.$$

The *formal dimension* of X is $\dim X = n$.

The non-empty connected components of $X_i \backslash X_{i-1}$ are the *strata* of X. Those of $X_n \backslash X_{n-1}$ are *regular strata*, while the others are *singular strata*. The family of strata of X is denoted by $\mathcal{S}_X$. The *singular set* is X_{n-1}, denoted by Σ_X or simply Σ. The *formal dimension* of a stratum $S \subset X_i \backslash X_{i-1}$ is $\dim S = i$. The *formal codimension* of S is $\operatorname{codim} S = \dim X - \dim S$.

An open subset U of X can be provided with the *induced filtration*, defined by $U_i = U \cap X_i$. If M is a topological manifold, the *product filtration* is defined by $(M \times X)_i = M \times X_i$ (see remarks about shifted filtrations of subsections 3.1 and 5.1).

The more restrictive concept of stratified space provides a better behavior of the intersection (co)homology with regard to continuous maps.

DEFINITION 1.2. A *stratified space* is a filtered space verifying the following frontier condition: for any two strata $S, S' \in \mathcal{S}_X$ such that $S \cap \overline{S'} \neq \varnothing$ then $S \subset \overline{S'}$.

In their work ([**16**], [**17**]), Goresky and MacPherson proved that the intersection (co)homology has richer properties on a singular space X with local conical structure: these are the classical stratified pseudomanifolds. The local structure is characterized by the fact that any stratum S is a manifold having a conical transversal structure over the link L. This link must be in turn a compact stratified pseudomanifold. Friedman [**14**] observed that we can suppose the link L to be only a compact filtered space to preserve the (co)homological properties of X. These are the CS sets of L. Siebenman, [**25**].

DEFINITION 1.3. A *CS set* of dimension n is a filtered space,

$$\varnothing \subset X_0 \subseteq X_1 \subseteq \cdots \subseteq X_{n-2} \subseteq X_{n-1} \subsetneq X_n = X,$$

such that, for each i, $X_i \backslash X_{i-1}$ is a topological manifold of dimension i or the empty set. Moreover, for each point $x \in X_i \backslash X_{i-1}$, $i \neq n$, there exist

(i) an open neighborhood V of x in X, endowed with the induced filtration,
(ii) an open neighborhood U of x in $X_i \backslash X_{i-1}$,

(iii) a compact filtered space L, with formal dimension $n-i-1$, where the open cone, $\mathring{c}L$, is provided with the *conical filtration*, $(\mathring{c}L)_j = \mathring{c}L_{j-1}$, with $\mathring{c}\varnothing = \mathrm{v}$ the apex of $\mathring{c}L$,

(iv) a homeomorphism, $\varphi\colon U \times \mathring{c}L \to V$, such that
  (a) $\varphi(u, \mathrm{v}) = u$, for each $u \in U$,
  (b) $\varphi(U \times \mathring{c}L_j) = V \cap X_{i+j+1}$, for each $j \in \{0, \dots, n-i-1\}$.

The pair (V, φ) is a *conical chart* of x and the filtered space L is a *link* of x. The CS set is called *normal* if the links are connected.

In the above definition, the links are always non-empty sets. Therefore, the open subset $X_n \backslash X_{n-1}$ is dense. Links are not necessarily CS sets but they are always filtered spaces. Note also that the links associated to points living in the same stratum may be not homeomorphic but they always have the same intersection homology, see for example [**14**, Chapter 5]. Finally, note that any open subset of a CS set is a CS set, that any CS set is a locally compact stratified space [**4**, Theorem G] and that a paracompact CS set is metrizable [**5**, Proposition 1.11].

The pseudomanifolds are special cases of CS sets. Their definition varies in the literature; we consider here the original definition of M. Goresky and R. MacPherson [**16**].

DEFINITION 1.4. An *n-dimensional pseudomanifold* (or simply pseudomanifold) is an n-dimensional CS set, where the link L of a point $x \in X_i \backslash X_{i-1}$ is an $(n-i-1)$-dimensional pseudomanifold. We refer to a pseudomanifold such that $X_{n-1} = X_{n-2}$ as a *classical n-dimensional pseudomanifold.*

There are several notions of stratified maps.

DEFINITION 1.5. Let $f\colon X \to Y$ be a continuous map between two stratified spaces. The map f is a *stratified map,* if it sends a stratum S of X in a stratum S^f of Y, $f(S) \subset S^f$, verifying codim $S \geqslant$ codim S^f.

The stratified map f is a *stratified homeomorphism,* if f and f^{-1} are stratified maps.

The map f is a *stratum preserving stratified map* if $n = \dim X = \dim Y$ and $f^{-1}(Y_{n-\ell}) = X_{n-\ell}$, for any $\ell \in \{0, \dots, n\}$.

This notion of stratified map does not match exactly with those found in [**14**] or [**17**].

There are two families of perversities: filtration depending (introduced by Goresky-MacPherson [**16**]) and the stratification depending (introduced by MacPherson [**22**]).

DEFINITION 1.6. A *Goresky-MacPherson perversity* (or *GM-perversity*) (see [**16**]) is a map $\overline{p}\colon \mathbb{N} \to \mathbb{Z}$, verifying $\overline{p}(0) = \overline{p}(1) = \overline{p}(2) = 0$ and $\overline{p}(i) \leqslant \overline{p}(i+1) \leqslant \overline{p}(i) + 1$ for each $i \geqslant 2$.

We use in this work the notion of perversity introduced by MacPherson [**22**] and also present in [**11**–**13**, **15**, **23**, **24**]. Unlike classic perversities, these perversities are not maps depending only on the codimension of the strata but are maps defined on the strata themselves.

DEFINITION 1.7. A *perversity on a filtered space* X is a map $\overline{p}\colon \mathcal{S}_X \to \mathbb{Z} \cup \{\pm\infty\}$ taking the value 0 on the regular strata. The pair $(X, \overline{p})$ is called a *perverse space.* If X is a CS set, we will say that $(X, \overline{p})$ is a *perverse CS set.*

The *top perversity* is the perversity defined by $\overline{t}(S) = \text{codim}\,(S) - 2$ on singular strata. The *complementary perversity* of a perversity $\overline{p}$ is the perversity $D\overline{p} = \overline{t} - \overline{p}$. The *zero perversity* is defined by $\overline{0}(S) = 0$.

A GM-perversity induces a perversity on a filtered space X, still denoted by $\overline{p}$, defined by $\overline{p}(S) = \overline{p}(\text{codim } S)$.

DEFINITION 1.8. Let $f\colon X \to Y$ be a stratified map. The *pull-back of a perversity* $\overline{q}$ of Y is the perversity $f^*\overline{q}$ of X defined by $f^*\overline{q}(S) = \overline{q}(S^f)$.

2. Blown-up complex of a weighted simplicial complex.

The blown-up complex is based on the blow up of a filtered simplex. This technique goes back to [**2**], (see also [**24**]). In this section, we introduce the notions and first properties associated to the blow up of a regular euclidean simplex (see Definition 2.2) and, more generally, of a weighted simplicial complex (see Definition 2.1).

The cone of an euclidean simplex $\Delta = [e_0, \ldots, e_m]$ is the simplicial simplex $\mathtt{c}\Delta = [e_0, \ldots, e_m, \mathtt{v}]$.

The notation $\nabla \lhd \Delta$ means that ∇ is a face of Δ.

2.1. Differential complexes associated to a simplicial complex. Let $\mathcal{L}$ be an oriented simplicial complex whose family of vertices is $\mathcal{V}(\mathcal{L}) = \{e_0, \ldots, e_m\}$. We denote $(N_*(\mathcal{L}), \partial)$ the R-complex of linear simplices of $\mathcal{L}$. The differential of $(N_*(\mathcal{L}), \partial)$ is defined by

$$\partial[e_{i_0}, \ldots, e_{i_p}] = \sum_{k=0}^{p} (-1)^k [e_{i_0}, \ldots, \hat{e}_{i_k}, \ldots, e_{i_p}].$$

For any vertex $e \in \mathcal{V}(\mathcal{L})$, we define the homomorphism $- * e\colon N_p(\mathcal{L}) \to N_{p+1}(\mathcal{L})$, by

$$[e_{i_0}, \ldots, e_{i_p}] * e = \begin{cases} [e_{i_0}, \ldots, e_{i_p}, e] & \text{if } [e_{i_0}, \ldots, e_{i_p}, e] \subset \mathcal{L}, \\ 0 & \text{if not.} \end{cases}$$

This operator is extended to the empty set by $\emptyset * e = [e]$. When $[e_0, \ldots, e_r]$ does not match an ordered simplex, by convention, it means $(-1)^{\varepsilon(\sigma)}[e_{\sigma(0)}, \ldots, e_{\sigma(r)}]$, where σ is the permutation for which $[e_{\sigma(0)}, \ldots, e_{\sigma(r)}]$ is an ordered simplex and $\varepsilon(\sigma)$ its signature. We also put $[e_0, \ldots, e_r] = 0$ if two vertices e_i are equal.

Let $(N^*(\mathcal{L}), \delta)$ be the R-dual, $\hom(N_*(\mathcal{L}), R)$, equipped with the transpose differential of ∂, defined by $(\delta f)(v) = -(-1)^{|f|} f(\partial v)$. Among the elements of $N^*(\mathcal{L})$, we consider the dual basis of the simplices of $\mathcal{L}$; i.e., if F is a p-simplex of $\mathcal{L}$, we denote by $\mathbf{1}_F$ the p-cochain taking the value 1 on F and 0 on the other simplices of $\mathcal{L}$. If $e \in \mathcal{V}(\mathcal{L})$, we also introduce a homomorphism, $- * e\colon N^p(\mathcal{L}) \to N^{p+1}(\mathcal{L})$, defined by

$$\mathbf{1}_{[e_{i_0}, \ldots, e_{i_p}]} * e = (-1)^p \mathbf{1}_{[e_{i_0}, \ldots, e_{i_p}, e]},$$

putting $\mathbf{1}_{[e_{i_0}, \ldots, e_{i_p} e]} = 0$ if $[e_{i_0}, \ldots, e_{i_p}, e] \not\subset \mathcal{L}$. On the elements of the dual basis, the differential $\delta\colon N^*(\mathcal{L}) \to N^{*+1}(\mathcal{L})$ takes the value
$\delta \mathbf{1}_{[e_{i_0}, \ldots, e_{i_p}]} = \sum_{k \notin \{i_0, \ldots, i_p\}} \mathbf{1}_{[e_{i_0}, \ldots, e_{i_p}, e_k]}$. Using $[e_{i_0}, \ldots, e_{i_k}] = 0$ when two of the vertices are equal, we can consider that the above sum is indexed by the set of vertices of $\mathcal{L}$; i.e.,

$$\delta \mathbf{1}_{[e_{i_0}, \ldots, e_{i_p}]} = \sum_{e \in \mathcal{V}(\mathcal{L})} \mathbf{1}_{[e_{i_0}, \ldots, e_{i_p}, e]} = (-1)^p \sum_{e \in \mathcal{V}(\mathcal{L})} \mathbf{1}_{[e_{i_0}, \ldots, e_{i_p}]} * e. \tag{2.1}$$

The cochain $\delta\mathbf{1}_F$ depends on the simplicial complex $\mathcal{L}$ in which the face F lives. We denote by $\delta^{\mathcal{L}}\mathbf{1}_F$ the differential of $\mathbf{1}_F$ in $\mathcal{L}$, when necessary.

DEFINITION 2.1. A *weighted simplicial complex* is an oriented simplicial complex, where each vertex e has a weight $w(e) \in \{0, \ldots, n\}$. The integer n is called the *formal dimension of $\mathcal{L}$.*

Let $\mathcal{L}_i$ be the union of the simplices of $\mathcal{L}$ whose vertices have a weight equal to i. Any simplex F of $\mathcal{L}$ is written (modulo orientation) as a join, $F = F_0 * \cdots * F_n$, with $F_i \subset \mathcal{L}_i$. They are the *filtered euclidean simplices.*

Let $\mathtt{c}\mathcal{L}_i$ be the cone on $\mathcal{L}_i$, whose apex $\mathtt{v}_i$ is called *virtual vertex*, chosen as the last vertex of the cone; i.e., if $F_i = [e_{i_0}, \ldots, e_{i_p}] \subset \mathcal{L}_i$, then $\mathtt{c}F_i = [e_{i_0}, \ldots, e_{i_p}, \mathtt{v}_i]$. The face $[\mathtt{v}_i]$ can be considered as the cone on the empty set. For a comprehensive treatment of all these cases, we consider the empty set as a simplex of $\mathcal{L}_i$.

The *full blow up* of a weighted simplicial complex $\mathcal{L}$ is the prismatic set $\widetilde{\mathcal{L}}^{all} = \mathtt{c}\mathcal{L}_0 \times \cdots \times \mathtt{c}\mathcal{L}_n$. For each $i \in \{0, \ldots, n\}$, we denote by L_i - a simplex F_i of $\mathcal{L}_i$ - or the cone $\mathtt{c}F_i$ of a simplex of $\mathcal{L}_i$, - either the singleton reduced to the virtual vertex $\mathtt{v}_i$. The products $L_0 \times \cdots \times L_n$ are called *faces of $\widetilde{\mathcal{L}}^{all}$* and represented by

$$(F, \varepsilon) = (F_0, \varepsilon_0) \times \cdots \times (F_n, \varepsilon_n),$$

satisfying the following conventions:

- the iterated join, $F = F_0 * \cdots * F_n$, is a simplex of $\mathcal{L}$;
- if $\varepsilon_i = 0$ and $F_i \neq \emptyset$, then $(F_i, 0) = F_i$ is a simplex of $\mathcal{L}_i$;
- if $\varepsilon_i = 1$ and $F_i \neq \emptyset$, then $(F_i, 1) = \mathtt{c}F_i$ is the cone over the simplex F_i of $\mathcal{L}_i$;
- if $F_i = \emptyset$, one must have $\varepsilon_i = 1$.

Let $(F, \varepsilon) = (F_0, \varepsilon_0) \times \cdots \times (F_n, \varepsilon_n)$ be a face of $\widetilde{\mathcal{L}}^{all}$ and let $\gamma \in \{0, \ldots, n\}$. We put,

- $|(F, \varepsilon)|_{<\gamma} = \sum_{i<\gamma}(\dim F_i + \varepsilon_i)$ and $|(F, \varepsilon)|_{>\gamma} = \sum_{i>\gamma}(\dim F_i + \varepsilon_i)$,
- $|(F, \varepsilon)|_{\leqslant\gamma} = \sum_{i\leqslant\gamma}(\dim F_i + \varepsilon_i)$ and $|(F, \varepsilon)|_{\geqslant\gamma} = \sum_{i\geqslant\gamma}(\dim F_i + \varepsilon_i)$,
- $|(F, \varepsilon)| = |(F, \varepsilon)|_{\leqslant n}$.

DEFINITION 2.2. The *blow up* of a weighted simplicial complex of dimension n, $\mathcal{L}$, is the sub-prism $\widetilde{\mathcal{L}}$ of $\widetilde{\mathcal{L}}^{all}$ consisting of the $(F, \varepsilon)'s$ such that $\varepsilon_n = 0$. Notice that the corresponding simplices, $F = F_0 * \cdots * F_n \subset \mathcal{L}$, must verify $F_n \neq \emptyset$. This kind of filtered simplices are called *regular.*

EXAMPLE 2.3. The generic case of a weighted simplicial complex is a regular simplex $\Delta = \Delta_0 * \cdots * \Delta_n$. Assume Δ is oriented by an order on its vertices, $(e_i)_{0\leqslant i\leqslant m}, e_i < e_{i+1}$, in a compatible way with the join decomposition. More precisely, expressing the maximal simplex by its vertices, and indicating by a vertical bar the filtration change, we have

$$\Delta = [\underbrace{e_0, \ldots, e_{k_0}}_{\Delta_0} \mid \underbrace{e_{k_0+1}, \ldots, e_{k_1}}_{\Delta_1} \mid \cdots \mid \underbrace{\emptyset}_{\Delta_i} \mid \cdots \mid \underbrace{e_{k_{n-1}+1}, \ldots, e_{k_n}}_{\Delta_n}], \tag{2.2}$$

with $k_n = m$. The vertices $\{e_0, \ldots, e_{k_\ell}\}$ generate the simplex $\Delta_0 * \cdots * \Delta_\ell$. A face Δ_i can be empty, as shown in the decomposition above. As we have already noticed, giving the filtration $\Delta = \Delta_0 * \cdots * \Delta_n$ of the simplex Δ is equivalent to giving a weight on each vertex of Δ, in the sense of Definition 2.1. The corresponding blow up is

$$\widetilde{\Delta} = \mathtt{c}\Delta_0 \times \cdots \times \mathtt{c}\Delta_{n-1} \times \Delta_n. \tag{2.3}$$

EXAMPLE 2.4. The coding of the faces of the blow up of a regular simplex under the form of a product $(F,\varepsilon) = (F_0,\varepsilon_0) \times \cdots \times (F_n,\varepsilon_n)$ is used throughout the text. To familiarize the reader with this representation, we specify the blow up $\widetilde{\Delta}$ of $\Delta = \Delta_0 * \Delta_1 = [e_0] * [e_1, e_2]$.

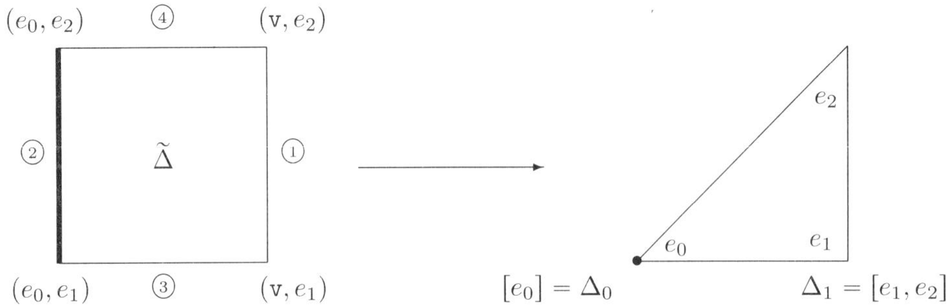

The four one dimensional faces of the blow up $\widetilde{\Delta} = \mathtt{c}\Delta_0 \times \Delta_1$ are encoded as

$$①=(\emptyset, 1) \times (\Delta_1, 0), \quad ②=([e_0], 0) \times (\Delta_1, 0),$$
$$③=([e_0], 1) \times ([e_1], 0), \quad ④=([e_0], 1) \times ([e_2], 0),$$

corresponding to

$$①=[\mathtt{v}] \times \Delta_1, \quad ②=\Delta_0 \times \Delta_1, \quad ③=\mathtt{c}\Delta_0 \times [e_1], \quad ④=\mathtt{c}\Delta_0 \times [e_2].$$

2.2. Blown-up complex associated to a weighted simplicial complex. We recall the notations of the previous paragraph. The element $\mathbf{1}_{(F_i,\varepsilon_i)}$ is the cochain on $\mathtt{c}\mathcal{L}_i$, taking the value 1 on the simplex (F_i, ε_i) and 0 on other simplices of $\mathtt{c}\mathcal{L}_i$. If $F = F_0 * \cdots * F_n \subset \mathcal{L}$ and $(F,\varepsilon) = (F_0,\varepsilon_0) \times \cdots \times (F_n,\varepsilon_n)$, we write

$$\mathbf{1}_{(F,\varepsilon)} = \mathbf{1}_{(F_0,\varepsilon_0)} \otimes \cdots \otimes \mathbf{1}_{(F_n,\varepsilon_n)}.$$

The R-module generated by these elements when F runs over the simplices of $\mathcal{L}$ is denoted by $\widetilde{N}^{all,*}(\mathcal{L}; R)$ or $\widetilde{N}^{all,*}(\mathcal{L})$ if there is no ambiguity. The sub-R-module generated by the elements such that $\varepsilon_n = 0$, endowed with the differential

$$\begin{aligned} \delta\mathbf{1}_{(F,\varepsilon)} &= \sum_{i=0}^{n-1} (-1)^{|(F,\varepsilon)|_{<i}} \mathbf{1}_{(F_0,\varepsilon_0)} \otimes \cdots \otimes \delta^{\mathtt{c}\mathcal{L}_i} \mathbf{1}_{(F_i,\varepsilon_i)} \otimes \cdots \otimes \mathbf{1}_{F_n} \\ &\quad + (-1)^{|(F,\varepsilon)|_{<n}} \mathbf{1}_{(F_0,\varepsilon_0)} \otimes \cdots \otimes \mathbf{1}_{(F_{n-1},\varepsilon_{n-1})} \otimes \delta^{\mathcal{L}_n} \mathbf{1}_{F_n}, \end{aligned} \tag{2.4}$$

is called the *blown-up complex* of $\mathcal{L}$ and denoted by $\widetilde{N}^*(\mathcal{L}; R)$ or simply $\widetilde{N}^*(\mathcal{L})$. (Recall that $\delta^{\mathtt{c}\mathcal{L}_i}$ is the differential of the cochain complex on the simplicial complex $\mathtt{c}\mathcal{L}_i$.)

The elements of the blown-up complex are provided with an additional degree which looks like the degree of a differential form along the fiber in a bundle.

DEFINITION 2.5. Let $\ell \in \{1, \ldots, n\}$. The *$\ell$-perverse degree of the cochain* $\mathbf{1}_{(F,\varepsilon)}$ is equal to

$$\|\mathbf{1}_{(F,\varepsilon)}\|_\ell = \begin{cases} -\infty & \text{if} \quad \varepsilon_{n-\ell} = 1, \\ |(F,\varepsilon)|_{>n-\ell} & \text{if} \quad \varepsilon_{n-\ell} = 0. \end{cases}$$

If $\omega \in \widetilde{N}^*(\mathcal{L};R)$ decomposes as $\omega = \sum_\mu \lambda_\mu \mathbf{1}_{(F_\mu \varepsilon_\mu)}$, with $\lambda_\mu \neq 0$, its *ℓ-perverse degree* is equal to

$$\|\omega\|_\ell = \max_\mu \|\mathbf{1}_{(F_\mu,\varepsilon_\mu)}\|_\ell. \tag{2.5}$$

By convention, we set $\|0\|_\ell = -\infty$.

EXAMPLE 2.6. We compute the perverse degree of some cochains of the blow up $\widetilde{\Delta} = c\Delta_0 \times c\Delta_1 \times \Delta_2$ of the regular simplex $\Delta = \Delta_0 * \Delta_1 * \Delta_2$.

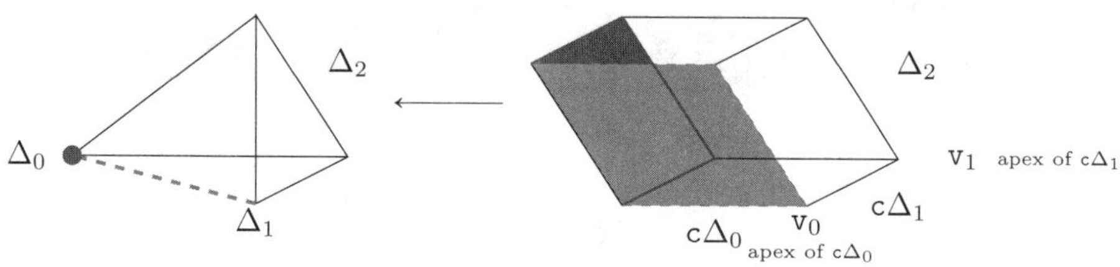

$$\begin{array}{ll}
\|\mathbf{1}_{\mathtt{v}_0\times\mathtt{v}_1\times\Delta_2}\|_2 = -\infty & \|\mathbf{1}_{\mathtt{v}_0\times\mathtt{v}_1\times\Delta_2}\|_1 = -\infty \\
\|\mathbf{1}_{c\Delta_0\times\Delta_1\times\Delta_2}\|_2 = -\infty & \|\mathbf{1}_{c\Delta_0\times\Delta_1\times\Delta_2}\|_1 = \dim\Delta_2 \\
\|\mathbf{1}_{\Delta_0\times\mathtt{c}\Delta_1\times\Delta_2}\|_2 = \dim\mathtt{c}\Delta_1 + \dim\Delta_2 & \|\mathbf{1}_{\Delta_0\times\mathtt{c}\Delta_1\times\Delta_2}\|_1 = -\infty \\
\|\mathbf{1}_{\Delta_0\times\Delta_1\times\Delta_2}\|_2 = \dim\Delta_1 + \dim\Delta_2 & \|\mathbf{1}_{\Delta_0\times\Delta_1\times\Delta_2}\|_1 = \dim\Delta_2
\end{array}$$

(We have used $\mathtt{v}_i$ the apex of the cone $\mathtt{c}\Delta_i$, $i = 0, 1$).

DEFINITION 2.7. In the case of a *regular simplex* $\Delta = \Delta_0 * \cdots * \Delta_n$, $\Delta_n \neq \emptyset$, the blown-up complex is the tensor product, $\widetilde{N}^*(\Delta) = N^*(\mathtt{c}\Delta_0) \otimes \cdots \otimes N^*(\mathtt{c}\Delta_{n-1}) \otimes N^*(\Delta_n)$ which corresponds to [**4**, Definition 1.31].

The *face operators* of a filtered simplex $\Delta = \Delta_0 * \cdots * \Delta_n$ are maps $\mu\colon \Delta' = \Delta'_0 * \cdots * \Delta'_n \to \Delta = \Delta_0 * \cdots * \Delta_n$, of the form $\mu = *_{i=0}^n \mu_i$ where μ_i is an injective map preserving the order. The face Δ' is a codimension 1 regular face. Each operator face of a factor of the join $\delta_\ell\colon \Delta'_i \to \Delta_i$ induces a map, still denoted by $\delta_\ell\colon N_*(\mathtt{c}\Delta'_0) \otimes \cdots \otimes N_*(\mathtt{c}\Delta'_{n-1}) \otimes N_*(\Delta'_n) \to N_*(\mathtt{c}\Delta_0) \otimes \cdots \otimes N_*(\mathtt{c}\Delta_i) \otimes \cdots \otimes N_*(\Delta_n)$, obtained by carrying out the join with the identity map on factors Δ_j, $j \neq \ell$. We call $\delta_\ell^*\colon N^*(\mathtt{c}\Delta_0) \otimes \cdots \otimes N^*(\mathtt{c}\Delta_{n-1}) \otimes N^*(\Delta_n) \to N^*(\mathtt{c}\Delta'_0) \otimes \cdots \otimes N^*(\mathtt{c}\Delta'_i) \otimes \cdots \otimes N^*(\Delta'_n)$, the transpose of the linear map δ_ℓ.

The blown-up complex of a weighted simplicial complex is expressed from the blown-up complex of regular simplices.

PROPOSITION 2.8. *Let $\mathcal{L}$ be a weighted simplicial complex. Then*

$$\widetilde{N}^*(\mathcal{L};R) \cong \varprojlim_{\Delta\subset\mathcal{L},\Delta \text{ regular}} \widetilde{N}^*(\Delta;R).$$

PROOF. An element ω of $\varprojlim \widetilde{N}^*(\Delta)$ is a family $\omega = (\omega_\Delta)_\Delta$, indexed by the regular simplices of $\mathcal{L}$, with $\omega_\Delta \in \widetilde{N}^*(\Delta)$ and satisfying the following face compatibility conditions: if $\delta_\ell\colon \nabla \to \Delta$ is a regular face of codimension 1 of a simplex Δ of $\mathcal{L}$, then $\delta_\ell^*\omega_\Delta = \omega_\nabla \in \widetilde{N}^*(\nabla)$. To $\omega \in \varprojlim \widetilde{N}^*(\Delta)$, we associate the cochain $\omega_\mathcal{L} \in \widetilde{N}^k(\mathcal{L})$ defined by

$$\omega_\mathcal{L}(F,\varepsilon) = \omega_F(F,\varepsilon).$$

Conversely, let $\omega_{\mathcal{L}} = \sum_F \alpha_{(F,\varepsilon)} \mathbf{1}_{(F,\varepsilon)} \in \widetilde{N}^k(\mathcal{L})$, where F runs over the regular simplices of $\mathcal{L}$, $|(F,\varepsilon)| = k$ and $\alpha_{(F,\varepsilon)} \in R$. For each regular simplex Δ of $\mathcal{L}$, we define $\omega_\Delta \in \widetilde{N}^k(\Delta)$ by

$$\omega_\Delta = \sum_{F \lhd \Delta} \alpha_{(F,\varepsilon)} \mathbf{1}_{(F,\varepsilon)}.$$

It remains to prove the compatibility with respect to the face operators. Let $\delta_\ell \colon \nabla \to \Delta$ be a regular face of Δ and F a simplex of Δ. By definition of the dual basis one has

$$\delta_\ell^*(\mathbf{1}_{(F,\varepsilon)}) = \begin{cases} \mathbf{1}_{(F,\varepsilon)} & \text{if } F \subset \nabla, \\ 0 & \text{if not.} \end{cases}$$

It follows

$$\delta_\ell^*(\omega_\Delta) = \sum_{F \lhd \nabla} \alpha_{(F,\varepsilon)} \mathbf{1}_{(F,\varepsilon)} = \omega_\nabla.$$

□

2.3. Adjunction of a vertex to a cochain of the blow up.

DEFINITION 2.9. Let $(F,\varepsilon) = (F_0,\varepsilon_0) \times \cdots \times (F_n,\varepsilon_n)$ be a face of $\widetilde{\mathcal{L}}^{all}$ and $\ell \in \{0,\ldots,n\}$. *The adjunction of a vertex $e \in \mathcal{L}_\ell$ to the cochain $\mathbf{1}_{(F,\varepsilon)}$ is the cochain*

$$\mathbf{1}_{(F,\varepsilon)} * e = (-1)^{|(F,\varepsilon)|_{>\ell}}\, \mathbf{1}_{(F_0,\varepsilon_0)} \otimes \cdots \otimes (\mathbf{1}_{(F_\ell,\varepsilon_\ell)} * e) \otimes \cdots \otimes \mathbf{1}_{(F_n,\varepsilon_n)}.$$

Likewise, for the virtual vertex $\mathtt{v}_\ell$, we set,

$$\mathbf{1}_{(F,\varepsilon)} * \mathtt{v}_\ell = (-1)^{|(F,\varepsilon)|_{>\ell}}\, \mathbf{1}_{(F_0,\varepsilon_0)} \otimes \cdots \otimes (\mathbf{1}_{(F_\ell,\varepsilon_\ell)} * \mathtt{v}_\ell) \otimes \cdots \otimes \mathbf{1}_{(F_n,\varepsilon_n)}.$$

We extend by linearity this adjunction to $\widetilde{N}^{all,*}(\mathcal{L};R)$.

The next property follows directly from the definition.

LEMMA 2.10. *Let $\mathcal{L}$ be a weighted simplicial complex and (F,ε) a face of $\widetilde{\mathcal{L}}^{all}$. Consider two vertices of $\mathcal{L}$, $e_\alpha \in \mathcal{L}_{\ell(\alpha)}$ and $e_\beta \in \mathcal{L}_{\ell(\beta)}$, and two virtual vertices $\mathtt{v}_\ell$ and $\mathtt{v}_{\ell'}$. Then the following properties are verified,*

$$\begin{aligned}
\mathbf{1}_{(F,\varepsilon)} * e_\alpha * e_\beta &= -\mathbf{1}_{(F,\varepsilon)} * e_\beta * e_\alpha, \\
\mathbf{1}_{(F,\varepsilon)} * e_\alpha * \mathtt{v}_\ell &= -\mathbf{1}_{(F,\varepsilon)} * \mathtt{v}_\ell * e_\alpha, \\
\mathbf{1}_{(F,\varepsilon)} * \mathtt{v}_\ell * \mathtt{v}_{\ell'} &= -\mathbf{1}_{(F,\varepsilon)} * \mathtt{v}_{\ell'} * \mathtt{v}_\ell.
\end{aligned}$$

PROPOSITION 2.11. *Let $\mathcal{L}$ be a weighted simplicial complex. The blown-up differential of an element $\mathbf{1}_{(F,\varepsilon)} \in \widetilde{N}^*(\mathcal{L};R)$ can be written as,*

$$\delta \mathbf{1}_{(F,\varepsilon)} = (-1)^{|(F,\varepsilon)|} \left(\sum_{e \in \mathcal{V}(\mathcal{L})} \mathbf{1}_{(F,\varepsilon)} * e + \sum_{i=0}^{n-1} \mathbf{1}_{(F,\varepsilon)} * \mathtt{v}_i \right). \tag{2.6}$$

PROOF. Let $\mathbf{1}_{(F,\varepsilon)} = \mathbf{1}_{(F_0,\varepsilon_0)} \otimes \cdots \otimes \mathbf{1}_{(F_{n-1},\varepsilon_{n-1})} \otimes \mathbf{1}_{F_n}$. By replacing $\delta^{c\mathcal{L}_i}$ and $\delta^{\mathcal{L}_n}$ by their values from (2.1) in the equality (2.4) and denoting

$|(F_i,\varepsilon_i)| = \dim F_i + \varepsilon_i$, we get,

$$\delta \mathbf{1}_{(F,\varepsilon)} = \sum_{i=0}^{n}(-1)^{|(F,\varepsilon)|_{<i}} \sum_{e_i\in\mathcal{V}(\mathcal{L}_i)} (-1)^{|(F_i,\varepsilon_i)|}\mathbf{1}_{(F_0,\varepsilon_0)}\otimes\cdots\otimes(\mathbf{1}_{(F_i,\varepsilon_i)} * e_i)\otimes\cdots\otimes\mathbf{1}_{(F_n,\varepsilon_n)}$$
$$+\sum_{i=0}^{n-1}(-1)^{|(F,\varepsilon)|_{\leqslant i}}\mathbf{1}_{(F_0,\varepsilon_0)}\otimes\cdots\otimes(\mathbf{1}_{(F_i,\varepsilon_i)} * \mathtt{v}_i)\otimes\cdots\otimes\mathbf{1}_{F_n}.$$

The wanted formula follows from the definition of the adjunction of a vertex. □

From Proposition 2.11 and Lemma 2.10, we directly deduce the behavior of the adjunction of a vertex with respect to the differential of the blown-up complex.

COROLLARY 2.12. *Let $\mathcal{L}$ be a weighted simplicial complex and let $\mathbf{1}_{(F,\varepsilon)} \in \widetilde{N}^*(\mathcal{L};R)$. For each vertex $e_\alpha \in \mathcal{V}(\mathcal{L})$ and each virtual vertex $\mathtt{v}_\ell$, one has $\delta(\mathbf{1}_{(F,\varepsilon)} * e_\alpha) = (\delta\mathbf{1}_{(F,\varepsilon)}) * e_\alpha$ and $\delta(\mathbf{1}_{(F,\varepsilon)} * \mathtt{v}_\ell) = (\delta\mathbf{1}_{(F,\varepsilon)}) * \mathtt{v}_\ell$.*

3. Blown-up intersection cohomology.

In this section, we define the blown-up intersection cohomology of a perverse space.

DEFINITION 3.1. Let X be a filtered space. A *filtered singular simplex* is a continuous map, $\sigma\colon \Delta \to X$, where the euclidean simplex Δ is endowed with a decomposition $\Delta = \Delta_0 * \Delta_1 * \cdots * \Delta_n$, called *$\sigma$-decomposition of* Δ, verifying

$$\sigma^{-1}X_i = \Delta_0 * \Delta_1 * \cdots * \Delta_i,$$

for each $i \in \{0,\dots,n\}$. The filtered singular simplex σ is *regular* if $\Delta_n \neq \emptyset$. To specify that the filtration of the euclidean simplex Δ is induced from that of X by the map σ, we sometimes write $\Delta = \Delta_\sigma$. This notation is particularly useful when a simplex carries two filtrations associated with two distinct maps.

The dimensions of the simplices Δ_i of the σ-decomposition measure the non-transversality of the simplex σ with respect to the strata of X. These simplices Δ_i may be empty, with the convention $\emptyset * Y = Y$, for any space Y. Note also that a singular simplex $\sigma\colon \Delta \to X$ is filtered if each $\sigma^{-1}(X_i)$, $i \in \{0,\dots,n\}$, is a face of Δ.

To any regular simplex, $\sigma\colon \Delta = \Delta_0 * \cdots * \Delta_n \to X$, we associate the cochain complex defined by

$$\widetilde{N}^*_\sigma = \widetilde{N}^*(\Delta) = N^*(\mathtt{c}\Delta_0)\otimes\cdots\otimes N^*(\mathtt{c}\Delta_{n-1})\otimes N^*(\Delta_n).$$

If $\delta_\ell\colon \Delta' = \Delta'_0 * \cdots * \Delta'_\ell * \cdots * \Delta'_n \to \Delta = \Delta_0 * \cdots * \Delta_\ell * \cdots * \Delta_n$ is a face operator, we denote $\partial_\ell\sigma$ the filtered simplex defined by $\partial_\ell\sigma = \sigma\circ\delta_\ell\colon \Delta' \to X$. The face operator δ_ℓ is regular if $\Delta'_n \neq \emptyset$.

DEFINITION 3.2. Let X be a filtered space. The *blown-up complex of X with coefficients in R*, $\widetilde{N}^*(X;R)$, is the cochain complex formed by the elements ω, associating to any regular filtered simplex, $\sigma\colon \Delta_0 * \cdots * \Delta_n \to X$, an element $\omega_\sigma \in \widetilde{N}^*_\sigma$, so that $\delta^*_\ell(\omega_\sigma) = \omega_{\partial_\ell\sigma}$, for any regular face operator, δ_ℓ. The differential $\delta\omega$ of $\omega \in \widetilde{N}^*(X;R)$ is defined by $(\delta\omega)_\sigma = \delta(\omega_\sigma)$ for any regular filtered simplex σ.

For any $\ell \in \{1, \ldots, n\}$, the element $\omega_\sigma \in \widetilde{N}^*_\sigma$ is endowed with the *perverse degree* $\|\omega_\sigma\|_\ell$, introduced in Definition 2.5. We extend this degree to the elements of $\widetilde{N}^*(X; R)$ as follows.

DEFINITION 3.3. Let X be a filtered space and $\omega \in \widetilde{N}^*(X; R)$. The *perverse degree of ω along a singular stratum*, S, is equal to

$$\|\omega\|_S = \sup \{\|\omega_\sigma\|_{\text{codim } S} \mid \sigma \colon \Delta \to X \text{ regular with } \sigma(\Delta) \cap S \neq \emptyset\}.$$

We denote by $\|\omega\|$ the map associating to any singular stratum S of X the element $\|\omega\|_S$ and 0 to a regular stratum.

Notice that, by definition, face operators δ^*_ℓ decrease the perverse degree.

DEFINITION 3.4. Let $(X, \overline{p})$ be a perverse space. A cochain $\omega \in \widetilde{N}^*(X; R)$ is *$\overline{p}$-allowable* if

$$\|\omega\| \leqslant \overline{p}. \tag{3.1}$$

A cochain ω is a *$\overline{p}$-intersection cochain* if ω and its coboundary, $\delta\omega$, are $\overline{p}$-allowable. We denote by $\widetilde{N}^*_{\overline{p}}(X; R)$ the complex of $\overline{p}$-intersection cochains and $\mathscr{H}^{\overline{p}}_*(X; R)$ its homology, called *blown-up intersection cohomology* of X with coefficients in R, for the perversity $\overline{p}$.

3.1. Shifted filtrations. Blown-up intersection cohomology does not depend on the dimension of the strata of X but on the codimension of these strata (see [**14**, Section 4.3.1]). Let us consider on X the shifted filtration Y, where $m \in \mathbb{N}^*$:

$$\begin{aligned}\emptyset = Y_0 = \cdots = Y_{m-1} \subset Y_m = X_0 \subset \cdots \subset Y_{n+m-1} = \\ X_{n-1} \subset Y_{n+m} = X_n = X.\end{aligned} \tag{3.2}$$

So, the formal dimension of Y is $n+m$. For example, on $\mathbb{R}^m \times X$ we have two natural shifted filtrations: $(\mathbb{R}^m \times X)_i = \mathbb{R}^m \times X_i$, with $i \in \{0, \ldots, n\}$, and $(\mathbb{R}^m \times X)_i = \mathbb{R}^m \times X_{i-m}$, with $i \in \{0, \ldots, n+m\}$.

The family of regular simplices is the same for both filtrations. The perverse degree of a filtered simplex $\sigma \colon \Delta \to X = Y$ is the same for both filtrations. Let us see that. If S is a singular stratum of X (and therefore of Y) with $\operatorname{Im} \sigma \cap S \neq \emptyset$, we have $\operatorname{codim}_X S = \operatorname{codim}_Y S$. Let ℓ be this codimension. On the other hand, if $\Delta^X_\sigma = \Delta_0 * \cdots * \Delta_n$ is the induced filtration by $\sigma \colon \Delta \to X$ then $\Delta^Y_\sigma = \underbrace{\emptyset * \cdots * \emptyset}_{m \text{ times}} * \Delta'_m * \cdots * \Delta'_{n+m}$, with $\Delta'_{m+r} = \Delta_r$, is the induced filtration by $\sigma \colon \Delta \to Y$. So, if $(F = F_0 * \cdots * F_n, \varepsilon = (\varepsilon_0 * \cdots * \varepsilon_n))$ is a face of the blow up of Δ^X_σ then $(F' = \underbrace{\emptyset, \ldots, \emptyset}_{m \text{ times}} * F_0 * \cdots * F_n, \varepsilon' = (\underbrace{1, \ldots, 1}_{m \text{ times}}, \varepsilon_0, \ldots, \varepsilon_n))$ is the corresponding face of the blow up of Δ^Y_σ. We have, for $\ell \in \{1, \ldots, n\}$

$$\begin{aligned}\|\mathbf{1}_{(F,\varepsilon)}\|^X_\ell &= \begin{cases} -\infty & \text{if } \varepsilon_{n-\ell} = 1 \\ |(F, \varepsilon)|_{>n-\ell} & \text{if } \varepsilon_{n-\ell} = 0 \end{cases} = \begin{cases} -\infty & \text{if } \varepsilon'_{m+n-\ell} = 1 \\ |(F', \varepsilon')|_{>m+n-\ell} & \text{if } \varepsilon'_{m+n-\ell} = 0 \end{cases} \\ &= \|\mathbf{1}_{(F',\varepsilon')}\|^Y_\ell,\end{aligned}$$

and $\|\mathbf{1}_{(F',\varepsilon')}\|^Y_\ell = -\infty$ if $\ell \in \{n+1, \ldots, m+n\}$. We get $\widetilde{N}^*_{\overline{p}}(X; R) = \widetilde{N}^*_{\overline{p}}(Y; R)$ and therefore $\mathscr{H}^*_{\overline{p}}(X; R) = \mathscr{H}^*_{\overline{p}}(Y; R)$.

We prove in Theorem A that any stratified map induces a morphism in blown-up intersection cohomology. The following Proposition, which is a weaker version of this result, is used in Part 2.

PROPOSITION 3.5. *Let $f\colon X \to Y$ be a stratum preserving stratified map. The induced map $f\colon \widetilde{N}^*(Y;R) \to \widetilde{N}^*(X;R)$, defined by $(f^*(\omega))_\sigma = \omega_{f\circ\sigma}$, is a well defined chain map.*

Consider a perversity $\overline{p}$ on X and a perversity $\overline{q}$ on Y verifying $\overline{p} \geqslant f^*\overline{q}$. The induced operator $f^*\colon \widetilde{N}^*_{\overline{q}}(Y;R) \to \widetilde{N}^*_{\overline{p}}(X;R)$, is a well defined chain map inducing the morphism $f^*\colon \mathscr{H}^*_{\overline{q}}(Y;R) \to \mathscr{H}^*_{\overline{p}}(X;R)$.

PROOF. The definition makes sense since $(\delta^*_\ell f^*(\omega))_\sigma = \delta^*_\ell(f^*(\omega)_\sigma) = \delta^*_\ell(\omega_{f\circ\sigma}) = \omega_{f\circ\sigma\circ\delta_\ell} = (f^*(\omega))_{\sigma\circ\delta_\ell}$, for each $\omega \in \widetilde{N}^*(Y;R)$, each regular simplex $\sigma\colon \Delta \to X$ and each regular face $\delta_\ell\colon \nabla \to \Delta$.

The induced morphism is a chain map since $(\delta f^*(\omega))_\sigma = \delta(f^*(\omega)_\sigma) = \delta(\omega_{f\circ\sigma}) = (\delta\omega)_{f\circ\sigma} = (f^*(\delta\omega))_\sigma$, for each $\omega \in \widetilde{N}^*(Y;R)$ and each regular simplex $\sigma\colon \Delta_\sigma \to X$.

We turn now to perversities and firstly to the filtrations. The euclidean simplex Δ has two filtrations, respectively induced by σ and $f\circ\sigma$, denoted by Δ_σ and $\Delta_{f\circ\sigma}$, following the conventions of Definition 3.1. For each $\ell \in \{0,\dots,n\}$, the hypothesis $f^{-1}(Y_{n-\ell}) = X_{n-\ell}$ implies equalities, $\sigma^{-1}(X_{n-\ell}) = \sigma^{-1}f^{-1}(Y_{n-\ell}) = (f\circ\sigma)^{-1}(Y_{n-\ell})$ and $\Delta_\sigma = \Delta_{f\circ\sigma}$.

Let S be an ℓ-codimensional singular stratum of X such that $S \cap \operatorname{Im}\sigma \neq \varnothing$. The stratum S^f of Y, characterized by $f(S) \subset S^f$, also has codimension ℓ and verifies $S^f \cap \operatorname{Im}(f\circ\sigma) \neq \varnothing$. Moreover, the definition of f^* and the previous observation on the two filtrations of Δ give $\|f^*(\omega)_\sigma\|_\ell = \|\omega_{f\circ\sigma}\|_\ell$. From Definition 3.3, we deduce $\|f^*(\omega)\|_S \leqslant \|\omega\|_{S^f} \leqslant \overline{q}(S^f)$. The assumption made on the perversities $\overline{p}$ and $\overline{q}$ allows us to conclude $\|f^*(\omega)\|_S \leqslant \overline{p}(S)$. The same argument applied to the $\overline{q}$-allowable form, $\delta\omega$, gives $\|f^*(\delta\omega)\|_S \leqslant \overline{p}(S)$. The compatibility of f^* with the differential δ implies $f^*(\omega) \in \widetilde{N}^*_{\overline{p}}(X;R)$. □

REMARK 3.6. If f is a stratum preserving stratified map, for any filtered simplex $\sigma\colon \Delta \to X$, the σ- and $f\circ\sigma$- decompositions of Δ are the same: $\Delta_\sigma = \Delta_{f\circ\sigma}$. This is not the case for a general stratified map (see [**4**, A.25]).

4. Cup product.

We have defined the notion of cup product in [**4**] for "filtered face sets". In [**8**], we also introduced the notion of cup_i-products and Steenrod squares on the blown-up intersection cohomology. We give below a definition of cup product for any coefficient ring, in our context of stratum depending perversities.

DEFINITION 4.1. Two ordered simplices, $F = [a_0,\dots,a_k]$ and $G = [b_0,\dots,b_\ell]$, of an euclidean simplex Δ, are said to be *compatible* if $a_k = b_0$. In this case, we set $F \cup G = [a_0,\dots,a_k,b_1,\dots,b_\ell] \in N_*(\Delta)$. The *cup product* is defined on the dual basis of $N^*(\Delta)$ by

$$\mathbf{1}_F \cup \mathbf{1}_G = (-1)^{k\cdot\ell}\,\mathbf{1}_{F\cup G}$$

if F and G are compatible and 0 if not.

In the case of the cone $\mathtt{c}\Delta$, this law appears on $N^*(\mathtt{c}\Delta)$ as:

$$(4.1)\quad \mathbf{1}_{(F,\varepsilon)} \cup \mathbf{1}_{(G,\kappa)} = \begin{cases} (-1)^{|F|\cdot|(G,\kappa)|}\,\mathbf{1}_{(F\cup G,\kappa)} & \text{if } F, G \text{ compatible and } \varepsilon = 0,\\ \mathbf{1}_{(F,\varepsilon)} & \text{if } (G,\kappa) = (\emptyset, 1) \text{ and } \varepsilon = 1,\\ 0 & \text{if not.}\end{cases}$$

If $\Delta = \Delta_0 * \cdots * \Delta_n$ is a regular euclidean simplex, this definition extends to $\widetilde{N}^*(\Delta)$ as a law defined on a tensor product. More precisely, if $\omega_0 \otimes \cdots \otimes \omega_n$ and $\eta_0 \otimes \cdots \otimes \eta_n$ belong to $N^*(\mathtt{c}\Delta_0) \otimes \cdots \otimes N^*(\Delta_n)$, we set

$$(4.2)\quad (\omega_0 \otimes \cdots \otimes \omega_n) \cup (\eta_0 \otimes \cdots \otimes \eta_n) = (-1)^{\sum_{i>j}|\omega_i|\,|\eta_j|}(\omega_0 \cup \eta_0) \otimes \cdots \otimes (\omega_n \cup \eta_n).$$

Given two perversities $\overline{p}, \overline{q}$ defined on a filtered set X we define the perversity $\overline{p}+\overline{q}$ by $(\overline{p}+\overline{q})(S) = \overline{p}(S)+\overline{q}(S)$, with the convention $-\infty+\overline{p}(S) = \overline{p}(S)-\infty = -\infty$.

PROPOSITION 4.2. *For each filtered space X endowed with two perversities $\overline{p}$ and $\overline{q}$, there exists an associative multiplication,*

$$- \cup - \colon \widetilde{N}^k_{\overline{p}}(X;R) \otimes \widetilde{N}^\ell_{\overline{q}}(X;R) \to \widetilde{N}^{k+\ell}_{\overline{p}+\overline{q}}(X;R),$$

defined by $(\omega\cup\eta)_\sigma = \omega_\sigma \cup \eta_\sigma$, for each pair of cochains $(\omega,\eta) \in \widetilde{N}^k_{\overline{p}}(X;R) \times \widetilde{N}^\ell_{\overline{q}}(X;R)$ and each regular simplex $\sigma\colon \Delta \to X$. It induces a graded commutative multiplication with unity called intersection cup product,

$$- \cup - \colon \mathscr{H}^k_{\overline{p}}(X;R) \otimes \mathscr{H}^\ell_{\overline{q}}(X;R) \to \mathscr{H}^{k+\ell}_{\overline{p}+\overline{q}}(X;R).$$

PROOF. Begin by verifying that the product locally defined on each regular simplex extends to the blown-up complex. For this, consider two cochains $\omega, \eta \in \widetilde{N}^*(X;R)$ and $\delta_\ell\colon \nabla \to \Delta$ a regular face of a simplex $\sigma\colon \Delta \to X$. The map induced by δ_ℓ at the cochain level, $\delta_\ell^*\colon \widetilde{N}^*(X;R) \to \widetilde{N}^*(X;R)$, is the identity on each factor of the tensor product, except for one where it is induced by a canonical inclusion. It is compatible with the cup product and we can write, $\delta_\ell^*(\omega\cup\eta)_\sigma = \delta_\ell^*(\omega_\sigma \cup \eta_\sigma) =_{(1)} \delta_\ell^*(\omega_\sigma) \cup \delta_\ell^*(\eta_\sigma) = \omega_{\sigma\circ\delta_\ell} \cup \eta_{\sigma\circ\delta_\ell} = (\omega\cup\eta)_{\sigma\circ\delta_\ell}$, where $=_{(1)}$ comes from the naturality of the usual cup product. Then the product $\omega\cup\eta$ of the statement is well defined.

Let us now study the behavior of the cup product with respect to the perverse degree. This is a local issue. Let $\omega \in \widetilde{N}^*_{\overline{p}}(X;R)$ and $\eta \in \widetilde{N}^*_{\overline{q}}(X;R)$. We need to prove that

$$\|\omega_\sigma \cup \eta_\sigma\|_\ell \leqslant (\overline{p}+\overline{q})(S),$$

where S is a singular stratum of X, $\ell = \operatorname{codim} S$, and $\sigma\colon \Delta = \Delta_0 * \cdots * \Delta_n \to X$ is a regular simplex with $\operatorname{Im}\sigma \cap S \neq \emptyset$.

Without loss of generality, we can suppose $\omega_\sigma = \mathbf{1}_{(F,\varepsilon)}$ and $\eta_\sigma = \mathbf{1}_{(G,\kappa)}$. The result is clear if $\|\omega_\sigma \cup \eta_\sigma\|_\ell = -\infty$. So, we can suppose that $\|\omega_\sigma \cup \eta_\sigma\|_\ell \neq -\infty$. If $\overline{p}(S) = -\infty$, condition $\|\omega_\sigma\|_\ell \leqslant -\infty$ implies $\varepsilon_{n-\ell} = 1$ and therefore $\|\omega_\sigma \cup \eta_\sigma\|_\ell = -\infty$. So, we can suppose that $\overline{p}(S) > -\infty$ and similarly $\overline{q}(S) > -\infty$, which imply $(\overline{p}+\overline{q})(S) = \overline{p}(S)+\overline{q}(S)$. Using (4.1) and Definition 2.5, we see that condition $\|\omega_\sigma \cup \eta_\sigma\|_\ell \neq -\infty$ gives $\varepsilon_{n-\ell} = \kappa_{n-\ell} = 0$. The usual degree of the cup product of two cochains gives $\|\omega_\sigma \cup \eta_\sigma\|_\ell \leqslant \|\omega_\sigma\|_\ell + \|\eta_\sigma\|_\ell \leqslant \overline{p}(S)+\overline{q}(S) = (\overline{p}+\overline{q})(S)$.

Then, the cup product of a $\overline{p}$-allowable cochain and of a $\overline{q}$-allowable cochain is $(\overline{p}+\overline{q})$-allowable. For the intersection cochains, the result follows from the formula $\delta(\omega\cup\eta) = \delta(\omega)\cup\eta + (-1)^{|\omega|}\omega\cup\delta(\eta)$.

Associativity is deduced from the associativity of the usual cup product. The unity element is the 0-cochain taking the value 1 on any face.

Let us verify commutativity. We have constructed the product $- \cup_1 -\colon \widetilde{N}^k_{\overline{p}}(X,\mathbb{Z}_2) \otimes \widetilde{N}^\ell_{\overline{q}}(X,\mathbb{Z}_2) \to \widetilde{N}^{k+\ell-1}_{\overline{p}+\overline{q}}(X,\mathbb{Z}_2)$ verifying the Leibniz condition $\delta(x_1 \cup_1 x_2) = x_1 \cup x_2 + x_2 \cup x_1 + \delta x_1 \cup_1 x_2 + x_1 \cup_1 \delta x_2$ (see [**8**]). For general coefficients, and following the same procedure, we construct the product $- \cup_1 -\colon \widetilde{N}^k_{\overline{p}}(X;R) \otimes \widetilde{N}^\ell_{\overline{q}}(X;R) \to \widetilde{N}^{k+\ell-1}_{\overline{p}+\overline{q}}(X;R)$, taking into account the signs (see for example [**3**, p. 414]). So, for each $\omega \in \widetilde{N}^k_{\overline{p}}(X;R)$ and $\eta \in \widetilde{N}^\ell_{\overline{q}}(X;R)$ we have

$$\delta(\omega \cup_1 \eta) = (-1)^{p+q-1}\omega \cup \eta + (-1)^{p+q+pq}\eta \cup \omega + \delta\omega \cup_1 \eta + (-1)^p \omega \cup_1 \delta\eta.$$

So, if $\delta\omega = \delta\eta = 0$, we get the commutativity $[\omega] \cup [\eta] = (-1)^{pq}[\eta] \cup [\omega]$. □

See [**1**, **15**] for other cup products.

REMARK 4.3. Given perversities $\overline{a} \leqslant \overline{p}$ and $\overline{b} \leqslant \overline{q}$, the cup product $- \cup -\colon \widetilde{N}^*_{\overline{a}}(X;R) \otimes \widetilde{N}^*_{\overline{b}}(X;R) \to \widetilde{N}^*_{\overline{a}+\overline{b}}(X;R)$ is the restriction of $- \cup -\colon \widetilde{N}^*_{\overline{p}}(X;R) \otimes \widetilde{N}^*_{\overline{q}}(X;R) \to \widetilde{N}^*_{\overline{p}+\overline{q}}(X;R)$.

5. Intersection and tame intersection (co)homology.

For a perversity $\overline{p}$ such that $\overline{p} \nleqslant \overline{t}$, we may have $\overline{p}$-allowable simplices that are not regular. This failure has bad consequences; for instance, Poincaré duality may be not satisfied in this case (see [**6**]). To overcome this problem the tame intersection homology $\mathfrak{H}^{\overline{p}}_*(X;R)$ has been introduced in [**24**]. In this work, we use the simpler presentation of G. Friedman (see [**5**, **9**, **14**]).

We begin by recalling the notions of intersection homology. As proved in [**4**, Proposition A29], it can be computed using filtered chains.

DEFINITION 5.1. Consider a perverse space $(X,\overline{p})$ and a filtered simplex $\sigma\colon \Delta = \Delta_0 * \cdots * \Delta_n \to X$.

(i) The *perverse degree of* σ is the $(n+1)$-tuple, $\|\sigma\| = (\|\sigma\|_0, \ldots, \|\sigma\|_n)$, where $\|\sigma\|_i = \dim \sigma^{-1}(X_{n-i}) = \dim(\Delta_0 * \cdots * \Delta_{n-i})$, with the convention $\dim \emptyset = -\infty$.

(ii) Given a stratum S of X, the *perverse degree of* σ *along* S is defined by

$$\|\sigma\|_S = \begin{cases} -\infty, & \text{if } S \cap \operatorname{Im} \sigma = \emptyset, \\ \|\sigma\|_{\operatorname{codim} S} & \text{if } S \cap \operatorname{Im} \sigma \neq \emptyset. \end{cases}$$

(iii) The filtered singular simplex σ is $\overline{p}$-*allowable* if $\|\sigma\|_S \leqslant \dim \Delta - \operatorname{codim} S + \overline{p}(S)$, for any stratum S.

(iv) A chain c is $\overline{p}$-*allowable* if it is a linear combination of $\overline{p}$-allowable simplices. The chain c is a $\overline{p}$-*intersection chain* if c and ∂c are $\overline{p}$-allowable chains.

DEFINITION 5.2. Consider a perverse space $(X,\overline{p})$. We denote by $C_*(X;R)$ the complex of filtered chains of X, generated by filtered simplices. The dual complex is $C^*(X;R) = \hom(C_*(X;R),R)$.

The complex of $\overline{p}$-intersection chains of X with the differential ∂ is denoted by $C^{\overline{p}}_*(X;R)$. Its homology $H^{\overline{p}}_*(X;R)$ is the $\overline{p}$-*intersection homology* of X (see [**5**, Théorème A]). The dual complex $C^*_{\overline{p}}(X;R) = \hom(C^{\overline{p}}_*(X;R),R)$ computes the $\overline{p}$-*intersection cohomology* $H^*_{\overline{p}}(X;R)$ (see [**14**]).

DEFINITION 5.3. Given a regular simplex $\Delta = \Delta_0 * \cdots * \Delta_n$ we denote by $\mathfrak{d}$ the regular part of the chain $\partial\Delta$. That is $\mathfrak{d}\Delta = \partial(\Delta_0 * \cdots * \Delta_{n-1}) * \Delta_n$, if $|\Delta_n| = 0$, or $\mathfrak{d}\Delta = \partial\Delta$, if $|\Delta_n| \geqslant 1$.

DEFINITION 5.4. Let $(X, \overline{p})$ be a perverse space. Given a regular simplex $\sigma\colon \Delta \to X$, we define the chain $\mathfrak{d}\sigma$ by $\sigma \circ \mathfrak{d}$. Notice that $\mathfrak{d}^2 = 0$. We denote by $\mathfrak{C}_*(X; R)$ the chain complex generated by the regular simplices, endowed with the differential $\mathfrak{d}$.

DEFINITION 5.5. Let $(X, \overline{p})$ be a perverse space. A $\overline{p}$-allowable filtered simplex $\sigma\colon \Delta \to X$ is *$\overline{p}$-tame* if σ is also a regular simplex. A chain ξ is *$\overline{p}$-tame* if is a linear combination of $\overline{p}$-tame simplices. The chain ξ is a *tame $\overline{p}$-intersection chain* if ξ and $\mathfrak{d}\xi$ are $\overline{p}$-tame chains.

We write $\mathfrak{C}^{\overline{p}}_*(X; R) \subset \mathfrak{C}_*(X; R)$ the complex of tame $\overline{p}$-intersection chains endowed with the differential $\mathfrak{d}$. Its homology $\mathfrak{H}^{\overline{p}}_*(X; R)$ is the *tame $\overline{p}$-intersection homology* of X

Main properties of this homology have been studied in [**5**, **14**]. We have proven in [**5**] that the homology $\mathfrak{H}^{\overline{p}}_*(X; R)$ coincides with those of [**10**, **24**] (see also [**14**, Chapter 6]). It is also proved that $\mathfrak{H}^{\overline{p}}_*(X; R) = H^{\overline{p}}_*(X; R)$ if $\overline{p} \leqslant \overline{t}$. The *tame $\overline{p}$-intersection cohomology* $\mathfrak{H}^*_{\overline{p}}(X; R)$ is defined by using the dual complex $\mathfrak{C}^*_{\overline{p}}(X; R) = \operatorname{Hom}(\mathfrak{C}^{\overline{p}}_*(X; R), R)$. It verifies $\mathfrak{H}^*_{\overline{p}}(X; R) = H^*_{\overline{p}}(X; R)$ if $\overline{p} \leqslant \overline{t}$.

5.1. Shifted filtrations. Intersection homology does not depend on the dimension of the strata of X but on the codimension of these strata (see [**14**, Section 4.3.1]). Let us consider on X the shifted filtration Y of (3.2) in subsection 3.1. So, the formal dimension of Y is $n + m$. Following subsection 3.1 we have:

$$\|\sigma\|^X_\ell = \dim \sigma^{-1}(X_{n-\ell}) = \dim \sigma^{-1}(Y_{n+m-\ell}) = \|\sigma\|^Y_\ell .$$

The allowability condition is the same. The perversity $\overline{p}$ is also a perversity on Y. If S is a stratum of X (and therefore of Y) with $\operatorname{Im} \sigma \cap S \neq \emptyset$ we have $\ell = \operatorname{codim}_X S = \operatorname{codim}_Y S$. So,

$$\|\sigma\|^X_\ell \leqslant \dim \Delta - \ell + \overline{p}(S) \Longleftrightarrow \|\sigma\|^Y_\ell \leqslant \dim \Delta - \ell + \overline{p}(S).$$

We get $C^{\overline{p}}_*(X; R) = C^{\overline{p}}_*(Y; R)$ and therefore $H^{\overline{p}}_*(X; R) = H^{\overline{p}}_*(Y; R)$. By the same reasons, we also have $H^*_{\overline{p}}(X; R) = H^*_{\overline{p}}(Y; R)$, $\mathfrak{H}^{\overline{p}}_*(X; R) = \mathfrak{H}^{\overline{p}}_*(Y; R)$ and $\mathfrak{H}^*_{\overline{p}}(X; R) = \mathfrak{H}^*_{\overline{p}}(Y; R)$.

6. Cap product.

We introduce the notion of cap product for the blown-up intersection (co)homology, already treated in [**7**] in a different context. First of all, we work at the simplex level.

Let Δ be an m-dimensional euclidean simplex. We denote by $[\Delta]$ its face of maximal dimension. The *classical cap product* $- \cap [\Delta]\colon N^*(\Delta) \to N_{m-*}(\Delta)$ is defined by

$$\mathbf{1}_F \cap [\Delta] = \begin{cases} G & \text{if } F \cup G = \Delta \text{ (where } \cup \text{ is the map of Definition 4.1)}, \\ 0 & \text{otherwise.} \end{cases} \tag{6.1}$$

Consider now the cone $\mathtt{c}\Delta$ whose apex is denoted by $\mathtt{v}$, which is the last vertex of the cone (see page 51). We have the formula:

$$\mathbf{1}_{(F,j)} \cap [\mathtt{c}\Delta] = \begin{cases} (G,1) & \text{if } j=0 \text{ or } (F,j)=(\Delta,1), \\ 0 & \text{if not,} \end{cases} \tag{6.2}$$

where the simplex G is the face of Δ with $F \cup G = \Delta$ (cf. Definition 4.1) if $j = 0$ and $G = \varnothing$ if $(F,j) = (\Delta,1)$.

We extend it to regular simplices $\Delta = \Delta_0 * \cdots * \Delta_n$ as follows. Denote $\widetilde{N}_*(\Delta) = N_*(\mathtt{c}\Delta_0) \otimes \cdots \otimes N_*(\mathtt{c}\Delta_{n-1}) \otimes N_*(\Delta_n)$.

DEFINITION 6.1. We define the *cap product* $- \,\tilde{\cap}\, \widetilde{\Delta} \colon \widetilde{N}^*(\Delta) \to \widetilde{N}_{*-m}(\Delta)$, linearly from

$$\mathbf{1}_{(F,\varepsilon)} \,\tilde{\cap}\, \widetilde{\Delta} = (-1)^{\nu(F,\varepsilon,\Delta)} (\mathbf{1}_{(F_0,\varepsilon_0)} \cap \mathtt{c}[\Delta_0]) \otimes \cdots \otimes (\mathbf{1}_{F_n} \cap [\Delta_n]), \tag{6.3}$$

where $\mathbf{1}_{(F,\varepsilon)} = \mathbf{1}_{(F_0,\varepsilon_0)} \otimes \cdots \otimes \mathbf{1}_{(F_{n-1},\varepsilon_{n-1})} \otimes \mathbf{1}_{F_n}$, $\widetilde{\Delta} = \mathtt{c}\Delta_0 \times \cdots \times \mathtt{c}\Delta_{n-1} \times \Delta_n$ (see (2.3)) and $\nu(F,\varepsilon,\Delta) = \sum_{j=0}^{n-1} (\dim \Delta_j + 1) \left(\sum_{i=j+1}^{n} |(F_i,\varepsilon_i)|\right)$, with the convention $\varepsilon_n = 0$.

At the filtered simplex level, the *local intersection cap product* $- \cap \widetilde{\Delta} \colon \widetilde{N}^*(\Delta) \to N_{m-*}(\Delta)$ is defined by

$$c \cap \widetilde{\Delta} = \mu_*(c \,\tilde{\cap}\, \widetilde{\Delta}),$$

where the map $\mu_* \colon \widetilde{N}_*(\Delta) \to N_*(\Delta)$ is defined by

(6.4)

$$\mu_* \left(\otimes_{k=0}^{n-1} (F_k, \varepsilon_k) \otimes F_n \right) = \begin{cases} F_0 * \cdots * F_\ell & \text{if } \dim(F,\varepsilon) = \dim(F_0 * \cdots * F_\ell), \\ 0 & \text{otherwise,} \end{cases}$$

where $(F,\varepsilon) = (F_0,\varepsilon_0) \otimes \cdots \otimes (F_{n-1},\varepsilon_{n-1}) \otimes F_n$ and ℓ is the smallest integer, j, such that $\varepsilon_j = 0$.

DEFINITION 6.2. Since $\widetilde{\Delta} = \mathtt{c}\Delta_0 \times \cdots \times \mathtt{c}\Delta_{n-1} \times \Delta_n$ (see (2.3)), we have the boundary chain

$$\begin{aligned} \partial \widetilde{\Delta} = & \sum_{\substack{i=0 \\ \Delta_i \neq \varnothing}}^{n} (-1)^{|\Delta|_{\leqslant i-1}+1} [\mathtt{c}\Delta_0] \otimes \cdots \otimes [\mathtt{c}\partial\Delta_i] \otimes \cdots \otimes [\Delta_n] \\ & + \sum_{\substack{i=0 \\ \Delta_i \neq \varnothing}}^{n-1} (-1)^{|\Delta|_{\leqslant i}+1} [\mathtt{c}\Delta_0] \otimes \cdots \otimes [\Delta_i] \otimes \cdots \otimes [\Delta_n]. \end{aligned}$$

Let $i \in \{0, \ldots, n-1\}$ such that $\Delta_i \neq \varnothing$, the face $\mathcal{H}_i = \mathtt{c}\Delta_0 \times \cdots \times \mathtt{c}\Delta_{i-1} \times \Delta_i \times \mathtt{c}\Delta_{i+1} \times \cdots \times \Delta_n$ is called a *hidden face* of $\widetilde{\Delta}$. This gives

$$\partial \widetilde{\Delta} = \widetilde{\mathfrak{d}\Delta} + \sum_{\substack{i=0 \\ \Delta_i \neq \varnothing}}^{n-1} (-1)^{|\Delta|_{\leqslant i}+1} \, \mathcal{H}_i, \tag{6.5}$$

where $\widetilde{\mathfrak{d}\Delta}$ is the blow up of the weighted simplicial complex $\mathfrak{d}\Delta$ (cf. Definition 2.2).

REMARK 6.3.

- For each $\omega \in \widetilde{N}^*(\Delta)$ the chain $\omega \cap \widetilde{\Delta}$ is regular or 0. Let us see that. Let $\mathbf{1}_{(F,\varepsilon)} = \mathbf{1}_{(F_0,\varepsilon_0)} \otimes \cdots \otimes \mathbf{1}_{F_n} \in \widetilde{N}^*(\Delta)$. From the definition of the cap product we have, up to a sign,
$$\mathbf{1}_{(F,\varepsilon)} \cap \widetilde{\Delta} = G_0 * \cdots * G_n,$$
if $\mathbf{1}_{(F,\varepsilon)} \cap \widetilde{\Delta} \neq 0$. The faces $G_\bullet$ have been defined in (6.1), (6.2) and (6.4). The simplex $\mathbf{1}_{(F,\varepsilon)} \cap \widetilde{\Delta}$ is regular: $F_n \neq \emptyset \Rightarrow G_n \neq \emptyset$.
- Given a hidden face $\mathcal{H}_i = \mathtt{c}\Delta_0 \times \cdots \times \Delta_i \times \cdots \times \Delta_n$, we have
$$\mu_*(\mathcal{H}_i) = \begin{cases} \Delta_0 * \cdots * \Delta_i & \text{if } \dim(\Delta_{i+1} * \cdots * \Delta_n) = 0 \\ 0 & \text{otherwise} \end{cases},$$
which is a non regular face or 0.

We define the intersection cap product for any filtered space X.

DEFINITION 6.4. Let X be a filtered space. The *intersection cap product*
$$- \cap -\colon \widetilde{N}^*(X;R) \otimes \mathfrak{C}_m(X;R) \to \mathfrak{C}_{m-*}(X;R)$$
is defined by linear extension of
$$\omega \cap \sigma = \sigma_*(\omega_\sigma \cap \widetilde{\Delta}),$$
where $\omega \in \widetilde{N}^*(X;R)$ and $\sigma\colon \Delta \to X$ is a regular simplex.

LEMMA 6.5. *Let Δ be a regular simplex. The map $\mu_*\colon \widetilde{N}_*(\Delta) \to N_*(\Delta)$ commutes with differentials.*

PROOF. Set $\Delta = \Delta_0 * \cdots * \Delta_n$. We proceed by induction on n. The result is clear when $n = 0$. We consider the regular simplex $\nabla = \Delta_1 * \cdots * \Delta_n$, et $\widetilde{N}_*(\nabla) = N_*(\mathtt{c}\Delta_1) \otimes \cdots \otimes N_*(\mathtt{c}\Delta_{n-1}) \otimes N_*(\Delta_n)$. The map $\mu_{\Delta,*}$ is the composition
$$\mu_{\Delta,*}\colon N_*(\mathtt{c}\Delta_0) \otimes \widetilde{N}_*(\nabla) \xrightarrow{\mathrm{id}\otimes\mu_{\nabla,*}} N_*(\mathtt{c}\Delta_0) \otimes N_*(\nabla) \xrightarrow{\nu} N_*(\Delta),$$
(see (6.4)), where
$$\nu((F_0,\varepsilon_0)\otimes G) = \begin{cases} F_0 & \text{if } \varepsilon_0 = 0 \text{ and } \dim G = 0, \\ 0 & \text{if } \varepsilon_0 = 0 \text{ and } \dim G > 0, \\ F_0 * G & \text{if } \varepsilon_0 = 1. \end{cases}$$
By induction hypothesis, the operator $\mu_{\nabla,*}$ is compatible with differentials. It remains to prove that ν is also compatible with differentials. We distinguish four cases.

- If $\varepsilon_0 = 0$ and $\dim G = 0$, we have $\nu(\partial((F_0,0)\otimes G)) = \nu((\partial F_0,0)\otimes G) = \partial F_0 = \partial\nu((F_0,0)\otimes G)$.
- If $\varepsilon_0 = 0$ and $\dim G = 1$, we have $\nu(\partial((F_0,0)\otimes G)) = \nu((\partial F_0,0)\otimes G) + (-1)^{|F_0|}\nu((F_0,0)\otimes \partial G) = 0 + F_0 - F_0 = 0 = \partial\nu((F_0,0)\otimes G)$.
- If $\varepsilon_0 = 0$ and $\dim G > 1$, all the terms vanish.

- If $\varepsilon_0 = 1$, the result comes from

$$\begin{aligned} \nu(\partial((F_0,1)\otimes G)) &= \\ \nu((\partial F_0,1)\otimes G) + (-1)^{|F_0|+1}\left(\nu((F_0,0)\otimes G) + \nu((F_0,1)\otimes \partial G)\right) &= \\ (\partial F_0) * G + (-1)^{|F_0|+1}\begin{cases} F_0 & \text{if } \dim G = 0, \\ F_0 * \partial G & \text{if } \dim G \geqslant 1, \end{cases} &= \\ \partial(F_0 * G) = \partial\nu((F_0,1)\otimes G). \end{aligned}$$

□

PROPOSITION 6.6. *The intersection cap product verifies the following properties*

(i) $\mathfrak{d}(\omega \cap \xi) = (\delta\omega) \cap \xi + (-1)^{|\omega|}\omega \cap \mathfrak{d}\xi$, *and*
(ii) $(\omega \cup \eta) \cap \xi = (-1)^{|\omega|\cdot|\eta|}\ \eta \cap \omega \cap \xi$,

where $\omega, \eta \in \widetilde{N}^*(X;R)$ *and* $\xi \in \mathfrak{C}_*(X;R)$.

PROOF. (i) It suffices to prove it for a regular simplex $\sigma\colon \Delta \to X$. We write $k = |\omega|$. Using formula (6.5), the classic Leibniz formula for the cap product gives

$$\begin{aligned} \partial(\omega_\sigma \cap \widetilde{\Delta}) = \partial\mu_*(\omega_\sigma \mathbin{\widetilde{\cap}} \widetilde{\Delta}) &\overset{Lemma\ 6.5}{=} \mu_*\partial(\omega_\sigma \mathbin{\widetilde{\cap}} \widetilde{\Delta}) = \\ &\mu_*\left((\delta\omega_\sigma) \mathbin{\widetilde{\cap}} \widetilde{\Delta} + (-1)^k \omega_\sigma \mathbin{\widetilde{\cap}} (\partial\widetilde{\Delta})\right) = \end{aligned}$$

$$\overset{(6.5)}{=} \mu_*\left((\delta\omega_\sigma) \mathbin{\widetilde{\cap}} \widetilde{\Delta} + (-1)^k \omega_\sigma \mathbin{\widetilde{\cap}} \widetilde{\mathfrak{d}\Delta} + \sum_{\substack{i=0 \\ \Delta_i \neq \emptyset}}^{n-1} (-1)^{(-1)^{k+|\Delta|_{\leqslant i}}} \omega_\sigma \mathbin{\widetilde{\cap}} \mathcal{H}_i\right) =$$

$$= \underbrace{\mu_*\left((\delta\omega_\sigma) \mathbin{\widetilde{\cap}} \widetilde{\Delta} + (-1)^k \omega_\sigma \mathbin{\widetilde{\cap}} \widetilde{\mathfrak{d}\Delta}\right)}_{\text{regular or } 0} + \sum_{\substack{i=0 \\ \Delta_i \neq \emptyset}}^{n-1} (-1)^{(-1)^{k+|\Delta|_{\leqslant i}}} \underbrace{\mu_*(\omega_\sigma \mathbin{\widetilde{\cap}} \mathcal{H}_i)}_{\text{non regular or } 0}$$

(cf. Remark 6.3). So, $\mathfrak{d}(\omega_\sigma \cap \widetilde{\Delta}) = \mu_*((\delta\omega_\sigma) \mathbin{\widetilde{\cap}} \widetilde{\Delta}) + (-1)^k \mu_*(\omega_\sigma \mathbin{\widetilde{\cap}} \widetilde{\mathfrak{d}\Delta}) = (\delta\omega_\sigma) \cap \widetilde{\Delta} + (-1)^k \omega_\sigma \cap \widetilde{\mathfrak{d}\Delta}$, and therefore $\sigma_*\mathfrak{d}(\omega_\sigma \cap \widetilde{\Delta}) = (\delta\omega) \cap \Delta + (-1)^k \omega \cap \mathfrak{d}\Delta$.

The regular simplex $\sigma\colon \Delta \to X$ is in fact a stratified map when one considers on Δ the filtration $\Delta_0 \subset \Delta_0 * \Delta_1 \subset \cdots \subset \Delta_0 * \cdots * \Delta_n$. So, $\sigma_* \circ \partial = \partial \circ \sigma_*$ [**4**, Theorem F]. Since $\Delta_0 * \cdots * \Delta_{n-1} = \sigma^{-1}(\Sigma_X)$ then $\sigma_* \circ \mathfrak{d} = \mathfrak{d} \circ \sigma_*$. We get property (i) since $\mathfrak{d}(\omega \cap \Delta) = \mathfrak{d}\sigma_*(\omega_\sigma \cap \widetilde{\Delta}) = \sigma_*\mathfrak{d}(\omega_\sigma \cap \widetilde{\Delta})$.

(ii) Without loss of generality we can suppose $\omega_\sigma = \mathbf{1}_{(F,\varepsilon)} = \mathbf{1}_{(F_0,\varepsilon_0)} \otimes \cdots \otimes \mathbf{1}_{(F_{n-1},\varepsilon_{n-1})} \otimes \mathbf{1}_{F_n}$ and $\eta_\sigma = \mathbf{1}_{(H,\kappa)} = \mathbf{1}_{(H_0,\kappa_0)} \otimes \cdots \otimes \mathbf{1}_{(H_{n-1},\kappa_{n-1})} \otimes \mathbf{1}_{H_n}$. It suffices to prove

$$(\mathbf{1}_{(F,\varepsilon)} \cup \mathbf{1}_{(H,\kappa)}) \mathbin{\widetilde{\cap}} \widetilde{\Delta} = (-1)^{|(F,\varepsilon)|\cdot|(H,\kappa)|}\ \mathbf{1}_{(H,\kappa)} \mathbin{\widetilde{\cap}} (\mathbf{1}_{(F,\varepsilon)} \mathbin{\widetilde{\cap}} \widetilde{\Delta}).$$

For $n = 0$ we find the classic property of cup/cap products. In the general case, with the convention $\varepsilon_n = \kappa_n = 0$, we have

$$(\mathbf{1}_{(F,\varepsilon)} \cup \mathbf{1}_{(H,\kappa)}) \,\widetilde{\cap}\, \widetilde{\Delta} = \left(\bigotimes_{a=0}^{n} \mathbf{1}_{(F_a,\varepsilon_a)} \cup \bigotimes_{a=0}^{n} \mathbf{1}_{(H_a,\kappa_a)} \right) \widetilde{\cap}\, \widetilde{\Delta} \overset{(4.2)}{=}$$

$$(-1)^{\ltimes_1} \cdot \left(\bigotimes_{a=0}^{n} \left(\mathbf{1}_{(F_a,\varepsilon_a)} \cup \mathbf{1}_{(H_a,\kappa_a)} \right) \right) \widetilde{\cap}\, \widetilde{\Delta} \overset{(6.3)}{=}$$

$$(-1)^{\ltimes_2} \cdot \bigotimes_{a=0}^{n-1} \left((\mathbf{1}_{(F_a,\varepsilon_a)} \cup \mathbf{1}_{(H_a,\kappa_a)}) \cap [\mathtt{c}\Delta_a] \right) \otimes \left((\mathbf{1}_{F_n} \cup \mathbf{1}_{H_n}) \cap [\Delta_n] \right) \overset{classic}{=}$$

$$(-1)^{\ltimes_3} \cdot \bigotimes_{a=0}^{n-1} \left(\mathbf{1}_{(H_a,\kappa_a)} \cap \mathbf{1}_{(F_a,\varepsilon_a)} \cap [\mathtt{c}\Delta_a] \right) \otimes \left(\mathbf{1}_{H_n} \cap \mathbf{1}_{F_n} \cap [\Delta_n] \right) \overset{(6.3)}{=}$$

$$(-1)^{\ltimes_4} \cdot \mathbf{1}_{(H,\kappa)} \,\widetilde{\cap} \left(\bigotimes_{a=0}^{n-1} \left(\mathbf{1}_{(F_a,\varepsilon_a)} \cap [\mathtt{c}\Delta_a] \right) \otimes \left(\mathbf{1}_{F_n} \cap [\Delta_n] \right) \right) \overset{(6.3)}{=}$$

$$(-1)^{|(F,\varepsilon)|\cdot|(H,\kappa)|} \cdot \mathbf{1}_{(H,\kappa)} \,\widetilde{\cap}\, \mathbf{1}_{(F,\varepsilon)} \,\widetilde{\cap}\, \widetilde{\Delta}.$$

where $\ltimes_1 = \sum_{i>j} |(F_i,\varepsilon_i)|\cdot|(H_j,\kappa_j)|$, $\ltimes_2 = \ltimes_1 + \sum_{j=0}^{n-1} (\dim \Delta_j + 1) \left(\sum_{i>j} |(F_i,\varepsilon_i)| + |(H_i,\kappa_i)| \right)$, $\ltimes_3 = \ltimes_2 + \sum_i |(F_i,\varepsilon_i)| \cdot |(H_i,\kappa_i)|$ and

$$\begin{aligned} \ltimes_4 &= \ltimes_3 + \sum_{j=0}^{n-1} (\dim \Delta_j + 1 - |(F_j,\varepsilon_j)|) \left(\sum_{i>j} |(H_i,\kappa_i)| \right) = \sum_{i \geqslant j} |(F_i,\varepsilon_i)| \cdot |(H_j,\kappa_j)| \\ &\quad + \sum_{j=0}^{n-1} (\dim \Delta_j + 1) \left(\sum_{i>j} |(F_i,\varepsilon_i)| \right) - \sum_{j=0}^{n-1} |(F_j,\varepsilon_j)| \left(\sum_{i>j} |(H_i,\kappa_i)| \right) \\ &= \sum_{j=0}^{n-1} (\dim \Delta_j + 1) \left(\sum_{i>j} |(F_i,\varepsilon_i)| \right) + |(F,\varepsilon)| \cdot |(H,\kappa)|. \end{aligned}$$

□

The cap product between a cochain $\omega \in \widetilde{N}^*(X;R)$ and a chain $\xi \in C_*(X;R)$ may not exist. This happens when a simplex σ of ξ lies in the singular part of X. In this case, σ is not regular and we cannot construct the blow up of Δ_σ. This is the reason to use the tame intersection homology in the definition of the cap product: all the involved simplices are regular.

Proposition 6.7. *For each filtered space, X, endowed with two perversities, $\overline{p}$ and $\overline{q}$, the intersection cap product*

$$- \cap -\colon \widetilde{N}^*_{\overline{p}}(X;R) \otimes \mathfrak{C}^{\overline{q}}_m(X;R) \to \mathfrak{C}^{\overline{p}+\overline{q}}_{m-*}(X;R),$$

is well defined and induces a morphism

$$- \cap -\colon \mathscr{H}^*_{\overline{p}}(X;R) \otimes \mathfrak{H}^{\overline{q}}_m(X;R) \to \mathfrak{H}^{\overline{p}+\overline{q}}_{m-*}(X;R).$$

Proof. Following Proposition 6.6 it remains to study the behavior of the cap product with respect to the perverse degree on a singular stratum S. Let us consider $\omega \in \widetilde{N}^*(X;R)$ with $\|\omega\|_S \leqslant \overline{p}(S)$, $\sigma\colon \Delta \to X$ a regular simplex with $\operatorname{Im} \sigma \cap S \neq \emptyset$ and $\|\sigma\|_S \leqslant |\Delta| - \ell + \overline{q}(S)$, where $\ell = \operatorname{codim} S$. We need to prove

$$\|\omega \cap \sigma\|_S \leqslant |\omega \cap \sigma| - \ell + (\overline{p} + \overline{q})(S). \tag{6.6}$$

Without loss of generality, we can suppose $\omega \cap \sigma \neq 0$.

When $\overline{p}(S) = -\infty$, the condition $\|\omega\|_S \leqslant -\infty$ implies $\varepsilon_{n-\ell} = 1$ and therefore the $(n-\ell)$-factor of $\omega_\sigma \,\widetilde{\cap}\, \widetilde{\Delta}$ is $[\mathtt{v}_{n-\ell}]$ which gives $\omega \cap \sigma \cap S = \varnothing$ and therefore (6.6). On the other hand, if $\overline{q}(S) = -\infty$, the condition $\|\sigma\|_S = -\infty$ gives $\Delta_{n-\ell} = \varnothing$ then $\omega \cap \sigma \cap S = \varnothing$ and therefore (6.6).

We can suppose $(\overline{p} + \overline{q})(S) = \overline{p}(S) + \overline{q}(S)$. Then we have (6.6) from

$$\|\omega\cap\sigma\|_S \overset{claim}{\leqslant} \|\sigma\|_S+\|\omega_\sigma\|_\ell-|\omega_\sigma| \leqslant |\Delta|-\ell+\overline{p}(S)+\overline{q}(S)-|\omega_\sigma| = |\omega\cap\sigma|-\ell+\overline{p}(S)+\overline{q}(S).$$

We verify the claim. It suffices to consider $\omega_\sigma = \mathbf{1}_{(F,\varepsilon)}$ with $\mathbf{1}_{(F,\varepsilon)} \cap \widetilde{\Delta} \neq 0$. We know, from Remark 6.3, that $\mathbf{1}_{(F,\varepsilon)} \cap \Delta = \pm G_0 * \cdots * G_n$ with $|\Delta_i| = |G_i| + |(F_i, \varepsilon_i)|$, for each $i \in \{0, \ldots, n\}$. Then, $\|\omega \cap \sigma\|_S$ is equal to

$$\begin{aligned} |G_0 * \cdots * G_{n-\ell}| &= |G_0| + \cdots + |G_{n-\ell}| + n - \ell \\ &= |\Delta_0| + \cdots + |\Delta_{n-\ell}| + n - \ell - |(F_0, \varepsilon_0)| - \cdots - |(F_{n-\ell}, \varepsilon_{n-\ell})| \\ &= \|\sigma\|_S - |(F, \varepsilon)| + |(F, \varepsilon)|_{>n-\ell} = \|\sigma\|_S - |\omega_\sigma| + \|\omega_\sigma\|_\ell. \end{aligned}$$

□

REMARK 6.8. The intersection cap product is natural respectively to perversities. Given perversities $\overline{a} \leqslant \overline{p}$ and $\overline{b} \leqslant \overline{q}$, the cap product $- \cap -\colon \widetilde{N}^*_{\overline{a}}(X) \otimes \mathfrak{C}^{\overline{b}}_m(X) \to \mathfrak{C}^{\overline{a}+\overline{b}}_{m-*}(X)$ is the restriction of $- \cap -\colon \widetilde{N}^*_{\overline{p}}(X) \otimes \mathfrak{C}^{\overline{q}}_m(X) \to \mathfrak{C}^{\overline{p}+\overline{q}}_{m-*}(X)$.

7. Stratified maps: The local level.

In this section and the next one, we prove that any stratified map induces a homomorphism between the blown-up intersection cohomologies if a certain compatibility of the involved perversities is satisfied. In particular, if $\overline{p}$ is a GM-perversity, any stratified map $f\colon X \to Y$ induces a homomorphism $f\colon \mathscr{H}^*_{\overline{p}}(X; R) \to \mathscr{H}^*_{\overline{p}}(Y; R)$.

Let $f\colon X \to Y$ be a stratified map as in Definition 1.5. If $\sigma\colon \Delta \to X$ is a filtered simplex then the composite $f \circ \sigma\colon \Delta \to Y$ is also a filtered simplex but, in general, the two filtrations induced on Δ by σ and $f \circ \sigma$ differ. We denote them by Δ_σ and $\Delta_{f\circ\sigma}$. In [**4**, Corollary A.25 and Lemma A.24], we prove that the two filtrations Δ_σ and $\Delta_{f\circ\sigma}$ (on the same euclidean simplex) match through a finite number of elementary amalgamations that we describe and study in this section.

DEFINITION 7.1. Let $k \in \{0, \ldots, n-1\}$. An *elementary k-amalgamation* of a regular simplex $\Delta = \Delta_0 * \cdots * \Delta_n$ is the regular simplex: $\Delta' = \Delta'_0 * \cdots * \Delta'_{n-1}$ with

$$\Delta'_i = \begin{cases} \Delta_i & \text{if } i \leqslant n-k-2, \\ \Delta_{n-k-1} * \Delta_{n-k} & \text{if } i = n-k-1, \\ \Delta_{i-1} & \text{if } i \geqslant n-k. \end{cases}$$

If $\Delta_{n-k-1} = \varnothing$ then the elementary amalgamation is *k-simple*.

We study the effect of these maps at the blow up level. First of all, we introduce a technical tool.

7.1. Stratification. The simplices $\Delta = \Delta_0 * \cdots * \Delta_n$ and $\Delta' = \Delta'_0 * \cdots * \Delta'_{n-1}$ are filtered spaces (see Definition 1.1) of respective dimension n and $n-1$. We set $\mathbb{N}_\Delta = \{i \in \{0, \ldots, n\} \;/\; \Delta_0 * \cdots * \Delta_{n-i-1} \neq \Delta_0 * \cdots * \Delta_{n-i}\}$. We observe that the family of strata of Δ is

$$\mathcal{S}_\Delta = \{S_i = \Delta_0 * \cdots * \Delta_{n-i} \backslash \Delta_0 * \cdots * \Delta_{n-i-1} \;/\; i \in \mathbb{N}_\Delta\}.$$

Notice that codim $S_i = i$ for each $i \in \mathbb{N}_\Delta$.

If $k \in \{0, \ldots, n-1\}$, we denote $\varepsilon_k \colon \{0, \ldots, n\} \to \{0, \ldots, n-1\}$ the map defined by $\varepsilon_k(i) = i$ if $i \leqslant k$ and $\varepsilon_k(i) = i-1$ if $i \geqslant k+1$. We check easily that this maps restricts as a map $a_k \colon \mathbb{N}_\Delta \to \mathbb{N}_{\Delta'}$, called the *index map*. If $S_i \in \mathcal{S}_\Delta$, we have $S_i \subset S'_{a_k(i)}$. Thus by definition, the identity map is a stratified map from Δ to Δ', we denote it

$$\mathcal{A}[k] \colon \Delta = \Delta_0 * \cdots * \Delta_n \to \Delta' = \Delta_0 * \cdots * \Delta'_{n-1}.$$

If $S \in \mathcal{S}_\Delta$, with the notations of Definition 1.5 we have $S^{\mathcal{A}[k]} = S'_{a_k(i)}$, with $i =$ codim S, and therefore

$$\text{codim } S^{\mathcal{A}[k]} = a_k(\text{codim } S). \tag{7.1}$$

We also say that $\mathcal{A}[k]$ is an elementary amalgamation. If the elementary k-amalgamation is simple, then a_k is a bijection and $\mathcal{A}[k]$ is a stratified homeomorphism (see Definition 1.5).

The elementary k-amalgamation has an effect only on two components . We first focus on them. For the sake of the simplicity we write $E_0 = \Delta_{n-k-1}$ and $E_1 = \Delta_{n-k}$, they can be empty.

DEFINITION 7.2. We define a map

$$\theta \colon N_*(\mathtt{c}E_0) \otimes N_*(\mathtt{c}E_1) \to N_*(\mathtt{c}(E_0 * E_1))$$

by

$$\theta((F_0, \varepsilon_0) \otimes (F_1, \varepsilon_1)) = \begin{cases} (F_0 * F_1, \varepsilon_1) & \text{if } \varepsilon_0 = 1, \\ (F_0, 0) & \text{if } \varepsilon_0 = 0 \text{ and } |(F_1, \varepsilon_1)| = 0, \\ 0 & \text{if not.} \end{cases}$$

Notice that the restriction of θ to $N_*(\mathtt{c}E_0) \otimes N_*(E_1)$ gives a map, still denoted θ,

$$\theta \colon N_*(\mathtt{c}E_0) \otimes N_*(E_1) \to N_*(E_0 * E_1).$$

By duality, we get a map

$$\Xi \colon N^*(\mathtt{c}(E_0 * E_1)) \to N^*(\mathtt{c}E_0) \otimes N^*(\mathtt{c}E_1)$$

which verifies

$$\begin{cases} \Xi(\mathbf{1}_{(F_0 * F_1, \varepsilon)}) = (-1)^{|(F_0,1)| \cdot |(F_1,\varepsilon)|} \mathbf{1}_{(F_0,1)} \otimes \mathbf{1}_{(F_1,\varepsilon)} & \text{if } (F_1, \varepsilon) \neq (\varnothing, 0). \\ \Xi(\mathbf{1}_{(F_0,0)}) = \mathbf{1}_{(F_0,0)} \otimes \lambda_{\mathtt{c}E_1}, & \text{with } \lambda_{\mathtt{c}E_1} = \mathbf{1}_{(\varnothing,1)} + \sum_{e \in \mathcal{V}(E_1)} \mathbf{1}_{(e,0)}. \end{cases}$$

We also denote $\Xi \colon N^*(E_0 * E_1) \to N^*(\mathtt{c}E_0) \otimes N^*(E_1)$ the restriction of Ξ.

LEMMA 7.3. *The morphisms θ and Ξ are chain maps. Moreover,*

(1) *if $E_0 = \varnothing$ then θ and Ξ are isomorphisms,*
(2) *Ξ is compatible with the cup product,*

(3) *θ and Ξ are compatible with the cap product, i.e.,*

$$\begin{aligned}\theta(\Xi\left(\mathbf{1}_{(F_0*F_1,\varepsilon)}\right)\,\tilde{\frown}\,([\mathtt{c}E_0]\otimes[\mathtt{c}E_1])) &= \mathbf{1}_{(F_0*F_1,\varepsilon)}\frown[\mathtt{c}(E_0*E_1)],\ \textit{and}\\ \theta(\Xi\left(\mathbf{1}_{F_0*F_1}\right)\,\tilde{\frown}\,([\mathtt{c}E_0]\otimes[E_1])) &= \mathbf{1}_{F_0*F_1}\frown[E_0*E_1]\end{aligned}$$

where $[\nabla]$ is the maximal simplex of an euclidean simplex ∇, (F_0*F_1,ε), is a face of $\mathtt{c}(E_0*E_1)$ and F_0*F_1, is a face of E_0*E_1.

We postpone the proof of this Lemma and deduce first some properties on amalgamations from it.

DEFINITION 7.4. Let us consider the elementary k-amalgamation $\mathcal{A}[k]\colon\Delta\to\Delta'$ with $k\in\{0,\dots,n-1\}$ (see Definition 7.1). We define two homomorphisms $\mathcal{A}[k]_*\colon\widetilde{N}_*(\Delta)\to\widetilde{N}_*(\Delta')$ and $\mathcal{A}[k]^*\colon\widetilde{N}^*(\Delta')\to\widetilde{N}^*(\Delta)$ by

$$\mathcal{A}[k]_* = \underbrace{\mathrm{id}\otimes\cdots\otimes\mathrm{id}}_{n-k-1\ times}\otimes\,\theta\,\otimes\underbrace{\mathrm{id}\otimes\cdots\otimes\mathrm{id}}_{k\ times} \tag{7.2}$$

and

$$\mathcal{A}[k]^* = \underbrace{\mathrm{id}\otimes\cdots\otimes\mathrm{id}}_{n-k-1\ times}\otimes\,\Xi\,\otimes\underbrace{\mathrm{id}\otimes\cdots\otimes\mathrm{id}}_{k\ times} \tag{7.3}$$

If there is not ambiguity, we use the notation $\mathcal{A}_*$ and $\mathcal{A}^*$ for these two maps and a for the index map. In order to get coherent notation, we write $\|-\|_0=0$.

PROPOSITION 7.5. *Let $\mathcal{A}\colon\Delta\to\Delta'$ be an elementary k-amalgamation. Then, the maps $\mathcal{A}_*$ and $\mathcal{A}^*$ defined above satisfy the following properties.*

(1) *They are chain maps and $\mathcal{A}^*$ is compatible with the cup product.*
(2) *Let $\omega\in\widetilde{N}^*(\Delta')$, then we have*

$$\mathcal{A}_*(\mathcal{A}^*(\omega)\,\tilde{\frown}\,\widetilde{\Delta}) = \omega\,\tilde{\frown}\,\widetilde{\Delta'}.$$

(3) *In the case of simple amalgamations, they are isomorphisms.*
(4) *For each regular face operator $\delta_\ell\colon\nabla\to\Delta$ we denote by $\mathcal{A}^*_\Delta$ the previous map and by $\mathcal{A}^*_\nabla$ the map corresponding to the amalgamation induced on ∇. Then we have*

$$\delta_\ell^*\circ\mathcal{A}^*_\nabla = \mathcal{A}^*_\Delta\circ\delta_\ell^*.$$

(5) *The map $\mathcal{A}^*$ decreases the perverse degree, i.e., for $\omega\in\widetilde{N}^*(\Delta')$ and $\ell\in\mathbb{N}_\Delta$, we have*

$$\|\mathcal{A}^*(\omega)\|_\ell\leqslant\|\omega\|_{a(\ell)}.$$

Moreover, if the amalgamation is simple, this inequality becomes an equality,

$$\|\mathcal{A}^*(\omega)\|_\ell=\|\omega\|_{a(\ell)}.$$

PROOF. The properties (1) and (3) are consequences of Lemma 7.3 (2),(1).

(2) From the definition of $\mathcal{A}_*$ and $\mathcal{A}^*$ it suffices to apply Lemma 7.3 (3).

(4) Direct from definitions.

(5) Let $\omega=\mathbf{1}_{(F_0,\varepsilon_0)}\otimes\cdots\otimes\mathbf{1}_{(F_{n-k-2},\varepsilon_{n-k-2})}\otimes\mathbf{1}_{(F_{n-k-1}*F_{n-k},\varepsilon)}\otimes\mathbf{1}_{(F_{n-k+1},\varepsilon_{n-k+1})}\otimes\cdots\otimes\mathbf{1}_{F_n}\in\widetilde{N}^*(\Delta')$. We distinguish two cases (see Definitions 7.2, 7.4):

- Suppose $(F_{n-k},\varepsilon)\neq(\emptyset,0)$. We have, up to a sign, $\mathcal{A}^*(\omega)=\pm\mathbf{1}_{(F_0,\varepsilon_0)}\otimes\cdots\otimes\mathbf{1}_{(F_{n-k-2},\varepsilon_{n-k-2})}\otimes\mathbf{1}_{(F_{n-k-1},1)}\otimes\mathbf{1}_{(F_{n-k},\varepsilon)}\otimes\mathbf{1}_{(F_{n-k+1},\varepsilon_{n-k+1})}\otimes\cdots\otimes\mathbf{1}_{(F_n,\varepsilon_n)}$. The result comes from

$$\|\mathcal{A}^*(\omega)\|_\ell=\begin{cases}\|\omega\|_{\ell-1} & \text{if } \ell\geqslant k+2\\ -\infty & \text{if } \ell=k+1\\ \|\omega\|_\ell & \text{if } \ell\leqslant k\end{cases}\leqslant\begin{cases}\|\omega\|_{\ell-1} & \text{if } \ell\geqslant k+1\\ \|\omega\|_\ell & \text{if } \ell\leqslant k\end{cases}=\|\omega\|_{a(\ell)},$$

for $\ell\in\mathbb{N}_\Delta$.
- Suppose $(F_{n-k},\varepsilon)=(\emptyset,0)$. We have the equality $\mathcal{A}^*(\omega)=\mathbf{1}_{(F_0,\varepsilon_0)}\otimes\cdots\otimes\mathbf{1}_{(F_{n-k-2},\varepsilon_{k-2})}\otimes\mathbf{1}_{(F_{n-k-1},0)}\otimes\mathbf{1}_{(\emptyset,1)}\otimes\mathbf{1}_{(F_{n-k+1},\varepsilon_{n-k+1})}\otimes\cdots\otimes\mathbf{1}_{(F_n,\varepsilon_n)}+\sum_{e\in\mathcal{V}(\Delta_k)}\mathbf{1}_{(F_0,\varepsilon_0)}\otimes\cdots\otimes\mathbf{1}_{(F_{n-k-2},\varepsilon_{n-k-2})}\otimes\mathbf{1}_{(F_{n-k-1},0)}\otimes\mathbf{1}_{(e,0)}\otimes\mathbf{1}_{(F_{n-k+1},\varepsilon_{n-k+1})}\otimes\cdots\otimes\mathbf{1}_{(F_n,\varepsilon_n)}$. Then we have

$$\|\mathcal{A}^*(\omega)\|_\ell=\begin{cases}\|\omega\|_{\ell-1} & \text{if } \ell\geqslant k+2\\ \|\omega\|_\ell & \text{if } \ell\leqslant k\end{cases}=\|\omega\|_{a(\ell)},$$

for $\ell\in\mathbb{N}_\Delta$. Since the amalgamation is simple, the equality comes from the fact that $k+1\notin\mathbb{N}_\Delta$.

□

Proof of Lemma 7.3. We begin with the compatibility with the differentials. Let $H=(F_0,\varepsilon_0)\otimes(F_1,\varepsilon_1)$ and suppose $\varepsilon_0=1$ and $|(F_1,\varepsilon_1)|\geqslant 2$. We have

$$\begin{aligned}\theta\partial H &= \theta((\partial F_0,1)\otimes(F_1,\varepsilon_1))+(-1)^{|(F_0,1)|}\theta((F_0,1)\otimes(\partial F_1,\varepsilon_1))\\ &\quad +(-1)^{|(F_0,1)+|F_1|+1}\begin{cases}0 & \text{if } \varepsilon_1=0\\ \theta((F_0,1)\otimes(F_1,0)) & \text{if } \varepsilon_1=1\end{cases}\\ &= (\partial F_0*F_1,\varepsilon_1)+(-1)^{|(F_0,1)|}(F_0*\partial F_1,\varepsilon_1)\\ &\quad +\begin{cases}0 & \text{if } \varepsilon_1=0\\ (-1)^{|(F_0,1)+|F_1|+1}(F_0*F_1,0) & \text{if } \varepsilon_1=1\end{cases}\\ &= \partial(F_0*F_1,\varepsilon_1)\end{aligned}$$

On the other hand, we get $\partial\theta H=\partial(F_0*F_1,\varepsilon_1)$ and we have established the equality $\theta\partial=\partial\theta$ in the case $\varepsilon_0=1$ and $|(F_1,\varepsilon_1)|\geqslant 2$. The proof is similar in the other cases. The compatibility of Ξ with the differentials follows by duality.

(1) If $E_0=\emptyset$, then $\varepsilon_0=1$ and, by definition of the map θ, we have $\theta((\emptyset,1)\otimes(F_1,\varepsilon_1))=(F_1,\varepsilon_1)$. Therefore, θ is an isomorphism and so is Ξ by duality.

(2) Let $\mathbf{1}_{(F_0*F_1,\varepsilon)},\mathbf{1}_{(F_0'*F_1',\varepsilon')}\in\widetilde{N}^*(\mathtt{c}(E_0*E_1))$. We suppose that (F_1,ε) and (F_1',ε') are different from $(\emptyset,0)$. Let us consider $A=\Xi\left(\mathbf{1}_{(F_0*F_1,\varepsilon)}\right)\cup\Xi\left(\mathbf{1}_{(F_0'*F_1',\varepsilon')}\right)$ and $B=\Xi\left(\mathbf{1}_{(F_0*F_1,\varepsilon)}\cup\mathbf{1}_{(F_0'*F_1',\varepsilon')}\right)$. A direct computation gives the following equalities from definitions.

$$A=(-1)^{|(F_0,1)|\cdot|(F_1,\varepsilon)|+|(F_0',1)|\cdot|(F_1',\varepsilon')|}\left(\mathbf{1}_{(F_0,1)}\otimes\mathbf{1}_{(F_1,\varepsilon)}\right)\cup\left(\mathbf{1}_{(F_0',1)}\otimes\mathbf{1}_{(F_1',\varepsilon')}\right)=$$
$$(-1)^{|(F_1,\varepsilon)|\cdot|(F_0',1)|+|(F_0,1)|\cdot|(F_1,\varepsilon)|+|(F_0',1)|\cdot|(F_1',\varepsilon')|}\left(\mathbf{1}_{(F_0,1)}\cup\mathbf{1}_{(F_0',1)}\right)\otimes\left(\mathbf{1}_{(F_1,\varepsilon)}\cup\mathbf{1}_{(F_1',\varepsilon')}\right)$$

If $(F_0',\varepsilon)=(\emptyset,0)$ and F_1,F_1' are compatibles (see Definition 4.1), we have

$$\begin{aligned}A &= (-1)^{|(F_1,0)|\left(|(F_0,1)|+|(F_1',\varepsilon')|\right)}\mathbf{1}_{(F_0,1)}\otimes\mathbf{1}_{(F_1\cup F_1',\varepsilon')}\\ &= (-1)^{|(F_0*F_1,0)|\cdot|(F_1',\varepsilon')|}\Xi(\mathbf{1}_{(F_0*F_1\cup F_1',\varepsilon')})=B.\end{aligned}$$

If $(F'_0,\varepsilon)=(\emptyset,1)$ and $(F'_1,\varepsilon')=(\emptyset,1)$, we have $A=(-1)^{|(F_0,1)|\cdot|(F_1,1)|}\mathbf{1}_{(F_0,1)}\otimes\mathbf{1}_{(F_1,1)}=\Xi(\mathbf{1}_{(F_0*F_1,1)})=B$. In the other cases we have $A=B=0$.

We have established the equality $A=B$ in the case where (F_1,ε) and (F'_1,ε') are different from $(\emptyset,0)$. The verification is similar in the other cases.

(3) Suppose $(F_1,\varepsilon)\neq(\emptyset,0)$. We have

$$\begin{aligned} C &= \theta(\Xi(\mathbf{1}_{(F_0*F_1,\varepsilon)})\,\tilde{\frown}\,([\mathtt{c}E_0]\otimes[\mathtt{c}E_1]))\\ &= (-1)^{|(F_0,1)|\cdot|(F_1,\varepsilon)|}\theta((\mathbf{1}_{(F_0,1)}\otimes\mathbf{1}_{(F_1,\varepsilon)})\,\tilde{\frown}\,([\mathtt{c}E_0]\otimes[\mathtt{c}E_1]))\\ &= (-1)^{|(F_1,\varepsilon)|\cdot|\mathtt{c}E_0|+|(F_0,1)|\cdot|(F_1,\varepsilon)|}\theta((\mathbf{1}_{(F_0,1)}\frown[\mathtt{c}E_0])\otimes(\mathbf{1}_{(F_1,\varepsilon)}\frown[\mathtt{c}E_1])). \end{aligned}$$

We apply the definition of cap product (6.2) and we consider three different cases.

- If $F_0=E_0, F_1=E_1$ and $\varepsilon=1$, we have
$$C=\theta([\mathtt{v}]\otimes[\mathtt{v}])=[\mathtt{v}]=\mathbf{1}_{(F_0*F_1,\varepsilon)}\frown[\mathtt{c}(E_0*E_1)].$$
- If $F_0=E_0, F_1\cup G_1=E_1$ and $\varepsilon=0$, we have
$$C=\theta([\mathtt{v}]\otimes(G_1,1))=(G_1,1)=\mathbf{1}_{(F_0*F_1,\varepsilon)}\frown[\mathtt{c}(E_0*E_1)].$$
- The other cases correspond to $C=\mathbf{1}_{(F_0*F_1,\varepsilon)}\frown[\mathtt{c}(E_0*E_1)]=0$.

We have established the property (3) in the case $(F_1,\varepsilon)\neq(\emptyset,0)$. The verification is similar in the other cases.

□

8. Stratified maps: The global level. Theorem A.

Let $f\colon X\to Y$ be a stratified map, $\sigma\colon\Delta_\sigma\to X$ and $f\circ\sigma\colon\Delta_{f\circ\sigma}\to Y$ filtered simplices of X and Y respectively. In [**4**, Corollary A.25] we have proved that the filtrations Δ_σ and $\Delta_{f\circ\sigma}$ (of the same euclidean simplex Δ) are connected by an amalgamation, more exactly, by a finite sequence $\mathcal{A}_1$ of elementary amalgamations and a finite sequence $\mathcal{A}_2^{-1}$ of inverse of simple amalgamations as follows.

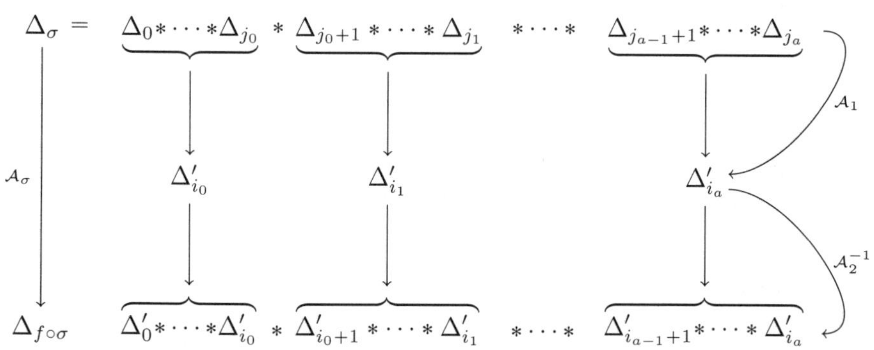

We denote the amalgamation $\mathcal{A}_\sigma=\mathcal{A}_2^{-1}\circ\mathcal{A}_1$, with $\mathcal{A}_1=\mathcal{A}_{1,1}\circ\cdots\circ\mathcal{A}_{1,u}$ and $\mathcal{A}_2=\mathcal{A}_{2,1}\circ\cdots\circ\mathcal{A}_{2,v}$. The elementary amalgamations $\mathcal{A}_{1,i}$ (resp. simple amalgamations $\mathcal{A}_{2,j}$) are written in a canonical way, going from the left to the right.

8.1. μ-amalgamation. Mention the amalgamation of Δ_σ collecting all the filtration in one factor. We call it the *μ-amalgamation* $\mathcal{A}''_\sigma \colon \Delta_\sigma \to \Delta$. The induced map $\mathcal{A}''_{\sigma,*} \colon \widetilde{N}_*(\Delta_\sigma) \to N_*(\Delta)$ is defined by

$$\bigotimes_{i=0}^{n}(F_i,\varepsilon_i) \rightsquigarrow (F_0 * \cdots * F_\ell, 0) \otimes \bigotimes_{i>\ell}(F_i,\varepsilon_i) \rightsquigarrow \begin{cases} F_0 * \cdots * F_\ell & \text{if } |(F,\epsilon)|_{>\ell} = 0 \\ 0 & \text{otherwise,} \end{cases}$$

with $\varepsilon_n = 0$ and ℓ is the smallest integer, j, such that $\varepsilon_j = 0$ (see Definition 6.1). We recover the map μ_* of (6.4). We also define $\mu^* = \mathcal{A}''^{*}_\sigma$.

With the previous notations, we notice the commutativity of the following diagram

$$\begin{array}{ccc} \Delta_\sigma & \xrightarrow{\mathcal{A}_\sigma} & \Delta_{f\circ\sigma} \\ {\scriptstyle \mathcal{A}''_\sigma}\big\downarrow & \swarrow {\scriptstyle \mathcal{A}''_{f\circ\sigma}} & \\ \Delta & & \end{array} \tag{8.1}$$

If $\overline{p}$ is a perversity on X and $\overline{q}$ is a perversity on Y with $\overline{p} \leqslant f^*\overline{q}$, we have proved in [**5**, Proposition 3.6] that the map f induces a homomorphism $f_* \colon H_*^{\overline{p}}(X;R) \to H_*^{\overline{p}}(Y;R)$. We define now the induced map in blown-up intersection cohomology and study its properties.

Definition 8.1. Let $f \colon X \to Y$ be a stratified map. The *induced map* $f \colon \widetilde{N}^*(Y;R) \to \widetilde{N}^*(X;R)$ is the map defined by $(f^*\omega)_\sigma = \mathcal{A}^*_\sigma \omega_{f\circ\sigma}$, for each $\omega \in \widetilde{N}^*(Y;R)$ and each regular simplex $\sigma \colon \Delta_\sigma \to X$.

With Proposition 7.5(4) this definition makes sense. It generalizes the notion of induced map of Proposition 3.5 (see Remark 3.6).

Theorem A. *Let $f \colon X \to Y$ be a stratified map.*

(1) *The induced morphism $f^* \colon \widetilde{N}^*(Y;R) \to \widetilde{N}^*(X;R)$ is a chain map.*
(2) *The map f^* is compatible with the cup product.*
(3) *The maps $f_* \colon \mathfrak{C}_*(X;R) \to \mathfrak{C}_*(Y;R)$ and $f^* \colon \widetilde{N}^*(Y;R) \to \widetilde{N}^*(X;R)$ verify*
$$f_*(f^*\omega \cap \xi) = \omega \cap f_*\xi.$$
for each $\omega \in \widetilde{N}^*(Y;R)$ and each $\xi \in \mathfrak{C}_*(X;R)$.
(4) *Let $\overline{p}$ be a perversity on X and $\overline{q}$ a perversity on Y such that $\overline{p} \geqslant f^*\overline{q}$. Then f induces a chain map $f^* \colon \widetilde{N}^*_{\overline{q}}(Y;R) \to \widetilde{N}^*_{\overline{p}}(X;R)$, thus a homomorphism $f^* \colon \mathscr{H}^*_{\overline{q}}(X;R) \to \mathscr{H}^*_{\overline{p}}(Y;R)$.*

As a GM-perversity verifies $\overline{p} \geqslant f^*\overline{p}$, the previous Theorem has an adaptation in this context.

Corollary 8.2. *Let $f \colon X \to Y$ be a stratified map and $\overline{p}$ a GM-perversity. Then f induces a homomorphism $f^* \colon \mathscr{H}^*_{\overline{p}}(Y;R) \to \mathscr{H}^*_{\overline{p}}(X;R)$ compatible with cup products.*

The maps induced for a GM-perversity satisfy also the condition (3) of Theorem A.

Proof of Theorem A. Properties (1) and (2) are direct consequences of Proposition 7.5.

(3) Notice first that $f_*: \mathfrak{C}_*(X;R) \to \mathfrak{C}_*(Y;R)$ is well defined since f_* preserves filtered simplices (see [**4**, Theorem F]) and $f(X\backslash\Sigma_X) \subset Y\backslash\Sigma_Y$, since f is stratified.

We can suppose that ξ is a regular simplex $\sigma\colon \Delta_\sigma \to X$. Applying Proposition 7.5(2) repeatedly we have $\mathcal{A}_{\sigma,*}(\mathcal{A}_\sigma^*\omega_{f\circ\sigma} \mathbin{\tilde{\frown}} \widetilde{\Delta}_\sigma) = \omega_{f\circ\sigma} \mathbin{\tilde{\frown}} \widetilde{\Delta}_{f\circ\sigma}$. The result follows from

$$\begin{aligned} f_*(f^*\omega \frown \sigma) &= f_*\sigma_*\mu_{\Delta_\sigma,*}((f^*\omega)_\sigma \mathbin{\tilde{\frown}} \widetilde{\Delta}_\sigma) = f_*\sigma_*\mu_{\Delta_\sigma,*}(\mathcal{A}_\sigma^*\omega_{f\circ\sigma} \mathbin{\tilde{\frown}} \widetilde{\Delta}_\sigma) \\ &\overset{(8.1)}{=} f_*\sigma_*\mu_{\Delta_{f\circ\sigma},*}\mathcal{A}_{\sigma,*}(\mathcal{A}_\sigma^*\omega_{f\circ\sigma} \mathbin{\tilde{\frown}} \widetilde{\Delta}_\sigma) = (f\circ\sigma)_*\mu_{\Delta_{f\circ\sigma},*}(\omega_{f\circ\sigma} \mathbin{\tilde{\frown}} \widetilde{\Delta}_{f\circ\sigma}) \\ &= \omega \frown f_*(\sigma). \end{aligned}$$

(4) Let $\omega \in \widetilde{N}^*_{\overline{q}}(Y;R)$. Let $\sigma\colon \Delta \to X$ be a regular simplex and S a singular stratum of X with $\operatorname{Im}\sigma \cap S \neq \emptyset$. We have to prove

$$\|\omega\|_{S^f} \leqslant \overline{q}(S^f) \Rightarrow \|f^*\omega\|_S \leqslant \overline{p}(S)$$

Since $\overline{p} \geqslant f^*\overline{q}$, it suffices to prove $\|\mathcal{A}_\sigma^*\omega_{f\circ\sigma}\|_{\operatorname{codim} S} \leqslant \|\omega_{f\circ\sigma}\|_{\operatorname{codim} S^f}$. Recall the decomposition, $\mathcal{A}_\sigma = \mathcal{A}_{2,v}^{-1} \circ \cdots \circ \mathcal{A}_{2,1}^{-1} \circ \mathcal{A}_{1,1} \circ \cdots \circ \mathcal{A}_{1,u}$, at the beginning of the section and denote by $a_{1,i}, a_{2,j}$ the associated index maps. We have

$$\operatorname{codim} S^f = (a_{2,v}^{-1} \circ \cdots \circ a_{2,1}^{-1} \circ a_{1,1} \circ \cdots \circ a_{1,u})(\operatorname{codim} S)$$

(see (7.1)). Now, the result follows from Proposition 7.5 (5). □

Part 2. Properties of the blown-up intersection cohomology

We present the properties of the blown-up intersection cohomology used in this work. We end this Part by comparing this cohomology with the intersection cohomology obtained from the dual cochains.

9. $\mathcal{U}$-small chains. Theorem B.

We establish a Theorem of $\mathcal{U}$-small filtered simplices, where $\mathcal{U}$ is an open cover of X (Theorem B).

Let Δ be an euclidean simplex whose vertices are $e_0, \ldots, e_m$. Let $e_{\{i_0\ldots i_p\}}$ be the barycenter of the simplex $[e_{i_0}, \ldots, e_{i_p}]$ and Sub Δ the simplicial complex given by the barycentric subdivision of Δ. The subdivision linear map, $\mathrm{Sub}_*\colon N_*(\Delta) \to N_*(\mathrm{Sub}\,\Delta)$, is defined by $\mathrm{Sub}_*[e_{i_0}, \ldots, e_{i_p}] = (-1)^p(\mathrm{Sub}_*\,\partial[e_{i_0}, \ldots, e_{i_p}]) * e_{\{i_0\ldots i_p\}}$. Any face of Sub Δ appears at most once in the expression of Sub_* and some faces do not.

Definition 9.1. An ℓ-simplex of Sub Δ is a *subdividing simplex* if it appears in the subdivision of an ℓ-simplex of Δ.

From the definition of Sub_* we know that the subdividing simplices of Sub Δ are the simplices of the form

$$\nabla = [e_{i_0}, e_{\{i_0 i_1\}}, e_{\{i_0 i_1 i_2\}}, \ldots, e_{\{i_0\ldots i_p\}}].$$

The vertex e_{i_0} is the *first vertex* and the vertex e_{i_p} is the *last vertex of* ∇. We write $e_{i_p} = \mathtt{der}\,\nabla$.

The transpose map of Sub_* is denoted by $\mathrm{Sub}^*\colon N^*(\mathrm{Sub}\,\Delta)\to N^*(\Delta)$. It is defined on the subdividing simplices by

$$\mathrm{Sub}^*(\mathbf{1}_{[e_{i_0},e_{\{i_0i_1\}},e_{\{i_0i_1i_2\}},\ldots,e_{\{i_0\ldots i_p\}}]})=\mathbf{1}_{[e_{i_0},\ldots,e_{i_p}]},$$

and by 0 on the other simplices. Since the linear map Sub_* is compatible with differentials, then Sub^* also is a cochain map.

To construct a homotopy between the barycentric subdivision operator and the identity, we consider the simplicial complex $K(\Delta)$, whose simplices are the joins, $F*G$, of a simplex F of Δ and of a simplex G of $\mathrm{Sub}\,\Delta$, such that: if $F=[e_{i_0},\ldots,e_{i_p}]$ and $G=[e_{J_0},\ldots,e_{J_k}]$, then $\{i_0,\ldots,i_p\}\subset J_\ell$, for each $\ell\in\{0,\ldots,k\}$. In addition, simplices F and G can be empty, but not simultaneously, making Δ and $\mathrm{Sub}\,\Delta$ two sub-complexes of $K(\Delta)$.

The homotopy operator, $T\colon N_*(\Delta)\to N_{*+1}(K(\Delta))$, defined by $T[e_{i_0},\ldots,e_{i_p}]=(-1)^{p+1}\left([e_{i_0},\ldots,e_{i_p}]-T\partial[e_{i_0},\ldots,e_{i_p}]\right)*e_{\{i_0\ldots i_p\}}$, verifies

$$(\partial T+T\partial)(F)=F-\mathrm{Sub}_*(F).$$

In particular, T takes the values, $T(e_0)=-e_0*e_{\{0\}}$ and $T[e_0,e_1]=[e_0e_1]*e_{\{01\}}-e_0*[e_{\{0\}},e_{\{0,1\}}]+e_1*[e_{\{1\}},e_{\{0,1\}}]$.

Definition 9.2. A *full simplex* of $K(\Delta)$ is a simplex of the form

$$F*G=[e_{i_0},\ldots,e_{i_p}]*[e_{\{i_0\ldots i_p\}},e_{\{i_0\ldots i_pi_{p+1}\}},\ldots,e_{\{i_0\ldots i_p\ldots i_{p+r}\}}].$$

The vertex $e_{i_{p+r}}$ is the *last vertex* of $F*G$ and denoted by $\mathtt{der}\,(F*G)$.

Note that if $F*G$ is full and $p=0$, then G is subdividing in the sense of Definition 9.1.

Endow the simplex Δ with a filtration, $\Delta=\Delta_0*\cdots*\Delta_n$, compatible with the order considered on the vertices, cf. (2.2). The simplicial complex $\mathrm{Sub}\,\Delta$ *is a weighted simplicial complex,* associating to the vertex $e_{\{i_0\ldots i_p\}}$ the weight ℓ if we have $[e_{i_0},\ldots,e_{i_p}]\triangleleft\Delta_0*\cdots*\Delta_\ell$ and $[e_{i_0},\ldots,e_{i_p}]\ntriangleleft\Delta_0*\cdots*\Delta_{\ell-1}$. Since the family of the vertices of $K(\Delta)$ is the union of those of Δ and those of $\mathrm{Sub}\,\Delta$, the complex $K(\Delta)$ also is a weighted simplicial complex. The chosen canonical basis of $\widetilde{N}^*(K(\Delta))$ is composed of elements of the form,

$$\mathbf{1}_{(F*G,\varepsilon)}=\mathbf{1}_{(F_0,\varepsilon_0)}\otimes\cdots\otimes\mathbf{1}_{(F_q*G_q,\varepsilon_q)}\otimes\cdots\otimes\mathbf{1}_{(G_{n-1},\varepsilon_{n-1})}\otimes\mathbf{1}_{G_n},$$

with $F=F_0*\cdots*F_q\triangleleft\Delta$ and $G=G_q*\cdots*G_n\triangleleft\mathrm{Sub}\,\Delta$.

Remark 9.3. The linear operator Sub^* does not necessarily respect the filter degree of simplices. Consider $\Delta=[e_0]*[e_1,e_2]$. Previous computations give $\mathrm{Sub}^*(\mathbf{1}_{[e_1,e_{\{0,1\}}]})=\mathbf{1}_{[e_1,e_0]}$. Taking into account the filtration defined on $\mathrm{Sub}\,\Delta$, one has $[e_1,e_{\{0,1\}}]=\varnothing*[e_1,e_{\{0,1\}}]$ and $[e_1,e_0]=-[e_0]*[e_1]$, from which one gets $\mathrm{Sub}^*(\mathbf{1}_{\varnothing*[e_1,e_{\{0,1\}}]})=-\mathbf{1}_{[e_0]*[e_1]}$.

More generally, let $\Delta=\Delta_0*\cdots*\Delta_n$ and let $\nabla=\nabla_0*\cdots*\nabla_n\triangleleft\mathrm{Sub}\,\Delta$. The previous example shows that the use of Sub^* for the construction of a cochain morphism,

$$\widetilde{\mathrm{Sub}}\colon N^*(\mathtt{c}\nabla_0)\otimes\cdots\otimes N^*(\mathtt{c}\nabla_{n-1})\otimes N^*(\nabla_n)\to N^*(\mathtt{c}\Delta_0)\otimes\cdots\otimes N^*(\mathtt{c}\Delta_{n-1})\otimes N^*(\Delta_n),$$

requires a reordering of the vertices according to the filtrations of Δ and $\mathrm{Sub}\,\Delta$. To avoid this difficulty we define below the morphism $\widetilde{\mathrm{Sub}}_\Delta\colon\widetilde{N}^*(\mathrm{Sub}\,\Delta)\to\widetilde{N}^*(\Delta)$ and the homotopy $\widetilde{\mathrm{T}}_\Delta\colon\widetilde{N}^*(K(\Delta))\to\widetilde{N}^{*-1}(\Delta)$ by induction.

The two canonical injections of Δ and Sub Δ into $K(\Delta)$ induce the chain complex epimorphisms,

- $\iota^*_\Delta\colon \widetilde{N}^*(K(\Delta)) \to \widetilde{N}^*(\Delta)$, defined by the identity on the elements $\mathbf{1}_{(F,\varepsilon)}$ with $F \triangleleft \Delta$ and 0 elsewhere,
- $\iota^*_{\mathrm{Sub}\,\Delta}\colon \widetilde{N}^*(K(\Delta)) \to \widetilde{N}^*(\mathrm{Sub}\,\Delta)$, defined by the identity on the elements $\mathbf{1}_{(G,\varepsilon)}$ with $G \triangleleft \mathrm{Sub}\,\Delta$ and 0 elsewhere.

LEMMA 9.4. *There exist a linear map,* $\widetilde{\mathrm{T}}_\Delta\colon \widetilde{N}^*(K(\Delta)) \to \widetilde{N}^{*-1}(\Delta)$, *and a complex cochain morphism,* $\widetilde{\mathrm{Sub}}_\Delta\colon \widetilde{N}^*(\mathrm{Sub}\,\Delta) \to \widetilde{N}^*(\Delta)$, *verifying*

$$\widetilde{\mathrm{T}}_\Delta \circ \widetilde{\delta}^{K(\Delta)} + \widetilde{\delta}^\Delta \circ \widetilde{\mathrm{T}}_\Delta = \iota^*_\Delta - \widetilde{\mathrm{Sub}}_\Delta \circ \iota^*_{\mathrm{Sub}\,\Delta}, \tag{9.1}$$

where $\widetilde{\delta}^{K(\Delta)}$ *and* $\widetilde{\delta}^\Delta$ *are the differentials on* $\widetilde{N}^*(K(\Delta))$ *and* $\widetilde{N}^*(\Delta)$.

9.1. Construction of $\widetilde{\mathrm{Sub}}_\Delta$ and $\widetilde{\mathrm{T}}_\Delta$. The homotopy $\widetilde{\mathrm{T}}_\Delta$ is constructed by induction at the level of the full blow up's, $\widetilde{N}^{all,*}(K(\Delta))$ and $\widetilde{N}^{all,*-1}(\Delta)$, specifying its value on the elements $\mathbf{1}_{(F*G,\varepsilon)}$, with $F \triangleleft \Delta$ and $G \triangleleft \mathrm{Sub}\,\Delta$.

We set $\widetilde{\mathrm{T}}_\Delta(\mathbf{1}_{(F*G,\varepsilon)}) = 0$ if $F * G$ is not full, in the sense of Definition 9.2. Now, let,

$$\mathbf{1}_{(F*G,\varepsilon)} = \mathbf{1}_{(F_0,\varepsilon_0)} \otimes \cdots \otimes \mathbf{1}_{(F_q*G_q,\varepsilon_q)} \otimes \cdots \otimes \mathbf{1}_{(G_{n-1},\varepsilon_{n-1})} \otimes \mathbf{1}_{(G_n,\varepsilon_n)},$$

with $F * G$ full. We use a double induction, on the length n of the filtration of Δ and on the dimension of the component (G_n, ε_n). We set $\Delta = \nabla * \Delta_n$.

a) If $\dim(G_n, \varepsilon_n) = 0$, we distinguish three cases

i) If $G_n = \varnothing$ and $\varepsilon_n = 1$, then $F * G \triangleleft \nabla$ and we set:

$$\widetilde{\mathrm{T}}_\Delta(\mathbf{1}_{(F*G,\varepsilon)} \otimes \mathbf{1}_{(\varnothing,1)}) = \widetilde{\mathrm{T}}_\nabla(\mathbf{1}_{(F*G,\varepsilon)}) \otimes \mathbf{1}_{(\varnothing,1)}. \tag{9.2}$$

ii) If $\dim G_n = 0$, $\varepsilon_n = 0$ and $q < n$, then $G_n = [e_J]$, where e_J is the barycenter of a face of Δ. Let $e_\alpha = \mathtt{der}\,(G_q * \cdots * G_n)$ the last vertex of the face G and $G' = G_q * \cdots * G_{n-1}$. (Notice $e_\alpha \in \Delta_n$, otherwise the weight of the vertex e_J of Sub Δ would be $n-1$.) We set:

$$\widetilde{\mathrm{T}}_\Delta(\mathbf{1}_{(F*G',\varepsilon)} \otimes \mathbf{1}_{([e_J],0)}) = \widetilde{\mathrm{T}}_\nabla(\mathbf{1}_{(F*G',\varepsilon)}) \otimes \mathbf{1}_{([e_\alpha],0)}. \tag{9.3}$$

iii) If $\dim G_n = 0$, $\varepsilon_n = 0$ and $q = n$, then $G = G_n = [e_{\{F\}}]$ is reduced to the barycenter of the face F. (Notice $F_n \neq \varnothing$ otherwise the weight of the barycenter of F could not be n in Sub Δ.) We set:

$$\widetilde{\mathrm{T}}_\Delta(\mathbf{1}_{(F,\varepsilon)} * e_{\{F\}}) = (-1)^{|(F,\varepsilon)|}\mathbf{1}_{(F,\varepsilon)}. \tag{9.4}$$

b) If $\dim(G_n, \varepsilon_n) \geqslant 1$, then $G_n \neq \varnothing$ and we set $e_\alpha = \mathtt{der}\,(G_q * \cdots * G_n)$ the last vertex of the face G. We write $G_n = G'_n * e_J$, so that $G' = G_q * \cdots * G_{n-1} * G'_n$ is subdividing and e_α is a vertex of the face J for which e_J is the barycenter. We set:

$$\widetilde{\mathrm{T}}_\Delta(\mathbf{1}_{(F*G',\varepsilon)} * e_J) = \widetilde{\mathrm{T}}_\Delta(\mathbf{1}_{(F*G',\varepsilon)}) * e_\alpha. \tag{9.5}$$

The construction of $\widetilde{\mathrm{T}}_\Delta$ does not change the value of ε_n; moreover, the image of an element $\mathbf{1}_{(F*G,\varepsilon)}$ with $\varepsilon_n = 0$ and $G_n \neq \varnothing$ also has a non-empty component in Δ_n. So we built a linear map

$$\widetilde{\mathrm{T}}_\Delta\colon \widetilde{N}^*(K(\Delta)) \to \widetilde{N}^{*-1}(\Delta).$$

We construct the map $\widetilde{\mathrm{Sub}}_\Delta\colon \widetilde{N}^*(\mathrm{Sub}\,\Delta) \to \widetilde{N}^*(\Delta)$ from $\widetilde{\mathrm{T}}_\Delta$, by

$$\widetilde{\mathrm{Sub}}_\Delta = -\widetilde{\mathrm{T}}_\Delta \circ \widetilde{\delta}^{K(\Delta)}.$$

REMARK 9.5. To any simplex, $F*G$, of $K(\Delta)$, we associate a family of vertices, $\mathcal{V}_\Delta(F*G)$, consisting of the vertices of F and the vertices of the faces of Δ having a vertex of G as barycenter. Adding the virtual vertices, we set $\mathcal{V}_\Delta(\mathbf{1}_{(F*G,\varepsilon)}) = \mathcal{V}_\Delta(F*G,\varepsilon)$. By construction of $\widetilde{\mathrm{T}}_\Delta$, we have

$$\mathcal{V}_\Delta(\mathbf{1}_{(F*G,\varepsilon)}) = \mathcal{V}_\Delta(\widetilde{\mathrm{T}}_\Delta(\mathbf{1}_{(F*G,\varepsilon)})).$$

The proof of Lemma 9.4 is postponed to the end of the proof of Lemma 9.4, page 78. First, we take advantage of this result in terms of the blown-up intersection cohomology.

Let $(X,\overline{p})$ be a perverse space. Recall that a cochain $\omega \in \widetilde{N}^*_{\overline{p}}(X;R)$ associates to any regular filtered simplex $\sigma\colon \Delta \to X$, a cochain $\omega_\sigma \in \widetilde{N}^*_\sigma$ and verifies $\|\omega\| \leqslant \overline{p}$, $\|\delta\omega\| \leqslant \overline{p}$ and $\delta^*_\ell(\omega_\sigma) = \omega_{\partial_\ell\sigma}$, for any regular face operator, δ_ℓ.

DEFINITION 9.6. Let $\mathcal{U}$ be an open cover of X. A *$\mathcal{U}$-small simplex* is a regular simplex, $\sigma\colon \Delta = \Delta_0 * \cdots * \Delta_n \to X$, such that there exists $U \in \mathcal{U}$ with $\mathrm{Im}\,\sigma \subset U$. The family of $\mathcal{U}$-small simplices is denoted by $\mathtt{Simp}_{\mathcal{U}}$.

The *blown-up complex of $\mathcal{U}$-small cochains of X with coefficients in R*, written $\widetilde{N}^{*,\mathcal{U}}(X;R)$, is the cochain complex made up of elements ω, associating to any $\mathcal{U}$-small simplex, $\sigma\colon \Delta = \Delta_0 * \cdots * \Delta_n \to X$, an element $\omega_\sigma \in \widetilde{N}^*(\Delta)$, so that $\delta^*_\ell(\omega_\sigma) = \omega_{\partial_\ell\sigma}$, for any face operator, $\delta_\ell\colon \Delta'_0 * \cdots * \Delta'_n \to \Delta_0 * \cdots * \Delta_n$, with $\Delta'_n \neq \emptyset$. If $\overline{p}$ is a perversity on X, we denote by $\widetilde{N}^{*,\mathcal{U}}_{\overline{p}}(X;R)$ the complex of $\mathcal{U}$-small cochains verifying $\|\omega\| \leqslant \overline{p}$ and $\|\delta\omega\| \leqslant \overline{p}$.

The following Theorem compares the complexes $\widetilde{N}^*_{\overline{p}}(X;R)$ and $\widetilde{N}^{*,\mathcal{U}}_{\overline{p}}(X;R)$.

THEOREM B. *Let $(X,\overline{p})$ be a perverse space endowed with an open cover, $\mathcal{U}$. Let $\rho_{\mathcal{U}}\colon \widetilde{N}^*_{\overline{p}}(X;R) \to \widetilde{N}^{*,\mathcal{U}}_{\overline{p}}(X;R)$ be the restriction map. The following properties are verified.*

(i) *There exist a cochain map, $\varphi_{\mathcal{U}}\colon \widetilde{N}^{*,\mathcal{U}}_{\overline{p}}(X;R) \to \widetilde{N}^*_{\overline{p}}(X;R)$, and a homotopy,*
$\Theta\colon \widetilde{N}^*_{\overline{p}}(X;R) \to \widetilde{N}^{*-1}_{\overline{p}}(X;R)$, *such that*

$$\rho_{\mathcal{U}} \circ \varphi_{\mathcal{U}} = \mathrm{id} \text{ and } \delta\circ\Theta + \Theta\circ\delta = \mathrm{id} - \varphi_{\mathcal{U}} \circ \rho_{\mathcal{U}}.$$

(ii) *Furthermore, if the cochain $\omega \in \widetilde{N}^{*,\mathcal{U}}_{\overline{p}}(X;R)$ is such that there exists a subset $K \subset X$ for which $\omega_\sigma = 0$ if $(\mathrm{Im}\,\sigma) \cap K = \emptyset$, then $\varphi_{\mathcal{U}}(\omega)$ also verifies $(\varphi_{\mathcal{U}}(\omega))_\sigma = 0$ if $(\mathrm{Im}\,\sigma) \cap K = \emptyset$.*

As an immediate consequence of property (i) we get the following corollary.

COROLLARY 9.7. *The restriction map, $\rho_{\mathcal{U}}\colon \widetilde{N}^*_{\overline{p}}(X;R) \to \widetilde{N}^{*,\mathcal{U}}_{\overline{p}}(X;R)$, is a quasi-isomorphism.*

First we establish the existence of a subdivision at the level of the blown-up complex.

PROPOSITION 9.8. *Consider X a filtered space. The maps $\widetilde{\mathrm{T}}_\Delta$ and $\widetilde{\mathrm{Sub}}_\Delta$ extend as $\widetilde{\mathrm{T}}\colon \widetilde{N}^*(X;R) \to \widetilde{N}^{*-1}(X;R)$ and as $\widetilde{\mathrm{Sub}} : \widetilde{N}^*(X;R) \to \widetilde{N}^*(X;R)$ verifying*

$$\widetilde{\mathrm{T}} \circ \delta + \delta \circ \widetilde{\mathrm{T}} = \mathrm{id} - \widetilde{\mathrm{Sub}}. \tag{9.6}$$

PROOF. We first construct a homotopy, $\widetilde{\mathrm{T}}\colon \widetilde{N}^*(X;R) \to \widetilde{N}^{*-1}(X;R)$, from the maps $\widetilde{\mathrm{T}}_\Delta$ previously defined.

Let $\sigma\colon \Delta \to X$ be a filtered simplex and $\omega \in \widetilde{N}^k(X;R)$. A simplex $F * G$ of $K(\Delta)$ can be described by its vertices, $F * G = [e_{i_0}, \ldots, e_{i_r}, e_{\{F_0\}}, \ldots, e_{\{F_s\}}]$, where e_j is a vertex of Δ and $e_{\{F_j\}} \in \mathcal{V}(\mathrm{Sub}\ \Delta)$ is the barycenter of the face $F_j \triangleleft \Delta$. It follows $e_{\{F_j\}} \in \Delta$ and we define a filtered simplex of X, $\sigma_{F*G} : F * G \to X$, using the barycentric coordinates by

$$\sigma_{F*G}\Big(\sum_j t_j e_j + \sum_\ell u_\ell e_{\{F_\ell\}}\Big) = \sigma\Big(\sum_j t_j e_j + \sum_\ell u_\ell e_{\{F_\ell\}}\Big). \tag{9.7}$$

Thanks to Proposition 2.8, we define $\omega_{K(\Delta)} \in \widetilde{N}^k(K(\Delta))$ by

$$\omega_{K(\Delta)} = \sum_{\substack{F*G \triangleleft K(\Delta) \\ |(F*G,\varepsilon)|=k}} \omega_{\sigma_{F*G}}(F*G,\varepsilon)\, \mathbf{1}_{(F*G,\varepsilon)}.$$

Write now

$$(\widetilde{\mathrm{T}}(\omega))_\sigma = \widetilde{\mathrm{T}}_\Delta(\omega_{K(\Delta)})$$

and verify that we obtain an element of $\widetilde{N}^*(X;R)$. Let $\sigma\colon \Delta = \Delta_0 * \cdots * \Delta_n \to X$ be a regular simplex and $\tau = \sigma \circ \delta_\ell \colon \nabla \to X$, where $\delta_\ell\colon \nabla \to \Delta$ is a regular face operator. Since $\omega_{K(\Delta)} \in \widetilde{N}^k(K(\Delta))$ then $\delta_\ell^* \omega_{K(\Delta)} = \omega_{K(\nabla)}$. The maps $\delta_\ell^*\colon \widetilde{N}^*(\Delta) \to \widetilde{N}^*(\nabla)$ and $\delta_\ell^*\colon \widetilde{N}^*(K(\Delta)) \to \widetilde{N}^*(K(\nabla))$ are defined by

$$\delta_\ell^*(\mathbf{1}_{(F,\varepsilon)}) = \begin{cases} \mathbf{1}_{(F,\varepsilon)} & \text{if } F \triangleleft \nabla, \\ 0 & \text{if not,} \end{cases} \quad \text{and } \delta_\ell^*(\mathbf{1}_{(F*G,\varepsilon)}) = \begin{cases} \mathbf{1}_{(F*G,\varepsilon)} & \text{if } F * G \triangleleft K(\nabla), \\ 0 & \text{if not.} \end{cases}$$

From these calculations and from Remark 9.5, directly follows the commutativity of the following diagram,

$$\begin{array}{ccc} \widetilde{N}^*(K(\Delta)) & \xrightarrow{\ \widetilde{\mathrm{T}}_\Delta\ } & \widetilde{N}^{*-1}(\Delta) \\ {\scriptstyle \delta_\ell^*}\downarrow & & \downarrow{\scriptstyle \delta_\ell^*} \\ \widetilde{N}^*(K(\nabla)) & \xrightarrow{\ \widetilde{\mathrm{T}}_\nabla\ } & \widetilde{N}^{*-1}(\nabla), \end{array}$$

from which we deduce, $\delta_\ell^* \widetilde{\mathrm{T}}(\omega)_\sigma = \delta_\ell^* \widetilde{\mathrm{T}}_\Delta(\omega_{K(\Delta)}) = \widetilde{\mathrm{T}}_\nabla(\delta_\ell^* \omega_{K(\Delta)}) = \widetilde{\mathrm{T}}_\nabla(\omega_{K(\nabla)}) = \widetilde{\mathrm{T}}(\omega)_\tau$. Thus the map $\widetilde{\mathrm{T}}\colon \widetilde{N}^*(X;R) \to \widetilde{N}^{*-1}(X;R)$ is well defined. Now we proceed to the construction of $\widetilde{\mathrm{Sub}} : \widetilde{N}^*(X;R) \to \widetilde{N}^*(X;R)$ from the operators $\widetilde{\mathrm{Sub}}_\Delta$ of Lemma 9.4. Let $\sigma\colon \Delta \to X$ be a regular simplex and let $\omega \in \widetilde{N}^k(X;R)$. If G is a face of $\mathrm{Sub}\ \Delta$, we denote by $\iota_G\colon G \to \Delta$ the canonical injection and we define (cf. Proposition 2.8)

$$\omega_{\mathrm{Sub}\ \Delta} = \sum_{\substack{G \triangleleft \mathrm{Sub}\ \Delta \\ |(G,\varepsilon)|=k}} \omega_{\sigma \circ \iota_G}(G,\varepsilon) \mathbf{1}_{(G,\varepsilon)} \quad \text{and} \quad (\widetilde{\mathrm{Sub}}\, \omega)_\sigma = \widetilde{\mathrm{Sub}}_\Delta(\omega_{\mathrm{Sub}\ \Delta}).$$

Notice

$$\begin{aligned}\iota^*_{\mathrm{Sub}\,\Delta}\left(\omega_{K(\Delta)}\right) &= \sum_{\substack{F*G\lhd K(\Delta)\\ |(F*G,\varepsilon)|=k}} \omega_{\sigma_{F*G}}(F*G,\varepsilon)\,\iota^*_{\mathrm{Sub}\,\Delta}\left(\mathbf{1}_{(F*G,\varepsilon)}\right)\\ &= \sum_{\substack{G\lhd \mathrm{Sub}\,\Delta\\ |(G,\varepsilon)|=k}} \omega_{\sigma\circ\iota_G}(G,\varepsilon)\mathbf{1}_{(G,\varepsilon)} = \omega_{\mathrm{Sub}\,\Delta}.\end{aligned}$$

We also have $\iota^*_\Delta\left(\omega_{K(\Delta)}\right) = \omega_\sigma$. It follows, by using Lemma 9.4,

$$\begin{aligned}\omega_\sigma - (\widetilde{\mathrm{T}}\delta\omega)_\sigma - (\delta\widetilde{\mathrm{T}}\omega)_\sigma &= \iota^*_\Delta\left(\omega_{K(\Delta)}\right) - \widetilde{\mathrm{T}}_\Delta\left(\widetilde{\delta}^{K(\Delta)}\omega_{K(\Delta)}\right) - \widetilde{\delta}^\Delta\widetilde{\mathrm{T}}_\Delta\left(\omega_{K(\Delta)}\right)\\ &= \widetilde{\mathrm{Sub}}_\Delta\left(\iota^*_{\mathrm{Sub}\,\Delta}(\omega_{K(\Delta)})\right) = \widetilde{\mathrm{Sub}}_\Delta\left(\omega_{\mathrm{Sub}\,\Delta}\right) = \left(\widetilde{\mathrm{Sub}}\,(\omega)\right)_\sigma\end{aligned}$$

We have shown $\widetilde{\mathrm{Sub}}\,(\omega) \in \widetilde{N}^*(X;R)$ and the equality $\widetilde{\mathrm{T}}\circ\delta + \delta\circ\widetilde{\mathrm{T}} = \mathrm{id} - \widetilde{\mathrm{Sub}}$. □

Proof of Theorem B. (i) We prove $\|\widetilde{\mathrm{T}}(\omega)\| \leqslant \|\omega\|$ and $\|\widetilde{\mathrm{Sub}}\,(\omega)\| \leqslant \|\omega\|$, by working on the chosen basis and the various possible cases. Making reference to Definition 2.5, since maps $\widetilde{\mathrm{T}}$ and $\widetilde{\mathrm{Sub}}$ do not modify the value of the parameter ε then we can limit ourselves to the case $\varepsilon_{n-\ell} = 0$, with $\ell \in \{1,\dots,n\}$. We proceed by induction on the complex $\widetilde{N}^{all}(\Delta)$, the case $n = 0$ being obvious. With the notations used in the construction of $\widetilde{\mathrm{T}}$, the following properties are verified.

- (9.2): if $G_n = \emptyset$ and $\varepsilon_n = 1$, the desired inequality comes from

$$\begin{aligned}\|\widetilde{\mathrm{T}}_\Delta(\mathbf{1}_{(F*G,\varepsilon)})\otimes\mathbf{1}_{(\emptyset,1)}\|_\ell &= \begin{cases} 0 & \text{if } \ell = 1,\\ \|\widetilde{\mathrm{T}}_\Delta(\mathbf{1}_{(F*G,\varepsilon)})\|_{\ell-1} & \text{if } \ell > 1,\end{cases}\\ &\leqslant \begin{cases} 0 & \text{if } \ell = 1,\\ \|\mathbf{1}_{(F*G,\varepsilon)}\|_{\ell-1} & \text{if } \ell > 1,\end{cases} \leqslant \|\mathbf{1}_{(F*G,\varepsilon)}\otimes\mathbf{1}_{(\emptyset,1)}\|_\ell.\end{aligned}$$

- The justifications of (9.3) and (9.4) are similar.
- (9.5): if $\dim(G_n,\varepsilon_n) \geqslant 1$, we denote by $\ell(\alpha)$ the weight of the generator e_α of Δ. By construction of $\widetilde{\mathrm{T}}_\Delta$, we have

$$\|\widetilde{\mathrm{T}}_\Delta(\mathbf{1}_{(F*G',\varepsilon)}) * e_\alpha\|_\ell = \begin{cases} \|\widetilde{\mathrm{T}}_\Delta(\mathbf{1}_{(F*G',\varepsilon)})\|_\ell \text{if } \ell(\alpha) \leqslant n-\ell,\\ \|\widetilde{\mathrm{T}}_\Delta(\mathbf{1}_{(F*G',\varepsilon)})\|_\ell + 1 \text{if } \ell(\alpha) > n-\ell,\end{cases} \leqslant \|\mathbf{1}_{(F*G,\varepsilon)}\|_\ell.$$

To study the perverse degree of $\widetilde{\mathrm{Sub}}_\Delta(-)$, consider a subdividing face $G\lhd \mathrm{Sub}\,\Delta$ with first term e_0. By definition of $\widetilde{\mathrm{Sub}}_\Delta$, one has, for each $\ell \in \{1,\dots,n\}$,

$$\|\widetilde{\mathrm{Sub}}_\Delta\mathbf{1}_{(G,\varepsilon)}\|_\ell = \|\widetilde{\mathrm{T}}_\Delta\widetilde{\delta}^{K(\Delta)}\mathbf{1}_{(G,\varepsilon)}\|_\ell = \|\widetilde{\mathrm{T}}_\Delta(\mathbf{1}_{(G,\varepsilon)}*e_0)\|_\ell \leqslant \|\mathbf{1}_{(G,\varepsilon)}*e_0\|_\ell = \|\mathbf{1}_{(G,\varepsilon)}\|_\ell.$$

(The last equality uses the fact that e_0, being the first vertex of G, lies in the smallest filtration of G.) Together with Proposition 9.8, this proves that $\widetilde{\mathrm{T}}$ and $\widetilde{\mathrm{Sub}}$ define two maps from the complex $\widetilde{N}^*_{\overline{p}}(X;R)$ to itself.

We construct a cochain map, $\varphi_{\mathcal{U}}\colon \widetilde{N}^{*,\mathcal{U}}_{\overline{p}}(X;R) \to \widetilde{N}^*_{\overline{p}}(X;R)$, and a homotopy, $\Theta\colon \widetilde{N}^*_{\overline{p}}(X;R) \to \widetilde{N}^{*-1}_{\overline{p}}(X;R)$, such that $\rho_{\mathcal{U}}\circ\varphi_{\mathcal{U}} = \mathrm{id}$ and $\delta\circ\Theta + \Theta\circ\delta = \mathrm{id} - \varphi_{\mathcal{U}}\circ\rho_{\mathcal{U}}$.

Write $\Psi_1 = \widetilde{\mathrm{T}}$ and, for $m \geqslant 2$, $\widetilde{\mathrm{Sub}}^m = \widetilde{\mathrm{Sub}}^{m-1} \circ \widetilde{\mathrm{Sub}}$ and $\Psi_m = \sum_{0 \leqslant i < m} \widetilde{\mathrm{T}} \circ \widetilde{\mathrm{Sub}}^i$. A simple calculation using (9.6) gives

$$\delta \circ \Psi_m + \Psi_m \circ \delta = \sum_{0 \leqslant i < m} \delta \circ \widetilde{\mathrm{T}} \circ \widetilde{\mathrm{Sub}}^i + \widetilde{\mathrm{T}} \circ \widetilde{\mathrm{Sub}}^i \circ \delta = \mathrm{id} - \widetilde{\mathrm{Sub}}^m. \tag{9.8}$$

If $\sigma \colon \Delta \to X$ is a regular simplex, we denote by $m(\sigma)$ the smallest integer such that the simplices of $\mathrm{Sub}^{m(\sigma)}\,\Delta$ have their image by σ included in an open subset of the cover $\mathcal{U}$. If F is a simplex of Δ, we denote by $\sigma_F \colon F \to X$ the restriction of σ and by $(F,\varepsilon) \lhd \widetilde{\Delta}$ the fact that (F,ε) is a face of the blow up $\widetilde{\Delta}$. Let $\omega \in \widetilde{N}^k(X;R)$. The homotopy Θ is defined by

$$(\Theta(\omega))_\sigma = \sum_{\substack{(F,\varepsilon)\lhd\widetilde{\Delta} \\ |(F,\varepsilon)|=k-1}} \left(\Psi_{m(\sigma_F)}(\omega)\right)_{\sigma_F}(F,\varepsilon)\,\mathbf{1}_{(F,\varepsilon)}.$$

Since the coefficient of $\mathbf{1}_{(F,\varepsilon)}$ depends only on σ_F and ε then we get the compatibility with the faces. In order to construct $\varphi_{\mathcal{U}}$, we calculate $\delta \circ \Theta$ and $\Theta \circ \delta$. Following (2.1), we have

$$\delta \mathbf{1}_{(F,\varepsilon)} = (-1)^k \sum_{(F,\varepsilon)\lhd\partial(\nabla,\kappa)} n_{(F,\varepsilon,\nabla,\kappa)} \mathbf{1}_{(\nabla,\kappa)}, \tag{9.9}$$

where $(F,\varepsilon)\lhd\partial(\nabla,\kappa)$ means that (F,ε) runs over the proper faces of (∇,κ) and

$$\partial(\nabla,\kappa) = \sum_{(F,\varepsilon)\lhd\partial(\nabla,\kappa)} n_{(F,\varepsilon,\nabla,\kappa)}(F,\varepsilon).$$

We deduce, for all $\omega \in \widetilde{N}^k(X;R)$,

$$(\delta\Theta(\omega))_\sigma = (-1)^k \sum_{\substack{(F,\varepsilon)\lhd\widetilde{\Delta} \\ |(F,\varepsilon)|=k-1}} \sum_{\substack{(F,\varepsilon)\lhd\partial(\nabla,\kappa) \\ (\nabla,\kappa)\lhd\widetilde{\Delta}}} n_{(F,\varepsilon,\nabla,\kappa)} \left(\Psi_{m(\sigma_F)}(\omega)\right)_{\sigma_F}(F,\varepsilon)\,\mathbf{1}_{(\nabla,\kappa)}.$$

On the other hand, the definition of Θ implies

$$\begin{aligned} (\Theta(\delta\omega))_\sigma &= \sum_{\substack{(\nabla,\kappa)\lhd\widetilde{\Delta} \\ |(\nabla,\kappa)|=k}} \left(\Psi_{m(\sigma_\nabla)}(\delta\omega)\right)_{\sigma_\nabla}(\nabla,\kappa)\,\mathbf{1}_{(\nabla,\kappa)} \\ &=_{(9.8)} \omega_\sigma - \sum_{\substack{(\nabla,\kappa)\lhd\widetilde{\Delta} \\ |(\nabla,\kappa)|=k}} \left(\widetilde{\mathrm{Sub}}^{m(\sigma_\nabla)}(\omega)\right)_{\sigma_\nabla}(\nabla,\kappa)\,\mathbf{1}_{(\nabla,\kappa)} \\ &\quad - \sum_{\substack{(\nabla,\kappa)\lhd\widetilde{\Delta} \\ |(\nabla,\kappa)|=k}} \left(\delta\Psi_{m(\sigma_\nabla)}(\omega)\right)_{\sigma_\nabla}(\nabla,\kappa)\,\mathbf{1}_{(\nabla,\kappa)}. \end{aligned}$$

It follows, again by using (9.9),

$$((\delta\Theta + \Theta\delta - \mathrm{id})(\omega))_\sigma = -(\varphi_{\mathcal{U}}(\omega))_\sigma$$

with

$$-(\varphi_{\mathcal{U}}(\omega))_\sigma = -\sum_{\substack{(\nabla,\kappa)\lhd\tilde{\Delta}\\ |(\nabla,\kappa)|=k}} \left(\widetilde{\mathrm{Sub}}^{m(\sigma_\nabla)}(\omega)\right)_{\sigma_\nabla} (\nabla,\kappa)\,\mathbf{1}_{(\nabla,\kappa)}+$$

$$(-1)^k \sum_{\substack{(F,\varepsilon)\lhd\tilde{\Delta}\\ |(F,\varepsilon)|=k-1}} \sum_{\substack{(F,\varepsilon)\lhd\partial(\nabla,\kappa)\\ (\nabla,\kappa)\lhd\tilde{\Delta}}} n_{(F,\varepsilon,\nabla,\kappa)} \left((\Psi_{m(\sigma_F)}-\Psi_{m(\sigma_\nabla)})(\omega)\right)_{\sigma_F}(F,\varepsilon)\,\mathbf{1}_{(\nabla,\kappa)}.$$

Observe that $(\varphi_{\mathcal{U}}(\omega))_\sigma$ is well defined for all $\omega\in\widetilde{N}^k(X;R)$ but we have to show that it is also well defined for every $\omega\in\widetilde{N}^{*,\mathcal{U}}(X;R)$ and any regular simplex $\sigma\colon\Delta\to X$. The first term of the sum above is defined since the cochains $(\widetilde{\mathrm{Sub}}^m(\omega))_{\sigma_\nabla}$ are well defined for any $m\geqslant m(\sigma_\nabla)$. The second term is also well defined since $m(\sigma_F)\leqslant m(\sigma_\nabla)$ and the cochain $(\widetilde{\mathrm{Sub}}^m(\omega))_{\sigma_F}$ is defined for any $m\geqslant m(\sigma_F)$. So we have $\varphi_{\mathcal{U}}(\omega)\in\widetilde{N}^*(X;R)$.

The equality $\delta\circ\Theta+\Theta\circ\delta=\mathrm{id}-\varphi_{\mathcal{U}}\circ\rho_{\mathcal{U}}$ is thus verified by construction of $\varphi_{\mathcal{U}}$ and Θ. If $\omega\in\widetilde{N}^{*,\mathcal{U}}(X;R)$ and if σ is $\mathcal{U}$-small, then $m(\sigma)=0$ and the family of indexes of the terms defining $\Psi_{m(\sigma)}$ is empty. Then

$$(\rho_{\mathcal{U}}(\varphi_{\mathcal{U}}(\omega)))_\sigma=\varphi_{\mathcal{U}}(\omega)_\sigma=\omega_\sigma-0=\omega_\sigma.$$

We have therefore established $\rho_{\mathcal{U}}\circ\varphi_{\mathcal{U}}=\mathrm{id}$. From $\mathrm{id}-\varphi_{\mathcal{U}}=\delta\circ\Theta+\Theta\circ\delta$, we deduce $\delta-\delta\circ\varphi_{\mathcal{U}}=\delta\circ\Theta\circ\delta=\delta-\varphi_{\mathcal{U}}\circ\delta$, and the compatibility of $\varphi_{\mathcal{U}}$ with differentials. It remains to study the behavior of Θ and $\varphi_{\mathcal{U}}$ relatively to the perverse degree. Since $\|\Psi_m(\omega)\|\leqslant\|\omega\|$ and $\|\widetilde{\mathrm{Sub}}\,(\omega)\|\leqslant\|\omega\|$, we have, for any singular stratum, S, $\|\Theta(\omega)\|_S\leqslant\|\omega\|_S$ and $\|\varphi_{\mathcal{U}}(\omega)\|_S\leqslant\|\omega\|_S$. This completes the proof of the assertion (i).

(ii) Let $\omega\in\widetilde{N}^{*,\mathcal{U}}_{\overline{p}}(X;R)$ verifying the hypothesis of (ii) for a subset $K\subset X$. Let $\sigma\colon\Delta\to X$ such that $(\mathrm{Im}\,\sigma)\cap K=\emptyset$. The element $(\varphi_{\mathcal{U}}(\omega))_\sigma$ is defined from cochains $\omega_{\sigma_{F*G}}$, where σ_{F*G} is the restriction of σ, as defined in (9.7). From this formula, we find that $\mathrm{Im}\,\sigma_{F*G}\subset\mathrm{Im}\,\sigma$, and then $\omega_{\sigma_{F*G}}=0$ and $(\varphi_{\mathcal{U}}(\omega))_\sigma=0$. □

Proof of Lemma 9.4. We have to show that $\widetilde{\mathrm{T}}_\Delta\colon\widetilde{N}^*(K(\Delta))\to\widetilde{N}^{*-1}(\Delta)$ verifies (9.1). For it, we use Proposition 2.11 to express the value of the differential δ in function of the adding of vertices.

(i) Start with $\mathbf{1}_{(F,\varepsilon)}\in\widetilde{N}^*(\Delta)$. By definition, the face F is not a full face (see Definition 9.2) on $K(\Delta)$, then we have $\widetilde{\mathrm{T}}_\Delta(\mathbf{1}_{(F,\varepsilon)})=0$. On the other hand, the only term of $\tilde{\delta}^{K(\Delta)}\mathbf{1}_{(F,\varepsilon)}$ having a non-zero image by $\widetilde{\mathrm{T}}_\Delta$ is $\mathbf{1}_{(F,\varepsilon)}*e_{\{F\}}$ where $e_{\{F\}}$is the barycenter of the face F. Therefore one has, using (9.4),

$$\widetilde{\mathrm{T}}_\Delta(\tilde{\delta}^{K(\Delta)}\mathbf{1}_{(F,\varepsilon)})=\widetilde{\mathrm{T}}_\Delta\left((-1)^{|(F,\varepsilon)|}\mathbf{1}_{(F,\varepsilon)}*e_{\{F\}}\right)=(-1)^{|(F,\varepsilon)|}(-1)^{|(F,\varepsilon)|}\mathbf{1}_{(F,\varepsilon)}=\mathbf{1}_{(F,\varepsilon)}.$$

It follows:

$$(\widetilde{\mathrm{T}}_\Delta\circ\tilde{\delta}^{K(\Delta)}+\tilde{\delta}^\Delta\circ\widetilde{\mathrm{T}}_\Delta)(\mathbf{1}_{(F,\varepsilon)})=(\iota^*_\Delta-\widetilde{\mathrm{Sub}}_\Delta\circ\iota^*_{\mathrm{Sub}\,\Delta})(\mathbf{1}_{(F,\varepsilon)}).$$

(ii) Continue with a simplex, $F*G\lhd K(\Delta)$, such that $F\neq\emptyset$ and $G\neq\emptyset$. In this case, we have to show

$$(\widetilde{\mathrm{T}}_\Delta\circ\tilde{\delta}^{K(\Delta)}+\tilde{\delta}^\Delta\circ\widetilde{\mathrm{T}}_\Delta)(\mathbf{1}_{(F*G,\varepsilon)})=0. \tag{9.10}$$

— We look first to the case of a *full simplex* distinguishing the various possibilities that appear in the construction of $\widetilde{\mathrm{T}}_\Delta$, located after the statement of Lemma 9.4. We use a first recurrence assuming equality (9.10) holds for any filtered euclidean simplex with formal dimension strictly less than n.

• If $G_n = \emptyset$ and $\varepsilon_n = 1$, we have $F * G \lhd K(\nabla)$, then:

$$\begin{aligned}
&\widetilde{\mathrm{T}}_\Delta\left(\widetilde{\delta}^{K(\Delta)}(\mathbf{1}_{(F*G,\varepsilon)}\otimes\mathbf{1}_{(\emptyset,1)})\right) =_{(1)}\\
&\widetilde{\mathrm{T}}_\nabla\left(\widetilde{\delta}^{K(\nabla)}\mathbf{1}_{(F*G,\varepsilon)}\right)\otimes\mathbf{1}_{(\emptyset,1)}\\
&+(-1)^{|(F*G,\varepsilon)|}\sum_{e\in\mathcal{V}(K(\Delta)_n)}\widetilde{\mathrm{T}}_\Delta\left(\mathbf{1}_{(F*G,\varepsilon)}\otimes\mathbf{1}_{(\emptyset,1)}*e\right) =_{(2)}\\
&\widetilde{\mathrm{T}}_\nabla\left(\widetilde{\delta}^{K(\nabla)}\mathbf{1}_{(F*G,\varepsilon)}\right)\otimes\mathbf{1}_{(\emptyset,1)}\\
&+(-1)^{|(F*G,\varepsilon)|}\sum_{e\in\mathcal{V}(\Delta_n)}\widetilde{\mathrm{T}}_\nabla\left(\mathbf{1}_{(F*G,\varepsilon)}\right)\otimes\mathbf{1}_{(\emptyset,1)}*e =_{(3)}\\
&-\widetilde{\delta}^\nabla\widetilde{\mathrm{T}}_\nabla\left(\mathbf{1}_{(F*G,\varepsilon)}\right)\otimes\mathbf{1}_{(\emptyset,1)}+(-1)^{|(F*G,\varepsilon)|}\widetilde{\mathrm{T}}_\nabla\left(\mathbf{1}_{(F*G,\varepsilon)}\right)\otimes\widetilde{\delta}^{c\Delta_n}\mathbf{1}_{(\emptyset,1)} =_{(1)}\\
&-\widetilde{\delta}^\Delta\widetilde{\mathrm{T}}_\Delta\left(\mathbf{1}_{(F*G,\varepsilon)}\otimes\mathbf{1}_{(\emptyset,1)})\right),
\end{aligned}$$

where $=_{(1)}$ uses (9.2), $=_{(2)}$ uses (9.5) and (9.2), $=_{(3)}$ is the induction hypothesis on ∇.

• The argument is similar when $\dim G_n = 0$ and $\varepsilon_n = 0$. A second induction on the dimension of G_n completes the proof in the case of a full simplex $F * G$.

— If $F * G$ is not full and if the differential $\widetilde{\delta}^{K(\Delta)}\mathbf{1}_{(F*G,\varepsilon)}$ only involves non full simplices, then the left hand side of (9.10) is zero and the result is true. So we have to consider the case of a non full simplex whose differential involves full simplices. Specifically, consider a full k-simplex $F' * G'$ with $F' = [e_{i_0}, \dots, e_{i_a}]$ and $G' = [e_{\{i_0\dots i_a\}}, \dots, e_{\{i_0\dots i_a\dots i_b\}}]$ and including a non full $(k-1)$-face $F * G$ with $F \neq \emptyset$ and $G \neq \emptyset$.

We need to establish

$$\widetilde{\mathrm{T}}_\Delta\widetilde{\delta}^{K(\Delta)}\left(\mathbf{1}_{(F*G,\varepsilon)}\right) = 0. \tag{9.11}$$

Consider the various possible cases.

(a) Suppose $a = b$, then $F = [e_{i_0}, \dots, \hat{e}_{i_x}, \dots, e_{i_a}]$ and $G = G' = [e_{\{F'\}}]$. The equality (9.11) is reduced to $\widetilde{\mathrm{T}}_\Delta\widetilde{\delta}^{K(\Delta)}\left(\mathbf{1}_{(F,\varepsilon)} * e_{\{F'\}}\right) = 0$. By keeping in the expression of the differential only the elements corresponding to full simplices, we have:

$$\widetilde{\mathrm{T}}_\Delta\widetilde{\delta}^{K(\Delta)}\left(\mathbf{1}_{(F,\varepsilon)} * e_{\{F'\}}\right) = (-1)^{(F,\varepsilon)+1}\widetilde{\mathrm{T}}_\Delta\left(\mathbf{1}_{(F,\varepsilon)} * e_{\{F'\}} * e_{i_x} + \mathbf{1}_{(F,\varepsilon)} * e_{\{F'\}} * e_{\{F\}}\right)$$

The definition of $\widetilde{\mathrm{T}}_\Delta$ gives,

$$\begin{aligned}
\widetilde{\mathrm{T}}_\Delta\left(\mathbf{1}_{(F,\varepsilon)} * e_{\{F'\}} * e_{i_x}\right) &= -\widetilde{\mathrm{T}}_\Delta\left(\mathbf{1}_{(F,\varepsilon)} * e_{i_x} * e_{\{F'\}}\right) = -(-1)^{|(F',\varepsilon)|}\mathbf{1}_{(F,\varepsilon)} * e_{i_x},\\
\widetilde{\mathrm{T}}_\Delta\left(\mathbf{1}_{(F,\varepsilon)} * e_{\{F'\}} * e_{\{F\}}\right) &= -\widetilde{\mathrm{T}}_\Delta\left(\mathbf{1}_{(F,\varepsilon)} * e_{\{F\}} * e_{\{F'\}}\right)\\
&= -\widetilde{\mathrm{T}}_\Delta\left(\mathbf{1}_{(F,\varepsilon)} * e_{\{F\}}\right) * e_{i_x} = -(-1)^{|(F,\varepsilon)|}\mathbf{1}_{(F,\varepsilon)} * e_{i_x}.
\end{aligned}$$

The conclusion comes from $|(F,\varepsilon)| + 1 = |(F',\varepsilon)|$.

For the last two cases, we use an induction on the dimension of the component G_n.

(b) Suppose $a<b$ and $F=[e_{i_0},\ldots,\hat{e}_{i_x},\ldots,e_{i_a}]$, then $G=G'=L*e_{\{i_0\ldots i_a\ldots i_b\}}$. The equality (9.11) is reduced to $\widetilde{\mathrm{T}}_\Delta\tilde{\delta}^{K(\Delta)}\left(\mathbf{1}_{(F*L,\varepsilon)}*e_{\{i_0\ldots i_a\ldots i_b\}}\right)=0$. By definition of $\widetilde{\mathrm{T}}_\Delta$ and following Corollary 2.12, one has:

$$\widetilde{\mathrm{T}}_\Delta\tilde{\delta}^{K(\Delta)}\left(\mathbf{1}_{(F*L,\varepsilon)}*e_{\{i_0\ldots i_a\ldots i_b\}}\right)=\widetilde{\mathrm{T}}_\Delta\left(\tilde{\delta}^{K(\Delta)}\mathbf{1}_{(F*L,\varepsilon)}\right)*e_{i_b}.$$

The simplex $F*L$ is not full and $L\neq\emptyset$ since $a<b$. We can apply the induction hypothesis, so $\widetilde{\mathrm{T}}_\Delta\tilde{\delta}^{K(\Delta)}\left(\mathbf{1}_{(F*L,\varepsilon)}\right)=0$.

(c) The argument is similar when $G=[e_{\{i_0\ldots i_a\}},\ldots,\hat{e}_{\{i_0\ldots i_a\ldots i_x\}},\ldots,e_{\{i_0\ldots i_a\ldots i_b\}}]$ and $a<b$, and thus $F=F'$.

(iii) It remains to verify the equality (9.1) for elements $\mathbf{1}_{(G,\varepsilon)}\in\widetilde{N}^*(\mathrm{Sub}\,\Delta)$. In this case, it is actually the *definition* of $\widetilde{\mathrm{Sub}}$ by

$$\widetilde{\mathrm{Sub}}\,\mathbf{1}_{(G,\varepsilon)}=-\left(\widetilde{\mathrm{T}}_\Delta\circ\tilde{\delta}^{K(\Delta)}\right)(\mathbf{1}_{(G,\varepsilon)}),$$

the other terms being zero.

To complete the proof of Lemma 9.4, it remains to verify the compatibility of $\widetilde{\mathrm{Sub}}$ with differentials,

$$\tilde{\delta}^{\Delta}\circ\widetilde{\mathrm{T}}_\Delta\circ\tilde{\delta}^{K(\Delta)}(\mathbf{1}_{(G,\varepsilon)})=\widetilde{\mathrm{T}}_\Delta\circ\tilde{\delta}^{K(\Delta)}\circ\tilde{\delta}^{\mathrm{Sub}\,\Delta}(\mathbf{1}_{(G,\varepsilon)}).\tag{9.12}$$

The two terms of this equation are zero except in the two following situations.

(i) $G=[e_{i_0},\ldots,e_{\{i_0\ldots i_a\}}]$,

(ii) $G=[e_{i_0},\ldots,\hat{e}_{\{i_0\ldots i_x\}},\ldots,e_{\{i_0\ldots i_x\ldots i_a\}}]$ with $x<a$.

Detail each case.

(i) The term $\tilde{\delta}^{K(\Delta)}(\mathbf{1}_{(G,\varepsilon)})$ has only one full term, $(-1)^{|(G,\varepsilon)|}\mathbf{1}_{(G,\varepsilon)}*e_{i_0}$. The term $\tilde{\delta}^{K(\Delta)}\tilde{\delta}^{\mathrm{Sub}\,\Delta}(\mathbf{1}_{(G,\varepsilon)})$ has the following full terms $-\sum_{e_{i_j}\in\mathcal{V}(\Delta)}\mathbf{1}_{(G,\varepsilon)}*e_{\{i_0\ldots i_a i_j\}}*e_{i_0}$. We can deduce,

$$\begin{aligned}\widetilde{\mathrm{T}}_\Delta\tilde{\delta}^{K(\Delta)}\tilde{\delta}^{\mathrm{Sub}\,\Delta}(\mathbf{1}_{(G,\varepsilon)}) &= -\sum_{e_{i_j}\in\mathcal{V}(\Delta)}\widetilde{\mathrm{T}}_\Delta(\mathbf{1}_{(G,\varepsilon)}*e_{\{i_0\ldots i_a i_j\}}*e_{i_0})\\ &= \sum_{e_{i_j}\in\mathcal{V}(\Delta)}\widetilde{\mathrm{T}}_\Delta(\mathbf{1}_{(G,\varepsilon)}*e_{i_0})*e_{i_j}\\ &= (-1)^{|(G,\varepsilon)|}\tilde{\delta}^{\Delta}\widetilde{\mathrm{T}}_\Delta(\mathbf{1}_{(G,\varepsilon)}*e_{i_0})=\tilde{\delta}^{\Delta}\widetilde{\mathrm{T}}_\Delta\tilde{\delta}^{K(\Delta)}(\mathbf{1}_{(G,\varepsilon)}).\end{aligned}$$

(ii) The term $\tilde{\delta}^{K(\Delta)}\mathbf{1}_{(G,\varepsilon)}$ does not have any full term. Also, the next term $\tilde{\delta}^{K(\Delta)}\tilde{\delta}^{\mathrm{Sub}\,\Delta}\mathbf{1}_{(G,\varepsilon)}$ has the following full terms $-\mathbf{1}_{(G,\varepsilon)}*e_{\{i_0\ldots i_x\}}*e_{i_0}-\mathbf{1}_{(G,\varepsilon)}*e_{\{i_0\ldots i_{x-1}i_{x+1}\}}*e_{i_0}$. As in the previous calculation, equation (9.12) follows from the definition of $\widetilde{\mathrm{T}}_\Delta$:

$$\widetilde{\mathrm{T}}_\Delta(\mathbf{1}_{(G,\varepsilon)}*e_{\{i_0\ldots i_x\}}*e_{i_0})=-\widetilde{\mathrm{T}}_\Delta(\mathbf{1}_{(G,\varepsilon)}*e_{\{i_0\ldots i_{x-1}i_{x+1}\}}*e_{i_0}).$$

(The sign comes from the permutation of e_{i_x} and $e_{i_{x+1}}$.) □

10. Mayer-Vietoris exact sequence. Theorem C.

If $U\subset V$ are two open subsets of a perverse space $(X,\overline{p})$, canonical inclusions $U\subset V\subset X$ induce cochain maps, $\widetilde{N}^*_{\overline{p}}(X;R)\to\widetilde{N}^*_{\overline{p}}(V;R)\to\widetilde{N}^*_{\overline{p}}(U;R)$. If there is no ambiguity, we keep the same notation for a cochain and its images by these maps.

THEOREM C (Mayer-Vietoris exact sequence). *Let $(X,\overline{p})$ be a paracompact perverse space, endowed with an open cover (U_1,U_2) and a subordinated partition of the unity, (f_1,f_2). For $i=1$, 2, we denote by $\mathcal{U}_i$ the cover of U_i consisting of the open subsets $(U_1\cap U_2, f_i^{-1}(]1/2,1]))$ and by $\mathcal{U}$ the cover of X, union of the covers $\mathcal{U}_i$. Then, the canonical inclusions, $U_i\subset X$ and $U_1\cap U_2\subset U_i$, induce a short exact sequence,*

$$0\to \widetilde{N}^{*,\mathcal{U}}_{\overline{p}}(X;R)\xrightarrow{\iota}\widetilde{N}^{*,\mathcal{U}_1}_{\overline{p}}(U_1;R)\oplus\widetilde{N}^{*,\mathcal{U}_2}_{\overline{p}}(U_2;R)\xrightarrow{\varphi}\widetilde{N}^{*}_{\overline{p}}(U_1\cap U_2;R)\to 0,$$

where $\varphi(\omega_1,\omega_2)=\omega_1-\omega_2$.

The following result is a direct consequence of Theorems B and C.

COROLLARY 10.1. *Let $(X,\overline{p})$ be a paracompact perverse space provided with an open cover (U_1,U_2). Then there is a long exact sequence for the blown-up intersection cohomology,*

$$\to\mathscr{H}^{i}_{\overline{p}}(X;R)\to\mathscr{H}^{i}_{\overline{p}}(U_1;R)\oplus\mathscr{H}^{i}_{\overline{p}}(U_2;R)\to\mathscr{H}^{i}_{\overline{p}}(U_1\cap U_2;R)\to\mathscr{H}^{i+1}_{\overline{p}}(X;R)\to$$

PROOF OF THEOREM C. Since the open cover $\mathcal{U}$ is the union of the covers $\mathcal{U}_1$ and $\mathcal{U}_2$, the morphism ι is injective.

We prove the surjectivity of the map φ. For $i=1$, 2, we define the fonction, $g_i\colon X\to\{0,1\}$, by

$$g_1(x)=\begin{cases}1 & \text{if } f_1(x)>1/2,\\ 0 & \text{if not,}\end{cases}\qquad\text{and}\qquad g_2(x)=\begin{cases}1 & \text{if } f_2(x)\geqslant 1/2,\\ 0 & \text{if not.}\end{cases}$$

The inequality $f_1(x)>1/2$ implies $f_2(x)<1/2$ and $g_2(x)=0$. The support of g_2 is therefore included in $U_2\backslash f_1^{-1}(]1/2,1])$. It is noted as well that the support of g_1 is included in $U_1\backslash f_2^{-1}(]1/2,1])$. On the other hand, by construction, one has $g_1(x)+g_2(x)=1$. We denote by $\tilde{g}_1$ and $\tilde{g}_2$ the two 0-cochains of perverse degree 0, respectively associated to g_1 and g_2, and defined in Lemma 10.2. If $\omega\in\widetilde{N}^{*}_{\overline{p}}(U_1\cap U_2;R)$, we denote by $\tilde{g}_i\cup\omega$ the cup product (cf. Section 4) of $\tilde{g}_i$ with ω, for $i=1$, 2. Since the cochain $\tilde{g}_1$ has a support included in $U_1\backslash f_2^{-1}(]1/2,1])$, then the cup product $\tilde{g}_1\cup\omega\in\widetilde{N}^{*,\mathcal{U}_1}_{\overline{p}}(U_1;R)$. Likewise, one has $\tilde{g}_2\cup\omega\in\widetilde{N}^{*,\mathcal{U}_2}_{\overline{p}}(U_2;R)$. We verify $\varphi(\tilde{g}_1\cup\omega,-\tilde{g}_2\cup\omega)=\omega$, which gives the surjectivity of φ.

The composition $\varphi\circ\iota$ is the zero map. It remains to consider an element $(\omega_1,\omega_2)\in\widetilde{N}^{*,\mathcal{U}_1}_{\overline{p}}(U_1;R)\oplus\widetilde{N}^{*,\mathcal{U}_2}_{\overline{p}}(U_2;R)$ such that $\varphi(\omega_1,\omega_2)=0$ and to construct $\omega\in\widetilde{N}^{*,\mathcal{U}}_{\overline{p}}(X;R)$ such that $\iota(\omega)=(\omega_1,\omega_2)$. If $\sigma\colon\Delta\to X$ is a regular simplex, we set,

- $\omega_\sigma=(\omega_1)_\sigma=(\omega_2)_\sigma$, if $\sigma(\Delta)\subset U_1\cap U_2$,
- $\omega_\sigma=(\omega_1)_\sigma$, if $\sigma(\Delta)\subset f_1^{-1}(]1/2,1])$,
- $\omega_\sigma=(\omega_2)_\sigma$, if $\sigma(\Delta)\subset f_2^{-1}(]1/2,1])$.

This definition makes sense because, on the one hand $f_1^{-1}(]1/2,1])\cap f_2^{-1}(]1/2,1])=\emptyset$ and on the other hand $U_1\cap U_2\cap f_1^{-1}(]1/2,1])\subset U_1\cap U_2$, where the two cochains $(\omega_1)_\sigma$ and $(\omega_2)_\sigma$ coincide. □

With the notation of the previous proof, the connecting morphism of the long exact sequence of Mayer-Vietoris is defined by

$$[\omega]\mapsto[(\delta\tilde{g}_1)\cup\omega].$$

LEMMA 10.2. *Let $(X,\overline{p})$ be a perverse space. Any map, $g\colon X \to R$, defines a 0-cochain $\tilde{g} \in \widetilde{N}^0_{\overline{0}}(X;R)$. Moreover, the association $g \mapsto \tilde{g}$ is R-linear.*

PROOF. Let $\sigma\colon \Delta = \Delta_0 * \dots * \Delta_n \to X$ be a regular simplex. We want to define $\tilde{g}_\sigma \in N^0(c\Delta_0)\otimes\dots\otimes N^0(c\Delta_{n-1})\otimes N^0(\Delta_n)$. If $b = (b_0,\dots,b_n) \in c\Delta_0 \times \dots \times c\Delta_{n-1} \times \Delta_n$, we denote by i_0 the smallest index for which b_i is not an apex of a cone (i.e., $i_0 = \min\{i \mid b_i \in \Delta_i\}$). Observe that the integer i_0 exists since $b_n \in \Delta_n$ and set $\tilde{g}_\sigma(b) = g(\sigma(b_{i_0}))$. This map is clearly compatible with the face operators and defines $\tilde{g} \in \widetilde{N}^0(X;R)$.

It remains to determine the perversity of the cochain $\tilde{g}_\sigma$. Since it is a 0-cochain, it is obviously $\overline{0}$-allowable and we only need to study its coboundary. To compute the ℓ-perverse degree of $\delta\tilde{g}_\sigma$, we consider $F = F_0\otimes\dots\otimes F_n = \{b_0\}\otimes\dots\otimes F_j\otimes\dots\otimes\{b_n\}$ with $\dim F_j = 1$ and $\partial F_j = b^1_j - b^0_j$. We have,

$$\delta\tilde{g}_\sigma(F) = \tilde{g}_\sigma(b_0,\dots,b^1_j,\dots,b_n) - \tilde{g}_\sigma(b_0,\dots,b^0_j,\dots,b_n).$$

If $n-\ell < j$, then $i_0 \leqslant n-\ell < j$ and $\delta\tilde{g}_\sigma(F) = g(\sigma(b_{i_0})) - g(\sigma(b_{i_0})) = 0$, and then $\|\delta\tilde{g}_\sigma\|_\ell = -\infty$.
Finally, $\tilde{g} \in \widetilde{N}^0_{\overline{0}}(X)$ and the association $g \mapsto \tilde{g}$ is R-linear by construction. □

11. Product with the real line. Theorem D.

Consider the product $X \times \mathbb{R}$ equipped with the product filtration $(X\times\mathbb{R})_i = X_i \times \mathbb{R}$ and a perversity $\overline{p}$. We also denote by $\overline{p}$ the perversity induced on X, that is, $\overline{p}(S) = \overline{p}(S\times\mathbb{R})$ for each stratum S of X. Let $I_0\ I_1\colon X \to X\times\mathbb{R}$ be the canonical injections, defined by $I_0(x) = (x,0)$ and $I_1(x) = (x,1)$. The canonical projections are denoted by $\mathrm{pr}\colon X\times\mathbb{R}\to X$ and $\mathrm{pr}_2\colon X\times\mathbb{R}\to\mathbb{R}$. Proposition 3.5 ensures that the maps I_0, I_1 and pr induce cochain maps between the blown-up complexes.

THEOREM D. *Let X be a filtered space and let $X\times\mathbb{R}$ equipped with the filtration product and a perversity $\overline{p}$. Denoting also $\overline{p}$ the perversity induced on X, the maps induced in the blown-up intersection cohomology by the projection $\mathrm{pr}\colon X\times\mathbb{R}\to X$ and the canonical injections, $I_0\ I_1\colon X\to X\times\mathbb{R}$, verify $(I_0\circ\mathrm{pr})^* = (I_1\circ\mathrm{pr})^* = \mathrm{id}$.*

The proof proceeds according to the scheme of Section 9 by constructing a homotopy Θ_Δ at the simplex level (cf. Proposition 11.1), then gluing them to get a homotopy at the level of blown-up complexes (see Proposition 11.3).

For any filtered simplex, $\Delta = \Delta_0 * \dots * \Delta_n$, one defines a weighted simplicial complex $\mathcal{L}_\Delta = \Delta\otimes[0,1]$, whose simplices are the joins $(F,\mathbf{0})*(G,\mathbf{1})$, with $F\lhd\Delta$, $G\lhd\Delta$, $(F,\mathbf{0})\subset\Delta\times\{0\}$ and $(G,\mathbf{1})\subset\Delta\times\{1\}$. Henceforth, we denote by $F*G$ such simplex, meaning that the first term, F, is identified with a simplex of $\Delta\times\{0\}$ and the second, G, with a simplex of $\Delta\times\{1\}$.

If F and G are compatible, in the sense of Definition 4.1, then, for the filtration induced by Δ, we have $F*G = F_0*\dots*(F_p*G_p)*\dots*G_n$, with $p\in\{0,\dots,n\}$. A face of the prismatic set $\widetilde{\mathcal{L}}_\Delta$, corresponding to compatible simplices, F and G of Δ, is denoted by,

$$(F*G,\varepsilon) = (F_0,\varepsilon_0)\times\dots\times(F_p*G_p,\varepsilon_p)\times\dots\times(G_{n-1},\varepsilon_{n-1})\times G_n.$$

In this writing, if $j<n$, $j\neq p$, one authorizes F_j and G_j to be the emptyset but the associated variable ε_j must be 1.

PROPOSITION 11.1. *There exists a linear map,* $\Theta_\Delta \colon \widetilde{N}^*(\Delta \otimes [0,1]) \to \widetilde{N}^{*-1}(\Delta)$, *such that*

$$(\Theta_\Delta \circ \delta + \delta \circ \Theta_\Delta)(\mathbf{1}_{(F*G,\varepsilon)}) = \begin{cases} 0 & \text{if } F \neq \emptyset \text{ and } G \neq \emptyset, \\ -\mathbf{1}_{(G,\varepsilon)} & \text{if } F = \emptyset, \\ \mathbf{1}_{(F,\varepsilon)} & \text{if } G = \emptyset, \end{cases} \tag{11.1}$$

where δ *denotes the differential of* $\widetilde{N}^*(\Delta \otimes [0,1])$ *and* $\widetilde{N}^{*-1}(\Delta)$.

PROOF. Define the map Θ_Δ on the elements of the dual basis. Let

$$\mathbf{1}_{(F*G,\varepsilon)} = \mathbf{1}_{(F_0,\varepsilon_0)} \otimes \cdots \otimes \mathbf{1}_{(F_p*G_p,\varepsilon_p)} \otimes \cdots \otimes \mathbf{1}_{(G_{n-1},\varepsilon_{n-1})} \otimes \mathbf{1}_{G_n}.$$

We set

$$\Theta_\Delta(\mathbf{1}_{(F*G,\varepsilon)}) = \begin{cases} (-1)^{|(F,\varepsilon)|_{<p}+|(G_p,\varepsilon_p)|}\, \mathbf{1}_{(F\cup G,\varepsilon)} & \text{if } F \text{ and } G \text{ are compatible,} \\ 0 & \text{if not,} \end{cases}$$

where $\mathbf{1}_{(F\cup G,\varepsilon)} = \mathbf{1}_{(F_0,\varepsilon_0)} \otimes \cdots \otimes \mathbf{1}_{(F_p\cup G_p,\varepsilon_p)} \otimes \cdots \otimes \mathbf{1}_{(G_{n-1},\varepsilon_{n-1})} \otimes \mathbf{1}_{G_n}$ and $F_p \cup G_p$ was introduced in Definition 4.1.

In $\mathbf{1}_{(F*G,\varepsilon)}$, we consider the faces F of $\Delta \times \{0\}$ and G of $\Delta \times \{1\}$. In the expression of $\Theta_\Delta(\mathbf{1}_{(F*G,\varepsilon)})$, we make an abuse of notation by keeping the letters F and G for faces of Δ. It remains to verify (11.1) and for that, we consider the following cases. We denote by $\mathcal{V}(\widetilde{\mathcal{L}}_\Delta)$ the union of $\mathcal{V}(\mathcal{L}_\Delta)$ with the family of virtual vertices.

- Suppose $F \neq \emptyset$, $G \neq \emptyset$, F and G compatible. Using equality (2.6) and Lemma 11.2, we obtain the equality (11.1) in this case:

$$\begin{aligned} \delta\Theta_\Delta(\mathbf{1}_{(F*G,\varepsilon)}) &= (-1)^{|(F*G,\varepsilon)|+1} \sum_{e\in\mathcal{V}(\tilde{\Delta})} \Theta_\Delta(\mathbf{1}_{(F*G,\varepsilon)}) * e \\ &= (-1)^{|(F*G,\varepsilon)|+1} \sum_{e\in\mathcal{V}(\tilde{\mathcal{L}}_\Delta)} \Theta_\Delta(\mathbf{1}_{(F*G,\varepsilon)} * e) = -\Theta_\Delta(\delta\mathbf{1}_{(F*G,\varepsilon)}). \end{aligned}$$

- Suppose $F \neq \emptyset$, $G \neq \emptyset$, F and G not compatible such that $\delta\mathbf{1}_{(F*G,\varepsilon)}$ contains compatible elements. This amounts to giving a k-simplex, $F'*G'$, of $\Delta \otimes [0,1]$, with F' and G' compatible and having a $(k-1)$-face, $F*G$, with F and G non compatible and non empty. Put $F' = [a_{i_0}, \ldots, a_{i_r}]$ and $G' = [b_{j_0}, \ldots, b_{j_s}]$ with $a_{i_r} = b_{j_0} \in \Delta_p$. (Recall that in the latter equality we identified $\Delta \cong \Delta \times \{0\} \cong \Delta \times \{1\}$.) We denote by $a_{j_t} \in \Delta \times \{0\}$ the vertex identified to $b_{j_t} \in \Delta \times \{1\}$, as well as $b_{i_t} \in \Delta \times \{1\}$ for the vertex identified to $a_{i_t} \in \Delta \times \{0\}$. For the simplices F and G, there are only two possibilities corresponding to compatible simplices F', G'. Equality (11.1) is deduced from the following calculations.

  + If $F = [a_{i_0}, \ldots, a_{i_{r-1}}]$ and $G = G'$, the expression $\Theta_\Delta(\delta\mathbf{1}_{(F*G,\varepsilon)})$ takes the values,

  $$(-1)^{|(G,\varepsilon)|+1+|(F,\varepsilon)|_{<p}+\varepsilon_p}\Theta_\Delta(\mathbf{1}_{(F*a_{i_r}*G,\varepsilon)} + \mathbf{1}_{(F*b_{i_{r-1}}*G,\varepsilon)}),$$

  if $a_{i_{r-1}} \in F_p$, and

  $$(-1)^{|(F,\varepsilon)|+|(G_p,\varepsilon_p)+1}\Theta_\Delta(\mathbf{1}_{(F*a_{i_r}*G,\varepsilon)}) + (-1)^{|(F,\varepsilon)|_{<a}+\varepsilon_a}\Theta_\Delta(\mathbf{1}_{(F*b_{i_{r-1}}*G,\varepsilon)}),$$

  if $a_{i_r-1} \in F_a$ with $a < p$, which implies

  $$\Theta_\Delta(\delta\mathbf{1}_{(F*G,\varepsilon)}) = -\mathbf{1}_{((F*a_{i_r})\cup G,\varepsilon)} + \mathbf{1}_{(F\cup(b_{i_{r-1}}*G),\varepsilon)} = 0.$$

  + The argument is the same for $F = F'$ and $G = [b_{j_1}, \ldots, b_{j_s}]$.

- Suppose $F = \emptyset$. Let $b_{i_0} \in \Delta_p \times \{1\}$ be the first vertex of G and a_{i_0} the vertex of $\Delta_p \times \{0\}$ whose projection on Δ_p is equal to the projection of b_{i_0}. We have $\Theta_\Delta(\mathbf{1}_{(G,\varepsilon)}) = 0$. To determine the second term, first note,

$$\begin{aligned}\mathbf{1}_{(G,\varepsilon)} * a_{i_0} &= (-1)^{|(G,\varepsilon)|_{>p}}(\mathbf{1}_{(G_p,\varepsilon_p)} * a_{i_0}) \otimes \dots \\ &= (-1)^{|(G,\varepsilon)|_{>p}+1}(\mathbf{1}_{(a_{i_0} * G_p,\varepsilon_p)}) \otimes \dots\end{aligned}$$

It follows

$$\begin{aligned}\Theta_\Delta \delta \mathbf{1}_{(G,\varepsilon)} &= (-1)^{|(G,\varepsilon)|}\Theta_\Delta(\mathbf{1}_{(G,\varepsilon)} * a_{i_0}) \\ &= (-1)^{|(G_p,\varepsilon_p)|+1}\Theta_\Delta(\mathbf{1}_{(a_{i_0} * G,\varepsilon)}) = -\,\mathbf{1}_{(G,\varepsilon)}.\end{aligned}$$

- The proof for the case $G = \emptyset$ is similar.

□

LEMMA 11.2. *Let $\mathbf{1}_{(F*G,\varepsilon)} \in \widetilde{N}^*(\Delta \otimes [0,1])$, with $F \neq \emptyset$ and $G \neq \emptyset$.*

1) *Let $\ell \in \{1, \dots, n\}$. For any vertex $e \in \Delta_\ell$, we denote $e_0 \in \Delta_\ell \times \{0\}$ and $e_1 \in \Delta_\ell \times \{1\}$ the vertices of $\Delta \otimes [0,1]$ corresponding to e. Then we have*

$$\Theta_\Delta(\mathbf{1}_{(F*G,\varepsilon)}) * e = \Theta_\Delta(\mathbf{1}_{(F*G,\varepsilon)} * e_0) + \Theta_\Delta(\mathbf{1}_{(F*G,\varepsilon)} * e_1).$$

2) *For any virtual vertex $\mathtt{v}_\ell$, $\ell < n$, we have*

$$\Theta_\Delta(\mathbf{1}_{(F*G,\varepsilon)} * \mathtt{v}_\ell) = (\Theta_\Delta(\mathbf{1}_{(F*G,\varepsilon)})) * \mathtt{v}_\ell. \tag{11.2}$$

PROOF. Let $\mathbf{1}_{(F*G,\varepsilon)} = \mathbf{1}_{(F_0,\varepsilon_0)} \otimes \dots \otimes \mathbf{1}_{(F_p*G_p,\varepsilon_p)} \otimes \dots \otimes \mathbf{1}_{(G_{n-1},\varepsilon_{n-1})} \otimes \mathbf{1}_{G_n}$.

1) • Let $\ell < p$. Using the Definition 2.9, we get $\Theta_\Delta(\mathbf{1}_{(F*G,\varepsilon)} * e_1) = 0$ and the following equalities

$$\begin{aligned}&\Theta_\Delta(\mathbf{1}_{(F*G,\varepsilon)} * e_0) = \\ &(-1)^{|(F*G,\varepsilon)|_{>\ell}}\Theta_\Delta(\mathbf{1}_{(F_0,\varepsilon_0)} \otimes \dots \otimes (\mathbf{1}_{(F_\ell,\varepsilon_\ell)} * e_0) \otimes \dots \otimes \mathbf{1}_{(F_p*G_p,\varepsilon_p)} \otimes \dots \otimes \mathbf{1}_{G_n}) = \\ &\quad (-1)^{|(F*G,\varepsilon)|_{>\ell}+|(F,\varepsilon)|_{<p}+1+|G_p|+\varepsilon_p} \dots \otimes (\mathbf{1}_{(F_\ell,\varepsilon_\ell)} * e) \otimes \dots \otimes \mathbf{1}_{(F_p \cup G_p,\varepsilon_p)} \otimes \dots = \\ &(-1)^{|(F,\varepsilon)|_{<p}+|G_p|+\varepsilon_p}(\mathbf{1}_{(F_0,\varepsilon_0)} \otimes \dots \otimes \mathbf{1}_{(F_p \cup G_p,\varepsilon_p)} \otimes \dots \otimes \mathbf{1}_{G_n}) * e = (\Theta_\Delta(\mathbf{1}_{(F*G,\varepsilon)})) * e.\end{aligned}$$

• The argument is the same if $\ell > p$.

• If $\ell = p$, these formulas become:

$$\begin{aligned}\Theta_\Delta(\mathbf{1}_{(F*G,\varepsilon)} * e_0) &= (-1)^{|(F*G,\varepsilon)|_{>p}}\Theta_\Delta(\mathbf{1}_{(F_0,\varepsilon_0)} \otimes \dots \otimes \mathbf{1}_{(F_p*G_p,\varepsilon_p)} * e_0 \otimes \dots \otimes \mathbf{1}_{G_n}). \\ &= (-1)^{|(F*G,\varepsilon)|_{\geqslant p}+\varepsilon_p}\Theta_\Delta(\mathbf{1}_{(F_0,\varepsilon_0)} \otimes \dots \otimes \mathbf{1}_{(F_p*G_p*e_0,\varepsilon_p)} \otimes \dots \otimes \mathbf{1}_{G_n}).\end{aligned}$$

Write $F_p = [a_{i_0}, \dots, a_{i_r}]$, $G_p = [b_{j_0}, \dots, b_{j_s}]$. If the simplices are not compatible, both members of the previous equality are zero. The only cases giving compatible simplices are one of the following cases:

$$(*)\ a_{i_t} < e < a_{i_{t+1}} \text{ with } t \in \{0, \dots, r-1\},\quad (*)\ e < a_{i_0} \text{ and } e \in \Delta_p. \tag{11.3}$$

We treat the first case, the second being identical. Notice:

$$\begin{aligned}F_p * G_p * e &= [a_{i_0}, \dots, a_{i_r}] * [b_{j_0}, \dots, b_{j_s}] * e \\ &= (-1)^{s+1+r-t}[a_{i_0}, \dots, a_{i_t}, e, a_{i_{t+1}}, \dots, a_{i_r}] * [b_{j_0}, \dots, b_{j_s}]\end{aligned}$$

and, setting $F_{p(e)} = [a_{i_0}, \dots, a_{i_t}, e, a_{i_{t+1}}, \dots, a_{i_r}]$,

$$F_p(e) \cup G_p = [a_{i_0}, \dots, a_{i_t}, e, a_{i_{t+1}}, \dots, a_{i_r} = b_{j_0}, \dots, b_{j_s}] = (-1)^{s+r-t}(F_p \cup G_p) * e.$$

It follows

$$\begin{aligned}
&\Theta_\Delta(\mathbf{1}_{(F*G,\varepsilon)} * e_0) = \\
&(-1)^{|(F*G,\varepsilon)|_{\geqslant p}+|(F,\varepsilon)|_{<p}+|G_p|+1}\mathbf{1}_{(F_0,\varepsilon_0)} \otimes \cdots \otimes \mathbf{1}_{((F_p\cup G_p)*e,\varepsilon_p)} \otimes \cdots \otimes \mathbf{1}_{G_n} = \\
&(-1)^{|(F,\varepsilon)|_{<p}+|G_p|+\varepsilon_p}(\mathbf{1}_{(F_0,\varepsilon_0)} \otimes \cdots \otimes (\mathbf{1}_{(F_p\cup G_p,\varepsilon_p)}) \otimes \cdots \otimes \mathbf{1}_{G_n}) * e = \\
&(\Theta_\Delta(\mathbf{1}_{(F_0,\varepsilon_0)} \otimes \cdots \otimes (\mathbf{1}_{(F_p*G_p,\varepsilon_p)}) \otimes \cdots \otimes \mathbf{1}_{G_n})) * e.
\end{aligned}$$

In the case (11.3), we have $\Theta_\Delta(\mathbf{1}_{(F*G,\varepsilon)} * e_1) = 0$, which gives the desired equality.

Similar argument gives the result in the symmetric case of the previous one: "$b_{j_t} < e < b_{j_{t+1}}$ with $t \in \{1, \ldots, s-1\}$ or $b_{j_s} < e$ with $e \in \Delta_p$."

2) We now consider a virtual vertex $\mathtt{v}_\ell$. If $\varepsilon_\ell = 1$, equality (11.2) is verified since its two members are zero. We therefore assume $\varepsilon_\ell = 0$. If $\ell < p$, we have:

$$\begin{aligned}
&\Theta_\Delta(\mathbf{1}_{(F*G,\varepsilon)} * \mathtt{v}_\ell) = \\
&(-1)^{|(F*G,\varepsilon)|_{>\ell}+|(F,\varepsilon)|_{<p}+1+|(G_p,\varepsilon_p)|} \cdots \otimes (\mathbf{1}_{(F_\ell,0)} * \mathtt{v}_\ell) \otimes \cdots \otimes \mathbf{1}_{(F_p\cup G_p,\varepsilon_p)} \otimes \cdots = \\
&(-1)^{|(F,\varepsilon)|_{<p}+|(G_p,\varepsilon_p)|}(\mathbf{1}_{(F_0,\varepsilon_0)} \otimes \cdots \otimes \mathbf{1}_{(F_p\cup G_p,\varepsilon_p)} \otimes \cdots \otimes \mathbf{1}_{G_n}) * \mathtt{v}_\ell = \\
&(\Theta_\Delta(\mathbf{1}_{(F*G,\varepsilon)})) * \mathtt{v}_\ell.
\end{aligned}$$

The argument is similar for $\ell \geqslant p$. □

PROPOSITION 11.3. *Let X be a filtered space and let $X \times \mathbb{R}$ endowed with the product filtration and a perversity $\overline{p}$. Denoting also by $\overline{p}$ the induced perversity on X, there exists a linear map $\Theta\colon \widetilde{N}^*_{\overline{p}}(X \times \mathbb{R}; R) \to \widetilde{N}^{*-1}_{\overline{p}}(X \times \mathbb{R}; R)$, such that*

$$\Theta \circ \delta + \delta \circ \Theta = (I_0 \circ \mathrm{pr})^* - \mathrm{id}\,. \tag{11.4}$$

PROOF. Let $\sigma\colon \Delta \to X \times \mathbb{R}$ be a regular simplex and $\omega \in \widetilde{N}^k(X \times \mathbb{R}; R)$. We define $\hat{\sigma}\colon \mathcal{L}_\Delta = \Delta \otimes [0,1] \to X \times \mathbb{R}$ by $\hat{\sigma}(x,t) = (\mathrm{pr}(\sigma(x)), t\,\mathrm{pr}_2(\sigma(x)))$ and we set

$$\omega_{\Delta\otimes[0,1]} = \sum_{(F*G,\varepsilon)\lhd\widetilde{\Delta\otimes[0,1]}} \omega_{\hat{\sigma}\circ\iota_{F*G}}(F*G,\varepsilon)\,\mathbf{1}_{(F*G,\varepsilon)},$$

where $\iota_{F*G}\colon F*G \hookrightarrow \Delta \otimes [0,1]$ is the canonical injection. By construction, $\omega_{\Delta\otimes[0,1]} \in \widetilde{N}^k(\Delta \otimes [0,1])$, and we can define

$$\Theta(\omega)_\sigma = \Theta_\Delta(\omega_{\Delta\otimes[0,1]}) \in \widetilde{N}^{k-1}(\Delta).$$

We need to verify $\Theta(\omega) \in \widetilde{N}^*(X \times \mathbb{R}; R)$. For this purpose, we consider a regular face operator $\delta_\ell\colon \nabla \to \Delta$ and $\tau = \sigma \circ \delta_\ell$. The property $\omega_{\Delta\otimes[0,1]} \in \widetilde{N}^*(\Delta \otimes [0,1])$ becomes $\delta_\ell^*\,\omega_{\Delta\otimes[0,1]} = \omega_{\nabla\otimes[0,1]}$. The map $\delta_\ell^*\colon \widetilde{N}^*(\Delta \otimes [0,1]) \to \widetilde{N}^*(\nabla \otimes [0,1])$ verifies

$$\delta_\ell^*(\mathbf{1}_{(F*G,\varepsilon)}) = \begin{cases} \mathbf{1}_{(F*G,\varepsilon)} & \text{if } F*G \lhd \nabla \otimes [0,1], \\ 0 & \text{if not,} \end{cases}$$

which implies the commutativity of the following diagram,

$$\begin{CD}
\widetilde{N}^*(\Delta \otimes [0,1]) @>{\Theta_\Delta}>> \widetilde{N}^{*-1}(\Delta \otimes [0,1]) \\
@V{\delta_\ell^*}VV @VV{\delta_\ell^*}V \\
\widetilde{N}^*(\nabla \otimes [0,1]) @>{\Theta_\nabla}>> \widetilde{N}^{*-1}(\nabla \otimes [0,1]).
\end{CD}$$

We deduce

$$\delta_\ell^*(\Theta(\omega)_\sigma) = \delta_\ell^*\Theta_\Delta(\omega_{\Delta\otimes[0,1]}) = \Theta_\nabla(\delta_\ell^*\omega_{\Delta\otimes[0,1]}) = \Theta_\nabla(\omega_{\nabla\otimes[0,1]}) = \Theta(\omega)_\tau.$$

The map $\Theta\colon \widetilde{N}^*(X\times\mathbb{R};R) \to \widetilde{N}^{*-1}(X\times\mathbb{R};R)$ is well defined. From Proposition 11.1, we can deduce

(11.5)

$$(\Theta\delta+\delta\Theta)(\omega)_\sigma = \sum_{(F,\varepsilon)\lhd\widetilde{\Delta\times\{0\}}} \omega_{\hat\sigma\circ\iota_F}(F,\varepsilon)\,\mathbf{1}_{(F,\varepsilon)} - \sum_{(G,\varepsilon)\lhd\widetilde{\Delta\times\{1\}}} \omega_{\hat\sigma\circ\iota_G}(G,\varepsilon)\,\mathbf{1}_{(G,\varepsilon)}.$$

Recall the canonical inclusions ι_0, $\iota_1\colon \Delta\to\Delta\otimes[0,1]$. If F is a face of $\Delta\times\{0\}$ identified to $\nabla\lhd\Delta$, one has $\iota_F=\iota_0\circ\iota_\nabla$, from where

$$\hat\sigma\circ\iota_F = \hat\sigma\circ\iota_0\circ\iota_\nabla = I_0\circ\mathrm{pr}\circ\sigma\circ\iota_\nabla$$

and

$$\sum_{(F,\varepsilon)\lhd\widetilde{\Delta\times\{0\}}} \omega_{\hat\sigma\circ\iota_F}(F,\varepsilon)\,\mathbf{1}_{(F,\varepsilon)} = (I_0\circ\mathrm{pr})^*\left(\sum_{(\nabla,\varepsilon)\lhd\tilde\Delta}\omega_{\sigma\circ\iota_\nabla}(\nabla,\varepsilon)\,\mathbf{1}_{(\nabla,\varepsilon)}\right) = (I_0\circ\mathrm{pr})^*(\omega)_\sigma.$$

Using $\iota_G=\iota_1\circ\iota_\nabla$ and $\hat\sigma\circ\iota_G=\hat\sigma\circ\iota_1\circ\iota_\nabla=\sigma\circ\iota_\nabla$, we show, as above, that the right-hand side of the second term of (11.5) is equal to ω_σ. In summary, we have shown,

$$(\Theta\delta+\delta\Theta)(\omega)_\sigma = (I_0\circ\mathrm{pr})^*(\omega)_\sigma - \omega_\sigma,$$

from which comes the equality (11.4).

As for the compatibility with the perverse degrees, we prove $\|\Theta(\omega)\|\leqslant\|\omega\|$, for each $\omega\in\widetilde{N}^*(X\times\mathbb{R};R)$, by arguing over the chosen basis. This inequality follows directly from the definition of Θ,

$$\|\Theta_\Delta(\mathbf{1}_{(F*G,\varepsilon)})\|_\ell = \|\mathbf{1}_{(F\cup G,\varepsilon)}\|_\ell \leqslant \|\mathbf{1}_{(F*G,\varepsilon)}\|_\ell,$$

for each $\ell\in\{1,\ldots,n\}$. With (11.4), we have constructed the homotopy $\Theta\colon \widetilde{N}^*_{\overline{p}}(X\times\mathbb{R};R)\to\widetilde{N}^{*-1}_{\overline{p}}(X\times\mathbb{R};R)$. □

Proof of Theorem D. This is a direct consequence of the equalities $\mathrm{pr}\circ I_0=\mathrm{pr}\circ I_1=\mathrm{id}$ and $\mathscr{H}^*_{\overline{p}}(I_0\circ\mathrm{pr})=\mathrm{id}$, proved in (11.4). □

The following result is used in Section 14.

Proposition 11.4. *Let $(X,\overline{p})$ be a paracompact perverse space. Let S^ℓ be the unit sphere of $\mathbb{R}^{\ell+1}$. The canonical projection $p_X\colon S^\ell\times X\to X$, $(z,x)\mapsto x$, induces an isomorphism $\mathscr{H}^k_{\overline{p}}(S^\ell\times X;R)\cong\mathscr{H}^k_{\overline{p}}(X;R)\oplus\mathscr{H}^{k-\ell}_{\overline{p}}(X;R)$.*

Proof. It suffices to use an induction on ℓ with the decomposition $S^\ell = S^\ell\backslash\{\text{Northpole}\}\cup S^\ell\backslash\{\text{Southpole}\}$ (see Theorem C and Theorem D). □

12. Blown-up intersection cohomology of a cone. Theorem E.

In this section, X is an n-dimensional *compact* filtered space and we represent the open cone as the quotient $\mathring{c}X = X\times[0,\infty[/X\times\{0\}$, whose apex is denoted by $\mathtt{w}$. The formal dimension of $\mathring{c}X$ is $n+1$ relatively to the conical filtration, $(\mathring{c}X)_i=\mathring{c}X_{i-1}$ if $i\geqslant 1$ and $(\mathring{c}X)_0=\{\mathtt{w}\}$. The purpose of this section is to prove the following proposition, cf. also [**4**, Corollary 1.47] and [**24**, Proposition 3.1.1].

THEOREM E. *Let X be a compact filtered space. Consider the open cone, $\mathring{c}X = X \times [0,\infty[/X \times \{0\}$, equipped with the conical filtration and a perversity $\overline{p}$. We also denote by $\overline{p}$ the perversity induced on X. The following properties are verified for any commutative ring, R.*

(a) *The inclusion $\iota\colon X \to \mathring{c}X$, $x \mapsto [x,1]$, induces an isomorphism, $\mathscr{H}^{k}_{\overline{p}}(\mathring{c}X;R) \xrightarrow{\cong} \mathscr{H}^{k}_{\overline{p}}(X;R)$, for each $k \leqslant \overline{p}(\mathtt{w})$.*

(b) *For each $k > \overline{p}(\mathtt{w})$, we have $\mathscr{H}^{k}_{\overline{p}}(\mathring{c}X;R) = 0$.*

12.1. Simplices on a filtered space and its cone. First we link the complexes of X and $\mathring{c}X$. The formal dimension of the cone being different from that of the original space, we introduce some operations which increase or decrease the length of filtrations.

- If $\Delta = \Delta_0 * \cdots * \Delta_{n+1}$ is a regular simplex, of formal dimension $n+1$, we define a regular simplex, of formal dimension n, by $\widehat{\Delta} = \Delta_1 * \cdots * \Delta_{n+1}$. Its filtration is characterized by $\widehat{\Delta}_i = \Delta_{i+1}$, for each $i \in \{0,\dots,n\}$.
- Let $\sigma\colon \Delta_\sigma = \Delta_0 * \cdots * \Delta_{n+1} \to \mathring{c}X$ be a regular simplex of $\mathring{c}X$. Since $\sigma(\widehat{\Delta}_\sigma) \subset X\times]0,\infty[$, we define the restriction
$$\hat{\sigma}\colon \Delta_{\hat{\sigma}} = \widehat{\Delta}_\sigma \xrightarrow{\sigma} X\times]0,\infty[.$$
- For each regular simplex of $\mathring{c}X$, $\sigma\colon \Delta_\sigma = \Delta_0 * \cdots * \Delta_{n+1} \to \mathring{c}X$, the image of a point $x \in \Delta_\sigma$ can be written as,
$$\sigma(x) = [\sigma_1(x), \sigma_2(x)] \in \mathring{c}X = X \times [0,\infty[/X \times \{0\}.$$
Associated to the simplex σ, there is the following regular simplex of $\mathring{c}X$,
$$\mathtt{c}\sigma\colon \Delta_{\mathtt{c}\sigma} = (\{\mathtt{p}\} * \Delta_0) * \cdots * \Delta_{n+1} \to \mathring{c}X,$$
defined by $\mathtt{c}\sigma((1-t)\mathtt{p} + tx) = [\sigma_1(x), t\sigma_2(x)]$. Moreover, if one considers $\hat{\sigma}\colon \widehat{\Delta}_\sigma \to X\times]0,\infty[\hookrightarrow \mathring{c}X$ as a filtered simplex of the cone, then $\mathtt{c}\hat{\sigma}$ is a face of $\mathtt{c}\sigma$.

The *truncation* of a cochain complex is defined for all positive integers s by

$$(\tau_{\leqslant s}C)^r = \begin{cases} C^r & \text{if } r < s, \\ \mathcal{Z}C^s & \text{if } r = s, \\ 0 & \text{if } r > s, \end{cases} \tag{12.1}$$

where $\mathcal{Z}C^s$ means the R-module of cocycles whose degree is s.

Construction of $f\colon \widetilde{N}^*(X\times]0,\infty[;R) \to \widetilde{N}^*(\mathring{c}X;R)$.

Let $\sigma\colon \Delta_\sigma = \Delta_0 * \cdots * \Delta_{n+1} \to \mathring{c}X$ and $\omega \in \widetilde{N}^*(X\times]0,\infty[;R)$. We denote by $\lambda_{\mathtt{c}\Delta_0}$ the cocycle $\mathbf{1}_{(\emptyset,1)} + \sum_{e\in\mathcal{V}(\Delta_0)} \mathbf{1}_{([e],0)} \in N^0(\mathtt{c}\Delta_0)$. We set

$$f(\omega)_\sigma = \lambda_{\mathtt{c}\Delta_0} \otimes \omega_{\hat{\sigma}}. \tag{12.2}$$

PROPOSITION 12.1. *Let $(\mathring{c}X,\overline{p})$ be a perverse space over the cone of the compact space X and let $(X\times]0,\infty[,\overline{p})$ be the induced perverse space. The correspondence defined above induces a cochain map,*

$$f\colon \tau_{\leqslant\overline{p}(\mathtt{w})}\widetilde{N}^*_{\overline{p}}(X\times]0,\infty[;R) \to \tau_{\leqslant\overline{p}(\mathtt{w})}\widetilde{N}^*_{\overline{p}}(\mathring{c}X;R).$$

PROOF. First, we check that the application f, defined locally at the level of simplices, extends globally to $\widetilde{N}^*(X\times]0,\infty[;R)$. For this, we must establish $\delta_k^* f(\omega)_\sigma = f(\omega)_{\sigma\circ\delta_k}$, for each $\omega \in \widetilde{N}^*(X\times]0,\infty[;R)$, each regular simplex, $\sigma\colon \Delta_\sigma = \Delta_0 * \cdots * \Delta_{n+1} \to \mathring{c}X$, and any regular face operator, $\delta_k\colon \nabla \to \Delta_\sigma$, with $k \in \{0,\dots,\dim\Delta_\sigma\}$. Let j_0 denote the dimension of Δ_0. To determine the effect of δ_k on the operation $\sigma \mapsto \hat{\sigma}$, we must distinguish $k > j_0$ of $k \leqslant j_0$. For the sake of convenience, we set $\delta_s = \mathrm{id}$ if $s < 0$. From the construction of $\hat{\sigma}$, we have

$$\widehat{\sigma\circ\delta_k} = \begin{cases} \hat{\sigma}\circ\delta_{k-j_0-1} & \text{if } \ k > j_0, \\ \hat{\sigma} & \text{if } \ k \leqslant j_0, \end{cases}$$

which implies $\widehat{\sigma\circ\delta_k} = \hat{\sigma}\circ\delta_{k-j_0-1}$, with the previous convention. We conclude

$$\delta_k^* f(\omega)_\sigma = \delta_k^*\left(\lambda_{\mathtt{c}\Delta_0}\otimes\omega_{\hat{\sigma}}\right) = \begin{cases} \lambda_{\mathtt{c}\Delta_0}\otimes\delta_{k-j_0-1}^*\omega_{\hat{\sigma}} & \text{if } \ k > j_0, \\ \lambda_{\mathtt{c}\nabla_0}\otimes\omega_{\hat{\sigma}} & \text{if } \ k \leqslant j_0. \end{cases}$$

It follows $\delta_k^* f(\omega)_\sigma = \lambda_{\mathtt{c}\nabla_0}\otimes\omega_{\widehat{\sigma\circ\delta_k}} = f(\omega)_{\sigma\circ\delta_k}$.

Since the 0-cochain $\lambda_{\mathtt{c}\Delta_0}$ is a cocycle, the compatibility with the differentials is immediate from the equalities

$$\delta\left(f(\omega)_\sigma\right) = \delta\left(\lambda_{\mathtt{c}\Delta_0}\otimes\omega_{\hat{\sigma}}\right) = \lambda_{\mathtt{c}\Delta_0}\otimes\delta\omega_{\hat{\sigma}} = f(\delta\,\omega)_\sigma.$$

The map f being compatible with the differentials, it remains to show that the image by f of a $\overline{p}$-allowable cochain, $\omega \in \widetilde{N}^*(X\times]0,\infty[;R)$, is a $\overline{p}$-allowable cochain in $\widetilde{N}^*(\mathring{c}X;R)$. We choose ω of degree less than or equal to $\overline{p}(\mathtt{w})$ and refer to Definition 3.4 for the property of $\overline{p}$-allowability. For the stratum reduced to $\mathtt{w}$, the allowability comes directly from $\|f(\omega)_\sigma\|_{n+1} \leqslant |\omega_{\hat{\sigma}}| \leqslant \overline{p}(\mathtt{w})$. Now consider a singular stratum S of X and a regular simplex $\sigma\colon \Delta_\sigma = \Delta_0 * \cdots * \Delta_{n+1} \to \mathring{c}X$, such that $\sigma(\Delta_\sigma)\cap(S\times]0,\infty[) \neq \emptyset$. Let $\ell = \operatorname{codim}_{X\times]0,\infty[}(S\times]0,\infty[)$ and notice the equivalence of the conditions $\sigma(\Delta_\sigma)\cap(S\times]0,\infty[) \neq \emptyset$ and $\hat{\sigma}(\Delta_{\hat{\sigma}})\cap(S\times]0,\infty[) \neq \emptyset$. For such stratum, we have $\ell \in \{1,\dots,n\}$ and $\|f(\omega)_\sigma\|_\ell = \|\lambda_{\mathtt{c}\Delta_0}\otimes\omega_{\hat{\sigma}}\|_\ell = \|\omega_{\hat{\sigma}}\|_\ell$. The result is a consequence of the inequality $\|\omega_{\hat{\sigma}}\|_\ell \leqslant \|\omega\|_{S\times]0,\infty[} \leqslant \overline{p}(S\times]0,\infty[)$, arising from the $\overline{p}$-allowability of ω. □

Construction of $g\colon \widetilde{N}^*(\mathring{c}X;R) \to \widetilde{N}^*(X\times]0,\infty[;R)$.

Let $\omega \in \widetilde{N}^*(\mathring{c}X;R)$ and $\tau\colon \Delta_\tau \to X\times]0,\infty[$ a regular simplex. We denote by $\mathtt{c}\tau\colon \Delta_{c\tau} = \{\mathtt{p}\} * \Delta_\tau \to \mathring{c}X$ the cone over τ defined above. Notice $\widetilde{\Delta}_{c\tau} = \mathtt{c}\{\mathtt{p}\}\times\widetilde{\Delta}_\tau$.

Let $\mathtt{v}_0$ be the apex of the cone over the component of filtration degree 0 of a filtered simplex. The cone $\mathtt{c}\{\mathtt{p}\}$ having two vertices $\mathtt{p}$ and $\mathtt{v}_0$, the cochain $\omega_{\mathtt{c}\tau}$ decomposes into

$$\omega_{\mathtt{c}\tau} = \mathbf{1}_{\mathtt{p}}\otimes\gamma_{\mathtt{p}} + \mathbf{1}_{\mathtt{v}_0}\otimes\gamma_{\mathtt{v}_0} + \mathbf{1}_{\mathtt{p}*\mathtt{v}_0}\otimes\gamma'_{\mathtt{v}_0}, \tag{12.3}$$

with $\gamma_{\mathtt{p}}, \gamma_{\mathtt{v}_0}, \gamma'_{\mathtt{v}_0} \in \widetilde{N}^*(\Delta_\tau)$. We set

$$g(\omega)_\tau = \gamma_{\mathtt{v}_0}.$$

PROPOSITION 12.2. *Let $(\mathring{c}X,\overline{p})$ be a perverse space with X compact and let $(X\times]0,\infty[,\overline{p})$ be the induced perverse space. The correspondence defined above induces a cochain map,*

$$g\colon \tau_{\leqslant\overline{p}(\mathtt{w})}\widetilde{N}^*_{\overline{p}}(\mathring{c}X;R) \to \tau_{\leqslant\overline{p}(\mathtt{w})}\widetilde{N}^*_{\overline{p}}(X\times]0,\infty[;R).$$

The proof follows the pattern of that of Proposition 12.1; we leave it to the reader.

Specify the compositions $f \circ g$ and $g \circ f$.

(a) Let $\omega \in \widetilde{N}^*(X;R)$. Consider a regular simplex, $\tau\colon \Delta_\tau \to X\times]0,\infty[$ and its associated map $\mathtt{c}\tau\colon \{\mathtt{p}\} * \Delta_\tau \to \mathring{\mathtt{c}}X$. Following (12.2), one has

$$f(\omega)_{\mathtt{c}\tau} = \lambda_{\mathtt{c}\{\mathtt{p}\}} \otimes \omega_\tau = \mathbf{1}_{\mathtt{p}} \otimes \omega_\tau + \mathbf{1}_{\mathtt{v}_0} \otimes \omega_\tau.$$

It follows, according to (12.3), $g(f(\omega))_\tau = \omega_\tau$ and $g \circ f = \mathrm{id}$.

(b) Let $\omega \in \widetilde{N}^*(\mathring{\mathtt{c}}X;R)$. Consider a regular simplex $\sigma\colon \Delta_\sigma = \Delta_0 * \cdots * \Delta_{n+1} \to \mathring{\mathtt{c}}X$, and its associated map $\mathtt{c}\sigma\colon (\{\mathtt{p}\} * \Delta_0) * \cdots * \Delta_{n+1} \to \mathring{\mathtt{c}}X$. The cochain $\omega_{\mathtt{c}\sigma}$ decomposes into

$$\omega_{\mathtt{c}\sigma} = \sum_{F \lhd \mathtt{c}\Delta_0} \mathbf{1}_F \otimes \gamma_F + \sum_{F \lhd \mathtt{c}\Delta_0} \mathbf{1}_{\mathtt{p}*F} \otimes \gamma'_F + \mathbf{1}_{\mathtt{p}} \otimes \gamma_\varnothing. \tag{12.4}$$

Since the cochain ω is globally defined and the simplex $\mathtt{c}\hat{\sigma}$ is a face of $\mathtt{c}\sigma$, we deduce $\omega_{\mathtt{c}\hat{\sigma}} = \mathbf{1}_{\mathtt{v}_0} \otimes \gamma_{\mathtt{v}_0} + \mathbf{1}_{\mathtt{p}*\mathtt{v}_0} \otimes \gamma'_{\mathtt{v}_0} + \mathbf{1}_{\mathtt{p}} \otimes \gamma_\varnothing$. It follows:

$$f(g(\omega))_\sigma = \lambda_{\mathtt{c}\Delta_0} \otimes g(\omega)_{\hat{\sigma}} = \lambda_{\mathtt{c}\Delta_0} \otimes \gamma_{\mathtt{v}_0}. \tag{12.5}$$

Construction of a homotopy $H\colon \widetilde{N}^*(\mathring{\mathtt{c}}X;R) \to \widetilde{N}^{*-1}(\mathring{\mathtt{c}}X;R)$.

If $\sigma\colon \Delta_\sigma = \Delta_0 * \cdots * \Delta_{n+1} \to \mathring{\mathtt{c}}X$ is a regular simplex, we define a map $H\colon \widetilde{N}^*(\Delta_{\mathtt{c}\sigma}) \to \widetilde{N}^{*-1}(\Delta_\sigma)$, i.e.,

$$H\colon N^*(\mathtt{c}(\mathtt{p} * \Delta_0)) \otimes N^*(\mathtt{c}\Delta_1) \otimes \cdots \otimes N^*(\Delta_{n+1}) \to N^{*-1}(\mathtt{c}\Delta_0) \otimes \cdots \otimes N^*(\Delta_{n+1}).$$

We decompose $\omega_{\mathtt{c}\sigma} \in \widetilde{N}^*(\Delta_{\mathtt{c}\sigma})$ as in the formula (12.4) and set:

$$(H(\omega))_\sigma = \sum_{\mathtt{v}_0 \neq F \lhd \mathtt{c}\Delta_0} (-1)^{|F|+1} \mathbf{1}_F \otimes \gamma'_F + \lambda_{\Delta_0} \otimes \gamma'_{\mathtt{v}_0}, \tag{12.6}$$

where $\mathtt{v}_0$ is the apex of the cone over the component filtration of degree 0 and λ_{Δ_0} the sum of 0-cochains on Δ_0.

PROPOSITION 12.3. *Let $(\mathring{\mathtt{c}}X, \overline{p})$ be a perverse space with X compact.*

(a) *The equality (12.6) induces a linear map, $H\colon \widetilde{N}^*(\mathring{\mathtt{c}}X;R) \to \widetilde{N}^{*-1}(\mathring{\mathtt{c}}X;R)$, verifying*

$$\delta \circ H + H \circ \delta = \mathrm{id} - f \circ g.$$

(b) *Using the notation introduced in (12.1), the application H induces a map,*

$$H\colon \tau_{\leqslant \overline{p}(\mathtt{w})} \widetilde{N}^*_{\overline{p}}(\mathring{\mathtt{c}}X;R) \to \tau_{\leqslant \overline{p}(\mathtt{w})} \widetilde{N}^{*-1}_{\overline{p}}(\mathring{\mathtt{c}}X;R).$$

PROOF. (a) We must establish the equality $\delta_k^* H(\omega)_\sigma = H(\omega)_{\sigma \circ \delta_k}$, for each cochain $\omega \in \widetilde{N}^*(\mathring{\mathtt{c}}X)$, each regular simplex $\sigma\colon \Delta_\sigma = \Delta_0 * \cdots * \Delta_{n+1} \to \mathring{\mathtt{c}}X$ and each regular face operator $\delta_k\colon D = D_0 * \cdots * D_{n+1} \to \Delta_\sigma$, with $k \in \{0, \ldots, \dim \Delta_\sigma\}$. Denoting by $\delta_*^{\mathtt{p}}$ the regular face operators of $\{\mathtt{p}\} * \Delta_\sigma$, we can write $\mathtt{c}(\sigma \circ \delta_k) = \mathtt{c}\sigma \circ \delta_{k+1}^{\mathtt{p}}$. If $k > \dim \Delta_0$, we set $k^\circ = k - \dim \Delta_0 - 1$ and $\delta_{k^\circ}^\circ\colon D_1 * \cdots * D_{n+1} \to \Delta_1 * \cdots * \Delta_{n+1}$ the induced face. Following (12.6), we have:

$$\delta_k^* H(\omega)_\sigma = \begin{cases} \sum_{\mathtt{v}_0 \neq F \lhd \mathtt{c}D_0} (-1)^{|F|+1} \mathbf{1}_F \otimes \gamma'_F + \lambda_{D_0} \otimes \gamma'_{\mathtt{v}_0} & \text{if } k \leqslant \dim \Delta_0, \\ \sum_{\mathtt{v}_0 \neq F \lhd \mathtt{c}\Delta_0} (-1)^{|F|+1} \mathbf{1}_F \otimes \delta_{k^\circ}^{\circ,*} \gamma'_F + \lambda_{\Delta_0} \otimes \delta_{k^\circ}^{\circ,*} \gamma'_{\mathtt{v}_0} & \text{if } k > \dim \Delta_0. \end{cases}$$

Using the equality $\omega_{\mathtt{c}(\sigma\circ\delta_k)} = \omega_{\mathtt{c}\sigma\circ\delta^{\mathtt{p}}_{k+1}} = \delta^{\mathtt{p},*}_{k+1}\omega_{\mathtt{c}\sigma}$ and (12.4), we get:

$$\omega_{\mathtt{c}(\sigma\circ\delta_k)} = \begin{cases} (\sum_{F\lhd \mathtt{c}D_0} \mathbf{1}_F\otimes\gamma_F + \mathbf{1}_{\mathtt{p}*F}\otimes\gamma'_F) + \mathbf{1}_{\mathtt{p}}\otimes\gamma_\emptyset & \text{if } k\leqslant \dim\Delta_0,\\ (\sum_{F\lhd \mathtt{c}D_0} \mathbf{1}_F\otimes\delta^{\circ,*}_{k^\circ}\gamma_F + \mathbf{1}_{\mathtt{p}*F}\otimes\delta^{\circ,*}_{k^\circ}\gamma'_F) + \mathbf{1}_{\mathtt{p}}\otimes\delta^{\circ,*}_{k^\circ}\gamma_\emptyset & \text{if } k> \dim\Delta_0,\end{cases}$$

which gives

$$\begin{aligned} H(\omega)_{\sigma\circ\delta_k} &= \begin{cases} (\sum_{\mathtt{v}_0\neq F\lhd \mathtt{c}D_0} (-1)^{|F|+1}\mathbf{1}_F\otimes\gamma'_F) + \lambda_{D_0}\otimes\gamma'_{\mathtt{v}_0}, & \text{if } k\leqslant \dim\Delta_0,\\ (\sum_{\mathtt{v}_0\neq F\lhd \mathtt{c}D_0} (-1)^{|F|+1}\mathbf{1}_F\otimes\delta^{\circ,*}_{k^\circ}\gamma'_F) + \lambda_{D_0}\otimes\delta^{\circ,*}_{k^\circ}\gamma'_{\mathtt{v}_0} & \text{if } k> \dim\Delta_0,\end{cases}\\ &= \delta_k^* H(\omega)_\sigma. \end{aligned}$$

Let us study the behavior of H towards the differentials. Let $\sigma\colon \Delta_\sigma = \Delta_0 * \dots * \Delta_{n+1} \to \mathring{\mathtt{c}}X$ be a regular simplex and $\omega\in\widetilde{N}^*(\mathring{\mathtt{c}}X;R)$. Let us start from the equality (12.4) and calculate the differential in eliminating terms having a zero image by H.

$$(12.7)\ (H(\delta\omega))_\sigma = H\left(\sum_{e\in\mathcal{V}(\mathtt{c}\Delta_0)} \mathbf{1}_{p*e}\otimes\gamma_\emptyset + \sum_{F\lhd\mathtt{c}\Delta_0}\mathbf{1}_{F*\mathtt{p}}\otimes\gamma_F + \sum_{\substack{F\lhd\mathtt{c}\Delta_0\\ e\in\mathcal{V}(\mathtt{c}\Delta_0)}} \mathbf{1}_{\mathtt{p}*F*e}\otimes\gamma'_F + \sum_{F\lhd\mathtt{c}\Delta_0}(-1)^{|F|+1}\mathbf{1}_{\mathtt{p}*F}\otimes\delta\gamma'_F\right).$$

Notice

$$\sum_{e\in\mathcal{V}(\mathtt{c}\Delta_0)} \mathbf{1}_{p*e}\otimes\gamma_\emptyset + \sum_{F\lhd\mathtt{c}\Delta_0}\mathbf{1}_{F*\mathtt{p}}\otimes\gamma_F = \sum_{e\in\mathcal{V}(\Delta_0)} \mathbf{1}_{p*e}\otimes\gamma_\emptyset + \sum_{\mathtt{v}_0\neq F\lhd\mathtt{c}\Delta_0}\mathbf{1}_{F*\mathtt{p}}\otimes\gamma_F + \mathbf{1}_{\mathtt{p}*\mathtt{v}_0}\otimes(\gamma_\emptyset-\gamma_{\mathtt{v}_0}).$$

Replacing in (12.7) and developing the definition of H, we get:

$$\begin{aligned}(H(\delta\omega))_\sigma &= -\sum_{e\in\mathcal{V}(\Delta_0)}\mathbf{1}_e\otimes\gamma_\emptyset + \sum_{\mathtt{v}_0\neq F\lhd\mathtt{c}\Delta_0}\mathbf{1}_F\otimes\gamma_F + \lambda_{\Delta_0}\otimes(\gamma_\emptyset-\gamma_{\mathtt{v}_0}-\delta\gamma'_{\mathtt{v}_0})\\ &+ \sum_{\substack{F\lhd\mathtt{c}\Delta_0\\ e\in\mathcal{V}(\mathtt{c}\Delta_0)}}(-1)^{|F|}\mathbf{1}_{F*e}\otimes\gamma'_F + \sum_{\mathtt{v}_0\neq F\lhd\mathtt{c}\Delta_0}\mathbf{1}_F\otimes\delta\gamma'_F.\end{aligned}$$

On the other hand, the quantity $\delta\circ H$ can be written

$$\begin{aligned}\delta(H(\omega))_\sigma &= \sum_{\substack{\mathtt{v}_0\neq F\lhd\mathtt{c}\Delta_0\\ e\in\mathcal{V}(\mathtt{c}\Delta_0)}}(-1)^{|F|+1}\mathbf{1}_{F*e}\otimes\gamma'_F - \sum_{\mathtt{v}_0\neq F\lhd\mathtt{c}\Delta_0}\mathbf{1}_F\otimes\delta\gamma'_F\\ &+ \sum_{e\in\mathcal{V}(\Delta_0)}\mathbf{1}_{e*\mathtt{v}_0}\otimes\gamma'_{\mathtt{v}_0} + \lambda_{\Delta_0}\otimes\delta\gamma'_{\mathtt{v}_0}.\end{aligned}$$

Using (12.5), the sum of the two expressions can be reduced to:

$$(12.8)\qquad H(\delta\omega)_\sigma + \delta H(\omega)_\sigma = \sum_{F\lhd\mathtt{c}\Delta_0}\mathbf{1}_F\otimes\gamma_F - \lambda_{\mathtt{c}\Delta_0}\otimes\gamma_{\mathtt{v}_0} = \omega_\sigma - (f\circ g)(\omega)_\sigma.$$

(b) As in the proof of Proposition 12.1, we are reduced to consider a singular stratum T of $\mathring{\mathtt{c}}X$, a $\overline{p}$-allowable cochain $\omega\in\widetilde{N}^*(\mathring{\mathtt{c}}X)$, of degree less than or equal to $\overline{p}(\mathtt{w})$, and a regular simplex, $\sigma\colon \Delta_\sigma \to \mathring{\mathrm{c}}X$, with $\sigma(\Delta_\sigma)\cap T\neq\emptyset$. Let $\ell = \operatorname{codim}_{\mathring{\mathrm{c}}X} T \in \{1,\dots,n+1\}$. Notice that $\mathtt{c}\sigma(\Delta_{\mathtt{c}\sigma})\cap T\neq\emptyset$. Thus, according to the definition of perverse degree (cf. Definition 2.5), we have

$$\begin{aligned}\overline{p}(T) &\geqslant \|\omega\|_T \geqslant \|\omega_{\mathtt{c}\sigma}\|_\ell = \max\{\|\mathbf{1}_F\otimes\gamma_F\|_\ell,\ \|\mathbf{1}_{\mathtt{p}*F}\otimes\gamma'_F\|_\ell,\ \|\mathbf{1}_{\mathtt{p}}\otimes\gamma_\emptyset\|_\ell \mid F\triangleleft \mathtt{c}\Delta_0\}\\ &\geqslant \max\{\|\mathbf{1}_{\mathtt{p}*F}\otimes\gamma'_F\|_\ell \mid F\triangleleft\mathtt{c}\Delta_0\},\end{aligned}$$

where the equality uses (2.5) and (12.4). We develop this expression by distinguishing two cases.

- Let $\ell\neq n+1$. By definition, for any face $F\triangleleft\mathring{\mathrm{c}}\Delta_0$, we have the equality $\|\mathbf{1}_{\mathtt{p}*F}\otimes\gamma'_F\|_\ell = \|\mathbf{1}_F\otimes\gamma'_F\|_\ell$, if $F\neq \mathtt{v}_0$, and $\|\mathbf{1}_{\mathtt{p}*\mathtt{v}_0}\otimes\gamma'_{\mathtt{v}_0}\|_\ell = \|\lambda_{\Delta_0}\otimes\gamma'_{\mathtt{v}_0}\|_\ell$. It follows:
$$\overline{p}(T)\geqslant \max\{\|\mathbf{1}_F\otimes\gamma'_F\|_\ell,\ \|\lambda_{\Delta_0}\otimes\gamma'_{\mathtt{v}_0}\|_\ell \mid \mathtt{v}_0\neq F\triangleleft\mathtt{c}\Delta_0\} = \|H(\omega)_\sigma\|_\ell,$$
where the last equality uses (2.5) and (12.6). We deduce $\overline{p}(T)\geqslant\|H(\omega)\|_T$.
- Let $\ell = n+1$. In this case we have
$$\begin{aligned}\|H(\omega)_\sigma\|_{n+1} &= \max\{\|\mathbf{1}_{\mathtt{p}*F}\otimes\gamma'_F\|_{n+1},\ \|\lambda_{\Delta_0}\otimes\gamma'_{\mathtt{v}_0}\|_{n+1} \mid \mathtt{v}_0\neq F\triangleleft\mathtt{c}\Delta_0\}\\ &= \max\{|\gamma'_F|,\ |\gamma'_{\mathtt{v}_0}| \mid F\triangleleft\Delta_0\}\leqslant |\omega|-1\leqslant\overline{p}(\mathtt{w}),\end{aligned}$$
where the first equality uses (2.5) and (12.6). It follows $\overline{p}(\mathtt{w})\geqslant\|H(\omega)\|_{\mathtt{w}}$.

It has been shown $\|H(\omega)\|\leqslant\overline{p}$. The property $\|\delta H(\omega)\|\leqslant\overline{p}$ is deduced using (12.8). □

The determination of the cohomology of a cone follows from the properties of f, g and H.

Proof of Theorem E. (a) From Propositions 12.1, 12.2 and 12.3, we deduce that the map $g\colon \tau_{\leqslant\overline{p}(\mathtt{w})}\widetilde{N}^*_{\overline{p}}(\mathring{\mathrm{c}}X;R)\to\tau_{\leqslant\overline{p}(\mathtt{w})}\widetilde{N}^*_{\overline{p}}(X\times]0,\infty[;R)$ is a quasi-isomorphism. We know from Theorem D, that the inclusion $I_1\colon X\to X\times]0,\infty[$ induces a quasi-isomorphism. It remains to prove $I_1^*\circ g=\iota^*$. The involved stratifications on $X\times]0,\infty[$ are different in these two quasi-isomorphisms but the cohomologies are the same (see subsection 3.1).

For this, consider $\omega\in\tau_{\overline{p}(\mathtt{w})}\widetilde{N}^*_{\overline{p}}(\mathring{\mathrm{c}}X;R)$ and $\tau\colon\Delta_\tau\to X$ a regular simplex. By definition, we have $(\iota^*\omega)_\tau=\omega_{\iota\circ\tau}$. Notice $\iota\circ\tau=\mathtt{c}(I_1\circ\tau)\circ\delta_0$, where $\delta_0(x)=0\cdot\mathtt{p}+1\cdot x$. It then follows

$$(\iota^*\omega)_\tau=\omega_{\mathtt{c}(I_1\circ\tau)\circ\delta_0}=\delta_0^*\,\omega_{\mathtt{c}(I_1\circ\tau)}=_{(12.3)}\gamma_{\mathtt{v}_0}=g(\omega)_{I_1\circ\tau}=I_1^*(g(\omega))_\tau.$$

To prove part (b), we consider a cocycle $\omega\in\widetilde{N}^k_{\overline{p}}(\mathring{\mathrm{c}}X;R)$ with $k>\overline{p}(\mathtt{w})$. Following Proposition 12.3, we have $\delta H(\omega)=\omega-f(g(\omega))$ and it suffices to establish the equality $f(g(\omega))=0$.

We prove it by contradiction, assuming that there is a regular simplex $\sigma\colon\Delta_\sigma\to\mathring{\mathrm{c}}X$, such that $f(g(\omega))_\sigma\neq0$. Following (12.5), this implies $\gamma_{\mathtt{v}_0}\neq0$. We get a contradiction,

$$\begin{aligned}k>\overline{p}(\mathtt{w}) &\geqslant \|f(g(\omega))\|_{\mathtt{w}}\geqslant\|f(g(\omega))_{\mathtt{c}\sigma}\|_{n+1}=\|\lambda_{\mathtt{c}(\mathtt{p}*\Delta_0)}\otimes\gamma_{\mathtt{v}_0}\|_{n+1}\\ &= \|\lambda_{\mathtt{p}*\Delta_0}\otimes\gamma_{\mathtt{v}_0}\|_{n+1}=|\gamma_{\mathtt{v}_0}|=k.\end{aligned}$$

□

The proof of the topological independence Theorem of Section 14 is based on the following excision theorem.

12.2. Relative cohomology. Let X be a filtered space and $Y \subset X$ a subspace endowed with the induced stratification. Consider a perversity $\overline{p}$ on X. We also denote by $\overline{p}$ the induced perversity on Y. The *complex of relative $\overline{p}$-intersection cochains* is the direct sum $\widetilde{N}^*_{\overline{p}}(X,Y;R) = \widetilde{N}^*_{\overline{p}}(X;R) \oplus \widetilde{N}^{*-1}_{\overline{p}}(Y;R)$, endowed with the differential $D(\alpha,\beta) = (d\alpha, \alpha - d\beta)$. Its homology is the *relative blown-up $\overline{p}$-intersection cohomology of the perverse pair* $(X,Y,\overline{p})$, denoted by $\mathscr{H}^*_{\overline{p}}(X,Y;R)$.

By definition, we have a long exact sequence associated to the perverse pair $(X,Y,\overline{p})$:

$$\ldots \to \mathscr{H}^i_{\overline{p}}(X;R) \xrightarrow{\imath^*} \mathscr{H}^i_{\overline{p}}(Y;R) \to \mathscr{H}^{i+1}_{\overline{p}}(X,Y;R) \xrightarrow{\mathrm{pr}^*} \mathscr{H}^{i+1}_{\overline{p}}(X;R) \to \ldots, \tag{12.9}$$

where $\mathrm{pr}\colon \widetilde{N}^*_{\overline{p}}(X,Y;R) \to \widetilde{N}^*_{\overline{p}}(X;R)$ is defined by $\mathrm{pr}(\alpha,\beta) = \alpha$ and $\imath^*$ comes from the restriction map $\imath\colon \widetilde{N}^*_{\overline{p}}(X;R) \to \widetilde{N}^*_{\overline{p}}(Y;R)$.

PROPOSITION 12.4. *Let $(X,\overline{p})$ be a paracompact perverse space. If F is a closed subset of X and U an open subset of X with $F \subset U$, then the natural inclusion $(X\backslash F, U\backslash F) \hookrightarrow (X,U)$ induces the isomorphism*

$$\mathscr{H}^*_{\overline{p}}(X,U;R) \cong \mathscr{H}^*_{\overline{p}}(X\backslash F, U\backslash F;R).$$

PROOF. For the sake of simplicity we write the complexes $\mathcal{A}^* = \widetilde{N}^{*,\mathcal{U}}_{\overline{p}}(X;R)$, $\mathcal{B}^* = \widetilde{N}^{*,\mathcal{U}_1}_{\overline{p}}(X\backslash F;R)$, $\mathcal{C}^* = \widetilde{N}^{*,\mathcal{U}_2}_{\overline{p}}(U;R)$ and $\mathcal{E}^* = \widetilde{N}^*_{\overline{p}}(U\backslash F;R)$, the terms of the Mayer-Vietoris exact sequence of Theorem C, applied to the cover $\mathcal{U} = \{X\backslash F, U\}$. Associated to this sequence we have the exact sequence

$$0 \to (\mathcal{A}^* \oplus \mathcal{C}^{*-1}, D) \xrightarrow{f} (\mathcal{B}^* \oplus \mathcal{C}^* \oplus \mathcal{C}^{*-1} \oplus \mathcal{E}^{*-1}, D') \xrightarrow{g} (\mathcal{E}^* \oplus \mathcal{E}^{*-1}, D) \to 0,$$

where $D(a,c) = (da, a - dc)$, $f(a,c) = (a,a,c,0)$, $D'(b,c_0,c_1,e) = (db, dc_0, c_0 - dc_1, b - c_0 - de)$, $g(b,c_0,c_1,e) = (b - c_0, e)$ and $D(e,e') = (de, e - de')$. The right-hand complex is acyclic since $D(e,e') = 0 \Rightarrow (e,e') = D(e',0)$. The chain map $h\colon (\mathcal{B}^* \oplus \mathcal{E}^{*-1}, D) \to (\mathcal{B}^* \oplus \mathcal{C}^* \oplus \mathcal{C}^{*-1} \oplus \mathcal{E}^{*-1}, D')$, defined by $h(b,e) = (b,0,0,e)$, where $D(b,e) = (db, b - de)$, is a quasi-isomorphism, since

$$\begin{aligned} 0 = D'(b,c_0,c_1,e) \quad &\Rightarrow \quad (b,c_0,c_1,e) = h(b, e + c_1) + D'(0,c_1,0,0) \\ h(b',e') = D'(b,c_0,c_1,e) \quad &\Rightarrow \quad (b',e') = D(b, e + c_1). \end{aligned}$$

Let $k\colon (\mathcal{A}^* \oplus \mathcal{C}^{*-1}, D) \to (\mathcal{B}^* \oplus \mathcal{E}^{*-1}, D)$, defined by $k(a,c) = (a,c)$. If $D(a,c) = 0$, then $h(k(a,c)) - f(a,c) = (0,-a,-c,c) = D'(0,-c,0,0)$. Therefore, k is a quasi-isomorphism. Applying Theorem B and the definition of relative blown-up $\overline{p}$-intersection cohomology we get the result. □

The following result comes directly from the Mayer-Vietoris sequence and the long exact sequence associated to a pair.

PROPOSITION 12.5. *Let X be a compact filtered space. Let us consider the cone, $\mathring{c}X$, with apex $\mathtt{w}$, endowed with the cone filtration. Consider a perversity $\overline{p}$ on $\mathring{c}X$ and denote also by $\overline{p}$ the induced perversity on X. Then*

$$\mathscr{H}^j_{\overline{p}}(\mathring{c}X, \mathring{c}X\backslash\{\mathtt{w}\};R) = \begin{cases} \mathscr{H}^{j-1}_{\overline{p}}(X;R) & \text{if } j \geqslant \overline{p}(\mathtt{w}) + 2, \\ 0 & \text{if } j \leqslant \overline{p}(\mathtt{w}) + 1. \end{cases}$$

13. Comparison between intersection cohomologies. Theorem F.

If R is a field, the existence of a quasi-isomorphism between the cochain complexes, $C^*_{D\overline{p}}(X;R)$ and $\widetilde{N}^*_{\overline{p}}(X;R)$, has been established in [**4**, Theorem B] for GM-perversities in the framework of filtered face sets. We generalize this result using general perversities without the condition $D\overline{p} \leqslant \overline{t}$. In order to eliminate this last constraint we use tame intersection cohomology and homology instead of intersection cohomology and homology.

Recall that a perverse CS set $(X,\overline{p})$ is *locally* $(D\overline{p},R)$*-torsion free* if, for each singular stratum S and each $x \in S$ with link L, we have that $\mathfrak{T}^{D\overline{p}}_{\overline{p}(S)}(L;R) = 0$, i.e., the torsion R-submodule of $\mathfrak{H}^{D\overline{p}}_{\overline{p}(S)}(L;R)$, vanishes (see [**18**]).

THEOREM F. *Let $(X,\overline{p})$ be a paracompact separable perverse CS set and R a Dedekind ring. The blown-up and the tame intersection complexes, $\widetilde{N}^*_{\overline{p}}(X;R)$ and $\mathfrak{C}^*_{D\overline{p}}(X;R)$, are related by a quasi-isomorphism if one of the following hypotheses is satisfied.*

1) *The ring R is a field.*
2) *The space X is a locally $(D\overline{p},R)$-torsion free pseudomanifold.*

Under any of these assumptions, there exists an isomorphism

$$\mathscr{H}^*_{\overline{p}}(X;R) \cong \mathfrak{H}^*_{D\overline{p}}(X;R).$$

This isomorphism exists for the top perversity $\overline{t}$, for any CS set and any Dedekind ring (see [**4**, Remark 1.51]). This Theorem has a particular form in the classical setting of stratified pseudomanifolds and GM-perversities.

COROLLARY 13.1. *Let $\overline{p}$ be a GM-perversity, with* complementary perversity *$\overline{q}$, i.e., $\overline{q}(i) = i-2-\overline{p}(i)$ for $i \geqslant 2$. We consider a Dedekind ring R. Let X be a paracompact separable classical pseudomanifold. The blown-up and the intersection complexes, $\widetilde{N}^*_{\overline{p}}(X;R)$ and $C^*_{\overline{q}}(X;R)$, are related by a quasi-isomorphism if X is locally $(\overline{q},R)$-torsion free. So, we have an isomorphism $\mathscr{H}^*_{\overline{p}}(X;R) \cong H^*_{\overline{q}}(X;R)$.*

The proof of Theorem F, similar to that used in [**5**, Théorème A], meets the scheme used in [**21**, Theorem 10], [**24**, Lemma 1.4.1], [**14**, Section 5.1].

PROPOSITION 13.2. *Let $\mathcal{F}_X$ be the category whose objects are (stratified homeomorphic to) open subsets of a given paracompact and separable CS set X and whose morphisms are stratified homeomorphisms and inclusions. Let $\mathcal{A}b_*$ be the category of graded abelian groups. Let $F^*, G^*\colon \mathcal{F}_X \to \mathcal{A}b$ be two functors and $\Phi\colon F^* \to G^*$ a natural transformation satisfying the conditions listed below.*

(i) *F^* and G^* admit exact Mayer-Vietoris sequences and the natural transformation Φ induces a commutative diagram between these sequences,*
(ii) *If $\{U_\alpha\}$ is a disjoint collection of open subsets of X and $\Phi\colon F_*(U_\alpha) \to G_*(U_\alpha)$ is an isomorphism for each α, then $\Phi\colon F^*(\bigsqcup_\alpha U_\alpha) \to G^*(\bigsqcup_\alpha U_\alpha)$ is an isomorphism.*
(iii) *If L is a compact filtered space such that X has an open subset stratified homeomorphic to $\mathbb{R}^i \times \mathring{c}L$ and, if $\Phi\colon F^*(\mathbb{R}^i \times (\mathring{c}L\backslash\{\mathrm{v}\})) \to G^*(\mathbb{R}^i \times (\mathring{c}L\backslash\{\mathrm{v}\}))$ is an isomorphism, then so is $\Phi\colon F^*(\mathbb{R}^i \times \mathring{c}L) \to G^*(\mathbb{R}^i \times \mathring{c}L)$. Here, v is the apex of the cone $\mathring{c}L$.*

(iv) *If U is an open subset of X contained within a single stratum and homeomorphic to an Euclidean space, then $\Phi\colon F^*(U) \to G^*(U)$ is an isomorphism.*

Then $\Phi\colon F^*(X) \to G^*(X)$ is an isomorphism.

PROOF. Let U be an open subset of X. We apply Lemma 13.3, taking for $P(U)$ the property

$$\text{“}\Phi\colon F^*(U) \to G^*(U) \text{ is an isomorphism.”}$$

Notice that the CS set X is a separable, locally compact and metrizable space [**5**, Proposition 1.11]. We proceed by induction on the dimension of X, the result is immediate when $\dim X = 0$, that is, a discrete space. The inductive step occurs in two stages.

• *We first show $P(Y)$ for any open subset Y of a fixed conical chart.* We can suppose $Y = \mathbb{R}^m \times \mathring{c}L$. If $L = \emptyset$, we apply (iv). Let us suppose $L \neq \emptyset$. We consider the basis $\mathcal{V}$ of open sets of Y, composed of

- open subsets V of Y, with $\mathtt{v} \notin V$, which are CS sets of strictly lower dimension than X,
- open subsets $V = B \times \mathring{c}_\varepsilon L$, where $B \subset \mathbb{R}^m$ is an open m-cube, $\varepsilon > 0$ and $\mathring{c}_\epsilon = (L \times [0,\varepsilon[)/(L \times \{0\})$.

This family is closed by finite intersections and it verifies the assumptions of Lemma 13.3. For a), it suffices to apply the induction hypothesis together with (iii) and (iv). The property b) derives from (i) and property c) from (ii).

• *We prove* $P(X)$ by considering the open basis of X consisting of open subsets of conical charts. This family is stable by finite intersections and verifies the assumptions of Lemma 13.3. The condition a) is a consequence of the first step. The two other conditions derive from (i) and (ii), as previously. □

LEMMA 13.3. *Let X be a locally compact topological space, metrizable and separable. We are given an open basis of X, $\mathcal{U} = \{U_\alpha\}$, closed by finite intersections, and a statement $P(U)$ on open subsets of X satisfying the following three properties.*

a) *The property $P(U_\alpha)$ is true for all α.*
b) *If U, V are open subsets of X for which properties $P(U)$, $P(V)$ and $P(U \cap V)$ are true, then $P(U \cup V)$ is true.*
c) *If $(U_i)_{i\in I}$ is a family of open subsets of X, pairwise disjoint, verifying the property $P(U_i)$ for all $i \in I$, then $P(\bigsqcup_i U_i)$ is true.*

Then the property $P(X)$ is true.

PROOF. Since the family $\mathcal{U}$ is closed by finite intersections, the properties a) and b) imply that the property P is true for any finite union of elements of $\mathcal{U}$.

Since the space X is separable, metrizable and locally compact, then its Alexandroff's compactification $\hat{X} = X \sqcup \{\infty\}$ is metrizable (cf. [**26**, Exercise 23C]) and there exists a proper map, $f\colon X \to [0,\infty[$, defined by $f(x) = 1/d(\infty,x)$. We associate to f a countable family of compact subsets, $A_n = f^{-1}([n,n+1])$. Each A_n possesses a finite cover consisting of open subsets $U_\alpha \in \mathcal{U}$, included in $f^{-1}(]n-1/2,n+3/2[)$; we denote by U_n the union of the elements of this cover. Since the open subset U_n is a finite union of elements of $\mathcal{U}$, then the property $P(U_n)$ is true.

Let $U_{\text{even}} = \bigsqcup_n U_{2n}$ and let $U_{\text{odd}} = \bigsqcup_n U_{2n+1}$. The hypothesis c) implies that $P(U_{\text{even}})$ and $P(U_{\text{odd}})$ are true. Furthermore, since the intersection $U_{2n} \cap U_{2n+1}$ is a finite union of elements U_α of $\mathcal{U}$, then the property $P(U_{2n} \cap U_{2n+1})$ is true. From $U_{\text{even}} \cap U_{\text{odd}} = \bigsqcup_n U_{2n} \cap U_{2n+1}$ and from the hypothesis c), we deduce that the property $P(U_{\text{even}} \cup U_{\text{odd}})$ is true. The conclusion then follows from $U_{\text{even}} \cup U_{\text{odd}} = \bigcup_n U_n \supset \bigcup_n A_n = X$. □

Let X be a filtered space. We construct the map

$$\chi\colon \widetilde{N}^*(X;R) \to \mathfrak{C}^*(X;R)$$

as follows. If $\omega \in \widetilde{N}^*(X;R)$ and if $\sigma\colon \Delta = \Delta_0 * \cdots * \Delta_n \to X$ is a regular simplex, we set:

$$\chi(\omega)(\sigma) = \omega_\sigma(\widetilde{\Delta}).$$

Proposition 13.4. *Let $(X,\overline{p})$ be a perverse space. The operator χ induces a chain map, $\chi\colon \widetilde{N}^*_{\overline{p}}(X;R) \to \mathfrak{C}^*_{D\overline{p}}(X;R)$.*

Proof. Recall that $\mathfrak{C}^*_{D\overline{p}}(X;R) = \operatorname{Hom}(\mathfrak{C}^{D\overline{p}}_*(X;R),R)$. Since the simplices coming from $\mathfrak{C}^{D\overline{p}}_*(X;R)$ are regular, then the definition of χ makes sense. This is the reason of considering tame intersection chains instead of intersection chains.

We need to prove $\chi \circ \delta = \mathfrak{d}^* \circ \chi$, where $\mathfrak{d}^*$ is the linear dual of $\mathfrak{d}$ (see Definition 5.4). Consider a cochain ω of $\widetilde{N}^*_{\overline{p}}(X;R)$ and a chain $\xi \in \mathfrak{C}^{D\overline{p}}_*(X;R)$. We prove $\chi(\delta\omega)(\xi) = (\mathfrak{d}^*\chi)(\omega)(\xi) = -(-1)^{|\omega|}\chi(\omega)(\mathfrak{d}\xi)$. In fact, it suffices to prove $\chi(\delta\omega)(\sigma) = -(-1)^{|\omega|}\chi(\omega)(\mathfrak{d}\sigma)$ for a regular simplex $\sigma\colon \Delta \to X$ $D\overline{p}$-allowable. Let us suppose that:

$$\omega_\sigma(\mathcal{H}_i) = 0 \text{ for each hidden face } \mathcal{H}_i \tag{13.1}$$

(see (6.5)). Then

$$\chi(\delta\omega)(\sigma) = (\delta\omega)_\sigma(\widetilde{\Delta}) = -(-1)^{|\omega|}\,\omega_\sigma(\partial\widetilde{\Delta}) = -(-1)^{|\omega|}\,\omega_\sigma(\widetilde{\mathfrak{d}\Delta}) = -(-1)^{|\omega|}\,\chi(\omega)(\mathfrak{d}\sigma).$$

Let us prove the claim (13.1). We suppose $\omega_\sigma(\mathcal{H}_i) \neq 0$ for some hidden face $\mathcal{H}_i$. By definition of a hidden face, notice that $\Delta_{n-i} \neq \emptyset$, thus there exists an i-dimensional stratum S of X with $S \cap \operatorname{Im}\sigma \neq \emptyset$. Then, since σ is $D\overline{p}$-allowable and ω is $\overline{p}$-allowable,

- $0 \leqslant |\Delta|_{\leqslant n-i} = \|\sigma\|_S \leqslant \dim\sigma - \operatorname{codim} S + D\overline{p}(S) = \dim\sigma - \overline{p}(S) - 2$, and
- $0 \leqslant |\Delta|_{>n-i} = |\mathcal{H}_i|_{>n-i} = \|\mathbf{1}_{\mathcal{H}_i}\|_{\operatorname{codim} S} \leqslant_{(1)} \|\omega_\sigma\|_{\operatorname{codim} S} \leqslant \overline{p}(S)$,

where $\leqslant_{(1)}$ comes from $\omega_\sigma(\mathcal{H}_i) \neq 0$. We conclude that $\overline{p}(S)$ is finite. Adding up these two inequalities, we get $\dim\sigma - 1 = \dim\Delta - 1 = |\Delta|_{\leqslant n-i} + |\Delta|_{>n-i} \leqslant \|\omega_\sigma\|_S + \|\sigma\|_S \leqslant \dim\sigma - 2$. This contradiction gives the claim (13.1) and ends the proof. □

Proof of Theorem F. By hypothesis, X is a separable paracompact CS set. Properties (i), (ii), (iv) of Proposition 13.2 are verified. For the item (iii), we consider a filtered compact space L, for which the chain map of Proposition 13.4 induces an isomorphism $\chi^*\colon \mathscr{H}^*_{\overline{p}}(\mathbb{R}^i \times L \times]0,\infty[;R) \xrightarrow{\cong} \mathfrak{H}^*_{\overline{q}}(\mathbb{R}^i \times L \times]0,\infty[;R)$. Since the space $\mathbb{R}^i \times L\times]0,\infty[$ is an open subset of $\mathbb{R}^i \times \mathring{c}L$, the following diagram commutes.

$$\begin{array}{ccc} \mathscr{H}^*_{\overline{p}}(\mathbb{R}^i \times L\times]0,\infty[;R)) & \xrightarrow{\chi^*} & \mathfrak{H}^*_{D\overline{p}}(\mathbb{R}^i \times L\times]0,\infty[;R) \\ \uparrow & & \uparrow \\ \mathscr{H}^*_{\overline{p}}(\mathbb{R}^i \times \mathring{c}L;R) & \xrightarrow{\chi^*_{\mathbb{R}^i\times \mathring{c}L}} & \mathfrak{H}^*_{D\overline{p}}(\mathbb{R}^i \times \mathring{c}L;R)) \end{array} \tag{13.2}$$

Recall the blown-up intersection cohomology computation (see Theorem D and Theorem E):

$$\begin{aligned} \mathscr{H}^k_{\overline{p}}(\mathbb{R}^i \times \mathring{c}L;R) &= \mathscr{H}^k_{\overline{p}}(\mathring{c}L;R) \\ &= \begin{cases} \mathscr{H}^k_{\overline{p}}(L;R) = \mathscr{H}^k_{\overline{p}}(\mathbb{R}^i \times L\times]0,\infty[;R) & \text{if} \quad k \leqslant \overline{p}(\mathtt{w}), \\ 0 & \text{if} \quad k > \overline{p}(\mathtt{w}). \end{cases} \end{aligned}$$

The tame intersection cohomology is determined in [**14**, Chapter 7]:

$$\begin{aligned} \mathfrak{H}^k_{D\overline{p}}(\mathbb{R}^i \times \mathring{c}L;R) &= \mathfrak{H}^k_{D\overline{p}}(\mathring{c}L;R) \\ &= \begin{cases} \mathfrak{H}^k_{D\overline{p}}(L;R) = \mathfrak{H}^k_{D\overline{p}}(\mathbb{R}^i \times L\times]0,\infty[;R) & \text{if} \quad k \leqslant \overline{p}(\mathtt{w}), \\ \operatorname{Ext}(\mathfrak{H}^{D\overline{p}}_{k-1}(L;R),R) & \text{if} \quad k = \overline{p}(\mathtt{w})+1, \\ 0 & \text{if} \quad k > \overline{p}(\mathtt{w})+1. \end{cases} \end{aligned}$$

The involved stratifications on L are different in these computations but the cohomologies are the same (see subsections 3.1, 5.1).

In case 1) we have $\operatorname{Ext}(\mathfrak{H}^{D\overline{p}}_{\overline{p}(\mathtt{w})}(L;R),R) = 0$ since R is a field. In case 2) the link L is a compact pseudomanifold. According to [**14**, Section 5.7], the tame intersection homology of L is finitely generated. Since the torsion module of $\mathfrak{H}^{D\overline{p}}_{\overline{p}(\mathtt{w})}(L;R)$ is zero by hypothesis, we deduce $\operatorname{Ext}(\mathfrak{H}^{D\overline{p}}_{\overline{p}(\mathtt{w})}(L;R),R) = 0$.

So, in both cases, the vertical maps of (13.2) are isomorphisms. Since χ^* is an isomorphism then $\chi^*_{\mathbb{R}^i\times\mathring{c}L}$ is also an isomorphism. ☐

For some particular perversities the blown-up intersection cohomology is described in terms of the ordinary cohomology, denoted by $H^*(-;R)$.

PROPOSITION 13.5. *Let X be a paracompact separable CS set. Then*

(a) $\mathscr{H}^*_{\overline{0}}(X;R) \cong H^*(X;R)$ *if X is normal.*

(b) $\mathscr{H}^*_{\overline{p}}(X;R) \cong H^*(X,\Sigma;R)$ *if $\overline{p} < 0$.*

(c) $\mathscr{H}^*_{\overline{p}}(X;R) \cong H^*(X\backslash\Sigma;R)$ *if $\overline{p} > \overline{t}$.*

These isomorphisms preserve the cup product.

PROOF. The complex of singular chains (resp. relative chains) with coefficients in R, is denoted by $S_*(X;R)$ (resp.$S_*(X,\Sigma;R)$). It computes the ordinary homology $H_*(X;R)$ (resp. $H_*(X,\Sigma;R)$).

(a) Let $F\colon S^*(X;R) \to \widetilde{N}^*_{\overline{0}}(X;R)$ be the differential operator defined by $F(\omega)_\sigma = \mu^*(\sigma^*(\omega))$, for any regular filtered simplex $\sigma\colon \Delta \to X$. The operator μ^* is associated to the μ-amalgamation of Δ (see 8.1.1). Applying Proposition 7.5 repeatedly, we conclude that F is a chain map commuting with the cup product.

By definition of μ^* we have $||F(\omega)_\sigma||_\ell \leqslant ||\sigma^*(\omega)||_0 = 0$. The operator F is therefore well defined.

We consider the statement: "The operator F is a quasi-isomorphism" and we prove it by using Proposition 13.2. We verify the four properties (i)-(iv) of this Proposition.

(i) It is well known that the functor $S^*(-;R)$ verifies the Mayer-Vietoris property. It has been proven in Theorem C that the functor $\widetilde{N}^*_{\overline{0}}(-;R)$ also verifies Mayer-Vietoris property. One easily checks that F induces a commutative diagram between these sequences.

(ii),(iv) Straightforward.

(iii) It is well known that $H^*(\mathbb{R}^i \times \mathring{c}L;R) = H^0(\mathbb{R}^i \times \mathring{c}L;R) = R$ (constant cochains). From Theorems D, E, we have $\mathscr{H}^*_{\overline{0}}(\mathbb{R}^i \times \mathring{c}L;R) = \mathscr{H}^0_{\overline{0}}(L;R)$. Let us consider a basis point $x_0 \in L\backslash\Sigma$, which is connected from [**14**, Lemma 2.6.3]. Also, for any point $x_1 \in L$ there exists a regular simplex $\sigma\colon [0,1] \to L$ going from x_1 to x_0. If $\omega \in \widetilde{N}^0_{\overline{0}}(L;R)$ is a cocycle, the cochain ω_σ takes the same value on x_1 and on x_0. This implies $\mathscr{H}^0_{\overline{0}}(L;R) = R$.

(b) Consider the previous operator $F\colon S^*(X;R) \to \widetilde{N}^*(X;R)$ defined by $F(\omega)_\sigma = \mu^*(\sigma^*(\omega))$, for any regular filtered simplex $\sigma\colon \Delta \to X$. We know that it is a chain map commuting with the cup product. If ω vanishes on Σ then $\|F(\omega)_\sigma\|_\ell = -\infty$ for any $\ell \in \{1,\dots,n\}$. So, $F\colon S^*(X,\Sigma;R) \to \widetilde{N}^*_{\overline{p}}(X;R)$ is well defined. We consider the statement: "The operator F is a quasi-isomorphism" and we prove it as in the previous case. In fact, the only item to prove is (iii). It comes from $H^*(\mathbb{R}^i \times \mathring{c}L, \mathbb{R}^i \times \mathring{c}\Sigma_L;R) = 0 = \mathscr{H}^*_{\overline{p}}(\mathbb{R}^i \times \mathring{c}L;R)$, induced by Theorems D, E.

(c) Consider the natural restriction $\gamma\colon \widetilde{N}^*_{\overline{p}}(X;R) \to \widetilde{N}^*_{\overline{p}}(X\backslash\Sigma;R) = S^*(X\backslash\Sigma;R)$. It is a chain map commuting with the cup product. We consider the statement: "The operator γ is a quasi-isomorphim" we proceed as in the previous case. In fact, the only property to prove is the item (iii) of Proposition 13.2. We know that $H^*(\mathbb{R}^i \times \mathring{c}L\backslash\Sigma_{\mathbb{R}^i\times\mathring{c}L};R) = H^*(\mathbb{R}^i \times (L\backslash\Sigma_L)\times]0,\infty[;R) = H^*(L\backslash\Sigma_L;R) = H^*(\mathbb{R}^i \times (\mathring{c}L\backslash\{\mathrm{v}\})\backslash\Sigma_{\mathbb{R}^i\times(\mathring{c}L\backslash\{\mathrm{v}\})};R)$. On the other hand, we have $\mathscr{H}^*_{\overline{p}}(\mathbb{R}^i \times \mathring{c}L;R) = \mathscr{H}^*_{\overline{p}}(L;R) = \mathscr{H}^*_{\overline{p}}(\mathbb{R}^i \times (\mathring{c}L\backslash\{\mathrm{v}\});R)$ (cf. Theorems D, E). It suffices to apply the hypothesis of (iii). □

14. Topological invariance. Theorem G.

We prove in this section that the blown-up intersection cohomology is a topological invariant, working with CS sets and GM-perversities. We follow the procedure of King [**21**] (see also [**14**]).

14.1. Intrinsic filtration. A key ingredient in the King's proof of the topological invariance is the intrinsic filtration of a CS set X. It was introduced in [**21**], crediting to Sullivan, see also [**20**]. We refer the reader to [**14**] where there is an exhaustive study of this notion.

Let X be an n-dimensional filtered space. Two points $x, y \in X$ are *equivalent* if there exists a homeomorphism $h\colon (U,x) \to (V,y)$, where U, V are neighborhoods of x, y respectively. The local structure of a CS set implies that two points of the same stratum are necessarily equivalent. Hence the equivalence classes are unions of components of strata of X.

Let Y_i be the union of the equivalence classes which only contain components of strata of dimension $\leqslant i$. The filtration $Y_0 \subset Y_1 \subset \cdots \subset Y_n = X$ is the *intrinsic filtration* of X, denoted by X^*. It is proved that X^* is an n-dimensional CS set [**21**]. Since the filtration X^* coarsens the original filtration of X then the identity map $\nu \colon X \to X^*$ is a stratified map, called the *intrinsic aggregation* (see Definition 1.5). Recall that, given a perversity $\overline{p}$, the map ν induces a morphism $\nu^* \colon \mathscr{H}^*_{\overline{p}}(X^*;R) \to \mathscr{H}^*_{\overline{p}}(X;R)$ (see Theorem A).

PROPOSITION 14.1. *Let $\overline{p}$ be a GM-perversity. Let X be a paracompact CS set with no codimension one strata. Let S be a stratum of X and (U,φ) a conical chart of a point $x \in S$. If the intrinsic aggregation $\nu \colon X \to X^*$ induces the isomorphism $\nu_* \colon \mathscr{H}^*_{\overline{p}}((U\backslash S)^*;R) \xrightarrow{\cong} \mathscr{H}^*_{\overline{p}}(U\backslash S;R)$ then it also induces the isomorphism,*

$$\nu_* \colon \mathscr{H}^*_{\overline{p}}(U^*;R) \xrightarrow{\cong} \mathscr{H}^*_{\overline{p}}(U;R).$$

PROOF. We analyze the local structure of X and X^*. Without loss of generality, we can suppose $U = \mathbb{R}^k \times \mathring{c}W$, where W is a compact filtered space (possibly empty) and $S \cap U = \mathbb{R}^k \times \{\mathtt{w}\}$, $\mathtt{w}$ being the apex of the cone $\mathring{c}W$. Following [**21**, Lemma 2 and Proposition 1], there exists a homeomorphism

$$h \colon (\mathbb{R}^k \times \mathring{c}W)^* \xrightarrow{\cong} \mathbb{R}^m \times \mathring{c}L, \tag{14.1}$$

which is also a stratified map, where L is a compact filtered space (possibly empty) and $m \geqslant k$. Moreover, h verifies

$$h(\mathbb{R}^k \times \{\mathtt{w}\}) \subset \mathbb{R}^m \times \{\mathtt{v}\} \text{ and } h^{-1}(\mathbb{R}^m \times \{\mathtt{v}\}) = \mathbb{R}^k \times \mathring{c}A, \tag{14.2}$$

where A is an $(m-k-1)$ sphere and $\mathtt{v}$ is the apex of the cone $\mathring{c}L$. Now, the hypothesis and the conclusion of the Proposition become

$$h \colon \mathscr{H}^*_{\overline{p}}(\mathbb{R}^m \times \mathring{c}L\backslash h(\mathbb{R}^k \times \{\mathtt{w}\});R) \xrightarrow{\cong} \mathscr{H}^*_{\overline{p}}(\mathbb{R}^k \times \mathring{c}W\backslash(\mathbb{R}^k \times \{\mathtt{w}\});R) \tag{14.3}$$

and

$$h \colon \mathscr{H}^*_{\overline{p}}(\mathbb{R}^m \times \mathring{c}L;R) \xrightarrow{\cong} \mathscr{H}^*_{\overline{p}}(\mathbb{R}^k \times \mathring{c}W;R). \tag{14.4}$$

The existence of the homeomorphism (14.1) implies $k+s = m+t$, and $s \geqslant t$, since $m \geqslant k$, where $s = \dim W$ and $t = \dim L$.

The result is clear when $s=-1$. So, we can suppose $s \geqslant 0$ which implies that the stratum $\mathbb{R}^k \times \{\mathtt{w}\}$ is singular. Since X has no codimension one strata, we indeed have $s \geqslant 1$. When $t=-1$ then $L = \emptyset$, $s = \dim A$ and therefore

$$\begin{aligned}\mathscr{H}^*_{\overline{p}}(\mathbb{R}^k \times \mathring{c}W;R) &\overset{(14.2)}{\cong} \mathscr{H}^*_{\overline{p}}(\mathbb{R}^k \times \mathring{c}A;R) \cong_{(2)} \mathscr{H}^*_{\overline{p}}(\mathring{c}A;R) \cong_{(3)} \left\{\begin{array}{ll} R & \text{if } *=0 \\ 0 & \text{if not}\end{array}\right. \\ &\cong \mathscr{H}^*_{\overline{p}}(\mathbb{R}^m \times \mathring{c}L;R).\end{aligned}$$

Here, $\cong_{(2)}$ is Theorem D and $\cong_{(3)}$ comes from

- Theorem E, and
- $0 \leqslant \overline{p}(s+1) \leqslant \overline{t}(s+1) = s-1 < \dim A$ since $\overline{p}$ is a GM-perversity and $s \geqslant 1$.

So, we can suppose $t \geqslant 0$ which implies that the stratum $\mathbb{R}^m \times \{\mathtt{v}\}$ is singular.

We establish (14.4) in two different cases.

- *First case*: $i > \overline{p}(s+1)$.

Theorem E gives $\mathscr{H}^i_{\overline{p}}(\mathbb{R}^k \times \mathring{c}W;R) = 0$ and then we have to prove that $\mathscr{H}^i_{\overline{p}}(\mathbb{R}^m \times \mathring{c}L;R) = 0$. Since the stratum $\mathbb{R}^m \times \{\mathtt{v}\}$ is singular and $\overline{p}$ is a GM-perversity then $i > \overline{p}(s+1) \geqslant \overline{p}(t+1)$. Now, Theorem E applied to $\mathscr{H}^i_{\overline{p}}(\mathbb{R}^m \times \mathring{c}L;R)$ gives the result.

• *Second case* : $i \leqslant \overline{p}(s+1)$. We have the isomorphisms

$$(14.5)\ \mathscr{H}^i_{\overline{p}}(\mathbb{R}^m \times \mathring{c}L\backslash h(\mathbb{R}^k \times \{\mathtt{w}\});R) \cong_{(1)} \mathscr{H}^i_{\overline{p}}(\mathbb{R}^k \times \mathring{c}W\backslash(\mathbb{R}^k \times \{\mathtt{w}\});R) \cong \mathscr{H}^i_{\overline{p}}(\mathbb{R}^k \times (\mathring{c}W\backslash\{\mathtt{w}\});R) \cong \mathscr{H}^i_{\overline{p}}(\mathbb{R}^k \times]0,1[\times W;R) \cong_{(2)} \mathscr{H}^i_{\overline{p}}(W;R),$$

where $\cong_{(1)}$ is the hypothesis (14.3) and $\cong_{(2)}$ is Theorem D. Write $h(\mathbb{R}^k \times \{\mathtt{w}\}) = B \times \{\mathtt{v}\} \subset \mathbb{R}^m \times \{\mathtt{v}\}$, with B closed. We obtain a new sequence of isomorphisms

$$(14.6)\quad \begin{aligned} &\mathscr{H}^i_{\overline{p}}(\mathbb{R}^m \times \mathring{c}L\backslash h(\mathbb{R}^k \times \{\mathtt{w}\}), \mathbb{R}^m \times \mathring{c}L\backslash\mathbb{R}^m \times \{\mathtt{v}\};R) \cong \\ &\mathscr{H}^i_{\overline{p}}((\mathbb{R}^m \times \mathring{c}L)\backslash(B \times \{\mathtt{v}\}), \mathbb{R}^m \times (\mathring{c}L\backslash\{\mathtt{v}\});R) \cong_{(1)} \\ &\mathscr{H}^i_{\overline{p}}((\mathbb{R}^m\backslash B) \times \mathring{c}L, (\mathbb{R}^m\backslash B) \times (\mathring{c}L\backslash\{\mathtt{v}\});R) \cong \mathscr{H}^i_{\overline{p}}((\mathbb{R}^m\backslash B) \times (\mathring{c}L, \mathring{c}L\backslash\{\mathtt{v}\});R) \cong_{(2)} \\ &\mathscr{H}^i_{\overline{p}}(\mathbb{R}^{k+1} \times A \times (\mathring{c}L, \mathring{c}L\backslash\{\mathtt{v}\});R) \cong_{(3)} \\ &\mathscr{H}^i_{\overline{p}}(\mathring{c}L, \mathring{c}L\backslash\{\mathtt{v}\};R) \oplus \mathscr{H}^{i-m+1+k}_{\overline{p}}(\mathring{c}L, \mathring{c}L\backslash\{\mathtt{v}\};R). \end{aligned}$$

The isomorphism $\cong_{(1)}$ is the excision of $B \times (\mathring{c}L\backslash\{\mathtt{v}\})$ (see Proposition 12.4), $\cong_{(2)}$ comes from (14.2) and $\cong_{(3)}$ from Theorem D and Proposition 11.4. Notice that $\mathbb{R}^m \times \mathring{c}L$, $\mathring{c}L$ and $\mathring{c}L\backslash\{\mathtt{w}\}$ are F_σ-subsets, [**26**, Exercise 3H], of X^* and therefore paracompact spaces [**26**, Theorem 20.12 a),b)]. The hypothesis on i implies $i - m + 1 + k \leqslant \overline{p}(s+1) - s + 1 + t \leqslant \overline{p}(t+1) + 1$, since $\overline{p}$ is a GM-perversity. Then $\mathscr{H}^{i-m+1+k}_{\overline{p}}(\mathring{c}L, \mathring{c}L\backslash\{\mathtt{v}\};R)$ vanishes (see Proposition 12.5). We have proved that $\mathscr{H}^i_{\overline{p}}(\mathbb{R}^m \times \mathring{c}L\backslash h(\mathbb{R}^k \times \{\mathtt{w}\}), \mathbb{R}^m \times \mathring{c}L\backslash\mathbb{R}^m \times \{\mathtt{v}\};R) \cong \mathscr{H}^i_{\overline{p}}(\mathring{c}L, \mathring{c}L\backslash\{\mathtt{v}\};R)$. Finally, from this isomorphism, from the long exact sequence of a pair (12.9) and from (14.5), we get that $\mathscr{H}^i_{\overline{p}}(W;R) \cong \mathscr{H}^i_{\overline{p}}(\mathring{c}L;R)$. Applying Theorem E and Theorem D we have $\mathscr{H}^i_{\overline{p}}(\mathbb{R}^m \times \mathring{c}L;R) \cong \mathscr{H}^i_{\overline{p}}(\mathring{c}L;R) \cong \mathscr{H}^i_{\overline{p}}(W;R) \cong \mathscr{H}^i_{\overline{p}}(\mathring{c}W;R) \cong \mathscr{H}^i_{\overline{p}}(\mathbb{R}^k \times \mathring{c}W;R)$. □

THEOREM G. *Suppose X is a separable paracompact CS set with no codimension one strata. Let $\overline{p}$ be a GM-perversity. Then, the intrinsic aggregation $\nu\colon X \to X^*$ induces an isomorphism $\mathscr{H}^*_{\overline{p}}(X;R) \cong \mathscr{H}^*_{\overline{p}}(X^*;R)$. It follows that $\mathscr{H}^*_{\overline{p}}(X;R)$ is independent (up to isomorphism) of the choice of a stratification of X as CS set with no codimension one strata.*

In particular, if X' is another CS set with no codimension one strata which is homeomorphic to X (not necessarily stratified homeomorphic), then $\mathscr{H}^*_{\overline{p}}(X;R) \cong \mathscr{H}^*_{\overline{p}}(X';R)$. These $\overline{p}$-depending isomorphisms can be chosen to preserve the cup product.

PROOF. We consider the statement: "The intrinsic aggregation $\nu\colon X \to X^*$ induces a quasi-isomorphism" and we prove it by using Proposition 13.2. We verify the four conditions of this Proposition.

(i) The functor $\widetilde{N}^*_{\overline{p}}(-;R)$ satisfies Mayer-Vietoris property (see Theorem C). Notice that we know from [**26**, Theorem 20.12 b)] that X^* is paracompact.

(ii),(iv) Straightforward.

(iii) See Proposition 14.1.

Last statement comes from the fact that ν is a stratified map and Theorem A. □

15. Decomposition of the ordinary cap product.

In this last section, we show how the ordinary cap product factorizes through the cap product we have defined for the blown-up intersection cohomology. This result extends the factorization developed in [**7**, Theorem A].

PROPOSITION 15.1. *Let X be a normal separable paracompact CS set endowed with two perversities $\overline{p},\overline{q}$ verifying $\overline{0} \leqslant \overline{p}$ and $\overline{p}+\overline{q} \leqslant \overline{t}$. Then, there exists a commutative diagram*

$$\begin{array}{ccc} H^*(X;R)\otimes H_m(X;R) & \xrightarrow{\ \cap\ } & H_{m-*}(X;R) \\ \Big\downarrow \Phi \qquad \phi \Big\uparrow & & \Big\uparrow \phi \\ \mathscr{H}^*_{\overline{p}}(X;R)\otimes H^{\overline{q}}_m(X;R) & \xrightarrow{\ \cap\ } & H^{\overline{p}+\overline{p}}_{m-*}(X;R), \end{array}$$

that is, $\phi(\Phi(\alpha)\cap\xi) = \alpha\cap\phi(\xi)$, for each $\alpha \in H^*(X;R)$ and $\xi \in H^{\overline{q}}_m(X;R)$. The map Φ preserves the $\cup$-product. Moreover, if $(\overline{p},\overline{q}) = (\overline{0},\overline{t})$ then Φ and ϕ can be chosen isomorphisms.

PROOF. Consider the natural inclusions $\widetilde{N}^*_{\overline{0}}(X;R) \hookrightarrow \widetilde{N}^*_{\overline{p}}(X;R)$, $C^{\overline{q}}_*(X;R) \hookrightarrow C^{\overline{t}}_*(X;R)$ and $C^{\overline{p}+\overline{q}}_*(X;R) \hookrightarrow C^{\overline{t}}_*(X;R)$. So, it suffices to prove the statement for $(\overline{p},\overline{q}) = (\overline{0},\overline{t})$ (see Remark 6.8).

The complex of singular chains (resp. cochains), with coefficients on R, is denoted by $S_*(X;R)$ (resp. $S^*(X;R)$), it computes the ordinary homology $H_*(X;R)$ (resp. ordinary cohomology $H^*(X;R)$). We proceed in three steps.

- *Construction of ϕ.* We consider the inclusion $\iota\colon C^{\overline{t}}_*(X;R) \hookrightarrow S_*(X;R)$. We prove the statement :"The map ι is a quasi-isomorphim" by using [**14**, Theorem 5.1.4]. Let us verify the four conditions (1)-(4) of this Theorem.
  (1) It is well known that the functor $S_*(-;R)$ verifies the Mayer-Vietoris property. It has been proven in [**5**, Proposition 4.1] that the functor $C^{\overline{t}}_*(-;R)$ also verifies Mayer-Vietoris property.
  (2) Since the support of chains are compact.
  (3) Since

$$\begin{aligned} H_*(\mathbb{R}^i\times \mathring{c}L;R) &= H_0(\mathbb{R}^i\times \mathring{c}L;R) = R \overset{Normal}{=} H_0(L;R) \\ &= H_0(\mathbb{R}^i\times(\mathring{c}L\backslash\{\mathtt{v}\});R) \\ &\overset{Hyp.\ (3)}{=} H^{\overline{t}}_0(\mathbb{R}^i\times(\mathring{c}L\backslash\{\mathtt{v}\});R) \overset{[5,\ \text{Prop. } 5.4]}{=} H^{\overline{t}}_*(\mathbb{R}^i\times \mathring{c}L;R) \end{aligned}$$

  (4) Straightforward.

  The differential operator ϕ is the isomorphism ι_*.

- *Construction of Φ.* Consider the operator F constructed in the proof of Proposition 13.5 (a) and set $\Phi = F_*$.

- *Diagram commutativity.* This is a local question. We can consider a cochain $\omega \in \widetilde{N}^*(X)$ and a regular simplex $\sigma\colon \Delta \to X$. We have: $\phi(\Phi(\omega) \cap \sigma) = \sigma_* \mu_*((\Phi(\omega))_\sigma \,\tilde{\cap}\, \widetilde{\Delta}) = \sigma_* \mu_*(\mu^*(\sigma^* \omega) \,\tilde{\cap}\, \widetilde{\Delta}) \overset{Prop.7.5(2)}{=} \sigma_*(\sigma^* \omega \cap \Delta) = \omega \cap \sigma = \omega \cap \phi(\sigma)$.

□

The case $\overline{q} = \overline{0}$ corresponds to the decomposition of [**7**, Theorem A]. If we work with non positive perversities, we have a decomposition of the cap product as follows.

PROPOSITION 15.2. *Let X be a separable paracompact CS set endowed with two perversities $\overline{p} < \overline{0}$ and $\overline{q} < \overline{0}$. Then, there exists a commutative diagram*

$$\begin{array}{ccc} H^*(X,\Sigma;R) \otimes H_m(X\backslash\Sigma;R) & \xrightarrow{\cap} & H_{m-*}(X\backslash\Sigma;R) \\ \downarrow \Phi \qquad \uparrow \phi & & \uparrow \phi \\ \mathscr{H}^*_{\overline{p}}(X;R) \otimes H^{\overline{q}}_m(X;R) & \xrightarrow{\cap} & H^{\overline{p}+\overline{q}}_{m-*}(X;R), \end{array}$$

that is, $\phi(\Phi(\alpha) \cap \xi) = \alpha \cap \phi(\xi)$, for each $\alpha \in H^*(X,\Sigma;R)$ and $\xi \in H^{\overline{q}}_m(X;R)$, where Φ, ϕ are isomorphisms and the first one preserves the $\cup$-product .

PROOF. We proceed in three steps.

- *Construction of ϕ.* We prove that the inclusion $C^{\overline{p}}_*(X;R) \overset{\iota}{\hookleftarrow} S_*(X\backslash\Sigma, R)$ is a quasi-isomorphim. We proceed as in the previous Proposition. In fact, the only item to prove is the property (3) of [**14**, Theorem 5.1.4]. The regular part of $\mathbb{R}^i \times \mathring{c}L$ is $\mathbb{R}^i \times (L\backslash\Sigma_L) \times]0,\infty[$. We know that $H_*(\mathbb{R}^i \times (L\backslash\Sigma_L) \times]0,\infty[;R) = H_*(L\backslash\Sigma_L;R)$. On the other hand, we have $H^{\overline{p}}_*(\mathbb{R}^i \times \mathring{c}L;R) = H_*(L\backslash\Sigma_L;R)$ (cf. [**5**, Prop. 5.4]). Since both isomorphisms are induced by the canonical projection then we get (3). The map ϕ is the isomorphism ι_*.
- *Construction of Φ.* Consider the operator F constructed in the proof of Proposition 13.5 (b) and set $\Phi = F_*$.
- *Diagram commutativity.* As in the previous Proposition.

□

References

[1] M. Banagl, *Rational generalized intersection homology theories*, Homology Homotopy Appl. **12** (2010), no. 1, 157–185. MR2607414

[2] J.-P. Brasselet, G. Hector, and M. Saralegi, *Théorème de de Rham pour les variétés stratifiées* (French), Ann. Global Anal. Geom. **9** (1991), no. 3, 211–243. MR1143404

[3] G. E. Bredon, *Topology and geometry*, Graduate Texts in Mathematics, vol. 139, Springer-Verlag, New York, 1997. Corrected third printing of the 1993 original. MR1700700

[4] D. Chataur, M. Saralegi-Aranguren, and D. Tanré, *Intersection Cohomology. Simplicial blow-up and rational homotopy*, ArXiv Mathematics e-prints. no. 1205.7057 (2012), To appear in Mem. Amer. Math. Soc.

[5] D. Chataur, M. Saralegi-Aranguren, and D. Tanré, *Homologie d'intersection. Perversités générales et invariance topologique*, ArXiv Mathematics e-prints. no. 1602.03009 (2016).

[6] D. Chataur, M. Saralegi-Aranguren, and D. Tanré, *Poincaré duality with cap products in intersection homology*, Adv. Math. **326** (2018), 314–351, DOI 10.1016/j.aim.2017.12.017. MR3758431

[7] D. Chataur, M. Saralegi-Aranguren, and D. Tanré, *Singular decompositions of a cap product*, Proc. Amer. Math. Soc. **145** (2017), no. 8, 3645–3656. MR3652815

[8] D. Chataur, M. Saralegi-Aranguren, and D. Tanré, *Steenrod squares on intersection cohomology and a conjecture of M Goresky and W Pardon*, Algebr. Geom. Topol. **16** (2016), no. 4, 1851–1904. MR3546453

[9] G. Friedman, *Superperverse intersection cohomology: stratification (in)dependence*, Math. Z. **252** (2006), no. 1, 49–70. MR2209151

[10] G. Friedman, *Singular chain intersection homology for traditional and super-perversities*, Trans. Amer. Math. Soc. **359** (2007), no. 5, 1977–2019. MR2276609

[11] G. Friedman, *Intersection homology Künneth theorems*, Math. Ann. **343** (2009), no. 2, 371–395. MR2461258

[12] G. Friedman, *Intersection homology with general perversities*, Geom. Dedicata **148** (2010), 103–135. MR2721621

[13] G. Friedman, *An introduction to intersection homology with general perversity functions*, Topology of stratified spaces, Math. Sci. Res. Inst. Publ., vol. 58, Cambridge Univ. Press, Cambridge, 2011, pp. 177–222. MR2796412

[14] G. Friedman, *Singular intersection homology*, Available at http://faculty.tcu.edu/gfriedman/IHbook.pdf (2017).

[15] G. Friedman and J. E. McClure, *Cup and cap products in intersection (co)homology*, Adv. Math. **240** (2013), 383–426. MR3046315

[16] M. Goresky and R. MacPherson, *Intersection homology theory*, Topology **19** (1980), no. 2, 135–162. MR572580

[17] M. Goresky and R. MacPherson, *Intersection homology. II*, Invent. Math. **72** (1983), no. 1, 77–129. MR696691

[18] M. Goresky and P. Siegel, *Linking pairings on singular spaces*, Comment. Math. Helv. **58** (1983), no. 1, 96–110. MR699009

[19] S. Halperin, *Lectures on minimal models*, Mém. Soc. Math. France (N.S.) **9-10** (1983), 261. MR736299

[20] M. Handel, *A resolution of stratification conjectures concerning CS sets*, Topology **17** (1978), no. 2, 167–175. MR0478169

[21] H. C. King, *Topological invariance of intersection homology without sheaves*, Topology Appl. **20** (1985), no. 2, 149–160. MR800845

[22] R. MacPherson, *Intersection homology and perverse sheaves*, Unpublished AMS Colloquium Lectures, San Francisco (1991).

[23] M. Saralegi, *Homological properties of stratified spaces*, Illinois J. Math. **38** (1994), no. 1, 47–70. MR1245833

[24] M. Saralegi-Aranguren, *de Rham intersection cohomology for general perversities*, Illinois J. Math. **49** (2005), no. 3, 737–758. MR2210257

[25] L. C. Siebenmann, *Deformation of homeomorphisms on stratified sets. I, II*, Comment. Math. Helv. **47** (1972), 123–136; ibid. 47 (1972), 137–163. MR0319207

[26] S. Willard, *General topology*, Addison-Wesley Publishing Co., Reading, Mass.-London-Don Mills, Ont., 1970. MR0264581

LAFMA, UNIVERSITÉ DE PICARDIE JULES VERNE, 33, RUE SAINT-LEU, 80039 AMIENS CEDEX 1, FRANCE

Email address: David.Chataur@u-picardie.fr

LABORATOIRE DE MATHÉMATIQUES DE LENS, UNIVERSITÉ D'ARTOIS, EA 2462, SP18, RUE JEAN SOUVRAZ, 62307 LENS CEDEX, FRANCE

Email address: martin.saraleguiaranguren@univ-artois.fr

DÉPARTEMENT DE MATHÉMATIQUES, UNIVERSITÉ DE LILLE 1, UMR 8524, 59655 VILLENEUVE D'ASCQ CEDEX, FRANCE

Email address: Daniel.Tanre@univ-lille1.fr

Contemporary Mathematics
Volume **708**, 2018
http://dx.doi.org/10.1090/conm/708/14270

Homotopically rigid Sullivan algebras and their applications

Cristina Costoya, David Méndez, and Antonio Viruel

ABSTRACT. In this paper we construct an infinite family of homotopically rigid spaces. These examples are then used as building blocks to forge highly connected rational spaces with prescribed finite group of self-homotopy equivalences. They are also exploited to provide highly connected inflexible and strongly chiral manifolds.

INTRODUCTION

The group of self-homotopy equivalences of a space X, $\mathcal{E}(X)$, is rarely trivial. An obvious example is $X = K(\mathbb{Z}/2, n)$ since $\mathcal{E}(K(\mathbb{Z}/2, n)) \cong \mathrm{Aut}(\mathbb{Z}/2) \cong *$. The first elaborated example with non trivial rational homology is constructed in [**17**] by D. Kahn. He expressed in [**18**] the belief that spaces with trivial group of self-homotopy equivalences, named homotopically rigid spaces, might play a role in some way of decomposing a space. Thus, the problem of determining X for which $\mathcal{E}(X)$ is trivial becomes of interest, [**18**, Problem 3].

A decade after, Arkowitz and Lupton came across an example of a minimal Sullivan algebra (equivalently, a rational homotopy type of a space) M with $\mathcal{E}(M)$ trivial, [**3**, Example 5.1]. This, let us say homotopically rigid algebra, M provides us with motivation and guidance: in our paper [**5**], we use M as a building block to construct minimal Sullivan algebras, [**5**, Definition 2.1], with the property that their group of self-homotopy equivalences are isomorphic to an arbitrary finite group G previously fixed, [**5**, Theorem 1.1]. In other words, every finite group G is realizable through minimal Sullivan algebras that are built upon M; this is a partial answer to Kahn's realizability problem and we refer to [**5**] for more details. Observe that as an application, for the trivial group $G \cong \{e\}$, M is then a building block for an infinite family of homotopically rigid minimal Sullivan algebras (equivalently,

2010 *Mathematics Subject Classification.* Primary 55P10, 55P62; Secondary 57N65.

Key words and phrases. Self-homotopy equivalences, rational homotopy theory, inflexible manifolds, Poincaré duality.

The first author was partially supported by Ministerio de Economía y Competitividad (Spain), grant MTM2016-79661-P, and by Xunta de Galicia grant EM2013/16.

The second author was partially supported by Ministerio de Educación, Cultura y Deporte grant FPU14/05137, by FEDER-MEC grants MTM2013-41768-P and MTM2016-78647-P, by Ministerio de Economía y Competitividad (Spain) grant MTM2016-79661-P, and by Xunta de Galicia grant EM2013/16.

The third author was partially supported by FEDER-MEC grants MTM2013-41768-P and MTM2016-78647-P, and by Xunta de Galicia grant EM2013/16.

homotopically rigid rational spaces). However, as interesting as it may seem, all the examples of the type above just mentioned are 7-connected since M is 7-connected itself.

This paper came to light in our attempt to: (i) decide how rare homotopically rigid spaces are, since very few examples are known in literature. In that sense, we construct an infinite family of such spaces as highly connected as wanted; and (ii) show that building blocks, other than Arkowitz-Lupton example M, can be chosen in the resolution of Kahn's realizability problem in such a way that we are able to enlarge the family of examples in literature, all of them with low connectivity.

The paper is organized as follows. First, in Section 1 we construct an infinite family of highly connected minimal Sullivan algebras (see Definition 1.1) and we prove that they are homotopically rigid (see Theorem 1.6). Second, in Section 2 we show that algebras from Definition 1.1 serve as building blocks for new highly connected minimal Sullivan algebras (see Definition 2.1) with the property that their group of self-homotopy equivalences are isomorphic to any finite group G that we fix (see Theorem 2.9 and Theorem 2.10). In other words, we are realizing the group G through minimal Sullivan algebras that can be chosen as highly connected as desired. Section 3 consists of applications of the previous sections to differential geometry: we enlarge the class of inflexible manifolds existing in literature (see Theorem 3.1), and create new strongly chiral manifolds (see Corollary 3.3, and Corollary 3.4). We end this paper with an Appendix by Pascal Lambrechts and Don Stanley that allows us to guarantee the existence of highly connected inflexible manifolds.

Acknowledgments. The authors would like to thank the referee for carefully reading our manuscript and for giving such valuable comments which helped improving the quality of the paper.

1. Highly connected homotopically rigid CDGA

In this section an infinite family of highly connected homotopically rigid minimal Sullivan algebras is constructed. They extend the family of examples in literature, all of them of low connectivity.

We refer to [**9**] for basic facts in rational homotopy theory. Only simply connected $\mathbb{Q}$-algebras of finite type are considered here. If W is a graded rational vector space, we write ΛW for the free commutative graded algebra on W. This is a symmetric algebra on W^{even} tensored with an exterior algebra on W^{odd}. A Sullivan algebra is a nilpotent commutative differential graded algebra (CDGA) which is free as commutative graded algebra on a simply connected graded vector space W of finite dimension in each degree. It is minimal if in addition $d(W) \subset \Lambda^{\geq 2}W$. A Sullivan algebra is pure if $d = 0$ on W^{even} and $d(W^{\text{odd}}) \subset \Lambda(W^{\text{even}})$. We recall that the geometric realization functor of Sullivan [**21**] gives the equivalence of categories between the homotopy category of minimal Sullivan algebras, and the homotopy category of rational simply connected spaces of finite type. Henceforward, by abuse of language, we sometimes refer to minimal Sullivan algebras as rational spaces.

Recall that the homotopy rigid algebra constructed by Arkowitz and Lupton [**3**, Example 5.1] is

$$M = \big(\Lambda(x_1, x_2, y_1, y_2, y_3, z), d\big)$$

where dimensions and differential are

$$\begin{array}{ll}
|x_1| = 8, & dx_1 = 0 \\
|x_2| = 10, & dx_2 = 0 \\
|y_1| = 33, & dy_1 = x_1^3 x_2 \\
|y_2| = 35, & dy_2 = x_1^2 x_2^2 \\
|y_3| = 37, & dy_3 = x_1 x_2^3 \\
|z| = 119, & dz = y_1 y_2 x_1^4 x_2^2 - y_1 y_3 x_1^5 x_2 + y_2 y_3 x_1^6 + x_1^{15} + x_2^{12}.
\end{array}$$

Trying to obtain highly connected homotopy rigid algebras by re-scaling the degree of the generators in M is hopeless: the differential in M leads to a system of linear equations, involving the degree of the generators, whose unique solution is the one given by M.

We now introduce the main building blocks of this paper.

DEFINITION 1.1. For any even integer, $k > 4$, we define the minimal Sullivan algebra

$$M_k = \big(\Lambda(x_1, x_2, y_1, y_2, y_3, z), d\big)$$

where

$$\begin{array}{ll}
|x_1| = 5k - 2 & dx_1 = 0, \\
|x_2| = 6k - 2 & dx_2 = 0, \\
|y_1| = 21k - 9 & dy_1 = x_1^3 x_2, \\
|y_2| = 22k - 9 & dy_2 = x_1^2 x_2^2, \\
|y_3| = 23k - 9 & dy_3 = x_1 x_2^3, \\
|z| = 15k^2 - 11k + 1 & dz = x_1^{3k-12}(x_1^2 y_2 y_3 - x_1 x_2 y_1 y_3 + x_2^2 y_1 y_2) + x_1^{\frac{6k-2}{2}} + x_2^{\frac{5k-2}{2}}.
\end{array}$$

REMARK 1.2. Observe that since k is even, only x_1, x_2 have even degree; the rest of generators have odd degree. This is a technicality which proves to be crucial in the proof of Theorem 1.6, the main result in this section.

For the sake of clarity, the proof of Theorem 1.6 is split into three lemmas. Recall that the algebras M_k are introduced in Definition 1.1.

Lemmas on generators of M_k. The first two lemmas measure how the degrees of the generators are, in some sense, far apart.

LEMMA 1.3. *Generators of M_k satisfy the following inequalities:*

$$|x_1| < |x_2| < |y_1| < |y_2| < |y_3| < |x_1 y_1| < |x_2 y_3| < |z|.$$

PROOF. All the inequalities are straightforward except for the last one, which follows from the fact that $x_2 y_3$ is a factor in a term of dz, and M_k has no generator in degree 1. □

LEMMA 1.4. *None of the following integers are divisible by $|x_1|$ or by $|x_2|$:*

TABLE 1

$\|z\| - \|y_1\|$	$\|z\| - \|y_2\|$	$\|z\| - \|y_3\|$
$\|z\| - \|y_1\| - \|x_1\| - \|x_2\|$	$\|z\| - \|y_2\| - \|x_1\| - \|x_2\|$	$\|z\| - \|y_3\| - \|x_1\| - \|x_2\|$

PROOF. We first show that they are not divisible by $|x_1|$, which is an even number. Let us write $|x_1| = 5k - 2 = 2n$. Then, for $k = \frac{2n+2}{5}$, TABLE 1 becomes:

$\frac{1}{5}(12n^2 - 40n - 2)$	$\frac{1}{5}(12n^2 - 42n - 4)$	$\frac{1}{5}(12n^2 - 44n - 6)$
$\frac{1}{5}(12n^2 - 62n - 4)$	$\frac{1}{5}(12n^2 - 64n - 6)$	$\frac{1}{5}(12n^2 - 66n - 8)$

Let f be any of the polynomials in the table above. If f is divisible by $|x_1| = 2n$, then $5f$ is so. Every monomial in $5f$ is divisible by $2n$ but the independent term, which is at most 8 thus not divisible by $2n > 18$. Therefore neither of $5f$ and f is divisible by $2n$.

In a similar way we prove that they are not divisible by $|x_2|$, which is also an even number. Let us write $|x_2| = 6k - 2 = 2n$. For $k = \frac{n+1}{3}$ TABLE 1 becomes:

$\frac{1}{3}(5n^2 - 22n + 3)$	$\frac{1}{3}(5n^2 - 23n + 2)$	$\frac{1}{3}(5n^2 - 24n + 1)$
$\frac{1}{3}(5n^2 - 33n + 4)$	$\frac{1}{3}(5n^2 - 34n + 3)$	$\frac{1}{3}(5n^2 - 35n + 2)$

Again, let f be any of the polynomials in the table above. If f is divisible by $|x_2| = 2n > 22$, then f and $3f$ are divisible by n. Every monomial in $3f$ is divisible by n but the independent term, which is at most 4 thus not divisible by $n > 11$. Therefore neither of $3f$ and f is divisible by n, and f is not divisible by $2n$. □

The third and final lemma describes some specific elements in M_k. By M_k^s we denote the elements of degree s.

LEMMA 1.5. *Let $A_i \in \mathbb{Q}[x_1, x_2]$ be such that $y_i A_i \in M_k^{|z|}$. Then A_i is a multiple of $(x_1x_2)^2$, for $i = 1, 2, 3$. Moreover, if $\sum_{i=1}^3 y_i A_i$ is a cocycle, then it is also a coboundary.*

PROOF. For $y_i A_i \in M_k^{|z|}$, then $|A_i| = |z| - |y_i|$ which, by Lemma 1.4, is not divisible by $|x_1|$ or by $|x_2|$. In this situation, A_i has not monomials that are pure powers of x_1 or x_2 and, as a consequence, A_i is a multiple of x_1x_2. Resulting from that $A_i = x_1x_2B_i$ for some $B_i \in \mathbb{Q}[x_1, x_2]$ with $|B_i| = |z| - |y_i| - |x_1| - |x_2|$ which, again by Lemma 1.4, is not divisible by $|x_1|$ or by $|x_2|$. The same argument as above shows that B_i is a multiple of x_1x_2, so $B_i = x_1x_2C_i$ for some $C_i \in \mathbb{Q}[x_1, x_2]$. Then, $A_i = (x_1x_2)^2C_i$.

Moreover, if $\Sigma_{i=1}^3 y_iA_i$ is a cocycle, $d(\Sigma_{i=1}^3 y_iA_i) = x_1^3x_2A_1 + x_1^2x_2^2A_2 + x_1x_2^3A_3 = (x_1x_2)^3(x_1^2C_1 + x_1x_2C_2 + x_2^2C_3) = 0$ leading to $x_1^2C_1 + x_1x_2C_2 + x_2^2C_3 = 0$. This equality forces C_3 and C_1 to be multiples of x_1 and x_2 respectively. Then, $C_1 = x_2\bar{C}_1$ and $C_3 = x_1\bar{C}_3$ for some $\bar{C}_1, \bar{C}_3 \in \mathbb{Q}[x_1, x_2]$ and the equality above becomes $x_1x_2(x_1\bar{C}_1 + C_2 + x_2\bar{C}_3) = 0$. Therefore $C_2 = -x_1\bar{C}_1 - x_2\bar{C}_3$ and

$$\begin{aligned} y_1A_1 + y_2A_2 + y_3A_3 &= y_1(x_1x_2)^2C_1 + y_2(x_1x_2)^2C_2 + y_3(x_1x_2)^2C_3 \\ &= y_1(x_1x_2)^2x_2\bar{C}_1 + y_2(x_1x_2)^2(-x_1\bar{C}_1 - x_2\bar{C}_3) + y_3(x_1x_2)^2x_1\bar{C}_3 \\ &= \bar{C}_1x_2(y_1x_1^2x_2^2 - y_2x_1^3x_2) + \bar{C}_3x_1(y_3x_1^2x_2^2 - y_2x_1x_2^3) \\ &= \bar{C}_1x_2d(y_2y_1) + \bar{C}_3x_1d(y_2y_3) \\ &= d(\bar{C}_1x_2y_2y_1 + \bar{C}_3x_1y_2y_3), \end{aligned}$$

which exhibits $\sum_{i=1}^3 y_iA_i$ as a coboundary. □

Main theorem in Section 1. We have all the ingredients to present and prove the main result in this section. Recall that M_k, for any even $k > 4$, are the $(5k-3)$-connected algebras introduced in Definition 1.1. The following result directly implies that M_k are homotopically rigid algebras:

THEOREM 1.6. *The monoid of the homotopy classes of self-maps of M_k has only two elements: the zero map and the identity.*

PROOF. Let $f \in \operatorname{End}(M_k)$. By degree reasons, following Lemma 1.3 we know that $f(x_i) = a_i x_i$, $f(y_j) = b_j y_j$, and $f(z) = cz + \Sigma_{i=1}^{3} A_i y_i + B y_1 y_2 y_3$, $a_i, b_j, c \in \mathbb{Q}$, $A_1, A_2, A_3, B \in \mathbb{Q}[x_1, x_2]$. We point out here that since $f(z)$ is of odd degree, no summand from $\Lambda(x_1, x_2) \otimes \Lambda^2(y_1, y_2, y_3)$ can be part of it.

While $df(y_i) = f(dy_i)$ allows us to immediately obtain the following relations

$$b_1 = a_1^3 a_2, \quad b_2 = a_1^2 a_2^2 \quad \text{and} \quad b_3 = a_1 a_2^3, \tag{1.1}$$

comparing $f(dz) = df(z)$ is more demanding. On the one hand we have that

$$\begin{aligned} f(dz) = a_1^{3k-12} x_1^{3k-12} (a_1^2 b_2 b_3 x_1^2 y_2 y_3 - a_1 a_2 b_1 b_3 x_1 x_2 y_1 y_3 + a_2^2 b_1 b_2 x_2^2 y_1 y_2) \\ + a_1^{\frac{6k-2}{2}} x_1^{\frac{6k-2}{2}} + a_2^{\frac{5k-2}{2}} x_2^{\frac{5k-2}{2}}, \end{aligned} \tag{1.2}$$

and on the other hand

$$\begin{aligned} df(z) = c \left[x_1^{3k-12} (x_1^2 y_2 y_3 - x_1 x_2 y_1 y_3 + x_2^2 y_1 y_2) + x_1^{\frac{6k-2}{2}} + x_2^{\frac{5k-2}{2}} \right] \\ + x_1^3 x_2 A_1 + x_1^2 x_2^2 A_2 + x_1 x_2^3 A_3 + B(x_1^3 x_2 y_2 y_3 - x_1^2 x_2^2 y_1 y_3 + x_1 x_2^3 y_1 y_2). \end{aligned} \tag{1.3}$$

Note that none of the summands in (1.2) is a multiple of $x_2 y_2 y_3$, whereas $B x_1^3 x_2 y_2 y_3$ appears as a summand in (1.3). This forces $B = 0$. Also, if we compare monomials of the form $x_1 x_2 P$, with $P \in \mathbb{Q}[x_1, x_2]$, in (1.2) and (1.3), we are led to the conclusion that $x_1^3 x_2 A_1 + x_1^2 x_2^2 A_2 + x_1 x_2^3 A_3 = 0$. This implies that $\sum_{i=1}^{3} y_i A_i$ is a cocycle and, by Lemma 1.5, a coboundary. Following the strategy of comparing the rest of the coefficients in both equations and considering (1.1) we get

$$\begin{cases} c = a_1^{3k-10} b_2 b_3 = a_1^{3k-7} a_2^5, \\ c = a_1^{3k-11} a_2 b_1 b_3 = a_1^{3k-7} a_2^5, \\ c = a_1^{3k-12} a_2^2 b_1 b_2 = a_1^{3k-7} a_2^5, \\ c = a_1^{3k-1}, \\ c = a_2^{\frac{5k-2}{2}}. \end{cases} \tag{1.4}$$

When $a_1 = 0$, it is easy to obtain that $b_1 = b_2 = b_3 = a_1 = a_2 = c = 0$. In other case, as $a_1^{3k-1} = a_2^{\frac{5k-2}{2}}$ and also $a_1^{3k-1} = a_1^{3k-7} a_2^5$, we get that $a_1^6 = a_2^5$, or equivalently, $a_1^3 = a_2^{\frac{5}{2}}$. Combining both identities, we have that $a_1^{3k-1} = a_1^{3k-2} a_2$. Therefore $a_1 = a_2$, and since $a_1^6 = a_2^5$, we deduce that $a_1 = a_2 = 1$. So, when $a_1 \neq 0$, we infer that $a_1 = a_2 = b_1 = b_2 = b_3 = c = 1$.

Summarizing all the steps, we have proved that for $f \in \operatorname{End}(M_k)$, there exists $s = 0, 1$, such that $f(x_i) = s x_i$, $f(y_j) = s y_j$, $f(z) = sz + d(m_z)$, $m_z \in M_k^{|z|-1}$. From there, it is only necessary to observe that $f(z) = sz + d(m_z)$ is homotopically equivalent to $\tilde{f}(z) = sz$ and we have succeeded in proving that the monoid of the homotopy classes of self-maps of M_k has only two elements: the zero map (for $s = 0$), and the identity (for $s = 1$). □

2. Realizing finite groups by highly connected CDGA

The purpose of this section is to prove that algebras from Definition 1.1 serve as building blocks in the resolution of Kahn's realizability problem. Recall that Kahn's problem asks, for an arbitrary group G, if $G \cong \mathcal{E}(X)$ for some space X; if so, G is said to be realized by X. We argue as in [**5**, §2]: for any finite group G, by Frucht's theorem [**11**], there exists a finite, connected and simple graph $\mathcal{G} = (V(\mathcal{G}), E(\mathcal{G}))$ whose automorphism group, $\mathrm{Aut}(\mathcal{G})$, is isomorphic to G. The landscape changes now from groups to graphs, and our approach consists on realizing the group $\mathrm{Aut}(\mathcal{G})$ by the following minimal Sullivan algebra:

Definition 2.1. Let $n \geq 1$ be an integer and $\mathcal{G}$ be a finite, connected and simple graph with more than one vertex. We define

$$M_n(\mathcal{G}) = \big(\Lambda(x_1, x_2, y_1, y_2, y_3, z) \otimes \Lambda\big(x_v, z_v \mid v \in V(\mathcal{G})\big), d\big)$$

where

$$\begin{array}{ll} |x_1| = 30n + 18, & dx_1 = 0, \\ |x_2| = 36n + 22, & dx_2 = 0, \\ |y_1| = 126n + 75, & dy_1 = x_1^3 x_2, \\ |y_2| = 132n + 79, & dy_2 = x_1^2 x_2^2, \\ |y_3| = 138n + 83, & dy_3 = x_1 x_2^3, \\ |x_v| = 180n^2 + 218n + 66, & dx_v = 0, \\ |z| = 540n^2 + 654n + 197, & dz = x_1^{18n}(x_1^2 y_2 y_3 - x_1 x_2 y_1 y_3 + x_2^2 y_1 y_2) \\ & \quad + x_1^{18n+11} + x_2^{15n+9}, \\ |z_v| = 540n^2 + 654n + 197, & dz_v = x_v^3 + \sum_{(v,w)\in E(\mathcal{G})} x_v x_w x_2^{5n+3}. \end{array}$$

Remark 2.2. Observe that $M_n(\mathcal{G})$ is described as a tensorial product where the left factor is the minimal Sullivan algebra M_{6n+4} from Definition 1.1, and the right factor codifies $\mathcal{G}$. Indeed, for every vertex $v \in V(\mathcal{G})$, v is represented by the generator x_v, while dz_v codifies the neighborhood of v, that is, the edges in $E(\mathcal{G})$ that are incident to the vertex v. Finally, notice that we are considering simple non oriented graphs, i.e. $E(\mathcal{G})$ consists of non ordered pairs of vertices, and therefore the pairs (v, w) and (w, v) represent the same element in $E(\mathcal{G})$.

Lemmas on generators of $M_n(\mathcal{G})$. We prove now three technical lemmas that are needed for the main result in this section, Theorem 2.9.

Lemma 2.3. *Let us consider $r_1 = |z| - |y_1|$, $r_2 = |z| - |y_2|$ and $r_3 = |z| - |y_3|$. The following integers are not divisible by either $|x_1|$ or $|x_2|$:*

Table 2

r_1	r_2	r_3
$r_1 - \lvert x_1\rvert - \lvert x_2\rvert$	$r_2 - \lvert x_1\rvert - \lvert x_2\rvert$	$r_3 - \lvert x_1\rvert - \lvert x_2\rvert$
$r_1 - \lvert x_v\rvert$	$r_2 - \lvert x_v\rvert$	$r_3 - \lvert x_v\rvert$
$r_1 - \lvert x_v\rvert - \lvert x_1\rvert - \lvert x_2\rvert$	$r_2 - \lvert x_v\rvert - \lvert x_1\rvert - \lvert x_2\rvert$	$r_3 - \lvert x_v\rvert - \lvert x_1\rvert - \lvert x_2\rvert$
$r_1 - 2\lvert x_v\rvert$	$r_2 - 2\lvert x_v\rvert$	$r_3 - 2\lvert x_v\rvert$
$r_1 - 2\lvert x_v\rvert - \lvert x_1\rvert - \lvert x_2\rvert$	$r_2 - 2\lvert x_v\rvert - \lvert x_1\rvert - \lvert x_2\rvert$	$r_3 - 2\lvert x_v\rvert - \lvert x_1\rvert - \lvert x_2\rvert$

PROOF. Nothing is to prove for the two top rows, it follows from Lemma 1.4. For the rest, we start by showing that they are not divisible by $|x_1| = 30n + 18$, which is an even number. Let us write $|x_1| = 2m$, and $n = \frac{m-9}{15}$. Then, the four bottom rows become:

$\frac{1}{15}(24m^2 - 122m - 6)$	$\frac{1}{15}(24m^2 - 128m - 12)$	$\frac{1}{15}(24m^2 - 134m - 18)$
$\frac{1}{15}(24m^2 - 188m - 12)$	$\frac{1}{15}(24m^2 - 194m - 18)$	$\frac{1}{15}(24m^2 - 200m - 24)$
$\frac{1}{15}(12m^2 - 124m - 6)$	$\frac{1}{15}(12m^2 - 130m - 12)$	$\frac{1}{15}(12m^2 - 136m - 18)$
$\frac{1}{15}(12m^2 - 190m - 12)$	$\frac{1}{15}(12m^2 - 196m - 18)$	$\frac{1}{15}(12m^2 - 202m - 24)$

Let f be any of the polynomials in the table above: if f is divisible by $|x_1| = 2m$, then $15f$ is so. Every monomial in $15f$ is divisible by $2m$ but the independent term, which is at most 24 thus not divisible by $|x_1| = 2m = 30n + 18 \geq 48$. Therefore neither of $15f$ nor f is divisible by $|x_1|$.

Similar arguments show that none of the integers from the table is divisible by $|x_2| = 36n + 22$, which is also an even number. Let us write $|x_2| = 2m$, and $n = \frac{m-11}{18}$. Then, the four bottom rows in TABLE 2 become:

$\frac{1}{9}(10m^2 - 65m + 9)$	$\frac{1}{9}(10m^2 - 68m + 6)$	$\frac{1}{9}(10m^2 - 71m + 3)$
$\frac{1}{9}(10m^2 - 98m + 12)$	$\frac{1}{9}(10m^2 - 101m + 9)$	$\frac{1}{9}(10m^2 - 104m + 6)$
$\frac{1}{9}(5m^2 - 64m + 9)$	$\frac{1}{9}(5m^2 - 67m + 6)$	$\frac{1}{9}(5m^2 - 70m + 3)$
$\frac{1}{9}(5m^2 - 97m + 12)$	$\frac{1}{9}(5m^2 - 100m + 9)$	$\frac{1}{9}(5m^2 - 103m + 6)$

Again, let f be any of the polynomials in the table above: if f is divisible by $|x_2| = 2m$, then f and $9f$ are divisible by m. Every monomial in $9f$ is divisible by m but the independent term, which is at most 12 thus not divisible by $m = 18n + 11 \geq 29$. Hence neither of $9f$ nor f is divisible by m. □

The description of some specific elements in the algebra $M_n(\mathcal{G})$ is also needed:

LEMMA 2.4. *Let $i = 1, 2, 3$ and $A_i \in \mathbb{Q}[x_1, x_2]$. If $y_i A_i \in \big(M_n(\mathcal{G})\big)^m$, for $m = |z|$, $|z| - |x_v|$, or $|z| - 2|x_v|$, then $A_i = (x_1x_2)^2 \bar{A}_i$, where $\bar{A}_i \in \mathbb{Q}[x_1, x_2]$. Moreover, if $\sum_{i=1}^3 y_i A_i$ is a cocycle, then it is a coboundary.*

PROOF. If $y_i A_i \in \big(M_n(\mathcal{G})\big)^m$, then $|A_i| = m - |y_i|$. When $m = |z|$, $|z| - |x_v|$, or $|z| - 2|x_v|$, $|A_i|$ is respectively the i-th term of the first, third or fifth row in TABLE 2. By Lemma 2.3, as $|A_i|$ is not divisible by neither $|x_1|$ nor $|x_2|$, we infer that A_i is not a pure power of x_1 nor x_2, so $A_i = x_1x_2B_i$, where $B_i \in \mathbb{Q}[x_1, x_2]$. Now, for $m = |z|$, $|z| - |x_v|$, or $|z| - 2|x_v|$, $|B_i|$ is, respectively, the i-th term of the second, fourth or sixth row in TABLE 2. Hence $|B_i|$ is not divisible by neither $|x_1|$ nor $|x_2|$, and the same arguments allow us to write $B_i = x_1x_2\bar{A}_i$. Thus, we finally obtain that $A_i = (x_1x_2)^2\bar{A}_i$, where $\bar{A}_i \in \mathbb{Q}[x_1, x_2]$.

Similar arguments as in the proof of Lemma 1.5 show that if $\sum_{i=1}^3 y_i A_i$ is a cocycle then it is a coboundary; details are omitted. □

The last technical result that we need is the following.

LEMMA 2.5. *There is no pair of positive integers* (α, β) *such that*

$$m = \alpha|x_1| + \beta|x_2|, \tag{2.1}$$

for $m = |x_v|$. *Moreover, there is no pair of non-negative integers* (α, β) *verifying* (2.1) *for* $m \in \{|z| - |y_1y_2y_3|, |z| - |x_vy_1y_2y_3|, |z| - |x_v^2y_1y_2y_3|\}$.

PROOF. To obtain the general solution of the diophantine equation (2.1), first notice that

$$\begin{aligned} -6|x_1| + 5|x_2| &= -6(30n + 18) + 5(36n + 22) \\ &= 2, \end{aligned}$$

so the greatest common divisor of $|x_1|$ and $|x_2|$ is 2. Thus it suffices to compute a particular solution (α, β), and express the general one as

$$\left(\alpha + k\frac{|x_2|}{2}, \beta - k\frac{|x_1|}{2}\right) = (\alpha + k(18n + 11), \beta - k(15n + 9)), \ k \in \mathbb{Z}.$$

For $m = |x_v| = 180n^2 + 218n + 66$, the pair $(0, 5n + 3)$ is a particular solution for (2.1) and the general solution is $\big(k(18n + 11), 5n + 3 - k(15n + 9)\big)$, $k \in \mathbb{Z}$. It is clear that for $k > 0$, the second component is negative, and for $k \leq 0$, the first component is non positive. Thus there is no solution where both integers are positive.

For the rest of cases we just write the general solution below here. It is straightforward to check that at least one of the two components of the pair is negative for all $k \in \mathbb{Z}$:

m	general solution
$\|z\| - \|y_1y_2y_3\|$	$\big(-12 + k(18n + 11), 15n + 8 - k(15n + 9)\big)$
$\|z\| - \|x_vy_1y_2y_3\|$	$\big(-12 + k(18n + 11), 10n + 5 - k(15n + 9)\big)$
$\|z\| - \|x_v^2y_1y_2y_3\|$	$\big(-12 + k(18n + 11), 5n + 2 - k(15n + 9)\big)$

□

Sullivan algebras $M_n(\mathcal{G})$ are elliptic. First, we recall that a Sullivan algebra $(\Lambda W, d)$ is said to be elliptic when both W and $H^*(\Lambda W)$ are finite-dimensional. The cohomology of an elliptic Sullivan algebra verifies Poincaré duality [**14**]. One can compute the degree of its fundamental class (a fundamental class of a Poincaré duality algebra $H = \Sigma_{i=0}^m H^i$, is a generator of H^m, m is said to be the formal dimension of the algebra) by the formula [**9**, Theorem 32.6]:

$$\textstyle\sum_{i=1}^p(\deg y_i) - \sum_{j=1}^q(\deg x_j - 1) \tag{2.2}$$

where $\deg y_i$ are the degrees of the elements of a basis of W^{odd} and $\deg x_j$ of a basis of W^{even}.

The existence of a fundamental class for elliptic Sullivan algebras gives rise to a notion of degree for self maps.

DEFINITION 2.6. Let M be an elliptic Sullivan algebra with fundamental class $c_M \in H^m(M)$. Given $f \in \operatorname{End}(M)$ we say that the degree of f is $k \in \mathbb{Q}$, $\deg(f) = k$, if $f^*(c_M) = kc_M$.

The following result is inspired by classical works from Hopf [**15, 16**]:

PROPOSITION 2.7. *Let M be an elliptic Sullivan algebra and $f \in [M, M]$. Then f is a self-homotopy equivalence if and only if* $\deg(f) \neq 0$.

PROOF. It is clear that a self homotopy equivalence has a non-zero degree since $\deg(\mathrm{Id}) = 1$ and the degree is a multiplicative invariant. Suppose now that $\deg(f) \neq 0$. Then, we are going to prove that $H^*(f)$ is injective. Indeed, assume that there exists a non trivial class $\alpha \in H^p(M)$ such that $f^*(\alpha) = 0$. Now, since $H^*(M)$ is a Poincaré duality algebra, consider $c_M \in H^m(M)$ the fundamental class and the (non-degenerate) bilinear form

$$\langle\,,\rangle\colon H^p(M) \otimes H^{m-p}(M) \to \mathbb{Q}$$

given by $xy = \langle x, y\rangle c_M$. This bilinear form induces an isomorphism

$$\theta\colon H^p(M) \to H^{m-p}(M)^\sharp$$

given by $\theta(x)(y) = \langle x, y\rangle$. Therefore, since $\alpha \neq 0$, there exists $\beta \in H^{m-p}(M)$ such that $\alpha\beta = c_M$. Then $0 = f^*(\alpha)f^*(\beta) = f^*(\alpha\beta) = f^*(c_M) = \deg(f)c_M$ which implies $0 = \deg(f)$, contradicting our initial assumption. Hence $H^*(f)$ is injective and, as we are considering finite type algebras, f^* is a quasi-isomorphism, which means that f is a self-homotopy equivalence. $\square$

Finally, we prove the following lemma.

LEMMA 2.8. *The algebra $M_n(\mathcal{G})$ is an elliptic Sullivan algebra of formal dimension* $540n^2 + 984n + 396 + |V(\mathcal{G})|(360n^2 + 436n + 132)$.

PROOF. We have to prove that the cohomology of $M_n(\mathcal{G})$ is finite-dimensional. Instead, we prove the equivalent condition that the pure Sullivan algebra associated to $M_n(\mathcal{G})$, [**9**, Proposition 32.4], is also elliptic. This pure algebra, denoted by $(M_n^\varsigma(\mathcal{G}), d_\varsigma)$, has the same generators as $M_n(\mathcal{G})$ and differentials

$$\begin{array}{ll} d_\varsigma(x_1) = 0, & d_\varsigma(x_2) = 0, \\ d_\varsigma(y_1) = x_1^3x_2, & d_\varsigma(y_2) = x_1^2x_2^2, \\ d_\varsigma(y_3) = x_1x_2^3, & d_\varsigma(z) = x_1^{18n+11} + x_2^{15n+9}, \\ d_\varsigma(x_v) = 0,\ v \in V(\mathcal{G}), & d_\varsigma(z_v) = x_v^3 + \Sigma_{(v,w)\in E(\mathcal{G})} x_vx_wx_2^{5n+3},\ v \in V(\mathcal{G}). \end{array}$$

The cohomology of $M_n^\varsigma(\mathcal{G})$ is finite-dimensional because

$$d_\varsigma(zx_1 - x_2^{15n+6}y_3) = x_1^{18n+12} \quad \text{and} \quad d_\varsigma(zx_2 - x_1^{18n+8}y_1) = x_2^{15n+10},$$

and the cohomology class

$$[x_v^3]^4 = \left[-\sum_{(v,w)\in E(\mathcal{G})} x_vx_wx_2^{5n+3}\right]^4 = 0.$$

Hence $M_n(\mathcal{G})$ is elliptic and its formal dimension is obtained by (2.2). $\square$

Main theorem in Section 2. We present the main result in this section. For $M_n(\mathcal{G})$, $n \geq 1$, the minimal Sullivan algebra from Definition 2.1, we prove:

THEOREM 2.9. *For any finite simple graph $\mathcal{G}$, the $(30n+17)$-connected algebra $M_n(\mathcal{G})$ realizes $\mathrm{Aut}(\mathcal{G})$, i.e.,*

$$\mathcal{E}(M_n(\mathcal{G})) \cong \mathrm{Aut}(\mathcal{G}).$$

Before getting into the details of the proof, one immediate consequence:

THEOREM 2.10. *Any finite group G can be realized by infinitely many highly connected rational spaces.*

PROOF. In [**11**], Frucht proves that given a finite group G is the group of automorphisms of a finite simple graph $\mathcal{G}$. It suffices then to apply Theorem 2.9. $\square$

For the proof of Theorem 2.9, two more lemmas are needed. Recall from Remark 2.2 that algebras $M_n(\mathcal{G})$ are constructed to codify the graph $\mathcal{G}$ in terms of vertices (generators x_v) and their neighborhoods (generators z_v). Since an automorphism of $\mathcal{G}$ is just a permutation of vertices that preserves neighborhoods, we deduce a first lemma. Explicitly:

LEMMA 2.11. *Every $\sigma \in \mathrm{Aut}(\mathcal{G})$ induces $f_\sigma \in \mathrm{Aut}(M_n(\mathcal{G}))$ defined by*

$$\begin{aligned} f_\sigma(\omega) &= \omega, && \text{for } \omega \in \{x_1, x_2, y_1, y_2, y_3, z\},\\ f_\sigma(x_v) &= x_{\sigma(v)}, && \text{for } v \in V(\mathcal{G}),\\ f_\sigma(z_v) &= z_{\sigma(v)}, && \text{for } v \in V(\mathcal{G}). \end{aligned}$$

Moreover, given $\sigma_1, \sigma_2 \in \mathrm{Aut}(\mathcal{G})$, then $f_{\sigma_1} \simeq f_{\sigma_2}$ if and only if $\sigma_1 = \sigma_2$.

PROOF. Given $\sigma \in \mathrm{Aut}(\mathcal{G})$, it is clear that $f_\sigma \in \mathrm{Aut}(M_n(\mathcal{G}))$. Finally, if $\sigma_1, \sigma_2 \in \mathrm{Aut}(\mathcal{G})$, and $f_{\sigma_1} \simeq f_{\sigma_2}$, then for every $v \in V(\mathcal{G})$ the elements $f_{\sigma_1}(x_v) = x_{\sigma_1(v)}$ and $f_{\sigma_2}(x_v) = x_{\sigma_2(v)}$ must be equal up to decomposable elements, that is $x_{\sigma_1(v)} = x_{\sigma_2(v)}$. Therefore $\sigma_1(v) = \sigma_2(v)$ for every $v \in V(\mathcal{G})$, and $\sigma_1 = \sigma_2$. $\square$

For the second lemma, we argue as in [**5**, Lemma 2.4]:

LEMMA 2.12. *For every $f \in \mathrm{End}\big(M_n(\mathcal{G})\big)$, one of the following holds:*

(1) *f is an automorphism, and there exists $\sigma \in \mathrm{Aut}(\mathcal{G})$ such that*

$$\begin{aligned} f(\omega) &= f_\sigma(\omega), && \text{for } \omega \in \big\{x_1, x_2, y_1, y_2, y_3, x_v \mid v \in V(\mathcal{G})\big\},\\ f(z) &= f_\sigma(z) + d(m_z), && m_z \in \big(M_n(\mathcal{G})\big)^{|z|-1},\\ f(z_v) &= f_\sigma(z_v) + d(m_{z_v}), && \text{for } v \in V(\mathcal{G}) \text{ and } m_{z_v} \in \big(M_n(\mathcal{G})\big)^{|z|-1}, \end{aligned}$$

where f_σ is the morphism given by Lemma 2.11.

(2) $\deg(f) = 0$.

In particular, we deduce that for $f \in \mathrm{End}\big(M_n(\mathcal{G})\big)$, either $f \simeq f_\sigma$, $\sigma \in \mathrm{Aut}(\mathcal{G})$, or $\deg(f) = 0$.

PROOF. We suppose that $\deg(f) \neq 0$, that is, f does not fall into case (2). We have to show that f falls into case (1). The element $f \in \mathrm{End}\big(M_n(\mathcal{G})\big)$ is uniquely determined by the image of generators of $M_n(\mathcal{G})$. First, we compute $f(x_1)$, $f(x_2)$, $f(y_1)$, $f(y_2)$, and $f(y_3)$. For degree reasons (see Lemma 1.3) we have:

$$f(x_1) = a_1x_1,\ f(x_2) = a_2x_2,\ f(y_1) = b_1y_1,\ f(y_2) = b_2y_2,\ f(y_3) = b_3y_3, \tag{2.3}$$

for $a_i, b_j \in \mathbb{Q}$. Since $df(y_i) = f(dy_i)$ for $i = 1, 2, 3$, we get that

$$b_1 = a_1^3 a_2, \quad b_2 = a_1^2 a_2^2, \quad \text{and} \quad b_3 = a_1 a_2^3. \tag{2.4}$$

Then, we continue with $f(x_v)$, for $v \in V(\mathcal{G})$. Observe that no pure power of x_1 can be a summand in $f(x_v)$. Indeed $|x_v| = |x_1|(6n+3) + (20n+12)$, with $20n + 12 < |x_1|$, for every $n \geq 1$, which means that $|x_1|$ does not divide $|x_v|$. Secondly, no summand of the form $x_1^\alpha x_2^\beta$, with α and β both positive can appear in $f(x_v)$, since $|x_v|$ cannot be expressed as $\alpha|x_1| + \beta|x_2|$, with α, β both positive, by Lemma 2.5. Then,

$$f(x_v) = \sum_{w \in V(\mathcal{G})} a(v,w) x_w + a_2(v) x_2^{5n+3} + A(v) y_1 y_2 + B(v) y_1 y_3 + C(v) y_2 y_3, \tag{2.5}$$

where $a(v,w), a_2(v) \in \mathbb{Q}$ and $A(v), B(v), C(v) \in \mathbb{Q}[x_1, x_2]$, for all $v, w \in V(\mathcal{G})$. Note that no term of the form $Ay_1y_2y_3$ or Ay_i, for $A \in \mathbb{Q}[x_1, x_2]$, can appear in the expression of $f(x_v)$ (see Remark 1.2). Now, since $df(x_v) = f(dx_v) = 0$, we get

$$0 = A(v)(x_1^3 x_2 y_2 - x_1^2 x_2^2 y_1) + B(v)(x_1^3 x_2 y_3 - x_1 x_2^3 y_1) + C(v)(x_1^2 x_2^2 y_3 - x_1 x_2^3 y_2).$$

In this equation the sum of the terms having y_1 as a factor, respectively y_2, y_3, must all be zero. We start by paying attention to coefficients in y_1. As

$$0 = A(v) x_1^2 x_2^2 + B(v) x_1 x_2^3 = x_1 x_2^2 (A(v) x_1 + B(v) x_2),$$

we infer that $A(v)x_1 + B(v)x_2 = 0$, and therefore that x_1 must be a factor in $B(v)$ and x_2 a factor in $A(v)$. Therefore $B(v) = \bar{B}(v)x_1$ for some $\bar{B}(v) \in \mathbb{Q}[x_1, x_2]$, $A(v) = \bar{A}(v)x_2$ for some $\bar{A}(v) \in \mathbb{Q}[x_1, x_2]$ and $\bar{A}(v)x_1x_2 + \bar{B}(v)x_1x_2 = 0$, so that $\bar{A}(v) = -\bar{B}(v)$. We continue by paying attention to coefficients in y_2. As

$$0 = A(v) x_1^3 x_2 - C(v) x_1 x_2^3 = x_1 x_2 (A(v) x_1^2 - C(v) x_2^2),$$

then $A(v)x_1^2 - C(v)x_2^2 = 0$. But $A(v) = \bar{A}(v)x_2$, so

$$0 = \bar{A}(v) x_1^2 x_2 - C(v) x_2^2 = x_2 (\bar{A}(v) x_1^2 - C(v) x_2).$$

The same kind of arguments as above imply now that $\bar{\bar{A}}(v)x_2 = \bar{A}(v)$ and $\bar{C}(v)x_1^2 = C(v)$, for some $\bar{\bar{A}}(v), \bar{C}(v) \in \mathbb{Q}[x_1, x_2]$. Then $\bar{\bar{A}}(v)x_1^2x_2 - \bar{C}(v)x_1^2x_2 = 0$, so $\bar{\bar{A}}(v) = \bar{C}(v)$. Finally, we pay attention to coefficients in y_3. As

$$0 = B(v) x_1^3 x_2 + C(v) x_1^2 x_2^2 = x_1^2 x_2 (B(v) x_1 + C(v) x_2),$$

we infer that $B(v)x_1 + C(v)x_2 = 0$. Since $B(v) = \bar{B}(v)x_1$ and $C(v) = \bar{C}(v)x_1^2$,

$$0 = \bar{B}(v) x_1^2 + \bar{C}(v) x_1^2 x_2 = x_1^2 (\bar{B}(v) + \bar{C}(v) x_2),$$

so $\bar{\bar{B}}(v)x_2 = \bar{B}(v)$, for some $\bar{\bar{B}}(v) \in \mathbb{Q}[x_1, x_2]$. Then $0 = \bar{\bar{B}}(v)x_2 + \bar{C}(v)x_2$, and so $\bar{\bar{B}}(v) = -\bar{C}(v)$.

Summarizing, we have $A(v) = \bar{C}(v)x_2^2$, $B(v) = -\bar{C}(v)x_1x_2$ and $C(v) = \bar{C}(v)x_1^2$, hence (2.5) can be rewritten as:

$$f(x_v) = \sum_{w \in V(\mathcal{G})} a(v,w) x_w + a_2(v) x_2^{5n+3} + \bar{C}(v)(x_2^2 y_1 y_2 - x_1 x_2 y_1 y_3 + x_1^2 y_2 y_3).$$

Let us now prove that $\bar{C}(v) = P(v)x_1x_2$, with $P(v) \in \mathbb{Q}[x_1, x_2]$. Notice that if we succeed, as $d(y_1y_2y_3) = x_1^3x_2y_2y_3 - x_1^2x_2^2y_1y_3 + x_1x_2^3y_1y_2$, the last summand in $f(x_v)$ can be expressed as $d(P(v)y_1y_2y_3)$. It suffices to prove that $|x_1|$ and $|x_2|$ are not divisors of the degree of $\bar{C}(v)$. This is clear as for $|\bar{C}(v)| = |x_v| - 2|x_2| - |y_1| - |y_2| = 180n^2 - 112n - 132 = |x_1|(6n-8) + (20n+12)$, with $0 < 20n + 12 < |x_1|$, for $n \geq 1$

and also for $|\bar{C}(v)| = |x_2|(5n-7)+(30n+22)$ with $0 < 30n+22 < |x_2|$, which lead us to the conclusion that $\bar{C}(v)$ cannot be a pure power of either x_1 or x_2. Then

$$f(x_v) = \sum_{w\in V(\mathcal{G})} a(v,w)x_w + a_2(v)x_2^{5n+3} + d\big(P(v)y_1y_2y_3\big). \tag{2.6}$$

Third, we determine $f(z)$. Remark that terms of the form $Py_1y_2y_3$ with $P \in \mathbb{Q}[x_1, x_2, x_v]$ cannot appear in the expression of $f(z)$, since by Lemma 2.5 there is no polynomial $Q \in \mathbb{Q}[x_1, x_2]$ such that $|Qx_v^m y_1y_2y_3| = |z|$, for $m = 0, 1, 2$. It is then clear that

$$\begin{aligned} f(z) = cz &+ \sum_{w\in V(\mathcal{G})} c(w)z_w + A_1y_1 + A_2y_2 + A_3y_3 \\ &+ \sum_{w\in V(\mathcal{G})} x_w\big(B_1(w)y_1 + B_2(w)y_2 + B_3(w)y_3\big) \\ &+ \sum_{\{v,w\}\subset V(\mathcal{G})} x_vx_w\big(C_1(v,w)y_1 + C_2(v,w)y_2 + C_3(v,w)y_3\big), \end{aligned} \tag{2.7}$$

where $c, c(w) \in \mathbb{Q}$, A_i, $B_i(w)$, $C_i(v,w) \in \mathbb{Q}[x_1, x_2]$ for all $v, w \in V(\mathcal{G})$, $i = 1, 2, 3$. Since $df(z)$ and $f(dz)$ must coincide, on the one hand using (2.3):

$$\begin{aligned} f(dz) =& b_1b_2a_1^{18n}a_2^2y_1y_2x_1^{18n}x_2^2 - b_1b_3a_1^{18n+1}a_2y_1y_3x_1^{18n+1}x_2 \\ &+ b_2b_3a_1^{18n+2}y_2y_3x_1^{18n+2} + a_1^{18n+11}x_1^{18n+11} + a_2^{15n+9}x_2^{15n+9}, \end{aligned} \tag{2.8}$$

and on the other hand:

$$\begin{aligned} df(z) =& c\big(x_1^{18n}(x_1^2y_2y_3 - x_1x_2y_1y_3 + x_2^2y_1y_2) + x_1^{18n+11} + x_2^{15n+9}\big) \\ &+ \sum_{w\in V(\mathcal{G})} c(w)\left[x_w^3 + \sum_{(w,u)\in E(\mathcal{G})} x_wx_ux_2^{5n+3}\right] \\ &+ A_1x_1^3x_2 + A_2x_1^2x_2^2 + A_3x_1x_2^3 \\ &+ \sum_{w\in V(\mathcal{G})} x_w\big[B_1(w)x_1^3x_2 + B_2(w)x_1^2x_2^2 + B_3(w)x_1x_2^3\big] \\ &+ \sum_{\{u,w\}\subset V(\mathcal{G})} x_ux_w\big[C_1(u,w)x_1^3x_2 + C_2(u,w)x_1^2x_2^2 + C_3(u,w)x_1x_2^3\big]. \end{aligned} \tag{2.9}$$

Now notice that in (2.8) there are no terms $x_1^\alpha x_2^\beta$ for $\alpha, \beta > 0$, whereas in (2.9) there are. This leads to $A_1x_1^3x_2 + A_2x_1^2x_2^2 + A_3x_1x_2^3 = 0$ and thus to $d(\sum_{i=1}^3 A_iy_i) = 0$. Moreover, in (2.8) there are no terms at all involving x_w, so we obtain $c(w) = 0$, $B_1(w)x_1^3x_2 + B_2(w)x_1^2x_2^2 + B_3(w)x_1x_2^3 = 0$, and $C_1(u,w)x_1^3x_2 + C_2(u,w)x_1^2x_2^2 + C_3(u,w)x_1x_2^3 = 0$ for all $u, w \in V(\mathcal{G})$ in (2.9). Hence $d\big(\sum_{i=1}^3 B_i(w)y_i\big) = 0$ and $d\big(\sum_{i=1}^3 C_i(u,w)y_i\big) = 0$, and by Lemma 2.4, we can finally reduce the expression to:

$$f(z) = cz + d(m_z), \quad m_z \in \big(M_n(\mathcal{G})\big)^{|z|-1}. \tag{2.10}$$

Before computing the last image $f(z_v)$, $v \in V(\mathcal{G})$, we work out here the value of the remaining constants. Compare again (2.8) to (2.9) and invoke (2.4) to obtain:

$$\begin{cases} c = a_1^{18n+2} b_2 b_3 = a_1^{18n+5} a_2^5, \\ c = a_1^{18n+1} a_2 b_1 b_3 = a_1^{18n+5} a_2^5, \\ c = a_1^{18n} a_2^2 b_1 b_2 = a_1^{18n+5} a_2^5, \\ c = a_1^{18n+11}, \\ c = a_2^{15n+9}, \end{cases}$$

which is a particular case of Equation (1.4) for $k = 6n + 4$. Therefore either $a_1 = a_2 = b_1 = b_2 = b_3 = c = 0$ or $a_1 = a_2 = b_1 = b_2 = b_3 = c = 1$. Since $\deg(f) \neq 0$, we know that $f \in \mathcal{E}(M_n(\mathcal{G}))$ (Proposition 2.7), and therefore $a_1 \neq 0$, from which it follows that $a_1 = a_2 = b_1 = b_2 = b_3 = c = 1$.

Finally, we end by computing $f(z_v)$, $v \in V(\mathcal{G})$. Since $|z_v| = |z|$, the degree reasons used to describe $f(z)$ in (2.7) apply, and $f(z_v)$ can be expressed as:

$$\begin{aligned} f(z_v) = e(v)z &+ \sum_{w \in V(\mathcal{G})} c(v,w) z_w + G_1(v)y_1 + G_2(v)y_2 + G_3(v)y_3 \\ &+ \sum_{w \in V(\mathcal{G})} x_w \big(H_1(v,w)y_1 + H_2(v,w)y_2 + H_3(v,w)y_3\big) \\ &+ \sum_{\{u,w\} \subset V(\mathcal{G})} x_u x_w \big(I_1(v,u,w)y_1 + I_2(v,u,w)y_2 + I_3(v,u,w)y_3\big), \end{aligned} \tag{2.11}$$

with $e(v), c(v,w) \in \mathbb{Q}$, and $G_i(v), H_i(v,w), I_i(v,u,w) \in \mathbb{Q}[x_1,x_2]$, for all $v,u,w \in V(\mathcal{G})$, $i = 1,2,3$.

Then,

(2.12)

$$\begin{aligned} df(z_v) = e(v) &\left[x_1^{18n}(x_1^2 y_2 y_3 - x_1 x_2 y_1 y_3 + x_2^2 y_1 y_2) + x_1^{18n+11} + x_2^{15n+9}\right] \\ &+ \sum_{w \in V(\mathcal{G})} c(v,w) \left(x_w^3 + \sum_{(w,u) \in E(\mathcal{G})} x_w x_u x_2^{5n+3} \right) \\ &+ G_1(v) x_1^3 x_2 + G_2(v) x_1^2 x_2^2 + G_3(v) x_1 x_2^3 \\ &+ \sum_{w \in V(\mathcal{G})} x_w \big(H_1(v,w) x_1^3 x_2 + H_2(v,w) x_1^2 x_2^2 + H_3(v,w) x_1 x_2^3\big) \\ &+ \sum_{\{w,u\} \subset V(\mathcal{G})} x_w x_u \big(I_1(v,u,w) x_1^3 x_2 + I_2(v,u,w) x_1^2 x_2^2 + I_3(v,u,w) x_1 x_2^3\big). \end{aligned}$$

On the other hand, using (2.6)

$$\begin{aligned} f(dz_v) = &\left(\sum_{w \in V(\mathcal{G})} a(v,w) x_w + a_2(v) x_2^{5n+3} + d\big(P(v) y_1 y_2 y_3\big) \right)^3 \\ + \sum_{(v,r) \in E(\mathcal{G})} &\left(\sum_{w \in V(\mathcal{G})} a(v,w) x_w + a_2(v) x_2^{5n+3} + d\big(P(v) y_1 y_2 y_3\big) \right) \\ &\times \left(\sum_{u \in V(\mathcal{G})} a(r,u) x_u + a_2(r) x_2^{5n+3} + d\big(P(r) y_1 y_2 y_3\big) \right) x_2^{5n+3}. \end{aligned} \tag{2.13}$$

First, notice that there is no summand in (2.12) of the form $x_v x_w x_u$, where u, v and w are pairwise distinc. This forces (2.13) to have at most two nontrivial coefficients $a(v,w)$. Furthermore, in (2.12) there is no factor of the form $x_w^2 x_v$, where $w \neq v$, which forces exactly one non trivial coefficient $a(v,w)$ in (2.13) (recall that $f \in \mathcal{E}(M_n(\mathcal{G}))$, hence at least one coefficient $a(v,w)$ must be nontrivial). Consequently, in (2.13) there is a unique summand containing x_w^3 and, in (2.12) there can only be one non trivial coefficient $c(v,w)$. This fact can be read as there is a self-map of $V(\mathcal{G})$, σ, such that $a\big(v,\sigma(v)\big)$ and $c\big(v,\sigma(v)\big)$ are the only non trivial coefficients. Second, notice that in (2.13) there is no factor that is a pure power of x_1, and thus, $e(v) = 0$ in (2.12). Also, since the graph has no loops, a term of type $x_{\sigma(v)}^2 x_2^{5n+3}$ does not appear in (2.12), and $a\big(v,\sigma(v)\big) \neq 0$ implies that $a_2(v) = 0$. We focus now on (2.13). Notice that

$$d\big(P(v)y_1y_2y_3\big) = P(v)(x_1^3x_2y_2y_3 - x_1^2x_2^2y_1y_3 + x_1x_2^3y_1y_2),$$

and if two terms of this form are multiplied, their product is zero ($y_i^2 = 0$). Thus,

$$\begin{aligned} f(dz_v) &= a\big(v,\sigma(v)\big)^3 x_{\sigma(v)}^3 + 3a\big(v,\sigma(v)\big)^2 x_{\sigma(v)}^2 d\big(P(v)y_1y_2y_3\big) \\ &\quad + \sum_{(v,r)\in E(\mathcal{G})} \Big[a\big(v,\sigma(v)\big)a\big(r,\sigma(r)\big)x_{\sigma(v)}x_{\sigma(r)} \\ &\quad + a\big(v,\sigma(v)\big)x_{\sigma(v)}d\big(P(r)y_1y_2y_3\big) \\ &\quad + a\big(r,\sigma(r)\big)x_{\sigma(r)}d\big(P(v)y_1y_2y_3\big)\Big] x_2^{5n+3}. \end{aligned} \tag{2.14}$$

Observe that only one term containing $x_{\sigma(v)}^2$ appears in (2.14), which is multiplied by $d\big(P(v)y_1y_2y_3\big) = P(v)(x_1^3x_2y_2y_3 - x_1^2x_2^2y_1y_3 + x_1x_2^3y_1y_2)$. However, in (2.12) there is no term containing both $x_{\sigma(v)}^2$ and y_i. Therefore, we can conclude that $P(v) = 0$, for every $v \in V(\mathcal{G})$, and hence

$$f(dz_v) = a\big(v,\sigma(v)\big)^3 x_{\sigma(v)}^3 + \sum_{(v,r)\in E(\mathcal{G})} a\big(v,\sigma(v)\big)a\big(r,\sigma(r)\big)x_{\sigma(v)}x_{\sigma(r)}x_2^{5n+3}. \tag{2.15}$$

Now, comparing terms containing x_1x_2, $x_1x_2x_w$ and $x_1x_2x_wx_u$ in (2.12) and (2.15), we get:

$$\begin{cases} G_1(v)x_1^3x_2 + G_2(v)x_1^2x_2^2 + G_3(v)x_1x_2^3 = 0, \\ H_1(v,w)x_1^3x_2 + H_2(v,w)x_1^2x_2^2 + H_3(v,w)x_1x_2^3 = 0, \\ I_1(v,u,w)x_1^3x_2 + I_2(v,u,w)x_1^2x_2^2 + I_3(v,u,w)x_1x_2^3 = 0. \end{cases}$$

In other words $\sum_{i=1}^3 y_iG_i(v)$, $\sum_{i=1}^3 y_iH_i(v,w)$ and $\sum_{i=1}^3 y_iI_i(v,u,w)$ are cocycles in degrees $|z|$, $|z| - |x_v|$ and $|z| - 2|x_v|$ respectively. We now invoke Lemma 2.4 to conclude that they are also coboundaries. Then, we can finally write

$$f(z_v) = c(v,\sigma(v))z_{\sigma(v)} + d(m_{z_v}),$$

for some $m_{z_v} \in \big(M_n(\mathcal{G})\big)^{|z|-1}$, and

$$df(z_v) = c(v,\sigma(v)) \left(x_{\sigma(v)}^3 + \sum_{(\sigma(v),u)\in E(\mathcal{G})} x_{\sigma(v)}x_u x_2^{5n+3} \right). \tag{2.16}$$

Comparing both (2.15) and (2.16), we get

$$\begin{aligned} c\big(v,\sigma(v)\big) &= a\big(v,\sigma(v)\big)^3, && \text{for all } v\in V(\mathcal{G}),\\ c\big(v,\sigma(v)\big) &= a\big(v,\sigma(v)\big)a\big(w,\sigma(w)\big), && \text{for all } (v,w)\in E(\mathcal{G}). \end{aligned}$$

Therefore $a\big(v,\sigma(v)\big)^2 = a\big(w,\sigma(w)\big)$, if $(v,w)\in E(\mathcal{G})$. Observe that since $\mathcal{G}$ is not a directed graph, (w,v) is also an edge of $\mathcal{G}$, then $a\big(w,\sigma(w)\big)^2 = a\big(v,\sigma(v)\big)$ and $a\big(v,\sigma(v)\big)^4 = a\big(v,\sigma(v)\big)$. Moreover, since $\mathcal{G}$ is connected, and $a\big(v,\sigma(v)\big)\neq 0$, then $a\big(v,\sigma(v)\big) = c\big(v,\sigma(v)\big) = 1$ for all $v\in V(\mathcal{G})$. Therefore we can write:

$$\begin{aligned} f(x_v) &= x_{\sigma(v)},\\ f(z_v) &= z_{\sigma(v)} + d(m_{z_v}). \end{aligned}$$

We shall see that in this case $\sigma\colon V(\mathcal{G})\to V(\mathcal{G})$ is, in fact, an element in $\mathrm{Aut}(\mathcal{G})$. We first show that σ is a full endomorphism of the graph $\mathcal{G}$, that is, $(v,w)\in E(\mathcal{G})$ if and only if $\big(\sigma(v),\sigma(w)\big)\in E(\mathcal{G})$. Indeed, $(v,w)\in E(\mathcal{G})$ if and only if there is a summand $x_v x_w x_2^{5n+3}$ in dz_v, and this is equivalent to having a summand $x_{\sigma(v)}x_{\sigma(w)}x_2^{5n+3}$ in $f(dz_v) = df(z_v) = d(z_{\sigma(v)})$, or equivalently, $\big(\sigma(v),\sigma(w)\big)\in E(\mathcal{G})$. Now, since for every $v\in V(\mathcal{G})$, $f(dz_v) = d(z_{\sigma(v)})$, σ is one-to-one on the neighborhood of every vertex. Thus $\sigma\in\mathrm{Aut}(\mathcal{G})$, which finally proves that if $\deg(f)\neq 0$, then f falls into case (1) of Lemma 2.12. □

Gathering Lemmas 2.11, and 2.12, we have succeeded in proving the following:

LEMMA 2.13. *The monoid of homotopy classes of self-maps of $M_n(\mathcal{G})$ is:*

$$[M_n(\mathcal{G}), M_n(\mathcal{G})] \cong \mathrm{Aut}(\mathcal{G}) \sqcup \{f\colon\ \deg(f)=0\}.$$

As a consequence, we obtain our main result, Theorem 2.9.

Proof of Theorem 2.9: From Proposition 2.7, we know that self-maps f with $\deg(f) = 0$ are not self-homotopy equivalences. Then, using Lemma 2.11, we conclude that $\mathrm{Aut}(\mathcal{G}) \cong \mathcal{E}(M_n(\mathcal{G}))$.

3. Inflexible manifolds

A closed, oriented and connected manifold M is said to be inflexible if the set of all the possible degrees of its continuous self-maps is finite. As the degree is multiplicative, this condition is equivalent to ask for the set of all the possible degrees to be a subset of $\{-1,0,1\}$. Inflexible manifolds naturally appear within the framework of functorial seminorms on singular homology developed by Gromov [**12**, **13**] and derived degree theorems (e.g. [**6**, Remark 2.6]): let M be a closed, oriented and connected manifold with fundamental class c_M. If there exists a functorial seminorm on singular homology $|\cdot|$ such that $|c_M|\neq 0$, then M is inflexible. In this way, oriented closed connected hyperbolic manifolds are shown to be inflexible; they do have non trivial simplicial volume, the value of ℓ^1-seminorm applied to the fundamental class [**12**, Section 0.3]. But ℓ^1-seminorm is trivial on simply connected manifolds [**12**, Section 3.1], which led Gromov to raise the question of whether every functorial seminorm on singular homology is trivial on all simply connected spaces [**13**, Remark (b) in 5.35]. This question is solved in the negative in [**6**] by constructing functorial seminorms associated to simply connected inflexible manifolds. Therefore, inflexible manifolds are extraordinary objects and still not many examples are known. Indeed, all the examples found in literature show low levels of connectivity when observing their minimal Sullivan models [**1**, **3**, **5**, **6**].

In this section, we closely follow the lines of [**5**] and provide new examples of inflexible manifolds whose Sullivan models are as highly connected as desired (see Corollary 3.2). To that purpose, we first need to recall some tools.

First, we recall that for elliptic Sullivan algebras, the Barge and Sullivan obstruction theory ([**4**, **21**]) decides if there exists a manifold (over $\mathbb{Z}$) of the same rational homotopy type of that algebra. Roughly speaking, the obstruction theory is trivial when the formal dimension of the algebra is not congruent to 0 (mod 4). Our examples will fall into this case.

Secondly, we recall a construction from [**5**, Proposition 3.1, Lemma 3.2]. Let A be a 1-connected elliptic Sullivan algebra of formal dimension $2m$. Then, we can construct a 1-connected elliptic Sullivan algebra of formal dimension $4m-1$ as follows: we choose a representative of the fundamental class of A, let us say x with $|x| = 2m$ and we define $\tilde{A} = (A \otimes \Lambda(z), dz = x)$. The algebra $\tilde{A}$ inherits properties of A: it is elliptic, the monoid of self-maps of $\tilde{A}$ coincides with the monoid of self-maps of A, and finally, the connectivity is preserved.

At this point, we wish to explain our strategy. We have proved that $M_n(\mathcal{G})$ from Definition 2.1 is an elliptic Sullivan algebra of even dimension, let us say $2m$, (see Lemma 2.8). By means of the construction above, we define a new elliptic Sullivan algebra $\tilde{M}_n(\mathcal{G})$ of formal dimension $4m-1$. The monoid of self-maps of both algebras coincide, hence $\mathcal{E}(\tilde{M}_n(\mathcal{G}))$ is finite which implies that $\tilde{M}_n(\mathcal{G})$ is inflexible. Also $\tilde{M}_n(\mathcal{G})$ and $M_n(\mathcal{G})$ have the same connectivity. Now, as we have already mentioned, the obstruction of Barge and Sullivan is trivial and $\tilde{M}_n(\mathcal{G})$ has the rational homotopy type of a manifold. Even more, by Proposition A.1, $\tilde{M}_n(\mathcal{G})$ has the rational homotopy type of $\mathcal{M}_n(\mathcal{G})$, a closed smooth manifold with the same connectivity as $\tilde{M}_n(\mathcal{G})$.

We can now prove the main theorem in this section.

THEOREM 3.1. *For any finite group G and any integer $n \geq 1$, G is the group of self-homotopy equivalences of the rationalization of an inflexible manifold which is $(30n+17)$-connected.*

PROOF. Take $\mathcal{G}$ a graph such that $\operatorname{Aut}(\mathcal{G}) \cong G$, [**11**]. For any $n \geq 1$, consider the algebra $M_n(\mathcal{G})$ from Definition 2.1. We know from Theorem 2.9 that $\mathcal{E}(M_n(\mathcal{G})) \cong \operatorname{Aut}(\mathcal{G})$. We also know from Lemma 2.8 that $M_n(\mathcal{G})$ is an elliptic minimal Sullivan algebra of formal dimension an even integer $2\delta = (18n+22)\deg x_1 + 2|V(\mathcal{G})|\deg x_v$. We apply [**5**, Proposition 3.1] to obtain an elliptic Sullivan algebra of formal dimension $\tilde{\delta} = 4\delta - 1$, which we denote $\tilde{M}_n(\mathcal{G})$, such that $\mathcal{E}(\tilde{M}_n(\mathcal{G})) \cong \mathcal{E}(M_n(\mathcal{G})) \cong G$, by Lemma 2.13. Since $\tilde{M}_n(\mathcal{G})$ is elliptic, every non zero degree map must be a self-homotopy equivalence by Proposition 2.7, and since $\mathcal{E}(\tilde{M}_n(\mathcal{G}))$ is finite, the set of degrees is finite as well and $\tilde{M}_n(\mathcal{G})$ is an inflexible Sullivan algebra. Now $\tilde{\delta} \not\equiv 0 \pmod 4$, so by Proposition A.1, $\tilde{M}_n(\mathcal{G})$ is the rationalization of a $(30n+17)$-connected manifold $\mathcal{M}_n(\mathcal{G})$. Observe that $\mathcal{M}_n(\mathcal{G})$ is also inflexible since so is $\tilde{M}_n(\mathcal{G})$. Thus

$$G \cong \operatorname{Aut}(\mathcal{G}) \cong \mathcal{E}(M_n(\mathcal{G})) \cong \mathcal{E}(\tilde{M}_n(\mathcal{G})) \cong \mathcal{E}(\mathcal{M}_n(\mathcal{G})_{\mathbb{Q}}),$$

where $\mathcal{M}_n(\mathcal{G})_{\mathbb{Q}}$ is the rational homotopy type of $\mathcal{M}_n(\mathcal{G})$. □

In particular, we immediately obtain the following.

COROLLARY 3.2. There exist infinitely many non-homotopically equivalent inflexible manifolds as highly connected as desired.

We end with some results on the existence of strongly chiral manifolds, that is, manifolds that do not admit orientation-reversing self-maps of degree -1 (see [**20**], [**1**] or [**5**]).

COROLLARY 3.3. For any $n \geq 1$ and $k > 1$, there exists a rationally $(30n+17)$-connected strongly chiral manifold M of dimension $(1080+720k)n^2+(1968+872k)n+264k+791$.

PROOF. Let $\mathcal{G}$ be a connected simple graph with k vertices, and let $\mathcal{M}_n(\mathcal{G})$ be the corresponding manifold obtained in the proof of Theorem 3.1 above, its minimal Sullivan algebra being $\tilde{M}_n(\mathcal{G})$. Any self-map of $\tilde{M}_n(\mathcal{G})$ is shown in [**5**, Lemma 3.2] to verify that $\deg(\tilde{f}) = \deg(\tilde{f}|_{M_n(\mathcal{G})})^2$, thus it has degree 0 or 1. □

Manifolds provided by Corollary 3.3 can be used to construct inflexible and strongly chiral product manifolds by exploiting techniques from [**19**]. In [**19**, Example 3.7], it is shown that given a simply connected inflexible manifold N one can construct an inflexible product manifold $M \times N$ just by considering M a closed oriented inflexible non-simply connected manifold that does not admit maps of non-zero degree from direct products and $\dim N < \dim M$. Here we prove an "inverse":

COROLLARY 3.4. Let M be a non necessarily simply connected closed oriented inflexible (resp. strongly chiral) manifold that does not admit maps of non-zero degree from direct products. Then there exist simply connected strongly chiral manifolds N such that $\dim N > \dim M$ and $M \times N$ is inflexible (resp. strongly chiral).

PROOF. Let $m = \dim M$, and $n \geq 1$ be an integer such that $m < 30n+17$. Let N be any of the $(30n+17)$-connected strongly chiral manifolds from Corollary 3.3. Since $H^m(N;\mathbb{Q}) = 0$, any continuous map $f\colon N \to M$ maps $H^m(M;\mathbb{Q})$ to 0. Thus the pair (M, N) is under the assumptions of [**19**, Theorem 1.4], and the result follows from [**19**, Corollary 1.5(c)] (resp. [**19**, Corollary 1.5(a)]). □

Appendix A. By Pascal Lambrechts and Don Stanley

Realization of rationally Poincaré duality complexes by highly connected bounding manifolds.

PROPOSITION A.1. *Let X be a k-connected rational space ($k \geq 1$) whose rational cohomology algebra satisfies Poincaré duality in dimension $n \geq 5$. When n is a multiple of 4, suppose furthermore that the quadratic form on $H^{n/2}(X;\mathbb{Q})$ has signature 0 and descends from a quadratic form over $\mathbb{Z}$. Then there exists a k-connected closed smooth manifold rationally equivalent to X and that is a boundary of a compact oriented manifold.*

PROOF. Let ε be some trivial vector bundle of rank $N > n$ over X. Then the Thom space of ε, $\mathrm{Th}(\varepsilon)$, is homotopy equivalent to $\Sigma^N(X_+)$ which is a wedge of rational spheres. Since X has Poincaré duality in dimension n there is a homotopy class $\alpha \in \pi_{n+N}(\mathrm{Th}(\varepsilon))$ whose image under the Hurewicz map is non trivial. One can pick a representative $f\colon S^{n+N} \to \mathrm{Th}(\varepsilon)$ of α transverse to X and then the preimage $M' := f^{-1}(X) \subset S^{n+N}$ is a smooth closed n-manifold with trivial stable normal bundle which comes with a map $f\colon M' \to X$ non trivial in $H^n(-;\mathbb{Q})$. Since X is k-connected, the classical surgery argument makes M' k-connected. Further surgeries of dimensions $\geq k$ can turn f into a rational homology equivalence up to the middle dimension. As explained in [**21**, Section 13] or [**4**] the obstruction to do surgery in the middle dimension in our case vanish if the dimension is not a multiple of 4 or if the dimension is a multiple of 4 and the quadratic form on $H^{n/2}(X;\mathbb{Q})$ is equivalent to $\sum_{i=1}^{l}(x_i^2 - y_i^2)$ (that is it descends from a quadratic form over $\mathbb{Z}$ and the signature vanishes). This gives a k-connected manifold M with a rational homotopy equivalence $M \simeq_{\mathbb{Q}} X$. Furthermore all the Pontrjagin and Stiefel-Whitney numbers of M' vanish because it has a trivial stable normal bundle, which implies that M', and hence M, is the boundary of an oriented compact manifold. □

REMARK A.2. Jim Fowler and Zhixu Su have proved that we cannot drop the hypothesis that the signature vanishes when the dimension is a multiple of 4, even if X is realizable by a smooth manifold. Indeed, consider the unique simply-connected rational homotopy type X with $H^*(X;\mathbb{Q}) \cong \mathbb{Q}[x]/(x^3)$ with $\deg(x) = 16$. This rational homotopy type is realizable by a closed smooth manifold but it cannot be 2-connected ([**10**, Theorem B]).

References

[1] M. Amann, *Degrees of self-maps of simply connected manifolds*, Int. Math. Res. Not. IMRN **18** (2015), 8545–8589. MR3417685

[2] M. Arkowitz, *Problems on self-homotopy equivalences*, Groups of homotopy self-equivalences and related topics (Gargnano, 1999), Contemp. Math., vol. 274, Amer. Math. Soc., Providence, RI, 2001, pp. 309–315. MR1817020

[3] M. Arkowitz and G. Lupton, *Rational obstruction theory and rational homotopy sets*, Math. Z. **235** (2000), no. 3, 525–539. MR1800210

[4] J. Barge, *Structures différentiables sur les types d'homotopie rationnelle simplement connexes*, Ann. Sci. École Norm. Sup. (4) **9** (1976), no. 4, 469–501. MR0440574

[5] C. Costoya and A. Viruel, *Every finite group is the group of self-homotopy equivalences of an elliptic space*, Acta Math. **213** (2014), no. 1, 49–62. MR3261010

[6] D. Crowley and C. Löh, *Functorial seminorms on singular homology and (in)flexible manifolds*, Algebr. Geom. Topol. **15** (2015), no. 3, 1453–1499. MR3361142

[7] Y. Félix, *La dichotomie elliptique-hyperbolique en homotopie rationnelle* (French, with English summary), Astérisque **176** (1989), 187. MR1035582

[8] Y. Félix, S. Halperin, and J.-C. Thomas, *Elliptic spaces*, Bull. Amer. Math. Soc. (N.S.) **25** (1991), no. 1, 69–73. MR1085825

[9] Y. Félix, S. Halperin, and J.-C. Thomas, *Rational homotopy theory*, Graduate Texts in Mathematics, vol. 205, Springer-Verlag, New York, 2001. MR1802847

[10] J. Fowler and Z. Su, *Smooth manifolds with prescribed rational cohomology ring*, Geom. Dedicata **182** (2016), 215–232. MR3500384

[11] R. Frucht, *Herstellung von Graphen mit vorgegebener abstrakter Gruppe* (German), Compositio Math. **6** (1939), 239–250. MR1557026

[12] M. Gromov, *Volume and bounded cohomology*, Inst. Hautes Études Sci. Publ. Math. **56** (1982), 5–99 (1983). MR686042

[13] M. Gromov, *Metric structures for Riemannian and non-Riemannian spaces*, Progress in Mathematics, vol. 152, Birkhäuser Boston, Inc., Boston, MA, 1999. Based on the 1981 French original [MR0682063 (85e:53051)]; With appendices by M. Katz, P. Pansu and S. Semmes; Translated from the French by Sean Michael Bates. MR1699320

[14] S. Halperin, *Finiteness in the minimal models of Sullivan*, Trans. Amer. Math. Soc. **230** (1977), 173–199. MR0461508

[15] H. Hopf, *On Some Properties of One-Valued Transformations of Manifolds*, Proc. Nat. Acad. Sci. U.S.A. **14** (1928), 206–214

[16] H. Hopf, *Zur Algebra der Abbildungen von Mannigfaltigkeiten* (German), J. Reine Angew. Math. **163** (1930), 71–88. MR1581232

[17] D. W. Kahn, *Realization problems for the group of homotopy classes of self-equivalences*, Math. Ann. **220** (1976), no. 1, 37–46. MR0402730

[18] D. W. Kahn, *Some research problems on homotopy-self-equivalences*, Groups of self-equivalences and related topics (Montreal, PQ, 1988), Lecture Notes in Math., vol. 1425, Springer, Berlin, 1990, pp. 204–207. MR1070586

[19] C. Neofytidis, *Degrees of self-maps of products*, Int. Math. Res. Not. (to appear).

[20] V. Puppe, *Simply connected 6-dimensional manifolds with little symmetry and algebras with small tangent space*, Prospects in topology (Princeton, NJ, 1994), Ann. of Math. Stud., vol. 138, Princeton Univ. Press, Princeton, NJ, 1995, pp. 283–302. MR1368664

[21] D. Sullivan, *Infinitesimal computations in topology*, Inst. Hautes Études Sci. Publ. Math. **47** (1977), 269–331 (1978). MR0646078

[22] A. T. White, *Graphs, groups and surfaces*, 2nd ed., North-Holland Mathematics Studies, vol. 8, North-Holland Publishing Co., Amsterdam, 1984. MR780555

Departamento de Computación, Universidade da Coruña, Campus de Elviña, s/n, 15071 - A Coruña, Spain

Email address: cristina.costoya@udc.es

Departamento de Álgebra, Geometría y Topología, Universidad de Málaga, E-29071-Málaga, Spain

Email address: davmenmar@uma.es

Departamento de Álgebra, Geometría y Topología, Universidad de Málaga, E-29071-Málaga, Spain

Email address: viruel@uma.es

IRMP, Université catholique de Louvain, 2 Chemin du Cyclotron, B-1348 Louvain-la-Neuve, Belgium

Email address: pascal.lambrechts@uclouvain.be

Department of Mathematics and Statistics, University of Regina, College West, Regina, Canada

Email address: stanley@math.uregina.ca

Contemporary Mathematics
Volume **708**, 2018
http://dx.doi.org/10.1090/conm/708/14262

A Dundas-Goodwillie-McCarthy theorem for split square-zero extensions of exact categories

Emanuele Dotto

ABSTRACT. Given a bimodule M over an exact category C, we define an exact category $C \ltimes M$ with a projection to C. This construction classifies certain split square-zero extensions of exact categories. We show that the trace map induces an equivalence between the relative K-theory of $C \ltimes M$ and its relative topological cyclic homology, after p-completion. When applied to the bimodule $\hom(-, - \otimes_A M)$ on the category of finitely generated projective modules over a ring A we recover the Dundas-Goodwillie-McCarthy theorem for split square-zero extensions of rings.

Introduction

The celebrated Dundas-Goodwillie-McCarthy theorem of [**McC97**], [**Dun97**], [**DGM13**] establishes a relationship between the K-theory and the topological cyclic homology of a nilpotent extension of ring spectra, and it led to several fundamental calculations of algebraic K-theory. The proof builds on the base case of split square-zero extensions of rings: if A is a ring and M is an A-bimodule, the cyclotomic trace map induces an equivalence between the reduced K-theory of the semi-direct product $A \ltimes M$ and its reduced topological cyclic homology. The methods employed in the proof involve objects which are defined in the more general framework of bimodules over linear categories. The aim of this paper is to introduce the notion of split square-zero extension of exact categories, and to show that the arguments of [**DGM13**] in fact prove a stronger result, which compares the K-theory and the TC of these more general extensions.

A bimodule over a category enriched in Abelian groups C is an enriched functor $M\colon C^{op} \otimes C \to Ab$ to the category of Abelian groups. We define the semi-direct product of C and M to be the category $C \ltimes M$ with the same objects as C, morphism groups

$$(C \ltimes M)(a,b) = C(a,b) \oplus M(a,b),$$

and where the composition of $(f,m)\colon b \to c$ and $(g,n)\colon a \to b$ is defined by the formula

$$(f,m) \circ (g,n) = \big(f \circ g, f_*n + g^*m \in M(a,c)\big)$$

(see §1). When the category C has one object this construction recovers the standard semi-direct product of rings and bimodules.

2010 *Mathematics Subject Classification.* Primary 19D55, 18E10.

If C has an exact structure and M is exact in both variables, we show in 1.8 that $C \ltimes M$ has an induced exact structure. Projecting off the module summand on morphisms gives an exact functor $p\colon C \ltimes M \to C$. The aim of this paper is to compare the relative K-theory of $C \ltimes M$ and its relative topological cyclic homology.

Main Theorem. *Let C be an exact category and $M\colon C^{op} \otimes C \to Ab$ an additive functor which is exact in both variables. Then the square of spectra*

$$\begin{array}{ccc} K(C \ltimes M) & \xrightarrow{\mathrm{Tr}} & \mathrm{TC}(C \ltimes M) \\ {\scriptstyle p_*}\downarrow & & \downarrow{\scriptstyle p_*} \\ K(C) & \xrightarrow[\mathrm{Tr}]{} & \mathrm{TC}(C) \end{array}$$

induced by the cyclotomic trace map is homotopy cartesian after completion at any prime.

Example. Suppose that $C = \mathcal{P}_A$ is the category of finitely generated projective modules over a ring A, and that M is an A-bimodule. There is an equivalence of exact categories

$$\mathcal{P}_A \ltimes \hom_A(-, - \otimes_A M) \simeq \mathcal{P}_{A \ltimes M}$$

over $\mathcal{P}_A$ (see 1.11), and our main theorem recovers the theorem of [**McC97**] for the split square-zero extension of rings $A \ltimes M \to A$.

Example. The exactness condition on the bimodule $M\colon C^{op} \otimes C \to Ab$ is automatically satisfied if the exact structure on C is split-exact. Thus our theorem always applies to direct-sum K-theory. One can for instance consider the category $C = \mathrm{Vect}_X$ of vector bundles over a scheme X and the bimodule $\hom(-, - \otimes V)$ for a fixed vector bundle V, and obtain the corresponding homotopy cartesian square for direct-sum K-theory.

Example. The theorem above applies also to categories that are not split-exact. Consider for example the category $\mathcal{M}_A$ of finitely generated modules over a ring A. Given a ring homomorphism $\phi\colon R \to A$, we equip $\mathcal{M}_A$ with the exact structure defined by the exact sequences in $\mathcal{M}_A$ which become split-exact after restricting scalars, that is whose image under the functor $\phi^*\colon \mathcal{M}_A \to Mod_R$ to the category of R-modules is split-exact (see [**Hel58**, §7]). Then any R-bimodule M defines a bimodule on the category $\mathcal{M}_A$ by sending two finitely generated A-modules P and Q to

$$\hom_R(\phi^* P, (\phi^* Q) \otimes_R M).$$

This bimodule is exact in each variable.

In §1 we show that the construction $C \ltimes M \to C$ classifies certain split square-zero extensions of exact categories: the exact functors $p\colon D \to C$ which admit an essentially surjective exact splitting, and whose kernel has zero composition and defines a bi-exact bimodule on C. In §2 we define an exact structure on $C \ltimes M$ and examine its K-theory. We employ the techniques of [**DGM13**] to determine a decomposition

$$K(C \ltimes M) \simeq \coprod_{c \in S_\bullet C} S_\bullet M(c, c),$$

where $S_\bullet M$ is an extension of the bimodule M to the $S_\bullet$-construction of C. The main ingredient of this decomposition is that the zero section induces a levelwise essentially surjective functor $S_\bullet C \to S_\bullet(C \ltimes M)$, which might be surprising when C is not split-exact. Finally in §3 we run the argument of [**McC97**] and use functor calculus on the decomposition of $K(C \ltimes M)$ above to prove the main theorem.

1. Split square-zero extensions of exact categories and bimodules

Let Ab be the symmetric monoidal category of Abelian groups with respect to the tensor product.

DEFINITION 1.1. A split extension of Ab-enriched categories is a triple (p, s, U), where $p\colon D \to C$ and $s\colon C \to D$ are Ab-enriched functors, $U\colon p \circ s \Rightarrow \mathrm{id}_C$ is a natural isomorphism, and the section s is essentially surjective.

We call (p, s, U) square-zero if for every pair of composable morphisms f and g in D with $p(f) = 0$ and $p(g) = 0$ in C, the composite $g \circ f$ is zero in D.

If C and D are exact categories we say that (p, s, U) is a split extension of exact categories if it is a split extension on the underlying Ab-enriched categories and p and s exact functors.

We observe that the isomorphism $U\colon p \circ s \Rightarrow \mathrm{id}_C$ insures that s is essentially injective. Since it is assumed to be essentially surjective, it follows that D and C have essentially the same objects.

EXAMPLE 1.2. Split square-zero extensions of Ab-enriched categories with one object are precisely the split square-zero extensions of rings $f\colon R \to A$. These are all of the form $A \ltimes M \to A$ for some A-bimodule M, where $A \ltimes M$ is the Abelian group $A \oplus M$ with multiplication

$$(a, m) \cdot (b, n) = (ab, an + mb).$$

What follows is the generalization of this classification for exact categories.

DEFINITION 1.3. A bimodule over an Ab-enriched category is an Ab-enriched functor $M\colon C^{op} \otimes C \to Ab$. For a morphism $f\colon a \to b$ and an object c in C we denote

$$f_* = M(\mathrm{id}_c \otimes f)\colon M(c, a) \to M(c, b) \quad \text{and} \quad f^* = M(f \otimes \mathrm{id}_c)\colon M(b, c) \to M(a, c)$$

the induced maps. If C has an exact structure we require $M\colon C^{op} \otimes C \to Ab$ to be exact in both variables. In particular $M(0, c) = M(c, 0) = 0$.

Given a bimodule M over an Ab-enriched category C we define an Ab-enriched category $C \ltimes M$ with the same objects of C, and morphism groups $(C \ltimes M)(a, b) = C(a, b) \oplus M(a, b)$. Composition is defined by

$$\big(f\colon b \to c, m \in M(b, c)\big) \circ \big(g\colon a \to b, n \in M(a, b)\big) = \big(f \circ g, f_* n + g^* m \in M(a, c)\big).$$

The projection onto the morphisms of C defines a functor $p\colon C \ltimes M \to C$. This functor is split (strictly) by the functor $s\colon C \to C \ltimes M$ which is the identity on objects and that sends a morphism f to $(f, 0)$.

REMARK 1.4. If C is a groupoid, there is a functor $\Delta\colon C \to C^{op} \times C$ that sends $f\colon c \to d$ to $(f^{-1}, f)\colon (c, c) \to (d, d)$. The category $C \ltimes M$ is then naturally isomorphic to the Grothendieck construction of the composite $M \circ \Delta\colon C \to C^{op} \times C \to Ab$, in symbols

$$C \ltimes M \cong C \wr (M \circ \Delta).$$

In particular by denoting iC the groupoid of isomorphisms of C we have that

$$i(C \ltimes M) = (iC) \ltimes M \cong (iC) \wr (M \circ \Delta),$$

where we denoted the restriction $(iC)^{op} \times iC \to Ab$ also by M.

EXAMPLE 1.5. Considering a ring A as an Ab-enriched category with one object, an Ab-enriched functor $M\colon A^{op} \otimes A \to Ab$ is precisely a bimodule over A in the classical sense. The construction $A \ltimes M$ above is the semi-direct product of rings and bimodules.

EXAMPLE 1.6. Let $\mathcal{P}_A$ be the exact category of finitely generated projective modules over a ring A, and let M be an A-bimodule. The functor

$$\hom_A(-, - \otimes_A M)\colon \mathcal{P}_A^{op} \otimes \mathcal{P}_A \longrightarrow Ab$$

is a bimodule on $\mathcal{P}_A$. It is exact in both variables since $\mathcal{P}_A$ is a split-exact category.

DEFINITION 1.7. Let C be an exact category. A sequence $a \xrightarrow{(i,m)} b \xrightarrow{(q,n)} c$ in $C \ltimes M$ is exact if the underlying sequence $a \xrightarrow{i} b \xrightarrow{q} c$ is exact in C, and if

$$q_* m + i^* n = 0 \quad \in M(a, c).$$

PROPOSITION 1.8. *Let C be an exact category and $M\colon C^{op} \otimes C \to Ab$ a functor which is exact in both variables. The exact sequences of* 1.7 *define an exact structure on $C \ltimes M$, and the triple* $(p\colon C \ltimes M \to C, s, \mathrm{id})$ *is a split square-zero extension of exact categories.*

PROOF. We verify the axioms of an exact category as presented in [**Kel90**, A1]. First we need to show that the sequences of 1.7 are closed under isomorphisms and that they are "exact". A morphism (f, m) is invertible in $C \ltimes M$ if and only if f is invertible in C. Moreover the condition $q_* m + i^* n = 0$ simply says that the second component of the composition $(q, n) \circ (i, m)$ is zero. It follows that the collection of exact sequences is closed under isomorphisms.

In the language of [**Kel90**] verifying exactness means showing that if $a \xrightarrow{(i,m)} b \xrightarrow{(q,n)} c$ is as in 1.7 then (i, m) is a kernel for (q, n), and that (q, n) is a cokernel for (i, m). The composite $(q, n) \circ (i, m) = (q \circ i, q_* m + i^* n)$ is zero by definition. We need to show that given any $(f, l)\colon d \to b$ with

$$(q, n) \circ (f, l) = (q \circ f, q_* l + f^* n) = 0$$

there is a unique map $(u, x)\colon d \to a$ with $(i, m) \circ (u, x) = (f, l)$. Since i is a kernel of q in C there is a unique $u\colon d \to b$ with the corresponding universal property in C. The element $x \in M(d, a)$ must be the unique solution of the equation

$$l = u^* m + i_* x.$$

Since $i_*\colon M(d, a) \to M(d, b)$ is injective by exactness of M, if a solution exists it must be unique. For the existence, $l - u^* m$ must belong to the image of i_* which is equal to the kernel of $q_*\colon M(d, b) \to M(d, c)$. We calculate that

$$q_*(l - u^* m) = q_* l - q_* u^* m = -f^* n - u^* q_* m = -f^* n + u^* i^* n = -f^* n + f^* n = 0$$

is indeed zero. The third equality holds because $-q_* m = i^* n$ by exactness. A dual proof shows that (q, n) is a cokernel for (i, m).

Next we show axiom Ex0 from [**Kel90**, A1]. The object 0 in C defines a zero-object in $C \ltimes M$ since the Abelian groups $C(0, c)$, $C(c, 0)$, $M(c, 0)$ and $M(0, c)$ are trivial for every object c of C. The identity map $(\mathrm{id}_0, 0)\colon 0 \to 0$ is an admissible epimorphism, since $(\mathrm{id}_0, 0) \circ (\mathrm{id}_0, 0) = (\mathrm{id}_0, 0)$ has null second component. This proves Ex0.

Now we show Ex2: pullbacks of admissible epimorphisms exist in $C \ltimes M$, and are admissible epimorphisms. Given a diagram $c \xrightarrow{(g,n)} b \overset{(f,m)}{\twoheadleftarrow} a$ in $C \ltimes M$, let

$$\begin{array}{ccc} p & \xrightarrow{\overline{g}} & a \\ {\scriptstyle \overline{f}}\downarrow & & \downarrow{\scriptstyle f} \\ c & \xrightarrow[g]{} & b \end{array}$$

be the pullback of the underlying diagram in C. First we find $\overline{m} \in M(p, c)$ and $\overline{n} \in M(p, a)$ such that

$$\begin{array}{ccc} p & \xrightarrow{(\overline{g},\overline{n})} & a \\ {\scriptstyle (\overline{f},\overline{m})}\downarrow & & \downarrow{\scriptstyle (f,m)} \\ c & \xrightarrow[(g,n)]{} & b \end{array}$$

commutes. Commutativity of this square is equivalent to the condition $f_*\overline{n} + \overline{g}^* m = g_*\overline{m} + \overline{f}^* n$. Choose $\overline{m} = 0$ and $\overline{n}$ in the preimage of $\overline{f}^* n - \overline{g}^* m$ by the surjective map $f_*\colon M(p, a) \twoheadrightarrow M(p, b)$. We show that p equipped with these maps satisfies the universal property

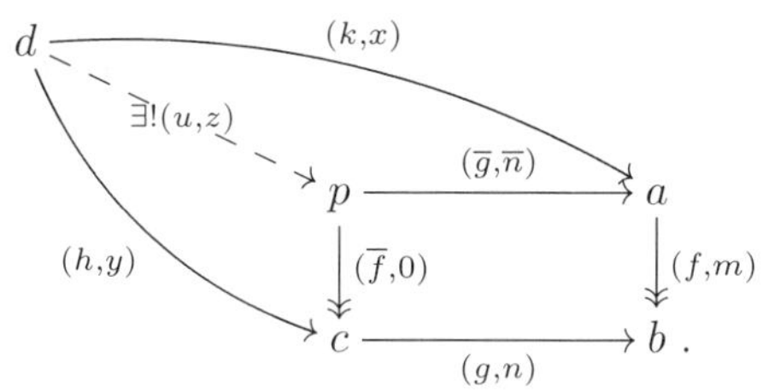

Let $u\colon d \to p$ be the universal map for the corresponding diagram in C. The element $z \in M(d, p)$ must be the unique solution of the system of equations

$$\begin{cases} \overline{g}_* z = x - u^*\overline{n} \\ \overline{f}_* z = y \end{cases} .$$

We observe that $x - u^*\overline{n}$ and y lie over the same element of $M(d, b)$, since

$$f_*(x - u^*\overline{n}) = f_*x - u^* f_*\overline{n} = f_*x - u^*(-\overline{g}^* m + \overline{f}^* n) = f_*x + k^*m - h^*n = g_*y.$$

The last equality holds by the commutativity of the outer diagram. Therefore a solution z to the equations above exists and is unique precisely when the map $(\overline{f}_*, \overline{g}_*)\colon M(d, p) \to M(d, c) \times_{M(d,b)} \times M(d, a)$ is an isomorphism, that is when

$$\begin{array}{ccc} M(d,p) & \xrightarrow{\overline{g}_*} & M(d,a) \\ {\scriptstyle \overline{f}_*}\downarrow & & \downarrow{\scriptstyle f_*} \\ M(d,c) & \xrightarrow[g_*]{} & M(d,b) \end{array}$$

is a pullback square. This follows from exactness of M in the second variable, and from the fact that kernels of f and $\overline{f}$ in C are isomorphic. To show that $(\overline{f},0)$ is an admissible epimorphism we choose an exact sequence $k \overset{i}{\rightarrowtail} p \overset{f}{\twoheadrightarrow} c$ in C. Then $k \overset{(i,0)}{\rightarrowtail} p \overset{(f,0)}{\twoheadrightarrow} c$ is clearly exact in $C \ltimes M$. A dual argument shows the analogous statement for pushouts of admissible monomorphisms, that is Ex2^{op} from [**Kel90**].

Axiom Ex1 states that admissible epimorphisms are closed under composition. That is, if

$$a \xrightarrow{(i,m)} b \xrightarrow{(q,n)} c \qquad\qquad a' \xrightarrow{(j,k)} c \xrightarrow{(p,l)} d$$

are exact, there exist an exact sequence $a'' \longrightarrow b \xrightarrow{(pq,p_*n+q^*l)} d$. By Ex2 we can form the pullback

$$\begin{array}{ccc} a'' & \xrightarrow{(\overline{j},\overline{k})} & b \\ {\scriptstyle (\overline{q},\overline{n})}\big\downarrow & & \big\downarrow{\scriptstyle (q,n)} \\ a' & \xrightarrow[(j,k)]{} & c \end{array}$$

in $C \ltimes M$, and the underlying square in C is also a pullback. Since C is an exact category the sequence $a'' \xrightarrow{\overline{j}} b \xrightarrow{pq} d$ is exact in C (see the third step in the proof of [**Kel90**, A1]). Thus we only need to verify that the composite $(p,l)\circ(q,n)\circ(\overline{j},\overline{k})$ is zero in $C \ltimes M$. By the commutativity of the square above this is the same as $(p,l)\circ(j,k)\circ(\overline{q},\overline{n})$, and $(p,l)\circ(j,k) = 0$ by exactness. This concludes the proof that the sequences of 1.7 form an exact structure.

The functors p and s are clearly exact. To see that (p,s,U) is square-zero, we need to show that the composition of morphisms of the form $(0,m)$ is zero. This is clear. □

Let $p\colon D \to C$ be the projection of a split extension of Ab-enriched categories (p,s,U). We let $\ker p\colon C^{op} \otimes C \to Ab$ be the bimodule defined by

$$\ker p(a,b) = \ker\big(D(s(a),s(b)) \xrightarrow{p} C(ps(a),ps(b))\big).$$

If the extension (p,s,U) is square-zero, there is a canonical functor $F\colon C \ltimes \ker p \to D$ that sends an object c to $s(c)$, and a morphism (f,m) to $s(f)+m$. The square-zero condition is used for showing that F preserves the composition of morphisms. Moreover the diagram

$$\begin{array}{ccc} C \ltimes \ker p & \xrightarrow{F} & D \\ \big\downarrow & & \big\downarrow{\scriptstyle p} \\ C & \underset{p\circ s}{\xrightarrow{\simeq}} & C \end{array}$$

commutes, and the functor $p \circ s$ is an equivalence of Ab-enriched categories. If (p,s,U) is a split square-zero extension of exact categories we observe that the bimodule $\ker p$ is automatically left exact in both variables.

PROPOSITION 1.9. *Let (p,s,U) be a split square-zero extension of Ab-enriched categories. The functor $F\colon C \ltimes \ker p \to D$ is an equivalence of Ab-enriched categories.*

Moreover, if p and s are exact and $\ker p$ *is right exact in each variable,* $F\colon C \ltimes \ker p \to D$ *is an equivalence of exact categories for the exact structure on* $C \ltimes \ker p$ *of* 1.8.

PROOF. The functor F is obviously essentially surjective, since s is by assumption. To see that it is fully faithful, we define an inverse of

$$F\colon C(a,b) \oplus \ker p(a,b) \longrightarrow D(s(a), s(b))$$

for every pair of objects a, b in C. Given a morphism $g\colon s(a) \to s(b)$ we let $U_*p(g)$ denote the composite

$$a \xrightarrow{U_a^{-1}} ps(a) \xrightarrow{p(g)} ps(b) \xrightarrow{U_b} b.$$

The inverse F^{-1} for the map above sends $g\colon s(a) \to s(b)$ to the pair

$$F^{-1}(g) = (U_*p(g), g - s(U_*p(g))).$$

For F^{-1} to be well-defined we need to verify that the second component lies in $\ker p(a,b)$, that is that $ps(U_*p(g)) = p(g)$. We start by remarking that by naturality of $U\colon ps \to \mathrm{id}$ the square

$$\begin{array}{ccc} psps(c) & \xrightarrow{U_{ps(c)}} & ps(c) \\ {\scriptstyle ps(U_c)}\downarrow & & \downarrow{\scriptstyle U_c} \\ ps(c) & \xrightarrow[U_c]{} & c \end{array}$$

commutes, and since U_c is an isomorphism this implies that $ps(U_c) = U_{ps(c)}$. Thus

$$ps(U_*p(g)) = ps(U_b) \circ psp(g) \circ ps(U_a^{-1}) = U_{ps(b)} \circ psp(g) \circ U_{ps(a)}^{-1} = p(g)$$

where the last equality holds by naturality of U. This shows that F^{-1} is well defined. It remains to verify that this is an inverse, namely that

$$FF^{-1}(g) = s(U_*p(g)) + g - s(U_*p(g)) = g,$$

and that

$$\begin{aligned} F^{-1}F(f,m) &= (U_*p(s(f)+m), s(f) + m - s(U_*p(s(f)+m))) \\ &= (U_*ps(f), s(f) + m - s(U_*ps(f))) \end{aligned}$$

equals (f, m). This is the case since by naturality $U_*ps(f) = f$.

Now we show that F is exact. By lemma 1.10 below one can choose an isomorphism between any exact sequence E in $C \ltimes \ker p$ and one of the form $a \xrightarrow{(i,0)} b \xrightarrow{(q,0)} c$. This gives an isomorphism between $F(E)$ and

$$F\big(a \xrightarrow{(i,0)} b \xrightarrow{(q,0)} c\big) = \big(s(a) \xrightarrow{s(i)} s(b) \xrightarrow{s(q)} s(c)\big)$$

which is exact by exactness of s.

It remains to show that if $E = \big(a \xrightarrow{(i,m)} b \xrightarrow{(q,n)} c\big)$ is any sequence of $C \ltimes \ker p$ whose image $F(E)$ is exact in D, then E is exact in $C \ltimes \ker p$. If $F(E)$ is exact, then the underlying sequence

$$\big(a \xrightarrow{i} b \xrightarrow{q} c\big) = pF(E)$$

is also exact. Moreover the composition $F(q,n) \circ F(i,m)$ is zero, and since F is faithful we must have $(q,n) \circ (i,m) = 0$ in $C \ltimes \ker p$, that is $q_*m + i^*n = 0$. $\square$

LEMMA 1.10. *Any exact sequence of $C \ltimes M$ is isomorphic (non canonically) to an exact sequence of the form* $a \xrightarrow{(i,0)} b \xrightarrow{(q,0)} c$.

PROOF. Let $a \xrightarrow{(i,m)} b \xrightarrow{(q,n)} c$ be an exact sequence of $C \ltimes M$. We find an isomorphism of the form

$$\begin{array}{ccccc} a & \xrightarrow{(i,m)} & b & \xrightarrow{(q,n)} & c \\ {\scriptstyle (\mathrm{id},0)}\downarrow & & \downarrow{\scriptstyle (\mathrm{id},x)} & & \downarrow{\scriptstyle (\mathrm{id},0)} \\ a & \xrightarrow[(i,0)]{} & b & \xrightarrow[(q,0)]{} & c \end{array}$$

for some $x \in M(b,b)$. The vertical maps are isomorphisms for any $x \in M(b,b)$. For the squares to commute, x needs to satisfy

$$i^*x = -m \qquad \text{and} \qquad q_*x = n.$$

This is the case precisely when the element $(-m,n)$ is in the image of the map

$$(i^*, q_*)\colon M(b,b) \longrightarrow M(a,b) \times_{M(a,c)} M(b,c).$$

Consider the diagram of Abelian groups

$$\begin{array}{ccccc} M(c,b) & \xrightarrow{\quad q_* \quad} & M(c,c) & = & M(c,c) \\ {\scriptstyle q^*}\downarrow & & \downarrow & & \downarrow{\scriptstyle q^*} \\ M(b,b) & \xrightarrow{(i^*,q_*)} & M(a,b) \times_{M(a,c)} M(b,c) & \longrightarrow & M(b,c) \\ {\scriptstyle j^*}\downarrow & & \downarrow & & \downarrow{\scriptstyle j^*} \\ M(a,b) & = & M(a,b) & \xrightarrow[q_*]{} & M(a,c)\,. \end{array}$$

The middle column is exact since it is the pullback of an exact sequence. Therefore (i^*, q_*) is surjective, and a lift of $(-m,n)$ exists. $\square$

EXAMPLE 1.11. Let A be a ring, M an A-bimodule, and $p\colon A \ltimes M \to A$ the projection with zero section $s\colon A \to A \ltimes M$. The induced functors $p_* = (-) \otimes_{A \ltimes M} A$ and $s_* = (-) \otimes_A (A \ltimes M)$ define a split square-zero extension of exact categories (s_* is essentially surjective by [**DGM13**, 1.2.5.4]). Therefore 1.9 provides an equivalence of exact categories

$$\mathcal{P}_{A \ltimes M} \xleftarrow{\simeq} \mathcal{P}_A \ltimes \ker p_*.$$

For $Q \in \mathcal{P}_A$ there is a natural isomorphism

$$s_*Q \cong Q \oplus (Q \otimes_A M)$$

where the right-hand side has module structure $(p', p \otimes n) \cdot (a, m) = (p'a, p \otimes na + p \otimes m)$. Under this isomorphism a module map $s_*Q \to s_*Q'$ is a matrix $\left(\begin{smallmatrix} \phi & 0 \\ f & \phi \otimes M \end{smallmatrix}\right)$ for module maps $\phi\colon Q \to Q'$ and $f\colon Q \to Q' \otimes_A M$ (see [**DGM13**, 1.2.5.1]). This gives a natural isomorphism of bimodules $\ker p_* \cong \hom_A(-, - \otimes_A M)$, and combined with 1.9 an equivalence of exact categories

$$\mathcal{P}_{A \ltimes M} \xleftarrow{\simeq} \mathcal{P}_A \ltimes \hom_A(-, - \otimes_A M).$$

2. The K-theory of $C \ltimes M$

Let C be an exact category and $M\colon C^{op} \otimes C \to Ab$ a bimodule over C which is exact in both variables. We use the techniques of [**DGM13**] to give a description of the K-theory $\mathrm{K}(C \ltimes M)$ in terms of the K-theory of bimodules of [**DGM13**, 1.3.4.2]. We recall from [**DGM13**, §1.3.3] that for every natural number p, the bimodule M extends to a bimodule $S_pM\colon S_pC^{op} \otimes S_pC \to Ab$ on S_pC, defined as the Abelian subgroup

$$S_pM(X,Y) \subset \bigoplus_{\theta \in Cat([1],[p])} M(X_\theta, Y_\theta)$$

of collections $\underline{m}$ satisfying the condition

$$X(a)^* m_\theta = Y(a)_* m_\rho$$

as elements of $M(X_\rho, Y_\theta)$, for all morphism $a\colon \rho \to \theta$ in $Cat([1],[p])$. In the terminology of §1 this condition expresses the commutativity of the diagram

$$\begin{array}{ccc} X_\rho & \xrightarrow{(0,m_\rho)} & Y_\rho \\ {\scriptstyle (X(a),0)}\downarrow & & \downarrow{\scriptstyle (Y(a),0)} \\ X_\theta & \xrightarrow[(0,m_\theta)]{} & Y_\theta \end{array}$$

in $C \ltimes M$. By iterating this construction we define, for all $\underline{p} = (p_1, \ldots, p_n)$, a bimodule on $S^{(n)}_{\underline{p}}C$

$$S^{(n)}_{\underline{p}}M = S_{p_1} \ldots S_{p_n} M.$$

A map $\sigma\colon \underline{p} \to \underline{q}$ induces a natural transformation $\sigma^*\colon S^{(n)}_{\underline{q}}M \Rightarrow S^{(n)}_{\underline{p}}M \circ ((\sigma^*)^{op} \otimes \sigma^*)$ defining an n-simplicial structure on $S^{(n)}_{\bullet}M$.

EXAMPLE 2.1. Consider the bimodule $\hom_A(-, - \otimes_A M)$ on the category $\mathcal{P}_A$ of finitely generated projective modules over a ring A, induced by an A-bimodule M. In this case $S^{(n)}_{\underline{p}}M$ is the Abelian group of natural transformations of diagrams

$$S^{(n)}_{\underline{p}}M(X,Y) = \hom_A(X, Y \otimes_A M),$$

where both the tensor product of a diagram of A-modules with M and the sum of natural transformations are pointwise.

Let $\coprod_C M$ be the following groupoid. It has the same objects of C, and only automorphisms. The automorphism group of c is $M(c,c)$, and composition is defined by addition in $M(c,c)$. It is the disjoint union of groups

$$\coprod\nolimits_C M = \coprod_{c \in ObC} M(c,c)$$

and its nerve is the disjoint union of the classifying spaces $\coprod_{c \in ObC} BM(c,c)$. This construction applied to $S^{(n)}_{\underline{p}}M$ defines a category

$$S^{(n)}(C;M)_{\underline{p}} := \coprod_{S^{(n)}_{\underline{p}}C} S^{(n)}_{\underline{p}}M.$$

The simplicial structure maps $\sigma^*\colon S^{(n)}_{\underline{q}}C \to S^{(n)}_{\underline{p}}C$ together with the natural transformations $\sigma^*\colon S^{(n)}_{\underline{q}}M \Rightarrow S^{(n)}_{\underline{p}}M \circ ((\sigma^*)^{op} \otimes \sigma^*)$ define an n-simplicial category $S^{(n)}_{\bullet}(C;M)$, whose realization is denoted

$$\mathrm{K}(C;M)_n = |\mathcal{N}_{\bullet}S^{(n)}_{\bullet}(C;M)| = |\coprod_{X \in ObS^{(n)}_{\bullet}C} S^{(n)}_{\bullet}BM(X,X)|.$$

There are structure maps

$$\mathrm{KR}(C;M)_n \wedge S^1 \longrightarrow \mathrm{K}(C;M)_{n+1}$$

induced by the inclusions of 1-simplices $S^{(n+1)}_{(1,p_1,\dots,p_n)} = S^{(n)}_{(p_1,\dots,p_n)}$, defined in a fashion analogous to the structure maps of $\mathrm{K}(C)$. Permuting the $S_{\bullet}$-constructions gives a Σ_n-action on $\mathrm{K}(C;M)_n$ compatible with the structure maps, defining a symmetric spectrum $\mathrm{K}(C;M)$.

REMARK 2.2. There is a category $C[M]$ with objects $\coprod_{c\in C} M(c,c)$ and morphism from m_c in the c-summand to $m_{c'}$ in c'-summand

$$C[M](m_c, m_{c'}) = \{f \in C(c,c') | f^* m_{c'} = f_* m_c\}.$$

This has an exact structure by defining a sequence to be exact if it is exact in C, and as simplicial sets

$$ObS_{\bullet}C[M] = \hom\left(\coprod_{S_{\bullet}C} S_{\bullet}M\right).$$

This induces an equivalence in K-theory $K(C[BM]) \simeq K(C;M)$, see [**DGM13**, §5.1.4].

REMARK 2.3. Our notation is off by a suspension factor compared to the notation of [**DM94**]. For $C = \mathcal{P}_A$ and $M = \hom_A(-, - \otimes_A M)$ what we denote $\mathrm{K}(\mathcal{P}_A;M)$ is $\mathrm{K}(\mathcal{P}_A;BM)$ in the notation of [**DM94**, 3.1].

Define a functor $\Psi\colon \coprod_C M \to i(C \ltimes M)$ by the identity on objects, and the inclusion at the identity $(\mathrm{id}_c, -)\colon M(c,c) \to C(c,c) \oplus M(c,c)$ on morphisms. This extends to a map of symmetric spectra $\mathrm{K}(C;M) \to \mathrm{K}(C \ltimes M)$.

THEOREM 2.4. *The map $\mathrm{K}(C;M) \to \mathrm{K}(C \ltimes M)$ is a levelwise equivalence of symmetric spectra above spectrum level one.*

PROOF. Since all our constructions are natural, $S_{\bullet}$ preserves equivalences, and the realization of bisimplicial sets preserves levelwise equivalences, it is enough to show that the map of the statement is an equivalence in spectrum degree one, that is that the functor

$$\Psi\colon \coprod_{S_{\bullet}C} S_{\bullet}M \longrightarrow iS_{\bullet}(C \ltimes M)$$

induces an equivalence on classifying spaces. We factor this map as

$$\Psi\colon \coprod_{S_{\bullet}C} S_{\bullet}M \longrightarrow tS_{\bullet}(C \ltimes M) \longrightarrow iS_{\bullet}(C \ltimes M),$$

where t is the set of isomorphisms of $S_{\bullet}(C \ltimes M)$ which lie in the image of Ψ. By the standard argument of [**Wal85**], the second map is an equivalence on realizations, as both spaces are equivalent to the realization of the objects of $S_{\bullet}(C \ltimes M)$ (see also [**DGM13**, §1.2.3.2]).

The first map is levelwise a fully faithful functor. We show that it is also essentially surjective, proving that it is a levelwise equivalence of categories. We show that any diagram $X \in S_p(C \ltimes M)$ is isomorphic to s_*p_*X, by a natural isomorphism $\phi\colon X \to s_*p_*X$ which at every object θ in $Cat([1],[p])$ is of the form $\phi_\theta = (\mathrm{id}, x_\theta)\colon X_\theta \to X_\theta$ for some $x_\theta \in M(X_\theta, X_\theta)$. This is a higher version of 1.10, where we proved this result for $p = 2$. For every morphism $\theta \leq \rho$ in $Cat([1],[p])$ let us denote the corresponding map in the diagram X by $(f_{\theta\rho}, m_{\theta\rho})\colon X_\theta \to X_\rho$. For ϕ to define a natural transformation, the elements x_θ need to satisfy the relations

$$f_{\theta\rho}^* x_\rho + m_{\theta\rho} = (f_{\theta\rho})_* x_\theta \qquad \in M(X_\theta, X_\rho) \tag{1}$$

for every $\theta \leq \rho$ in $Cat([1],[p])$. We represent an injective map $\theta\colon [1] \to [p]$ by listing its two values in the form $\theta = (j, j+k)$, where $1 \leq k \leq p$ and $0 \leq j \leq p-k$. We construct the elements $x_{(j,j+k)}$ inductively on k.

For $k = 1$ the elements $x_{(j,j+1)}$ lie on the diagonal of X, and we define $x_{(j,j+1)} = 0$. Since all the maps $X_{(j,j+1)} \to X_{(i,i+1)}$ in the diagram X are zero, these elements satisfy condition (1) for θ and ρ of the form $(j, j+1)$.

Now suppose that $x_{(i,i+l)}$ is defined for all $1 \leq l < k$ and $0 \leq i \leq p-l$, and let us define $x_{(j,j+k)}$. We observe that every non-identity map $X_{(i,i+l)} \to X_{(j,j+k)}$ with $l \leq k$ factors through $X_{(j,j+k-1)}$, where $x_{(j,j+k-1)}$ has already been inductively defined. Similarly every non-identity map $X_{(j,j+k)} \to X_{(i,i+l)}$ with $l \leq k$ factors through $X_{(j+1,j+k)}$, where $x_{(j+1,j+k)}$ has already been inductively defined. Thus for condition (1) to be satisfied among θ and ρ of the form $(j, j+l)$ with $l \leq k$, we only need to guarantee that the new elements $x_{(j,j+k)}$ are compatible with $x_{(j,j+k-1)}$ and $x_{(j+1,j+k)}$. Let us denote the corresponding maps by $(i,m)\colon X_{(j,j+k-1)} \to X_{(j,j+k)}$ and $(p,n)\colon X_{(j,j+k)} \to X_{(j+1,j+k)}$. The compatibility conditions that $x_{(j,j+k)}$ needs to satisfy are

$$i^* x_{(j,j+k)} = i_* x_{(j,j+k-1)} - m \qquad \text{and} \qquad p_* x_{(j,j+k)} = n + p^* x_{(j+1,j+k)}.$$

The elements $i_* x_{(j,j+k-1)} - m$ and $n + p^* x_{(j+1,j+k)}$ lie over the same point of $M(X_{(j,j+k-1)}, X_{(j+1,j+k)})$, and therefore a solution $x_{(j,j+k)}$ exists if the map (i^*, p_*) surjects onto the pullback P in the diagram

$$\begin{array}{ccccc}
M(X_{(j+k,j+k-1)}, X_{(j,j+k)}) & \xrightarrow{p_*} \!\!\!\!\to & M(X_{(j+k,j+k-1)}, X_{(j+1,j+k)}) & = & M(X_{(j+k,j+k-1)}, X_{(j+1,j+k)}) \\
\downarrow & & \downarrow & & \downarrow \\
M(X_{(j,j+k)}, X_{(j,j+k)}) & \xrightarrow{(i^*,p_*)} & P & \longrightarrow & M(X_{(j,j+k)}, X_{(j+1,j+k)}) \\
\downarrow{\scriptstyle i^*} & & \downarrow & & \downarrow{\scriptstyle i^*} \\
M(X_{(j,j+k-1)}, X_{(j,j+k)}) & = & M(X_{(j,j+k-1)}, X_{(j,j+k)}) & \xrightarrow[p_*]{} & M(X_{(j,j+k-1)}, X_{(j+1,j+k)})\,.
\end{array}$$

The vertical sequences are exact by exactness of M applied to the exact sequence induced by $(j, j+k-1, j+k)\colon [2] \to [p]$. Moreover the top left map p_* is surjective since p is an admissible epimorphism. Therefore (i^*, p_*) is surjective. $\square$

REMARK 2.5. Theorem 2.4 can be interpreted in terms of Grothendieck constructions. The equivalence of the statement factors as

$$\Psi\colon \coprod_{S_\bullet C} S_\bullet M \longrightarrow (iS_\bullet C) \ltimes S_\bullet M \xrightarrow{\simeq} iS_\bullet(C \ltimes M),$$

where the second map is an equivalence by 1.9, as $S_\bullet(C \ltimes M) \to S_\bullet C$ is a levelwise split square-zero extension of exact categories (the section is essentially surjective by the proof of 2.4). We saw in 1.4 that $(iS_\bullet C) \ltimes S_\bullet M$ is levelwise the Grothendieck construction of the composite

$$iS_pC \xrightarrow{\Delta} (iS_pC)^{op} \times iS_pC \xrightarrow{S_pM} Ab.$$

Similarly the category $\coprod_{S_\bullet C} S_\bullet M$ is levelwise the Grothendieck construction of the restriction to the discrete category of objects

$$ObS_pC \xrightarrow{\Delta} (ObS_pC)^{op} \times ObS_pC \xrightarrow{S_pM} Ab.$$

Theorem 2.4 can be rephrased as saying that the inclusion of objects $ObS_\bullet C \to iS_\bullet C$ induces an equivalence after geometric realization between the Grothendieck constructions

$$(ObS_\bullet C) \wr (S_\bullet M \circ \Delta) \xrightarrow{\simeq} (iS_\bullet C) \wr (S_\bullet M \circ \Delta).$$

Theorem 2.4 provides the following description of the homotopy fiber spectrum

$$\widetilde{\mathrm{K}}(C \ltimes M) = \mathrm{hof}\left(\mathrm{K}(C \ltimes M) \longrightarrow \mathrm{K}(C)\right).$$

Let $\widetilde{\mathrm{K}}(C; M)$ be the symmetric spectrum of [**DGM13**, 1.3.4.5], defined in degree n by the space

$$\widetilde{\mathrm{K}}(C; M)_n = \bigvee_{X \in ObS_\bullet^{(n)}C} BS_\bullet^{(n)}M(X, X)$$

with structure maps and symmetric actions defined similarly to the ones of $\mathrm{K}(C; M)$.

Corollary 2.6. *There is a natural equivalence of symmetric spectra*

$$\widetilde{\mathrm{K}}(C; M) \simeq \widetilde{\mathrm{K}}(C \ltimes M).$$

Proof. Projecting off the bimodule defines maps of $(n+1)$-simplicial sets

$$p_n \colon \mathcal{N}_\bullet S_\bullet^{(n)}(C; M) \longrightarrow ObS_\bullet^{(n)}C,$$

where the target has a trivial simplicial direction. The collection of realizations of n-simplicial sets $|ObS_\bullet^{(n)}C|$ forms a symmetric spectrum in the standard way, and there is a commutative diagram

$$\begin{array}{ccc} \mathrm{K}(C; M) & \xrightarrow{\simeq} & \mathrm{K}(C \ltimes M) \\ {\scriptstyle p}\downarrow & & \downarrow \\ \{|ObS_\bullet^{(n)}C|\} & \xrightarrow[\simeq]{} & \mathrm{K}(C) \end{array}.$$

This gives an equivalence from the homotopy fiber of p to $\widetilde{\mathrm{K}}(C \ltimes M)$. The inclusion at zero in $M(c, c)$ induces a splitting s for p. As a general fact, the homotopy fiber of p is equivalent to the homotopy cofiber of s. Indeed if $p \colon E \to B$ is a map of spectra split by a levelwise cofibration s, the square

$$\begin{array}{ccc} E & \longrightarrow & \mathrm{hoc}(s) \\ {\scriptstyle p}\downarrow & & \downarrow \\ B & \longrightarrow & * \end{array}$$

is cartesian since the horizontal homotopy fibers are equivalent ($B \to E \to \mathrm{hoc}(s)$ is a fiber sequence). The map on vertical homotopy fibers $\mathrm{hof}(p) \to \mathrm{hoc}(s)$ is therefore also an equivalence. In our case s is a levelwise cofibration and the homotopy cofiber of s is equivalent to its cofiber, which is precisely $\widetilde{\mathrm{K}}(C; M)$. $\square$

REMARK 2.7. When $C = \mathcal{P}_A$ with bimodule $\hom_A(-, - \otimes_A M)$ we recover the equivalence

$$\widetilde{\mathrm{K}}(\mathcal{P}_{A\ltimes M}) \underset{1.11}{\simeq} \widetilde{\mathrm{K}}(\mathcal{P}_A \ltimes \hom_A(-, - \otimes_A M)) \simeq \bigvee_{X \in ObS^{(n)}_{\bullet}\mathcal{P}_A} \hom_A(X, X \otimes_A BM)$$

of [**DM94**].

3. The categorical Dundas-Goodwillie-McCarthy theorem

We refer to [**DM96**] for the definition of the topological Hochschild homology space $\mathrm{THH}(C)$ of an exact category C. We let the topological Hochschild homology spectrum of C be the spectrum defined in degree n by the space $\mathrm{THH}(S^{(n)}_{\bullet}C)$, and we write $\mathrm{TC}(C)$ for the associated topological cyclic homology. For concreteness, we choose the model of [**DM96**] for TC. For many purposes this construction does not have the right rational homotopy type, but it has the right homotopy type after p-completion. This model admits a convenient definition of the trace map $\mathrm{Tr}\colon \mathrm{K} \to \mathrm{TC}$, see [**DM96**].

MAIN THEOREM. *Let C be an exact category, and $M\colon C^{op} \otimes C \to Ab$ an additive functor which is exact in both variables. The square of spectra*

$$\begin{array}{ccc} \mathrm{K}(C \ltimes M) & \xrightarrow{\mathrm{Tr}} & \mathrm{TC}(C \ltimes M) \\ \downarrow & & \downarrow \\ K(C) & \xrightarrow[\mathrm{Tr}]{} & \mathrm{TC}(C) \end{array}$$

is homotopy cartesian after completion at any prime, where $C \ltimes M$ has the exact structure defined in 1.7.

Using the description of the homotopy fiber $\widetilde{K}(C \ltimes M)$ of 2.6, the proof of [**McC97**] for the functor $\mathcal{P}_{A\ltimes M} \to \mathcal{P}_A$ adapts to our situation with minor changes. We recall the argument.

PROOF. For an Abelian group G and a finite pointed set X, let $G(X)$ be the Abelian group

$$G(X) = (\bigoplus_{x \in X} G \cdot x)/G \cdot *.$$

This construction is functorial in G, and precomposing it with $M\colon C^{op} \otimes C \to Ab$ defines a bimodule $M(X)\colon C^{op} \otimes C \to Ab$, with corresponding exact category $C \ltimes M(X)$. Any functor F from exact categories to the category Sp^{Σ} of symmetric spectra induces a functor $F(C \ltimes M(-))\colon Set^f_* \to \mathrm{Sp}^{\Sigma}$ on the category of finite sets. This extends to a functor from finite pointed simplicial sets to simplicial symmetric spectra. For any finite pointed simplicial set $X \in sSet^f_*$ define $F(C \ltimes M(X))$ as the realization of the simplicial spectrum

$$F(C \ltimes M(X)) = |[k] \longmapsto F(C \ltimes M(X_k))|.$$

The projection functor $C \ltimes M(X) \to C$ induces a map $F(C \ltimes M(X)) \to F(C)$, and we define

$$\widetilde{F}(C \ltimes M(X)) = \operatorname{hof}\left(F(C \ltimes M(X)) \longrightarrow F(C)\right)$$

as a functor $\widetilde{F}(C \ltimes M(-)) \colon sSet_*^f \to \mathrm{Sp}^\Sigma$.

Let $\underline{\mathrm{THH}}(C)$ denote the spectrum defined by the sequence of spaces $\mathrm{THH}(S_\bullet^{(n)}C)$. In [**McC97**, 4.2] the author constructs a commutative diagram

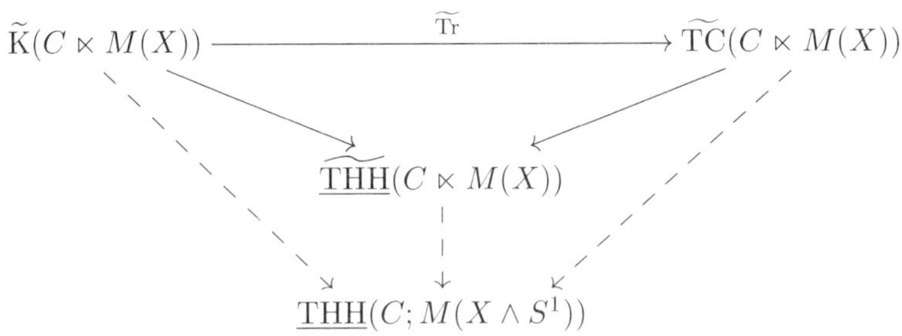

where the dashed maps are maps in the homotopy category. We denote for simplicity $S_\bullet^{(n)}M = M$. The left dashed map corresponds under the equivalence of 2.6 to the inclusion of wedges into sums

$$\bigvee_{c \in ObS_\bullet^{(n)}C} BM(c,c)(X) \longrightarrow \bigoplus_{c \in ObS_\bullet^{(n)}C} BM(c,c)(X) \xrightarrow{\simeq} \mathrm{THH}(S_\bullet^{(n)}C; BM(X)),$$

where the last map is the inclusion of the 0-simplices, which is an equivalence by [**DGM13**, 1.3.3.1, §4.3.4]. The composite is then roughly twice as connected as the connectivity of X, and therefore it induces an equivalence on Goodwillie differentials

$$D_*\widetilde{\mathrm{K}}(C \ltimes M(-)) \simeq D_*\underline{\mathrm{THH}}(C; BM(-)).$$

Now let us analyze the right-hand dashed map. Let H denote the Eilenberg-MacLane functor from categories enriched in Abelian groups to categories enriched in symmetric spectra. The inclusion of wedges into products defines an equivalence of spectrally enriched categories $HC \vee HM \to H(C \ltimes M)$ which induces an equivalence

$$\mathrm{THH}(HS_\bullet^{(n)}C \vee HM) \xrightarrow{\simeq} \mathrm{THH}(S_\bullet^{(n)}C \ltimes M) \xrightarrow{\simeq} \mathrm{THH}(S_\bullet^{(n)}(C \ltimes M))$$

over $\mathrm{THH}(S_\bullet^{(n)}C)$. The second map is the equivalence of remark 2.5. Moreover $\mathrm{THH}(HC \vee HM)$ splits as

$$\mathrm{THH}(HC \vee HM) \simeq \bigvee_{a \geq 0} \mathrm{THH}_a(C; M),$$

where $\mathrm{THH}_a(C; M)$ is the subspace of $\mathrm{THH}(HC \vee HM)$ defined as in [**Hes94**, 2.1], by distributing the smash products defining $\mathrm{THH}(HC \vee HM)$ onto the wedges of the mapping spectra of $HC \vee HM$, and by keeping only the wedge components in which the bimodule factors appear a times. By replacing C with the iterates of $S_\bullet$ we obtain a decomposition

$$\mathrm{THH}(S_\bullet^{(n)}(C \ltimes M)) \simeq \bigvee_{a \geq 0} \mathrm{THH}_a(S_\bullet^{(n)}C; M).$$

Hesselholt's connectivity estimates from [**Hes94**, §2.2] for the fixed-points of the spectra of $\mathrm{THH}_a(L \oplus P)$ hold just as well for our spectra $\underline{\mathrm{THH}}_a(C; M)$. The only difference is that our category C is allowed to have more than one object, and our construction of THH picks up an extra wedge sum over the set of objects of C which does not affect connectivity. Thus by Proposition [**Hes94**, 2.2] there is an equivalence

$$D_*\widetilde{\mathrm{TC}}(C \ltimes M(-);p) \simeq D_* \operatorname*{holim}_r \underline{\mathrm{THH}}_1(C;M(-))^{C_{p^r}}.$$

By [**Hes94**, 3.2,3.3] the fixed-points of $\underline{\mathrm{THH}}_1$ are equivalent to

$$\begin{aligned}\underline{\mathrm{THH}}_1(C;M(-))^{C_{p^r}} &\simeq (\underline{\mathrm{THH}}(C;M(-)) \wedge S^1_+)^{C_{p^r}} \\ &\simeq (\underline{\mathrm{THH}}(C;M(-)) \vee \Sigma\underline{\mathrm{THH}}(C;M(-)))\end{aligned}$$

and the fixed-points inclusions of the homotopy limit correspond to the map $p \vee \mathrm{id}$. Thus after p-completion this is equivalent to $\Sigma\underline{\mathrm{THH}}(C;M(-)) \simeq \underline{\mathrm{THH}}(C;BM(-))$. Combining these equivalences we have shown that the right dashed map induces an equivalence on differentials at a point

$$D_*\widetilde{\mathrm{TC}}(C \ltimes M(-)) \simeq D_*\underline{\mathrm{THH}}(C;BM(-))$$

after completion at a prime p.

By the commutativity of the diagram above the trace map induces an equivalence $D_*\widetilde{\mathrm{Tr}}\colon D_*\widetilde{\mathrm{K}}(C \ltimes M(-)) \xrightarrow{\simeq} D_*\widetilde{\mathrm{TC}}(C \ltimes M(-))$ after p-completion. We further observe that for any pointed set B there is an isomorphism of exact categories

$$C \ltimes M(B \vee X) \cong C \ltimes (M(B) \oplus M(X)) \cong (C \ltimes M(B)) \ltimes M(X),$$

where $M(X)$ is a bimodule on $C \ltimes M(B)$ via the projection $C \ltimes M(B) \to C$. This shows that $D_*\widetilde{\mathrm{K}}(C \ltimes M(B \vee -)) \simeq D_*\widetilde{\mathrm{TC}}(C \ltimes M(B \vee -))$, and therefore by [**Goo90**, 1.3(iv)] that the differential of the trace at any finite pointed simplicial set B

$$D_B\widetilde{\mathrm{Tr}}\colon D_B\widetilde{\mathrm{K}}(C \ltimes M(-)) \xrightarrow{\simeq} D_B\widetilde{\mathrm{TC}}(C \ltimes M(-))$$

is an equivalence after p-completion. Both functors $\widetilde{\mathrm{K}}(C \ltimes M(-))$, $\widetilde{\mathrm{TC}}(C \ltimes M(-))\colon sSet^f_* \to \mathrm{Sp}^\Sigma$ are (-1)-analytic in the sense of [**Goo90**],[**Goo92**]. The first one by the argument of [**McC97**, §3], which just uses the decomposition $\widetilde{\mathrm{K}}(C \ltimes M(-)) \simeq \widetilde{\mathrm{K}}(C;M(-))$. The second one by [**McC97**, §1] using the wedge decomposition of $\underline{\mathrm{THH}}(C \ltimes M(X))$ above. By [**Goo92**, 5.3] the trace map must have already been a p-completed equivalence before taking the differential

$$\widetilde{\mathrm{Tr}}\colon \widetilde{\mathrm{K}}(C \ltimes M(X)) \xrightarrow{\simeq} \widetilde{\mathrm{TC}}(C \ltimes M(X))$$

for any pointed ((−1)-connected) space X. For $X = S^0$ this shows that the square of the statement is cartesian. □

References

[DGM13] Bjørn Ian Dundas, Thomas G. Goodwillie, and Randy McCarthy, *The local structure of algebraic K-theory*, Algebra and Applications, vol. 18, Springer-Verlag London, Ltd., London, 2013. MR3013261

[DM94] Bjørn Ian Dundas and Randy McCarthy, *Stable K-theory and topological Hochschild homology*, Ann. of Math. (2) **140** (1994), no. 3, 685–701. MR1307900

[DM96] Bjørn Ian Dundas and Randy McCarthy, *Topological Hochschild homology of ring functors and exact categories*, J. Pure Appl. Algebra **109** (1996), no. 3, 231–294. MR1388700

[Dun97] Bjørn Ian Dundas, *Relative K-theory and topological cyclic homology*, Acta Math. **179** (1997), no. 2, 223–242. MR1607556

[Goo90] Thomas G. Goodwillie, *Calculus. I. The first derivative of pseudoisotopy theory*, K-Theory **4** (1990), no. 1, 1–27. MR1076523

[Goo92] Thomas G. Goodwillie, *Calculus. II. Analytic functors*, K-Theory **5** (1991/92), no. 4, 295–332. MR1162445

[Hel58] Alex Heller, *Homological algebra in abelian categories*, Ann. of Math. (2) **68** (1958), 484–525. MR0100622

[Hes94] Lars Hesselholt, *Stable topological cyclic homology is topological Hochschild homology*, Astérisque **226** (1994), 8–9, 175–192. K-theory (Strasbourg, 1992). MR1317119

[Kel90] Bernhard Keller, *Chain complexes and stable categories*, Manuscripta Math. **67** (1990), no. 4, 379–417. MR1052551

[McC97] Randy McCarthy, *Relative algebraic K-theory and topological cyclic homology*, Acta Math. **179** (1997), no. 2, 197–222. MR1607555

[Wal85] Friedhelm Waldhausen, *Algebraic K-theory of spaces*, Algebraic and geometric topology (New Brunswick, N.J., 1983), Lecture Notes in Math., vol. 1126, Springer, Berlin, 1985, pp. 318–419. MR802796

Mathematical Institute, University of Bonn, Endenicher Allee 60, D-53115 Bonn, Germany

Email address: `dotto@math.uni-bonn.de`

Contemporary Mathematics
Volume **708**, 2018
http://dx.doi.org/10.1090/conm/708/14261

Four approaches to cohomology theories with reality

J. P. C. Greenlees

Abstract. We give an account of the calculations of the $RO(Q)$-graded coefficient rings of some of the most basic Q-equivariant cohomology theories, where Q is a group of order 2. One purpose is to advertise the effectiveness of the Tate square, showing it has advantages over the slice spectral sequences in algebraically simple cases. A second purpose is to give a single account showing how to translate between the languages of different approaches.

1. Introduction

Historically, complex orientable cohomology theories (complex K-theory KU, complex cobordism MU, ...) have been the first to be exploited, partly because they tend to be easier to calculate. Real analogues of these (real K-theory KO, real cobordism MO, ...) contain new information but can be hard to approach directly. It is well known that representations over the reals can be viewed as complex representations with an additional conjugate-linear involution: real representations are complex representations with an action of the Galois group $Q = Gal(\mathbb{C}|\mathbb{R})$ of order 2. Atiyah used this technique [**2**] to define K-theory 'with Reality' (or simply *Real* K-theory with the initial letter capitalized), thereby obtaining a very practical approach to real K-theory by descent. This was then extended to cobordism by Landweber [**19**], Araki [**1**] and others. The resulting theories are Q-equivariant cohomology theories represented by Q-spectra.

The same discussion applies if we start with Q-equivariant theories: the complex orientable theories (Atiyah-Segal equivariant K-theory, tom Dieck complex bordism) are much more accessible. Many techniques were developed in the 1990s with applications to equivariant complex orientable theories most clearly in mind.

The point of this article is to describe some well known calculations for Real equivariant theories in the simplest cases (ordinary cohomology with constant integer coefficients and Real connective K-theory), giving several different methods in each case. In particular, despite the fact that the Q-equivariant Real theories are not complex orientable (even though the underlying non-equivariant theories are), some of the complex orientable methods are still extremely effective.

1.A. Notation. This subsection introduces our notation for some standard machinery. I am grateful to the referee for patiently and objectively pointing out ways in which notational conventions vary. Encouraged by the referee to clarify my

2010 *Mathematics Subject Classification.* Primary 55P91, 55N91, 55Q91.

conventions, the length of the subsection has increased about fivefold: I hope this will make the article more accessible, and ease communication between different notational tribes. On the other hand, those familiar with the customs of my tribe can skip directly to Subsection 1.C now.

1.A.1. *Grading.* We write σ for the sign representation of Q on $\mathbb{R}$, 1 for the trivial representation, and $\rho = 1 + \sigma$ for the real regular representation. We will be grading our groups over the real representation ring $RO(Q) = \{x + y\sigma \mid x, y \in \mathbb{Z}\}$. We write M_* to indicate grading over $\mathbb{Z}$ and $M_\bigstar$ to denote grading over $RO(Q)$. When we draw pictures, they will be displayed in the plane with the $\mathbb{Z}$ axis horizontal and the σ axis vertical, so that $x + y\sigma$ gets displayed at the point with cartesian coordinates (x, y).

Alternatives: We have followed [**15**] in using σ for the non-trivial real representation of Q. In the past each author seems to have used a different letter. For example, the present author previously used ξ [**3, 4, 9, 12**], the letter α is used by Hu and Kriz, and the letter τ is used elsewhere. Perhaps we can unite around σ?

We have used $RO(Q)$ grading both in notation and displays. Atiyah [**2**], Landweber [**19**], Araki [**1**] use $\mathbb{R}^{p,q}$ for $q \oplus p\sigma$ (sic). The Tate twist notation from algebraic geometry used by Dugger [**8**] is related to $RO(Q)$-grading by $\mathbb{R}^{p+q,q} = p+q\sigma$.

1.A.2. *Mackey functors.* We will use Dress's formulation of Mackey functors for a finite group G [**7**] as given by a covariant and a contravariant functor on finite G-sets subject to the Mackey condition. A Mackey functor M for a group G is therefore determined by its values on the transitive G-sets G/H together with certain structure maps. If $K \subseteq H$ the projection $\pi^H_K : G/K \longrightarrow G/H$ induces a restriction map $(\pi^H_K)^* = \mathrm{res}^H_K : M(G/H) \longrightarrow M(G/K)$ and an induction map $(\pi^H_K)_* = \mathrm{ind}^H_K : M(G/K) \longrightarrow M(G/H)$, and right multiplication $R_g : G/H \longrightarrow G/H^g$ induces an action of $W_G(H) = N_G(H)/H$ on $M(G/H)$.

We write $\underline{\mathbb{Z}}$ for the constant Mackey functor at $\mathbb{Z}$ (which is to say that the values on all orbits are $\mathbb{Z}$, restriction is the identity and induction is multiplication by the index of the smaller subgroup in the larger one).

1.A.3. *Spectra.* We will be working with cohomology theories $E^*_G(\cdot)$, which are represented by *genuine* G-spectra E (i.e., G-spectra indexed on a complete G-universe) in the sense that $E^*_G(X) = [X, E]^*_G$. Accordingly the integer grading can be extended to an $RO(G)$-grading.

For an orthogonal representation V we write $S(V)$ for the unit sphere and S^V for the one point compactification of V with ∞ as the basepoint. We write $X^G_V := \pi^G_V(X) = [S^V, X]^G$ for the Vth G-equivariant homotopy group of X. As usual $X_V := \pi_V(X) = [S^V, X]$ denotes the Vth non-equivariant homotopy group. The notation extends to virtual representations.

The subcategory of G-spectra of the form G/H_+ is called the *stable orbit category*, and calculation of the maps between them shows that they are generated by the maps $(\pi^H_K)_+$ and their duals, and that a Mackey functor as described above is equivalent to a contravariant functor on the stable orbit category. We write $\underline{\pi}^G_*(X)$ for the Mackey functor $[(\cdot)_+, X]^G_*$ whose value on G/H is $\pi^H_*(X) = [G/H_+, X]^G_*$.

Alternatives: The following is inserted in recognition of the fact that alternative views are held very strongly by some (for example by the referee!).

Some authors (including the present author in the 1980s) prefer to omit the group when writing $RO(Q)$-grading and Mackey functors (writing $\underline{\pi}_*(X)$ where we write $\underline{\pi}^Q_*(X)$ and $\pi_\bigstar(X)$ where we write $\pi^Q_\bigstar(X)$). This is presumably on the grounds of brevity, but it is at the cost of forcing the adoption of a more complicated notation for non-equivariant homotopy. The present choice is based on the ideas (i) that working equivariantly requires explicit acknowledgement whilst working non-equivariantly does not and (ii) that one should be able to substitute a particular grading for $\bigstar$ (so that considering grading zero for instance, $\pi^Q_\bigstar(X)$ becomes $\pi^Q_0(X)$).

1.A.4. *Ordinary cohomology.* A cohomology theory $E^*_G(\cdot)$ satisfying the dimension axiom (i.e., one whose values on cells G/H_+ is entirely in degree 0) is called *ordinary* (or *Bredon*). It is determined by the values on these cells, which define a Mackey functor $M = \underline{\pi}^G_0(E)$. We write $H^*_G(\cdot; M)$ for the cohomology theory and HM for the representing G-spectrum.

1.A.5. *Proto-Euler and Proto-Thom classes.* For any ring spectrum E, the unit map provides, for any representation V a class $\Sigma^V \iota \in E^V_G(S^V)$. A Thom class would be an element $\tau_V \in E^{|V|}_G(S^V)$ which generates a rank one free module. The Euler class (of degree $|V| - V$) would be the pullback of τ_V along the zero section $a_V : S^0 \longrightarrow S^V$.

With $G = Q$ some related classes are important for us. We write $a = a_\sigma : S^0 \longrightarrow S^\sigma$ for the inclusion, and think of it as a homotopy element of degree $-\sigma$.

There are Euler-like classes λ (of degree $1 - \sigma$, which does not survive to homotopy), u (of degree $2(1 - \sigma)$, important for ordinary cohomology), and U (of degree $4(1 - \sigma)$, important for $k\mathbb{R}$).

Alternatives: We follow [**15**] in using a for the degree $-\sigma$ Euler-like class. The letters e and i are used by some authors.

In [**16**, Definition 3.4] the class u is called $u_{2\sigma}$, and the class $2u^{-1}$ is called $e_{2\sigma}$. In the more general study of bordism with reality, a class u (of which ours is the image) plays a significant role, and its 2^nth power is an almost-periodicity at the nth chromatic level. This means that $U = u^2$, and that we should think of λ as a square root of u.

1.A.6. *Basepoints.* Our Q spaces will be equipped with a Q-fixed basepoint and cohomology will be reduced.

1.B. What's the point of this paper? The paper describes several different methods of calculating the $RO(Q)$-graded homotopy of two well known Real spectra: the integral Eilenberg-MacLane spectrum $H\underline{\mathbb{Z}}$ and connective K-theory with reality $k\mathbb{R}$. The results and the methods are all known. The intention is to describe several different methods in a way that lets one compare them, to provide more details and more pictures than is usually done, and to record the notations used by different authors. The author has found the need to make these comparisons quite often, and that it takes time to pin details down precisely from the existing literature. It is hoped the present summary may be useful to others.

1.C. Geography of the coefficients. It is worth highlighting some features of the coefficient rings $X^Q_\bigstar$. The background is that their non-equivariant counterparts are in even degrees. Next the reader should look at the pictures at the the start of Sections 2 and 3) (or at the picture of $tmf_1(3)^Q_\bigstar$ in [**13**, Subsection 13.C]).

The crudest feature is that the non-zero entries are mostly above the diagonal line of slope -1. Indeed the only exceptions to this are the a-multiple tails in the right half-plane. This is characteristic of theories which are equivariantly connective.

LEMMA 1.1. *If X is a connective Q-spectrum then $X^Q_{x+y\sigma}$ is zero below the antidiagonal $x+y=0$ in the half-plane $x<0$, and multiplication by a is an isomorphism below the antidiagonal in the half-plane $x \geq 0$.*

PROOF. Non-equivariant connectivity and the cofibre sequence $Q_+ \longrightarrow S^0 \longrightarrow S^\sigma$ shows multiplication by a is surjective from the antidiagonal, and isomorphic below it. The connectivity of X^Q_* shows there are zeros along the negative x axis. □

However, the most striking feature is that the coefficients of the theories described here have 'The Gap':

$$X^Q_{*\rho-i} = 0 \text{ for } i = 1,2,3$$

so that in the display of the $RO(Q)$-graded coefficients, the three diagonals above the leading diagonal are all zero. It is worth commenting on the implications of this.

LEMMA 1.2. *(i) The vanishing of the diagonals $X^Q_{*\rho-i}$ with $i=1,2$ is equivalent to X being* strongly even *in the sense of Meier and Hill* [**17**] *(i.e., that $X^Q_{*\rho-1}=0$ and the restriction maps $X^Q_{k\rho} \longrightarrow X_{2k}$ are isomorphic for all k).*

(ii) The vanishing of the three diagonals $X^Q_{*\rho-i}$ with $i=1,2,3$ is equivalent to being strongly even and having X_* even.

PROOF. Consider the long exact sequence in X-cohomology induced by the sequence $Q_+ \longrightarrow S^0 \longrightarrow S^\sigma$. □

1.D. The Tate square. We will make considerable use of the Tate square, so we briefly describe the construction from [**10, 12**]. This is based on the contractible free Q-space EQ and the join $\tilde{E}Q$ which are related by the cofibre sequence

$$EQ_+ \longrightarrow S^0 \longrightarrow \tilde{E}Q$$

brought to prominence by Carlsson's use of it in his proof of the Segal conjecture [**5**]. Since Q is a periodic group we use the model $EQ = S(\infty\sigma)$ for a contractible free Q-space and $\tilde{E}Q = S^{\infty\sigma}$ for the join with S^0 to give periodic resolutions as in [**10**].

We will use Scandinavian notation:

$$X^{hQ} = (X^h)^Q, \text{ where } X^h = F(EQ_+, X)$$

$$X_{hQ} \simeq (X_h)^Q, \text{ where } X_h = EQ_+ \wedge X$$

$$X^{tQ} = (X^t)^Q, \text{ where } X^t = F(EQ_+, X) \wedge \tilde{E}Q$$

$$X^{\Phi Q} = (X^\Phi)^Q, \text{ where } X^\Phi = X \wedge \tilde{E}Q.$$

The notation $X^{\Phi Q}$ is justified by the fact that since Q is of prime order, it is the geometric Q-fixed point spectrum.

It is easy to see that there are two linked cofibre sequences

$$\begin{array}{ccccc} X_h & \longrightarrow & X & \longrightarrow & X^\Phi \\ \simeq\downarrow & & \downarrow & & \downarrow \\ X_h & \longrightarrow & X^h & \longrightarrow & X^t \end{array}$$

and the indicated equivalence arises since $X \longrightarrow X^h$ is a non-equivariant equivalence. Accordingly the Tate square (the square on the right) is a homotopy pullback, and it is a homotopy pullback of rings if X is a ring.

Typically we proceed as follows

- Calculate $X^{hQ}_\bigstar$ using the homotopy fixed point spectral sequence described in Subsection 1.E.
- Infer $X^{tQ}_\bigstar$ by inverting a (since $\tilde{E}Q = S^{\infty\sigma}$, and evidently $\pi^Q_\bigstar(Y \wedge S^{\infty\sigma}) = \pi^Q_\bigstar(Y)[1/a]$).
- Infer $X_{hQ}^\bigstar$ from the lower cofibre sequence (which amounts to a simple Local Cohomology Theorem).
- Infer $X^{\Phi Q}_\bigstar$ from $X^{tQ}_\bigstar$ using Lemma 1.3; since both are a-periodic this follows from the integer graded homotopy groups.
- Infer $X^Q_\bigstar$ from the Tate square and the top sequence.

The following is immediate since the fibres of the two verticals in the Tate square are equivalent. The hypothesis on the non-equivariant spectrum shows X_{hQ} is connective and hence it follows that $X^{\Phi Q}$ is connective.

LEMMA 1.3. *(*[**13**, Lemma 11.2]*) Suppose X is a Q-spectrum which is non-equivariantly connective. If $X^Q \longrightarrow X^{hQ}$ is the connective cover then $X^{\Phi Q} \longrightarrow X^{tQ}$ is too.* □

1.E. The $RO(Q)$-graded homotopy fixed point spectral sequence. We note that we are trying to calculate the groups

$$X^{hQ}_{a+b\sigma} = [S^{a+b\sigma}, X^h]^Q = [S^{a+b\sigma}, F(EQ_+, X)]^Q$$
$$= [S^{a+b\sigma} \wedge EQ_+, X]^Q = [S^a, F(S^{b\sigma}, X)^{hQ}].$$

We could consider the separate homotopy fixed point spectral sequences for the various spectra $F(S^{b\sigma}, X)$. To avoid confusion later, we arrange this homologically from the start, so that it is in a (vertically shifted) second quadrant:

$$E^2_{s,t}(b) = H^{-s}(Q; X^{-t}(S^{b\sigma}))) \Rightarrow X_Q^{-(s+t)}(EQ_+ \wedge S^{b\sigma}) = \pi_{s+t}(F(S^{b\sigma}, X)^{hQ})$$

with differentials

$$d^r_{s,t} : E^r_{s,t}(b) \longrightarrow E^r_{s-r,t+r-1}(b).$$

Note that in the E^2-term Q acts on $S^{b\sigma}$ by a degree $(-1)^b$ map, and it is this that makes $X^*(S^{b\sigma})$ a Q-module.

However there is a great advantage to combining these spectral sequences as b varies, especially if X is a ring spectrum, since the pairings

$$S^{b_1\sigma} \wedge S^{b_2\sigma} \longrightarrow S^{(b_1+b_2)\sigma}$$

give the whole construction a multiplicative structure. Although the differentials all lie within the spectral sequence for a single value of b, the spectral sequences all come from the same filtration, so that the Leibniz rule applies across all the spectral sequences.

DEFINITION 1.4. The $RO(Q)$-graded homotopy fixed point spectral sequence for X is the trigraded spectral sequence

$$E^2_{s,t,b} = H^{-s}(Q; X_{t+b\sigma}) \Rightarrow [S^{s+t+b\sigma} \wedge EQ_+, X]^Q =: X^{hQ}_{s+t+b\sigma}$$

with differentials

$$d^r_{s,t,b} : E^r_{s,t,b} \longrightarrow E^r_{s-r,t+r-1,b}.$$

2. Ordinary cohomology

We may depict the $RO(Q)$-graded coefficients of the Eilenberg-MacLane spectrum, $H\underline{\mathbb{Z}}^Q_\bigstar$ as follows, where squares are copies of $\mathbb{Z}$; circles are also copies of $\mathbb{Z}$, but the generator is twice what one might expect. Dots (red online) are copies of $\mathbb{F}_2$. This was first calculated by Stong [**22, 23**] and Waner [**24**] and first published by Caruso [**6**], but it has appeared in several other places. The Mackey structure is made explicit in [**21**], [**16**, Figure 2, Page 405].

One notable feature is 'The Gap', which constitutes the three lines of slope 1 above the unit (i.e., the diagonals intersecting the x axis in $x = -3, -2, -1$).

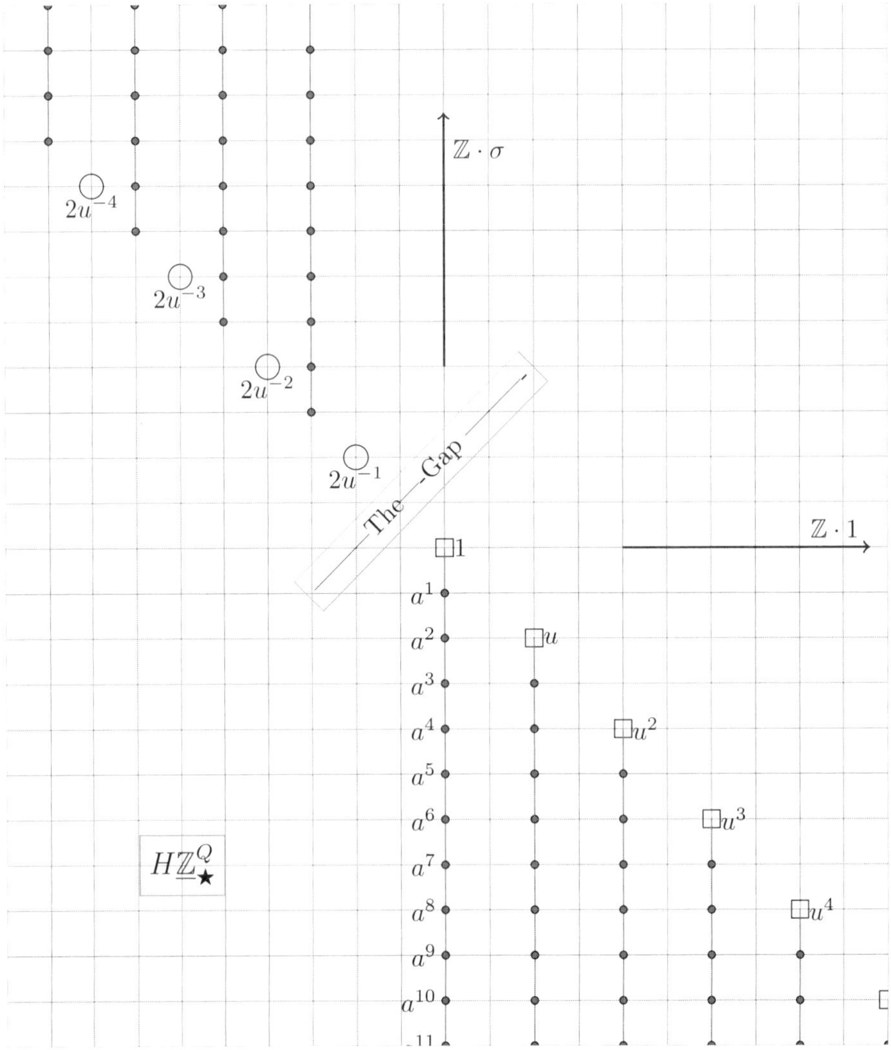

2.A. Cell approach. Note first that

$$H\underline{\mathbb{Z}}^Q_{a+b\sigma} = [S^{a+b\sigma}, H\underline{\mathbb{Z}}] = H_Q^{-a}(S^{b\sigma}; H\underline{\mathbb{Z}}).$$

Since $H\underline{\mathbb{Z}}$ satisfies the dimension axiom, it is natural to calculate this by finding a cell structure for $S^{b\sigma}$.

Indeed, we start from the cofibre sequence

$$Q_+ \longrightarrow S^0 \longrightarrow S^\sigma.$$

Suspending by $n\sigma$ and using the untwisting isomorphism $S^{n\sigma} \wedge Q_+ \simeq S^n \wedge Q_+$, this gives

$$S^n \wedge Q_+ \longrightarrow S^{n\sigma} \longrightarrow S^{(n+1)\sigma}.$$

First we consider a row in the lower half plane, $H\underline{\mathbb{Z}}^Q_{*-n\sigma}$, when $n \geq 0$. We argue as follows. We begin with $H^Q_*(S^0; \underline{\mathbb{Z}}) = \mathbb{Z}$ (by the dimension axiom). Adding an $(n+1)$-cell to deduce $H^Q_*(S^{(n+1)\sigma}; \underline{\mathbb{Z}})$ from $H^Q_*(S^{n\sigma}; \underline{\mathbb{Z}})$ we only ever have to determine the map

$$\mathbb{Z} = H^Q_n(S^n \wedge Q_+; \underline{\mathbb{Z}}) \longrightarrow H^Q_n(S^{n\sigma}; \underline{\mathbb{Z}}) = \mathbb{Z}$$

when n is even. It is always multiplication by 2: for $n = 0$ it is $\pi_* = \mathrm{ind}_1^Q$: $\underline{\mathbb{Z}}(Q/1) \longrightarrow \underline{\mathbb{Z}}(Q/Q)$. For larger even n we note inductively that

$$H^Q_n(S^{n\sigma}; \underline{\mathbb{Z}}) \xrightarrow{\cong} H^Q_n(S^{n\sigma}/S^{(n-1)\sigma}; \underline{\mathbb{Z}})$$

is an isomorphism. The composite

$$S^n \wedge Q_+ \longrightarrow S^{n\sigma} \longrightarrow S^{n\sigma}/S^{(n-1)\sigma} \simeq S^n \wedge Q_+$$

is the nth suspension of the case $n = 0$, where it is ind_1^Q.

Next, consider a row in the upper half plane, $H\underline{\mathbb{Z}}^Q_{*+n\sigma}$ when $n \geq 0$. We argue as follows. We begin with $H^*_Q(S^0; \underline{\mathbb{Z}}) = \mathbb{Z}$ (by the dimension axiom). Adding an $(n+1)$-cell to deduce $H^*_Q(S^{(n+1)\sigma}; \underline{\mathbb{Z}})$ from $H^*_Q(S^{n\sigma}; \underline{\mathbb{Z}})$ we only ever have to decide on the map

$$\mathbb{Z} = H^n_Q(S^{n\sigma}; \underline{\mathbb{Z}}) \longrightarrow H^n_Q(S^n \wedge Q_+; \underline{\mathbb{Z}}) = \mathbb{Z}$$

when n is even. For $n = 0$ it is $\pi^* = \mathrm{res}_1^Q : \mathbb{Z}(Q/Q) \xrightarrow{\cong} \mathbb{Z}(Q/1)$. For larger even n it is always multiplication by 2: we note inductively that

$$H^n_Q(S^{n\sigma}/S^{(n-1)\sigma}) \longrightarrow H^n_Q(S^{n\sigma}; \underline{\mathbb{Z}})$$

is multiplication by 2. The composite

$$S^n \wedge Q_+ \longrightarrow S^{n\sigma} \longrightarrow S^{n\sigma}/S^{(n-1)\sigma} \simeq S^n \wedge Q_+$$

is the nth suspension of the case $n = 0$, where it is the sum $N = 1 + x$ over multiplication by group elements, where $Q = \langle x \rangle$.

2.B. Quotient approach. The cohomology of a *space* only depends on the restriction maps in the Mackey coefficients, and if these form a constant coefficient system (i.e., restriction maps are the identity) it follows that it is the nonequivariant cohomology of the quotient: for any Q-space X

$$H^*_Q(X; \underline{\mathbb{Z}}) = H^*(X/Q; \mathbb{Z}).$$

For any $n \geq 0$ we have $S^{(n+1)\sigma} = S^0 * S(n\sigma)$, and hence

$$H^*_Q(S^{(n+1)\sigma}; \underline{\mathbb{Z}}) = H^*(S^0 * \mathbb{R}P^n; \mathbb{Z}).$$

This gives $H\underline{\mathbb{Z}}^Q_{*+n\sigma}$ for $n \geq 0$ from the well known cohomology of projective spaces. This fills in the upper half plane.

For the lower half plane we use the fact that for *free* Q-spectra

$$H^Q_*(X;\underline{\mathbb{Z}}) = H_*(X/Q;\mathbb{Z}).$$

This is not true for arbitrary Q-spectra (or even for Q-spaces). We then use the cofibre sequence

$$S(n\sigma)_+ \longrightarrow S^0 \longrightarrow S^{n\sigma}$$

We have

$$H^Q_*(S(n\sigma)_+;\underline{\mathbb{Z}}) = H_*(\mathbb{R}P^n_+;\mathbb{Z});$$

since S^0 only has homology in degree 0, we need only check that

$$S(n\sigma)_+ \longrightarrow S^0$$

is multiplication by 2 in degree 0. When $n = 0$ this is part of the definition of $\underline{\mathbb{Z}}$, and for higher dimensions, we use the fact that $S(\sigma)_+ \longrightarrow S(n\sigma)_+$ is an isomorphism in degree 0 since only higher cells have been attached. This fills in the lower half plane.

We note in passing that this also shows $H^Q_*(S^{n\sigma};\underline{\mathbb{Z}}) \not\cong H^Q_*(S^{n\sigma}/Q);\mathbb{Z})$ for $n \geq 1$.

2.C. Tate approach. The cleanest approach is to use the Tate diagram

$$\begin{array}{ccccc} H\underline{\mathbb{Z}}_h & \longrightarrow & H\underline{\mathbb{Z}} & \longrightarrow & H\underline{\mathbb{Z}}^\Phi \\ \simeq\downarrow & & \downarrow & & \downarrow \\ H\underline{\mathbb{Z}}_h & \longrightarrow & H\underline{\mathbb{Z}}^h & \longrightarrow & H\underline{\mathbb{Z}}^t \end{array}$$

The procedure is to start with homotopy fixed points. From this we can immediately deduce Tate cohomology by inverting a. This then gives homotopy orbits by using the bottom row. These terms only depend on the underlying spectrum (in this case $H\underline{\mathbb{Z}}$) made free. The geometric fixed points depends on the spectrum itself, so needs further input. Then the answer follows from the Tate square.

LEMMA 2.1.

$$H\underline{\mathbb{Z}}^{hQ}_{\bigstar} = BB[u, u^{-1}],$$

where $BB = \mathbb{Z}[a]/(2a)$, $|a| = -\sigma$, $|u| = 2 - 2\sigma$ *(a is the Euler class of σ and u is a Thom class).*

REMARK 2.2. (i) Notice how this is related to ordinary group cohomology, which occurs along the negative x-axis:

$$H^*(BQ_+;\mathbb{Z}) = H\underline{\mathbb{Z}}^{hQ}_* = \mathbb{Z}[y]/(2y), \text{ where } y = a^2u^{-1} \text{ is of codegree 2.}$$

(ii) The notation BB stands for Basic Block, and the usefulness of the notation will emerge as we consider other examples.

PROOF. We describe the $RO(Q)$-graded spectral sequence of Subsection 1.E for $H\underline{\mathbb{Z}}$. It is convenient to describe it from the E_1-term if we choose a good cell structure on EQ_+. Indeed we use the filtration

$$S(\sigma)_+ \subset S(2\sigma)_+ \subset S(3\sigma)_+ \subset \cdots \subset S(\infty\sigma)_+ = EQ_+.$$

There is one free Q-cell in each degree and in cellular homology this gives the standard 2-periodic resolution

$$0 \longleftarrow \mathbb{Z} \overset{\epsilon}{\longleftarrow} \mathbb{Z}Q \overset{1-x}{\longleftarrow} \mathbb{Z}Q \overset{1+x}{\longleftarrow} \mathbb{Z}Q \overset{1-x}{\longleftarrow} \cdots$$

of $\mathbb{Z}$ over the group ring $\mathbb{Z}Q$.

We may then describe the E^1 term conveniently as

$$E^1_{*,*,*} = \mathbb{Z}[a][\lambda, \lambda^{-1}]$$

where the (s,t,b) gradings are

$$|a| = (-1,1,-1), |\lambda| = (0,1,-1).$$

The differential is given by taking $d_1 a = 0$ and

$$d_1\lambda = 2a.$$

This is immediate from the algebraic resolution since Q acts non-trivially on $H^b(S^{b\sigma})$ for b odd.

It follows that $u = \lambda^2$ is a cycle and that there are no further differentials. □

We now immediately deduce the coefficients of the Tate theory by inverting a, since $S^{\infty\sigma} \simeq S^0[1/a]$, and then the homotopy orbits by taking local cohomology.

COROLLARY 2.3. *The Tate cohomology is given by*

$$H\underline{\mathbb{Z}}^{tQ}_{\bigstar} = \mathbb{F}_2[a, a^{-1}][u, u^{-1}],$$

and the coefficients of the homotopy orbits by

$$H\underline{\mathbb{Z}}^{\bigstar}_{hQ} = NB[u, u^{-1}]$$

where, $NB = \mathbb{Z} \oplus \Sigma^{-1}\mathbb{F}_2[a]^\vee$. *We note also that* $NB \simeq R\Gamma_{(a)}(BB)$, *i.e., it is the sum of right derived functors of the a-power torsion functor* $\Gamma_{(a)}$.

REMARK 2.4. The notation NB stands for Negative Block, and as for BB the usefulness of the notation should emerge once we have other examples.

For geometric fixed points we apply Lemma 1.3.

LEMMA 2.5. *The geometric fixed points are given by* $H\underline{\mathbb{Z}}^{\Phi Q}_{\bigstar} = \mathbb{F}_2[a, a^{-1}][u]$.

PROOF. By definition of the Eilenberg-MacLane spectrum, $(H\underline{\mathbb{Z}})^Q = H\mathbb{Z}$. It is immediate that $(H\underline{\mathbb{Z}})^{hQ}_*$ is zero in positive degrees, and the algebraic resolution shows that $(H\underline{\mathbb{Z}})^Q_0 \xrightarrow{\cong} (H\underline{\mathbb{Z}})^{hQ}_0$ is an isomorphism. The conclusion follows from Lemma 1.3. □

The calculation of $H\underline{\mathbb{Z}}^Q_{\bigstar}$ now concludes by using the Tate square. The only difference between $H\underline{\mathbb{Z}}^Q_{\bigstar}$ and the more regular $H\underline{\mathbb{Z}}^{hQ}_{\bigstar}$ comes from the negative Tate a-columns. The Tate square therefore induces an actual pullback square in gradings with even integer-part. The a-divisible columns in odd integer degrees ≤ -3 come from the cokernel of $(H\underline{\mathbb{Z}})^{hQ}_{\bigstar} \longrightarrow (H\underline{\mathbb{Z}})^{tQ}_{\bigstar}$.

COROLLARY 2.6.

$$H\underline{\mathbb{Z}}^Q_{\bigstar} = BB[u] \oplus u^{-1} \cdot NB[u^{-1}],$$

where $BB = \mathbb{Z}[a]/(2a)$ *as before and*

$$NB = \mathbb{Z} \oplus \Sigma^{-\rho}\mathbb{F}_2[a]^\vee.$$

The BB-module structure of NB is as suggested by the notation, and multiplication by u relates factors of NB and BB according to the notation, with the map $u^{-1}\cdot NB \longrightarrow BB$ being the composite $NB \longrightarrow \mathbb{Z} \longrightarrow BB$.

3. K-theory with reality

We now repeat the exercise with the connective version of Atiyah's K-theory with Reality: $k\mathbb{R}$. Again squares are copies of $\mathbb{Z}$ and dots are copies of $\mathbb{F}_2$; circles are also copies of $\mathbb{Z}$, but the generator is twice what one might expect. This calculation first appears in [**8**], and there are alternative approaches in [**4**, **13**].

One notable feature is 'The Gap', which constitutes the three lines of slope 1 above the unit (i.e., the diagonals intersecting the x axis in $x = -3, -2, -1$.). In the periodic form of the theory (i.e., with the degree ρ class $\overline{v}$ inverted) the a-columns vanish and The Gap is copied across the page with the U-periodicity.

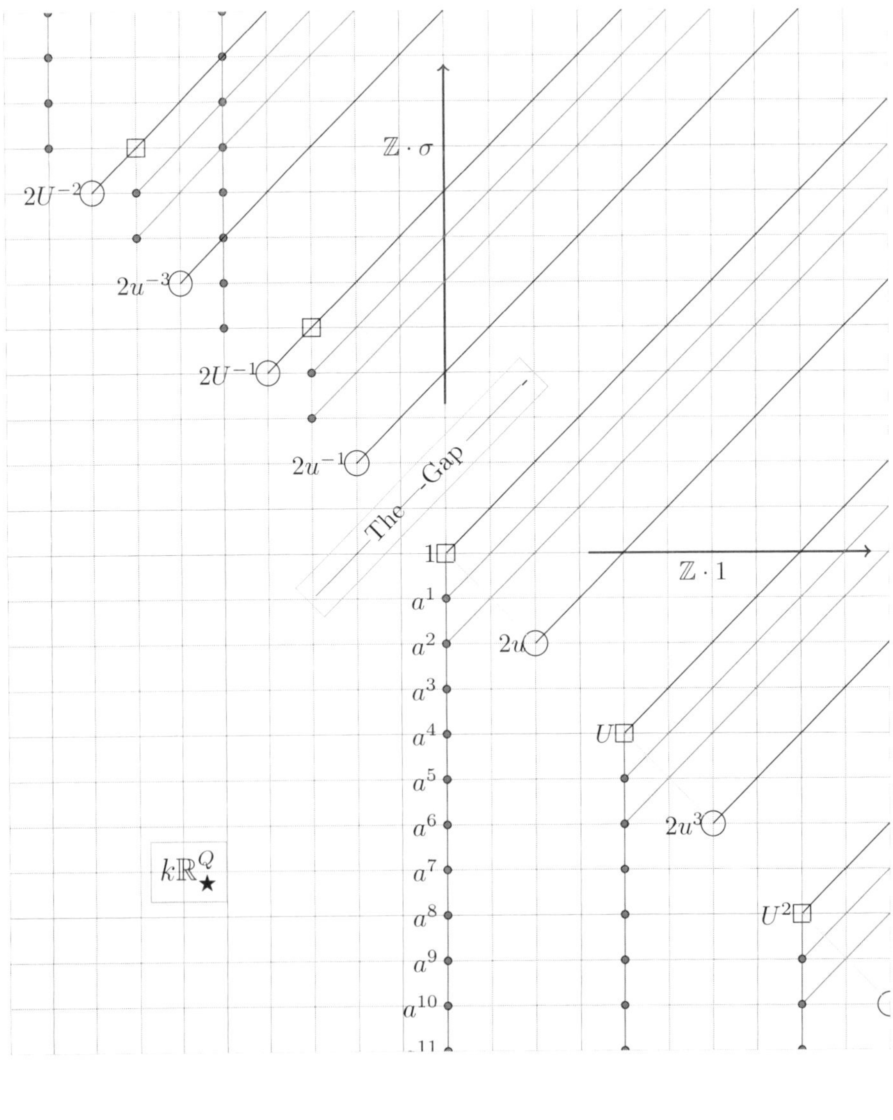

3.A. Adding cells. The approach through a cell structure of $S^{b\sigma}$ is a fairly unattractive since the value of the theory on cells is complicated: $k\mathbb{R}^Q_* = ko_*$. On the other hand, it is effectively what was done in [**4**], by working up the Tate filtration, so we will not repeat it here.

3.B. Quotient approach. The good behaviour on quotients is only easy to see on free spectra, so the content can be extracted from the Tate approach by truncating appropriately. Again, it does not seem worth repeating here.

3.C. Tate approach. The cleanest approach is to use the Tate diagram

$$\begin{array}{ccccc} k\mathbb{R}_h & \longrightarrow & k\mathbb{R} & \longrightarrow & k\mathbb{R}^\Phi \\ \simeq\downarrow & & \downarrow & & \downarrow \\ k\mathbb{R}_h & \longrightarrow & k\mathbb{R}^h & \longrightarrow & k\mathbb{R}^t \end{array}$$

The only real calculational input is the homotopy fixed points.

LEMMA 3.1. *The homotopy fixed point coefficients are*

$$k\mathbb{R}^{hQ}_\bigstar = BB[U, U^{-1}],$$

where $BB = \mathbb{Z}[a, \overline{v}]/(2a, \overline{v}a^3) \oplus 2U \cdot \mathbb{Z}[\overline{v}]$ *is the Basic Block for* $k\mathbb{R}$. *Here* $|a| = -\sigma$ *as before and* $|\overline{v}| = 1 + \sigma$; *the basic block* BB *is copied across the page by the periodicity operator* U *of degree* $|U| = 4 - 4\sigma$.

REMARK 3.2. Notice how this is related to ordinary group cohomology, which occurs along the negative x-axis:

$$k\mathbb{R}^{hQ}_{\leq 0} = \mathbb{Z}[Y]/(2Y), \text{ where } Y = a^4U^{-1} \text{ is of codegree 4.}$$

The element Y corresponds to y^2 from $H\underline{\mathbb{Z}}^{hQ}_*$ in Remark 2.2.

Similarly $ko_* = (k\mathbb{R})^{hQ}_{\geq 0}$ occurs along the positive x-axis, with $\eta = a\overline{v}$ generating ko_1, $2u\overline{v}^2$ generating ko_4 and the Bott element $U\overline{v}^4$ generating ko_8.

PROOF. We describe the $RO(Q)$-graded spectral sequence of Subsection 1.E for $k\mathbb{R}$. In fact it is convenient to describe it from the E_1-term if we choose a good cell structure on EQ_+. Indeed we use the filtration

$$S(\sigma)_+ \subset S(2\sigma)_+ \subset S(3\sigma)_+ \subset \cdots \subset S(\infty\sigma)_+ = EQ_+.$$

There is one free Q-cell in each degree and in cellular homology this gives the standard 2-periodic resolution

$$0 \longleftarrow \mathbb{Z} \stackrel{\epsilon}{\longleftarrow} \mathbb{Z}Q \stackrel{1-x}{\longleftarrow} \mathbb{Z}Q \stackrel{1+x}{\longleftarrow} \mathbb{Z}Q \stackrel{1-x}{\longleftarrow} \cdots$$

of $\mathbb{Z}$ over the group ring $\mathbb{Z}Q$.

We may then describe the E^1 term conveniently as

$$E^1_{*,*,*} = \mathbb{Z}[a, \overline{v}][\lambda, \lambda^{-1}]$$

where the (s, t, b) gradings are

$$|a| = (-1, 1, -1), |\lambda| = (0, 1, -1), |\overline{v}| = (0, 1, 1).$$

The differential is given by making $a, \overline{v}$ into d_1 cycles and

$$d_1\lambda = 2a.$$

This is clear from the algebraic resolution since we know Q acts to negate odd powers of v and is of degree -1 on $S^{b\sigma}$ for b odd.

This means that $u = \lambda^2$ is a d_1-cycle and we may calculate

$$E^2_{*,*,*} = \{\mathbb{Z}[a]/(2a)\}\,[\overline{v}][u, u^{-1}].$$

The next differential is the one piece of topological input. It is forced by the fact that $\eta^4 = 0$ (from the stable homotopy of S^0). The differential is given by making $a, \overline{v}$ into cycles again and

$$d_2 u = \overline{v}a^3.$$

This makes $U = u^2$ into a d_2-cycle and

$$E^3 = \left\{\mathbb{Z}[a, \overline{v}]/((2a, \overline{v}a^3) \oplus (2u)\cdot \mathbb{Z}[\overline{v}]\right\}[U, U^{-1}],$$

where the relations $(2u)\cdot a = 0$ and $(2u)^2 = 4u^2 = 4U$ are left implicit. Now we find that all subsequent differentials on $a, \overline{v}$ or U have target in a zero group, so the spectral sequence has collapsed. $\square$

Because $S^{\infty\sigma} \simeq S^0[1/a]$ it is then easy to calculate Tate cohomology by inverting a and homotopy orbits as local cohomology.

COROLLARY 3.3.

$$k\mathbb{R}^{tQ}_{\bigstar} = \mathbb{F}_2[a, a^{-1}][U, U^{-1}],$$

$$k\mathbb{R}^{\bigstar}_{hQ} = NB[U, U^{-1}],$$

where $NB = (2, \overline{v}) \oplus \Sigma^{-1}\mathbb{F}_2[a]^{\vee}$ is the Negative Block for $k\mathbb{R}$, more conceptually described as the right derived functors of the a-power torsion functor $\Gamma_{(a)}$:

$$NB = R\Gamma_{(a)}(BB).$$

For geometric fixed points we apply Lemma 1.3.

LEMMA 3.4. *The geometric fixed points are given by $k\mathbb{R}^{\Phi Q}_{\bigstar} = \mathbb{F}_2[a, a^{-1}][U]$.*

PROOF. By construction $KO = (K\mathbb{R})^Q$. Since $(K\mathbb{R})^{tQ} \simeq *$ (by nilpotence of a in $(K\mathbb{R})^{hQ}_{\bigstar}$) we see $(K\mathbb{R})^Q \simeq (K\mathbb{R})^{hQ}$. Now the behaviour of the homotopy fixed point spectral sequence for $K\mathbb{R}$ shows that of $k\mathbb{R}$ and hence that $(k\mathbb{R})^Q \longrightarrow (k\mathbb{R})^{hQ}$ is the connective cover. We may therefore apply Lemma 1.3 to obtain the conclusion. $\square$

Finally, we return to the Tate square, and read off the conclusion: the only difference between $k\mathbb{R}^Q_{\bigstar}$ and the more regular $k\mathbb{R}^{hQ}_{\bigstar}$ comes from the negative Tate a-columns.

COROLLARY 3.5. *The $RO(Q)$-graded homotopy of $k\mathbb{R}$ is as follows*

$$k\mathbb{R}^Q_{\bigstar} = BB[U] \oplus U^{-1}\cdot NB[U^{-1}],$$

where the action of U across the boundary

$$U^{-1}\cdot NB \longrightarrow BB$$

factors out the $H^1_{(a)}$ towers and includes the a-power torsion in BB.

3.D. The Bockstein spectral sequence. Dugger [**8**] has shown that there is a cofibre sequence

$$\Sigma^{\rho} k\mathbb{R} \xrightarrow{\overline{v}} k\mathbb{R} \longrightarrow H\underline{\mathbb{Z}}$$

where $\rho = 1+\sigma$. (This can be proved by noting the homotopy fixed point spectral sequence gives an element $\overline{v}$, and slice filtration methods easily show the mapping cone is $H\underline{\mathbb{Z}}$). We use the Dugger sequence to calculate $k\mathbb{R}^Q_\bigstar$ from $H\underline{\mathbb{Z}}^Q_\bigstar$ by the Bockstein spectral sequence. This spectral sequence happens to coincide with the slice spectral sequence for $k\mathbb{R}$, as we make explicit in Subsection 3.E, but we will not make use of this.

Accordingly (using the conventions of [**4**, Section 4.1]) we consider the spectral sequence obtained by taking $RO(Q)$-graded homotopy of the diagram

$$\begin{array}{ccccccccc} k\mathbb{R} & \longleftarrow & \Sigma^{\rho} k\mathbb{R} & \longleftarrow & \Sigma^{2\rho} k\mathbb{R} & \longleftarrow & \Sigma^{3\rho} k\mathbb{R} & \longleftarrow & \cdots \\ \downarrow & & \downarrow & & \downarrow & & \downarrow & & \\ \mathbb{Z} & & \Sigma^{\rho} H\underline{\mathbb{Z}} & & \Sigma^{2\rho} H\underline{\mathbb{Z}} & & \Sigma^{3\rho} H\underline{\mathbb{Z}} & & \end{array}$$

to obtain a spectral sequence

$$E^1 = H\underline{\mathbb{Z}}^Q_\bigstar[\hat{\overline{v}}] \Rightarrow k\mathbb{R}^Q_\bigstar$$

where $\hat{\overline{v}}$ is a formal variable of bidegree $(1, \rho)$. Since we have already calculated $H\underline{\mathbb{Z}}^Q_\bigstar$ this is straightforward once we have found a reasonable way to display the information. The standard thing to do in the $\mathbb{Z}$-graded case (for $\Sigma ko \longrightarrow ko \longrightarrow ku$ for example) is that for each row of constant s we write ku_* horizontally, shifted to the right by s, so that the differential d_r takes on the Adams form of going up r rows and one column to the left (i.e., subtracting 1).

In our present context, for each fixed s we have a whole $RO(Q)$-plane. Accordingly we will refer to this counterpart of the sth row as the sth *floor*. Once again we arrange that d^r goes up from the sth floor to the $(s+r)$th floor. If we place $\pi^Q_\bigstar(\Sigma^{s\rho} H\underline{\mathbb{Z}})$ on the sth floor, d^r again subtracts 1: symbolically, for any $\alpha \in RO(Q)$,

$$d^r : H\underline{\mathbb{Z}}^Q_\alpha \longrightarrow (\Sigma^{r\rho} H\underline{\mathbb{Z}})^Q_{\alpha-1} = H\underline{\mathbb{Z}}^Q_{\alpha-r\rho-1} = H\underline{\mathbb{Z}}^Q_{\alpha-r-1-r\sigma}$$

The value on x is defined by finding the image of x in $k\mathbb{R}^Q_{\alpha-1}$, dividing by $\overline{v}^{r-1}$ and looking at the image in $H\underline{\mathbb{Z}}^Q_\bigstar$ again. We need only determine it in a few cases and then propogate this using the multiplicative structure.

For those with old-fashioned habits, it is natural to arrange the calculation conventionally in the upper half-plane with the box at (n, s) containing $H\underline{\mathbb{Z}}^Q_{n-s+*\sigma}$, then d^r relates (n, s) to $(n-1, s+r)$ and reduces the σ-grading by r. We then find the value of $ku^Q_{n+*\sigma}$ by adding up the terms for fixed n

On the other hand this places a strain on visualization, and an alternative is to pile copies of $H\underline{\mathbb{Z}}^Q_\bigstar$ vertically above each other and look from above. Each $H\underline{\mathbb{Z}}^Q_{a+b\sigma}$ that you see then represents a whole pile, $H\underline{\mathbb{Z}}^Q_{a+b\sigma}[\hat{\overline{v}}]$, of copies. Then the differential d^r raises the floor in the pile by r and as indicated above it reduces $a+b\sigma$ by $(r+1)+r\sigma$. Finding the answer then involves adding up entries from different stacks with total degree constant.

We illustrate this for d^1, which has degree $-2-\sigma$:

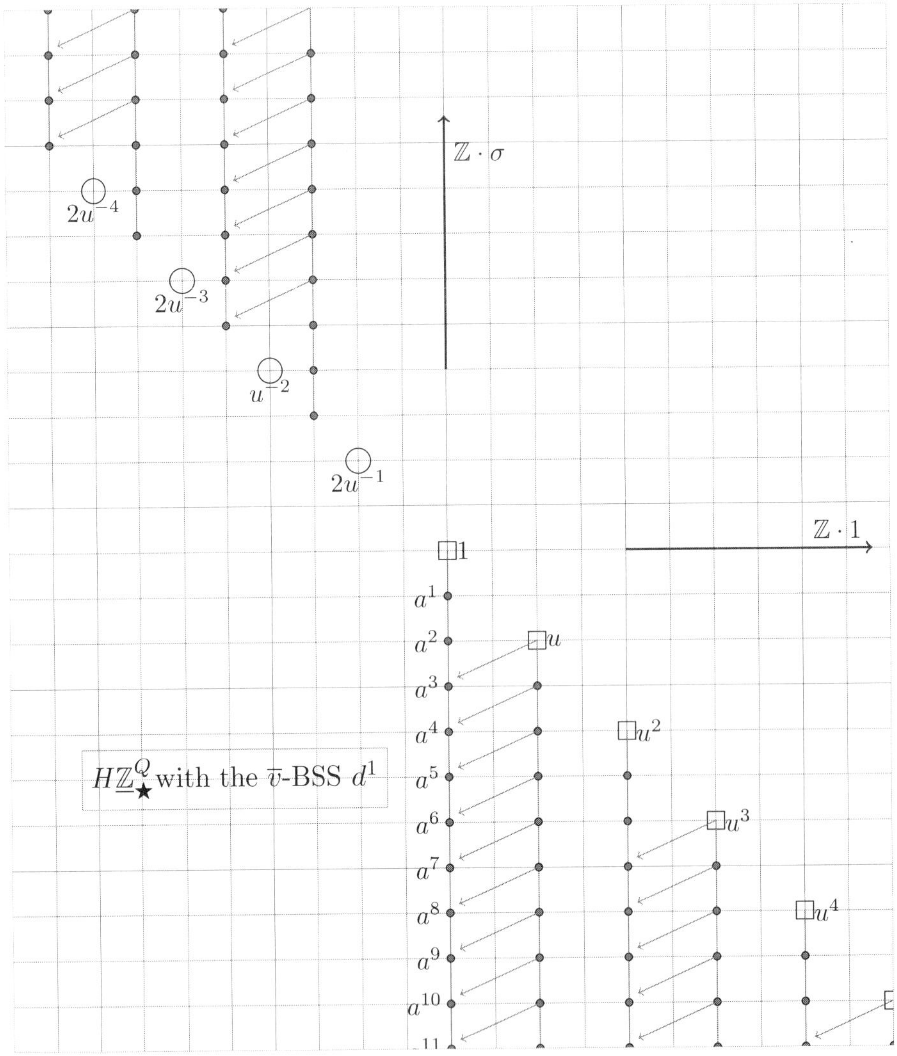

To justify the displayed d^1, note that it is induced by a map $\theta : H\underline{\mathbb{Z}} \longrightarrow \Sigma^{2+\sigma}H\underline{\mathbb{Z}}$, which is the analogue of the integral lift of Sq^3 in the non-equivariant setting. The stable operations of $H\underline{\mathbb{Z}}$ are relatively complicated, but in our case it suffices to observe that θ induces a map $\overline{\theta} : b \longrightarrow \Sigma^3 b$ where $b = F(EQ_+, H\mathbb{Z}/2)$ represents mod 2 Borel cohomology. The stable operations in this case are easy to determine: we immediately see $b^*_Q b = H^*(BQ_+; \mathbb{Z}/2) \otimes \mathcal{A}^*$ (the operations are studied further in [**11**]). In any case, we see that for representations V

$$\theta : H^n_Q(S^V; \underline{\mathbb{Z}}) = H\underline{\mathbb{Z}}^Q_{V-n} \longrightarrow H\underline{\mathbb{Z}}^Q_{V-n-2-\sigma} = H^{n+2}_Q(S^{V-\sigma})$$

is

$$\theta : H_Q^{n-1}(S(V);\underline{\mathbb{Z}}) \longrightarrow H_Q^{n+1}(S(V-\sigma);\underline{\mathbb{Z}})$$

for $n \geq 2$ and provided V contains σ. When $n-1$ is even, reduction mod 2 is an isomorphism and this agrees with

$$\overline{\theta} : b_Q^{n-1}(S(V)) \longrightarrow b_Q^{n+1}(S(V-\sigma)) = b_Q^{n+2}(S(V)).$$

Such an operation is a linear combination $\lambda Sq^3 + \mu Sq^2 Sq^1$. We need only argue that $\lambda \neq 0$, since this forces $\mu = 0$ (so as to get $(d^1)^2 = 0$). There are several approaches to understanding this operation. The first is to argue from the fact that $\eta^4 = 0$ (this involves looking at the whole BSS as displayed in Figure 1 below). The second is to study operations sufficiently to see the behaviour of d^1 is forced by the non-equivariant situation. Perhaps the most elementary is to begin by understanding the situation in geometric fixed points (i.e., after inverting a). After the fact, we know that the cofibre sequence

$$\Sigma k\mathbb{R}^{\Phi Q} = (\Sigma^\rho k\mathbb{R})^{\Phi Q} \longrightarrow k\mathbb{R}^{\Phi Q} \longrightarrow H\underline{\mathbb{Z}}^{\Phi Q}$$

has $H\underline{\mathbb{Z}}_*^{\Phi Q} = \mathbb{F}_2[x]$ where x is of degree 2 and $k\mathbb{R}_*^{\Phi Q} = \mathbb{F}_2[x^2]$. However, we do not need to use this full analysis of the cofibre sequence to see that the operation is non-trivial from the 2 column to the 0 column: we only need to know that $k\mathbb{R}_1^{\Phi Q} = 0$, which is easily checked from the Dugger sequence and known values of $H\underline{\mathbb{Z}}_\bigstar^Q$. This then determines the operation, which gives all the values. Using any one of these methods shows that d^1 is as displayed.

We could now jump straight to displaying $E^2 = E^\infty$. One simply notes that the $s = 0$ line consists of kernels of d^1 and each of the higher rows consists of the homology (with the sth row shifted s steps to the right).

Alternatively, instead of dealing with d^1 in piles, we could have used the traditional display, and we will describe this next. Each box for constant (n, s) contains a group graded on multiples of σ. Because of the almost-periodicity of $H\underline{\mathbb{Z}}_\bigstar^Q$ of $2(1-\sigma)$ and for typographical reasons we have arranged that $n(1-\sigma)$ is the mid-point of the (n, s)-box. In other words, the element x of (n, s)-degree $(2, 0)$ is represented by horizontal translation and corresponds to u of degree $2(1-\sigma)$.

We will now display the E^1-term in Figure 1 and then the $E^2 = E^\infty$ term in Figure 2. The squares and circles denote copies of $\mathbb{Z}$ and the dots denote copies of $\mathbb{F}_2$. The larger (pale blue online) blobs pick out the slice spectral sequence as detailed in the next section.

3.E. The slice spectral sequence (SSS). The SSS is usually presented as a mechanism for calculating the integer graded part $\pi_*^Q(X)$, with $RO(Q)$-graded part obtained by looking at suspensions of X [**14**]. Since suspension by ρ just shifts the slices, we may combine the suspensions and view it as a method for calculating the $RO(Q)$-graded homotopy. We will not give a detailed account of the SSS here since in this special case it is contained in the BSS. This short section is designed so that those already familiar with the SSS can pick it out from the BSS as described in the previous subsection.

For $k\mathbb{R}$, the slice filtration is precisely the same as the $\overline{v}$-Bockstein filtration, so that the $RO(Q)$-graded BSS and SSS are the same. The one piece of added value in the SSS is the way that the usual ($\mathbb{Z}$-graded) display highlights vanishing lines and 'The Gap'. This information is present in the BSS displays above, but it is not so easy to spot.

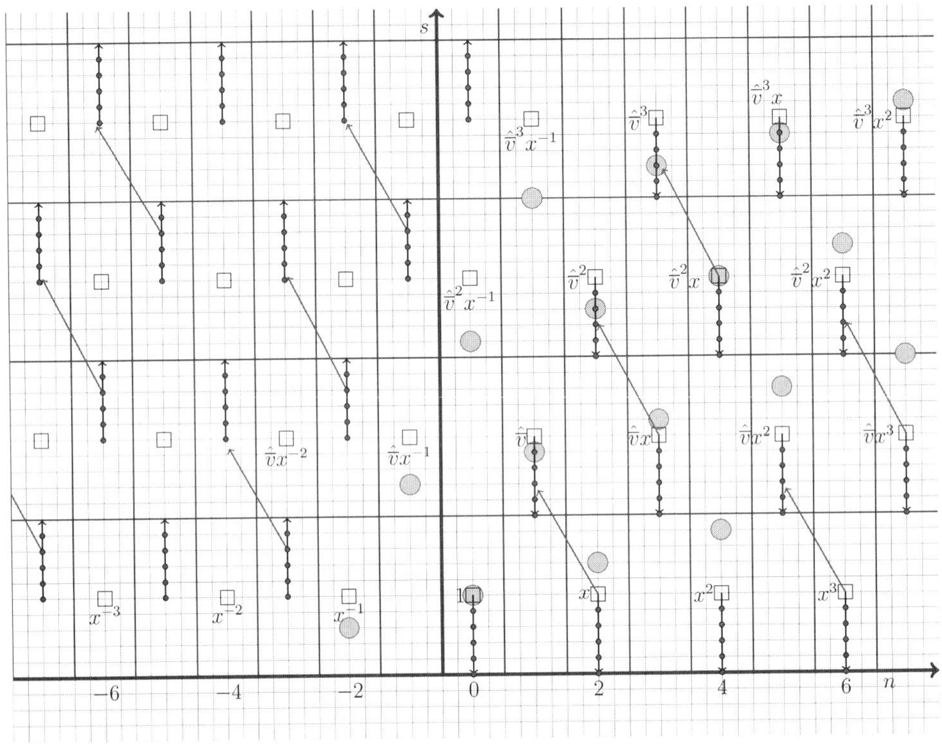

FIGURE 1. The $\overline{v}$-Bockstein spectral sequence: $E_1 = H\underline{\mathbb{Z}}^Q_\bigstar[\hat{\overline{v}}] \Rightarrow k\mathbb{R}^Q_\bigstar$. Squares are copies of $\mathbb{Z}$, small (red online) dots are copies of $\mathbb{F}_2$. The d^1 differentials are indicated by displaying the top or bottom degree component between the relevant boxes. The larger (blue online) blobs select out the degrees occurring in one slice spectral sequence.

Just to be completely explicit, for $k\mathbb{R}$ the negative slices are all zero and the odd slices are all zero. The positive even slices are suspensions of $H\underline{\mathbb{Z}}$: in the notation of [**15**] $P^{2k}_{2k}k\mathbb{R} = \Sigma^{k\rho}H\underline{\mathbb{Z}}$ for $k \geq 0$. In Adams grading $(n, s) = (2k - s, s)$ the usual SSS has

$$\pi^Q_{2k-s}(P^{2k}_{2k}k\mathbb{R}) = H^Q_{2k-s}(S^{k\rho}; \underline{\mathbb{Z}}) = H^Q_{k-s}(S^{k\sigma}; \underline{\mathbb{Z}}) = H\underline{\mathbb{Z}}^Q_{k-s-k\sigma},$$

so that the homotopy groups of one particular slice occur along a line of slope -1. The appropriate chart is displayed clearly as Figure 7 in [**16**, Page 414].

Note that the homotopy group above, at coordinates $(x_S, y_S) = (2k - s, s)$ in the SSS, appears at coordinates $(x_B, y_B) = (2k - s, k)$ in the BSS. Thus $(x_B, y_B) = (x_S, (x_S + y_S)/2)$ and $(x_S, y_S) = (x_B, 2y_B - x_B)$. The BSS differential d^1_B corresponds to the SSS differential d^3_S. The slice spectral sequence simply selects a subspectral sequence given by one particular σ-grading in each box.

The larger (pale blue online) blobs in the two $\overline{v}$-**B**SS charts are the potentially nonzero entries in the **S**SS for $\pi^Q_*(k\mathbb{R})$ (i.e., the integer graded part). We have displayed five diagonal rows of blobs (in adding $(1, 1)$ to (n, s), σ is subtracted). The

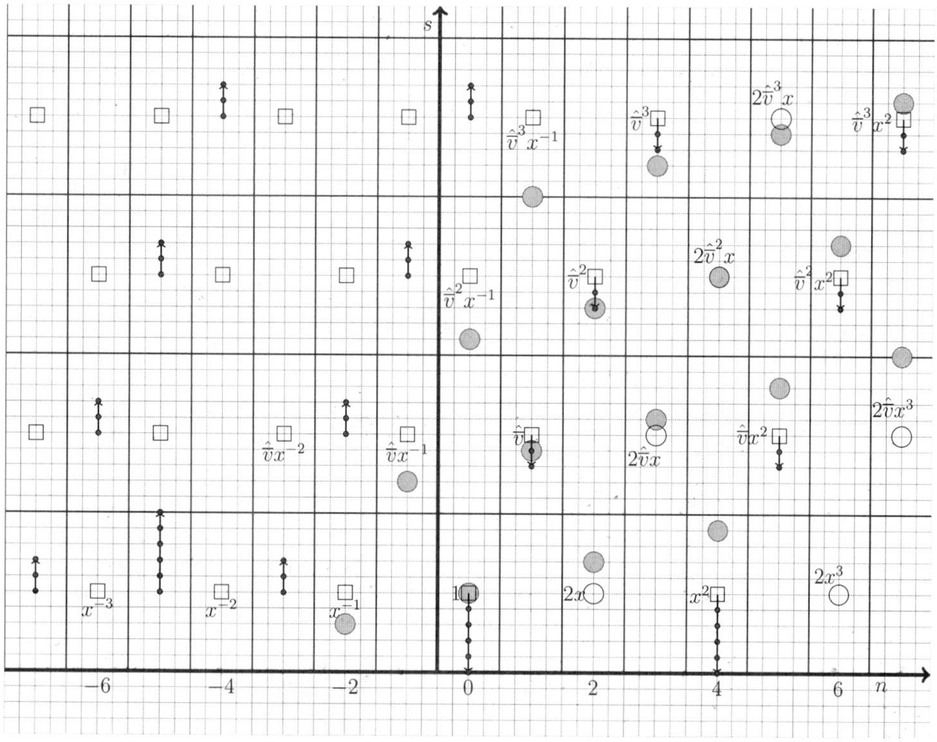

FIGURE 2. $E_2 = E_\infty$ of the $\overline{v}$-Bockstein spectral sequence: $E_1 = H\underline{\mathbb{Z}}^Q_\bigstar[\hat{\overline{v}}] \Rightarrow k\mathbb{R}^Q_\bigstar$. Squares are copies of $\mathbb{Z}$. Large circles are copies of $\mathbb{Z}$ of index 2 in E_1. Small (red online) dots are copies of $\mathbb{F}_2$. The larger (blue online) blobs select out the degrees occurring in one slice spectral sequence. The Gap occurs in this picture most prominently as the three zeroes below the diagonal squares emanating from $(n, s) = (-2, 0)$, but it also requires three zeroes along the diagonal emanating from $(n, 0)$ for any $n \leq -1$.

diagonal emanating from $(n, s) = (-2, 0)$ does not contain any non-zero SSS entries. The diagonal emanating from $(n, s) = (0, 0)$ contains 1, and $a^k\hat{\overline{v}}^k$ (corresponding to η^k) for $k \geq 1$. Of course $1, a\hat{\overline{v}}$ and $a^2\hat{\overline{v}}^2$ survive to E^∞ to give η and η^2 and higher powers are the images of differentials. The first non-zero entry on the diagonal emanating from $(n, s) = (2, 0)$ is on the $s = 2$ line, where it is $\hat{\overline{v}}^2 x$, and after that $a^k\hat{\overline{v}}^{2+k} x$ for $k \geq 1$. These all support nonzero differentials, so only $2\hat{\overline{v}}^2 x$ survives (to give the generator of $\pi_4^Q(k\mathbb{R}) = \pi_4(ko)$). In our display, only one blob is visible on diagonal emanating from $(n, s) = (4, 0)$, and it is zero.

Acknowledgments

I am grateful to R. R. Bruner, J. P. May and L. Meier for collaborations which fed into this account, to the referee for objectivity about conventions, and to D.Dugger, D.C.Ravenel and Mingcong Zeng for pointing out an error in the old version of the chart in Section 2.

References

[1] S. Araki, *Orientations in τ-cohomology theories*, Japan. J. Math. (N.S.) **5** (1979), no. 2, 403–430. MR614829

[2] M. F. Atiyah, *K-theory and reality*, Quart. J. Math. Oxford Ser. (2) **17** (1966), 367–386. MR0206940

[3] R. Bruner and J. Greenlees, *The Bredon-Löffler conjecture*, Experiment. Math. **4** (1995), no. 4, 289–297. MR1387694

[4] R. R. Bruner and J. P. C. Greenlees, *Connective real K-theory of finite groups*, Mathematical Surveys and Monographs, vol. 169, American Mathematical Society, Providence, RI, 2010. MR2723113

[5] G. Carlsson, *Equivariant stable homotopy and Segal's Burnside ring conjecture*, Ann. of Math. (2) **120** (1984), no. 2, 189–224. MR763905

[6] J. L. Caruso, *Operations in equivariant* $\mathbf{Z}/p$*-cohomology*, Math. Proc. Cambridge Philos. Soc. **126** (1999), no. 3, 521–541. MR1684248

[7] A. W. M. Dress, *Contributions to the theory of induced representations*, Algebraic K-theory, II: "Classical" algebraic K-theory and connections with arithmetic (Proc. Conf., Battelle Memorial Inst., Seattle, Wash., 1972), Springer, Berlin, 1973, pp. 183–240. Lecture Notes in Math., Vol. 342. MR0384917

[8] D. Dugger, *An Atiyah-Hirzebruch spectral sequence for KR-theory*, K-Theory **35** (2005), no. 3-4, 213–256 (2006). MR2240234

[9] J. P. C. Greenlees "Adams spectral sequences in equivariant topology." (Cambridge University Thesis, 1985)

[10] J. P. C. Greenlees, *Representing Tate cohomology of G-spaces*, Proc. Edinburgh Math. Soc. (2) **30** (1987), no. 3, 435–443. MR908451

[11] J. P. C. Greenlees, *Stable maps into free G-spaces*, Trans. Amer. Math. Soc. **310** (1988), no. 1, 199–215. MR938918

[12] J. P. C. Greenlees and J. P. May, *Generalized Tate cohomology*, Mem. Amer. Math. Soc. **113** (1995), no. 543, viii+178. MR1230773

[13] J. P C. Greenlees and L. Meier, *Gorenstein duality for real spectra*, Algebr. Geom. Topol. **17** (2017), no. 6, 3547–3619. MR3709655

[14] M. A. Hill, *The equivariant slice filtration: a primer*, Homology Homotopy Appl. **14** (2012), no. 2, 143–166. MR3007090

[15] M. A. Hill, M. J. Hopkins, and D. C. Ravenel, *On the nonexistence of elements of Kervaire invariant one*, Ann. of Math. (2) **184** (2016), no. 1, 1–262. MR3505179

[16] M. A. Hill, M. J. Hopkins, and D. C. Ravenel, *The slice spectral sequence for the* C_4 *analog of real K-theory*, Forum Math. **29** (2017), no. 2, 383–447. MR3619120

[17] M. A. Hill and L. Meier, *The* C_2*-spectrum* $\mathrm{Tmf}_1(3)$ *and its invertible modules*, Algebr. Geom. Topol. **17** (2017), no. 4, 1953–2011. MR3685599

[18] P. Hu and I. Kriz, *Real-oriented homotopy theory and an analogue of the Adams-Novikov spectral sequence*, Topology **40** (2001), no. 2, 317–399. MR1808224

[19] P. S. Landweber, *Conjugations on complex manifolds and equivariant homotopy of MU*, Bull. Amer. Math. Soc. **74** (1968), 271–274. MR0222890

[20] L. G. Lewis Jr., J. P. May, M. Steinberger, and J. E. McClure, *Equivariant stable homotopy theory*, Lecture Notes in Mathematics, vol. 1213, Springer-Verlag, Berlin, 1986. With contributions by J. E. McClure. MR866482

[21] N. Ricka, *Equivariant Anderson duality and Mackey functor duality*, Glasg. Math. J. **58** (2016), no. 3, 649–676. MR3530492

[22] R. Stong Letter to J. P. May, dated 4 September 1980.

[23] R. Stong, "Ordinary cohomology of points" 1980 13pp.

[24] S. Waner, "Notes on the cohomology of points" 1980.

School of Mathematics and Statistics, Hicks Building, Sheffield S3 7RH, United Kingdom

Email address: `j.greenlees@sheffield.ac.uk`

Contemporary Mathematics
Volume **708**, 2018
http://dx.doi.org/10.1090/conm/708/14264

Topological Hochschild homology and the Hasse-Weil zeta function

Lars Hesselholt

Introduction

In this paper, we consider the Tate cohomology of the circle group acting on the topological Hochschild homology of schemes. We show that in the case of a scheme smooth and proper over a finite field, this cohomology theory naturally gives rise to the cohomological interpretation of the Hasse-Weil zeta function by regularized determinants envisioned by Deninger [**15**]. In this case, the periodicity of the zeta function is reflected by the periodicity of said cohomology theory, whereas neither is periodic in general.

Let X be a scheme. Connes' periodic cyclic homology of X may be defined to be the anticommutative graded ring $\mathrm{HP}_*(X)$ given by the Tate cohomology groups

$$\mathrm{HP}_i(X) = \hat{H}^{-i}(\mathbb{T}, \mathrm{HH}(X))$$

in the sense of Greenlees [**20, 32**] of the circle group $\mathbb{T}$ acting on the Hochschild spectrum $\mathrm{HH}(X)$. By analogy, we consider the anticommutative graded ring $\mathrm{TP}_*(X)$ given by the Tate cohomology groups

$$\mathrm{TP}_i(X) = \hat{H}^{-i}(\mathbb{T}, \mathrm{THH}(X))$$

of $\mathbb{T}$ acting on Bökstedt's topological Hochschild spectrum $\mathrm{THH}(X)$ [**7, 23**]. The latter is an implementation of Waldhausen's philosophy to replace the initial ring of algebra $\mathbb{Z}$ by the initial ring of higher algebra $\mathbb{S}$. The graded ring $\mathrm{TP}_*(X)$ is the abutment of the Tate spectral sequence

$$E^2_{i,j} = H^{-i}(\mathbb{P}^{\infty}_{-\infty}(\mathbb{C}), \mathrm{THH}_j(X)) \Rightarrow \mathrm{TP}_{i+j}(X),$$

which is multiplicative and converges conditionally in the sense of Boardman [**6**]. The E^2-term is periodic with periodicity operator given by multiplication by

$$t = c_1(\mathcal{O}(1)) \in H^2(\mathbb{P}^{\infty}(\mathbb{C}), \mathbb{Z}) = H^2(\mathbb{P}^{\infty}_{-\infty}(\mathbb{C}), \mathbb{Z}).$$

In Connes' theory, the periodicity element t is an infinite cycle and is represented by a unique homotopy class $v \in \mathrm{HP}_{-2}(\mathbb{Z})$, the multiplication by which is Connes' S-operator. But in the topologically refined situation, the element t does not in general survive the spectral sequence, and, if it does, is typically not represented by a unique homotopy class. In particular, the graded ring $\mathrm{TP}_*(X)$ is generally

2010 *Mathematics Subject Classification.* Primary 11S40, 19D55; Secondary 14F30.

Assistance from DNRF Niels Bohr Professorship is gratefully acknowledged.

not periodic. However, by contrast to Connes' theory, the graded ring $\mathrm{TP}_*(X)$ is, in favorable situations, equipped with meromorphic Frobenius operators

$$\mathrm{TP}_*(X) \xrightarrow{\varphi_p} \mathrm{TP}_*(X).$$

The operator φ_p is obtained from the *cyclotomic* structure of $\mathrm{THH}(X)$ by a process that bears some resemblance to analytic continuation. In particular, it does not always exist, and, when it does, its construction requires some work.

Suppose now that X is defined over a finite field k of order $q = p^r$. In this situation, the periodicity element t is an infinite cycle and is represented by a preferred homotopy class $v \in \mathrm{TP}_{-2}(k)$, the definition of which is given in Section 4 below. Moreover, there is a canonical identification of the ring $\mathrm{TP}_0(k)$ with the ring W of p-typical Witt vectors in k. It follows that $\mathrm{TP}_*(X)$ is a 2-periodic anticommutative graded W-algebra. In this situation, the Frobenius operator φ_p is defined after inverting p. Moreover, the W-linear operator φ_p^r and the weight filtration of $\mathrm{TP}_*(X)$ determines the action of the geometric Frobenius $\mathrm{Fr}_q\colon X \to X$ on $\mathrm{TP}_*(X) \otimes_{W,\iota} \mathbb{C}$.

THEOREM A. *Let k be a finite field of order $q = p^r$, let W be its ring of p-typical Witt vectors, and let $\iota\colon W \to \mathbb{C}$ be a choice of embedding. If $f\colon X \to \mathrm{Spec}(k)$ is a smooth and proper morphism of schemes, then, as meromorphic functions on $\mathbb{C}$,*

$$\zeta(X,s) = \frac{\det_\infty(s\cdot \mathrm{id} - \Theta \mid \mathrm{TP}_{\mathrm{od}}(X) \otimes_{W,\iota} \mathbb{C})}{\det_\infty(s\cdot \mathrm{id} - \Theta \mid \mathrm{TP}_{\mathrm{ev}}(X) \otimes_{W,\iota} \mathbb{C})},$$

where Θ is a $\mathbb{C}$-linear graded derivation such that $q^\Theta = \mathrm{Fr}_q^*$ and $\Theta(v) = \frac{2\pi i}{\log q}\cdot v$.

The graded derivation Θ is determined by the specified value on the periodicity class v and by its values on $\mathrm{TP}_0(X) \otimes_{W,\iota} \mathbb{C}$ and $\mathrm{TP}_1(X) \otimes_{W,\iota} \mathbb{C}$, which are both finite dimensional $\mathbb{C}$-vector spaces. To specify the latter, we follow Deninger [**15**, Section 2.12] and decompose the Frobenius $\mathrm{Fr}_q^* = \mathrm{Fr}_{q,s}^* \circ \mathrm{Fr}_{q,u}^*$ in its semisimple and unipotent parts and define

$$\Theta = \log_q(\mathrm{Fr}_q^*) = \log_q(\mathrm{Fr}_{q,s}^*) + \log_q(\mathrm{Fr}_{q,u}^*).$$

Here we use some choice of principal branch of the logarithm on the semisimple part and the series $\frac{1}{\log q}\log(-)$ on the unipotent part. Since, by naturality, $\mathrm{Fr}_q^*(v) = v$, the operator Θ may be seen as the most general solution to $q^\Theta = \mathrm{Fr}_q^*$, which is in accordance with Deninger's philosophy that Θ should be the infinitesimal generator of a Frobenius flow Fr_t^*. We will recall the notions of regularized determinant and anomalous dimension in the final section of the paper.

The operator $s\cdot\mathrm{id} - \Theta$ has anomalous dimension zero. Therefore, we may fix any choice of a positive real scaling factor δ and state Theorem A as

$$\zeta(X,s) = \frac{\det_\infty(\delta(s\cdot \mathrm{id} - \Theta) \mid \mathrm{TP}_{\mathrm{od}}(X) \otimes_{W,\iota} \mathbb{C})}{\det_\infty(\delta(s\cdot \mathrm{id} - \Theta) \mid \mathrm{TP}_{\mathrm{ev}}(X) \otimes_{W,\iota} \mathbb{C})}.$$

In particular, we may take $\delta = \frac{1}{2\pi}$ as in the archimedean case [**14**, Theorem 1.1], where the scaling factor is essential.

We outline the proof of Theorem A. The present proof[1] uses the cohomological interpretation of the Hasse-Weil zeta function by crystalline cohomology established

[1] The recent proof of the Künneth formula for TP by Blumberg-Mandell [**4**, Theorem A] has facilitated a direct proof given by Tabuada [**35**, Theorem 3.27].

by Berthelot in his thesis [**2**, Théorème VII.3.2.3],

$$\zeta(X,s) = \frac{\det(\mathrm{id} - q^{-s}\,\mathrm{Fr}_q^* \mid H^{\mathrm{od}}_{\mathrm{crys}}(X/W)\otimes_{W,\iota}\mathbb{C})}{\det(\mathrm{id} - q^{-s}\,\mathrm{Fr}_q^* \mid H^{\mathrm{ev}}_{\mathrm{crys}}(X/W)\otimes_{W,\iota}\mathbb{C})}.$$

The comparison theorem of Bloch-Deligne-Illusie [**29**, Théorème II.1.4] shows that, in this formula, the crystalline cohomology $H^i_{\mathrm{crys}}(X/W)$ may be replaced by the hypercohomology $H^i(X, W\Omega^{\cdot}_X)$ with coefficients in the de Rham-Witt complex. Therefore, we have the two spectral sequences of hypercohomology, one of which gives rise to a natural spectral sequence of finite dimensional $\mathbb{C}$-vector spaces

$$E_2^{i,j} = (\lim_{n,F} H^i(X, W_n\Omega^j_X)) \otimes_{W,\iota}\mathbb{C} \Rightarrow H^{i+j}_{\mathrm{crys}}(X/W)\otimes_{W,\iota}\mathbb{C}$$

called the *conjugate* spectral sequence. Here, the identification of the E_2-term also uses the higher Cartier isomorphism of Illusie-Raynaud [**30**, Proposition III.1.4]. We prove in Theorem 6.8 below that, as a consequence of the definition of $\mathrm{THH}(X)$ in [**5, 17**] and the calculation of the equivariant homotopy groups of $\mathrm{THH}(\mathcal{O}_X)$ in [**21**, Theorem B], the same finite dimensional $\mathbb{C}$-vector spaces appear in the E^2-term of a natural spectral sequence

$$E^2_{i,j} = \bigoplus_{m\in\mathbb{Z}} (\lim_{n,F} H^{-i}(X, W_n\Omega^{j+2m}_X)) \otimes_{W,\iota}\mathbb{C} \Rightarrow \mathrm{TP}_{i+j}(X)\otimes_{W,\iota}\mathbb{C}$$

converging to the periodic topological cyclic homology of X. We call this spectral sequence the Hodge spectral sequence. The differentials in the spectral sequence preserve the direct sum decomposition of the E^2-term, and we call the number $w = j + m$ the weight.[2] By comparing the two spectral sequences and using that determinants are multiplicative on exact sequences, we conclude from Berthelot's cohomological interpretation that

$$\zeta(X,s) = \frac{\det(\mathrm{id} - q^{-s}\,\mathrm{Fr}_q^* \mid \mathrm{TP}_1(X)\otimes_{W,\iota}\mathbb{C})}{\det(\mathrm{id} - q^{-s}\,\mathrm{Fr}_q^* \mid \mathrm{TP}_0(X)\otimes_{W,\iota}\mathbb{C})}.$$

We wish to instead express the zeta function by using the full infinite dimensional graded $\mathbb{C}$-algebra $\mathrm{TP}_*(X)\otimes_{W,\iota}\mathbb{C}$. Indeed, we contend that this theory precisely is the infinite dimensional theory produced by Deninger [**15**] through an algebraic modification of the finite dimensional crystalline cohomology theory employing the Riemann-Hilbert correspondance on $\mathbb{G}_m$. Accordingly, as a consequence of op. cit., Corollary 2.8, we conclude that with Θ as in Theorem A,

$$\zeta(X,s) = \frac{\det_\infty(s\cdot\mathrm{id} - \Theta \mid \mathrm{TP}_{\mathrm{od}}(X)\otimes_{W,\iota}\mathbb{C})}{\det_\infty(s\cdot\mathrm{id} - \Theta \mid \mathrm{TP}_{\mathrm{ev}}(X)\otimes_{W,\iota}\mathbb{C})}.$$

This completes the outline of the proof of Theorem A. We remark that this result gives a cohomological interpretation of the zeta function $\zeta(X,s)$ itself as opposed to one of the function $Z(X,t)$ with $\zeta(X,s) = Z(X,q^{-s})$.

We close with a remark concerning periodicity. If k is a commutative ring, then the Hochschild homology $\mathrm{HH}_*(k)$ is an anticommutative graded k-algebra and the graded ideal $I \subset \mathrm{HH}_*(k)$ spanned by the homogeneous elements of positive even

[2] Bhatt-Morrow-Scholze [**3**] have now constructed a filtration of TP by weight, the filtration quotients of which are given by crystalline cohomology.

degree has a natural divided power structure [**12**, Exposé 7]. In particular, it is not possible for the canonical k-algebra homomorphism from the symmetric algebra

$$S_k(\mathrm{HH}_2(k)) \longrightarrow \mathrm{HH}_*(k)$$

to be an isomorphism, unless k is a $\mathbb{Q}$-algebra. The topological Hochschild homology $\mathrm{THH}_*(k)$ also is an anticommutative graded k-algebra, but there is no longer a divided power structure on graded ideal $I \subset \mathrm{THH}_*(k)$ spanned by homogeneous elements of positive even degree. Hence, the canonical k-algebra homomorphism

$$S_k(\mathrm{THH}_2(k)) \longrightarrow \mathrm{THH}_*(k)$$

can be an isomorphism, and Bökstedt's periodicity theorem [**8**] shows that this is indeed the case for $k = \mathbb{F}_p$; see also [**26**, Theorem 5.2]. It is this basic periodicity theorem in conjunction with the cyclotomic structure of topological Hochschild homology that makes Theorem A possible.

1. The Tate spectrum

In this section, we recall Greenlees' generalization of Tate cohomology [**20**]. We refer the reader to [**26**, Section 4] for more details and will follow the conventions therein. A more modern and comprehensive account, which in particular treats the multiplicative properties of the construction in detail, is given in [**32**].

Let G be a compact Lie group, let E be a free left G-CW-complex whose underlying space is contractible, and let $f\colon E_+ \to S^0$ be the map that collapses E onto $0 \in \{0, \infty\} = S^0$. We consider the following cofibration sequence

$$E_+ \xrightarrow{f} S^0 \xrightarrow{i} \tilde{E} \xrightarrow{\partial} \Sigma E_+$$

in the homotopy category of pointed left G-spaces. The sequence is unique, up to unique isomorphism, and the induced sequence of suspension G-spectra is a cofibration sequence in the G-stable homotopy category. We abuse notation and denote the latter sequence by the same symbols. Now, for X a G-spectrum, we consider the diagram

$$\begin{array}{ccccccc} (E_+ \otimes X)^G & \longrightarrow & X^G & \longrightarrow & (\tilde{E} \otimes X)^G & \longrightarrow & \Sigma(E_+ \otimes X)^G \\ \downarrow & & \downarrow & & \downarrow & & \downarrow \\ (E_+ \otimes [E_+, X])^G & \longrightarrow & [E_+, X]^G & \longrightarrow & (\tilde{E} \otimes [E_+, X])^G & \longrightarrow & \Sigma(E_+ \otimes [E_+, X])^G \end{array}$$

where we write "$\otimes$" and "$[-,-]$" to indicate the symmetric monoidal product and internal hom object in the homotopy category of G-spectra, respectively. The left-hand vertical morphism is an isomorphism. We write the lower sequence as

$$H_\cdot(G, S^{\mathfrak{g}} \otimes X) \xrightarrow{N} H^\cdot(G, X) \xrightarrow{i} \hat{H}^\cdot(G, X) \xrightarrow{\partial} \Sigma H_\cdot(G, S^{\mathfrak{g}} \otimes X)$$

and call it the Tate cofibration sequence. Here $S^{\mathfrak{g}}$ is the one-point compactification of the adjoint representation of G on its Lie algebra $\mathfrak{g}$.

We are interested mainly in the case, where G is either the full circle group $\mathbb{T}$ of complex numbers of modulus 1 or one of its finite subgroups $C_r \subset \mathbb{T}$. We record the following result, referring the reader to [**11**] for background on p-completion.

LEMMA 1.1. *Let p be a prime number and let X be a $\mathbb{T}$-spectrum. If the underlying spectrum of X is p-complete and bounded below, then the canonical morphism*

$$\hat{H}^{\cdot}(\mathbb{T}, X) \longrightarrow \operatorname{holim}_n \hat{H}^{\cdot}(C_{p^n}, X)$$

is a weak equivalence.

PROOF. It is proved in [**24**, Lemma 2.1.1] that the morphism in question becomes a weak equivalence after p-completion. Hence, it suffices to show that the domain and target of this morphism are p-complete. We prove the statement for the domain; the proof for the target is analogous. We wish to show that for every integer i, the canonical morphism

$$\pi_i(\hat{H}^{\cdot}(\mathbb{T}, X)) \longrightarrow \pi_i(\hat{H}^{\cdot}(\mathbb{T}, X), \mathbb{Z}_p)$$

is an isomorphism. The assumption that the underlying spectrum of X is bounded below implies that there exists a non-negative integer s depending on i such that the morphism

$$(\mathrm{Fil}_s(\tilde{E}) \otimes [E_+, X])^{\mathbb{T}} \longrightarrow (\tilde{E} \otimes [E_+, X])^{\mathbb{T}}$$

induced by the inclusion of the s-skeleton induces an isomorphism of homotopy groups in degrees less than or equal to i. We may further assume that the $\mathbb{T}$-CW-complex $\mathrm{Fil}_s(\tilde{E})$ is finite. Therefore, replacing X by $\mathrm{Fil}_s(\tilde{E}) \otimes X$, the underlying spectrum of which again is p-complete, it suffices to show that the group cohomology spectrum $H^{\cdot}(\mathbb{T}, X)$ is p-complete. In this case, the completion morphism factors as the composition

$$H^{\cdot}(\mathbb{T}, X) \longrightarrow H^{\cdot}(\mathbb{T}, X_p) \longrightarrow H^{\cdot}(\mathbb{T}, X)_p$$

of the morphism induced by the completion morphism for X and a canonical isomorphism. But the left-hand morphism is a weak equivalence by the assumption that the underlying non-equivariant spectrum of X is p-complete. □

The skeleton filtrations of the pointed G-CW-complexes E_+ and $\tilde{E}$ give rise to spectral sequences converging conditionally to the homotopy groups of the three terms in the Tate cofibration sequence. We call the spectral sequence converging to the homotopy groups of the Tate spectrum the Tate spectral sequence. In the situation of Lemma 1.1, we take $E = S(\mathbb{C}^\infty)$ and $\tilde{E} = S^{\mathbb{C}^\infty}$ with the G-CW-structures defined in [**27**, Section 4.4]. For the circle group $G = \mathbb{T}$, the spectral sequence takes the form

$$E^2 = S\{t^{\pm 1}\} \otimes \pi_*(X) \Rightarrow \hat{H}^{-*}(\mathbb{T}, X)$$

with the generator $t = c_1(\mathcal{O}(1)) \in E^2_{-2,0}$ specified below. For $G = C_{p^n}$, we have

$$E^2 = S\{t^{\pm 1}\} \otimes \pi_*(X)/p^n\pi_*(X) \Rightarrow \hat{H}^{-*}(C_{p^n}, X),$$

if the homotopy groups $\pi_*(X)$ are p-torsion free, and

$$E^2 = \Lambda\{u\} \otimes S\{t^{\pm 1}\} \otimes \pi_*(X) \Rightarrow \hat{H}^{-*}(C_{p^n}, X),$$

if the homotopy groups $\pi_*(X)$ are annihilated by p. We define the generators $t \in E^2_{-2,0}$ and $u \in E^2_{-1,0}$ to be the classes of the cycles specified in [**27**, Lemma 4.2.1]. We now define the generator t for $G = \mathbb{T}$ to be the unique class that restricts to the generator t for $G = C_{p^n}$. If X is a ring $\mathbb{T}$-spectrum, then the spectral sequences are multiplicative with the indicated bi-graded ring structure on the E^2-term, except that if $p = 2$ and $n = 1$, then $u^2 = t$. The restriction maps in Tate cohomology

$$\hat{H}^{\cdot}(\mathbb{T}, X) \longrightarrow \hat{H}^{\cdot}(C_{p^n}, X) \longrightarrow \hat{H}^{\cdot}(C_{p^{n-1}}, X)$$

induce multiplicative maps of Tate spectral sequences that, on the E^2-terms, are given by the tensor product of the canonical projections

$$\pi_*(X) \longrightarrow \pi_*(X)/p^n\pi_*(X) \longrightarrow \pi_*(X)/p^{n-1}\pi_*(X)$$

and the graded ring homomorphisms that take t to t and u to 0.

2. Cyclotomic spectra

We briefly recall the notion of a p-cyclotomic spectrum, which was introduced in [**26**, Section 2] and codifies the structure available on the topological Hochschild spectrum in addition to the circle action discovered by Connes. A better and more flexible version of this notion was introduced by Nikolaus and Scholze [**32**].

The group $\mathbb{T}$ has the special property that there exists a group isomorphism

$$\mathbb{T} \xrightarrow{\ \rho\ } \mathbb{T}/C_p,$$

which we will always choose to be the pth root. A p-cyclotomic structure on a $\mathbb{T}$-spectrum T was defined in [**26**, Definition 2.2] to be a morphism of $\mathbb{T}$-spectra

$$\rho^*(\tilde{E} \otimes T)^{C_p} \xrightarrow{\ r\ } T$$

with the property that for all $n \geqslant 0$, the induced map of C_{p^n}-fixed point spectra is a weak equivalence. The diagram at the beginning of Section 1 takes the form

$$\begin{array}{ccccccc}
H_{\cdot}(C_{p^n}, T) & \longrightarrow & \mathrm{TR}^{n+1}(T;p) & \xrightarrow{\ R\ } & \mathrm{TR}^n(T;p) & \xrightarrow{\ \partial\ } & \Sigma H_{\cdot}(C_{p^n}, T) \\
\Vert & & \downarrow{\scriptstyle \tilde{\gamma}} & & \downarrow{\scriptstyle \gamma} & & \Vert \\
H_{\cdot}(C_{p^n}, T) & \xrightarrow{\ N\ } & H^{\cdot}(C_{p^n}, T) & \xrightarrow{\ i\ } & \hat{H}^{\cdot}(C_{p^n}, T) & \xrightarrow{\ \partial\ } & H_{\cdot}(C_{p^n}, T)
\end{array}$$

where $\mathrm{TR}^n(T;p)$ is the $C_{p^{n-1}}$-fixed point spectrum of T, and where the morphism φ is the map of $C_{p^{n-1}}$-fixed point spectra induced by the composition

$$T \xleftarrow{\ r\ } \rho^*(\tilde{E} \otimes T)^{C_p} \xrightarrow{\ f^*\ } \rho^*(\tilde{E} \otimes [E_+, T])^{C_p}$$

of an inverse of the morphism r and the morphism f^* in the original diagram. We remark that, recently, Nikolaus and Scholze have made the remarkable observation that it is only the composite ρ-equivariant morphism

$$T \xrightarrow{\ \gamma\ } \hat{H}^{\cdot}(C_p, T)$$

that is of importance and that the factorization of this map inherent in the original definition of a p-cyclotomic structure is immaterial. As demonstrated in [**32**], this approach makes it possible to develop the theory of cyclotomic spectra without the use of equivariant stable homotopy theory. In the following, we will therefore refer to φ as the cyclotomic structure morphism. We call the morphism

$$\mathrm{TR}^{n+1}(T;p) \xrightarrow{\ R\ } \mathrm{TR}^n(T;p)$$

in the diagram above the restriction morphism; and we call the morphism

$$\mathrm{TR}^{n+1}(T;p) \xrightarrow{F} \mathrm{TR}^n(T;p)$$

given by the canonical inclusion the Frobenius morphism. We write

$$\mathrm{TF}(T;p) = \operatorname*{holim}_{n,F} \mathrm{TR}^n(T;p)$$

for the homotopy limit with respect to the Frobenius morphisms. Using the p-cyclotomic structure on T, we now define the inverse Frobenius operator

$$\mathrm{TF}(T;p) \xrightarrow{\varphi^{-1}} \mathrm{TF}(T;p)$$

to be the composition

$$\operatorname{holim}_{n,F} \mathrm{TR}^n(T;p) \longrightarrow \operatorname{holim}_{n,F} \mathrm{TR}^{n+1}(T;p) \longrightarrow \operatorname{holim}_{n,F} \mathrm{TR}^n(T;p)$$

of the restriction along the successor functor and the morphism induced by the restriction morphisms. The cyclotomic structure morphism induces a morphism

$$\mathrm{TF}(T;p) = \operatorname{holim}_{n,F} \mathrm{TR}^n(T;p) \xrightarrow{\gamma} \operatorname{holim}_{n,F} \hat{H}^{\cdot}(C_{p^n}, T),$$

and Lemma 1.1 identifies the target with $\hat{H}^{\cdot}(\mathbb{T}, T)$, provided that T is p-complete and bounded below. The inverse Frobenius operator φ^{-1} does typically not extend to an operator defined on $\hat{H}^{\cdot}(\mathbb{T}, T)$, even at the level of homotopy groups.

3. Topological Hochschild homology

In this section, we recall the structure of topological Hochschild homology for schemes smooth over a perfect field of characteristic $p > 0$ from [**21**]. We also recall Bökstedt's periodicity theorem from [**8**].

Topological Hochschild homology is defined in the same generality as algebraic K-theory and assigns to an exact category with weak equivalences a cyclotomic spectrum. In particular, for a scheme X, applying this theory to the exact category of perfect complexes of quasi-coherent $\mathcal{O}_X$-modules with quasi-isomorphisms as weak equivalences, defines a cyclotomic spectrum $\mathrm{THH}(X)$. The derived tensor product of perfect complexes gives rise to a pairing on $\mathrm{THH}(X)$, which makes it a commutative cyclotomic ring spectrum.[3] Moreover, for every morphism of schemes $f\colon X \to Y$, left derived extension of scalars gives rise to a morphism of cyclotomic ring spectra $f^*\colon \mathrm{THH}(Y) \to \mathrm{THH}(X)$; and for every proper morphism of schemes $f\colon X \to Y$, right derived restriction of scalars gives rise to a morphism of cyclotomic $\mathrm{THH}(Y)$-module spectra $f_*\colon \mathrm{THH}(X) \to \mathrm{THH}(Y)$.

The homotopy groups $\mathrm{THH}_*(X)$ form an anticommutative graded ring and we define $\mathrm{THH}_*(\mathcal{O}_X)$ to be the sheaf of anticommutative graded rings on the small étale site of X associated to the presheaf that to an étale morphism $f\colon U \to X$ assigns $\mathrm{THH}_*(U)$. There is a natural ring isomorphism

$$\mathcal{O}_X \xrightarrow{\eta} \mathrm{THH}_0(\mathcal{O}_X)$$

[3] It follows from [**32**, Section IV.2] that $\mathrm{THH}(X)$ is an $\mathbb{E}_\infty$-algebra in cyclotomic spectra.

given by the inclusion of the zero-skeleton, and we will consider $\mathrm{THH}_*(\mathcal{O}_X)$ as a sheaf of anticommutative graded $\mathcal{O}_X$-algebras via this map. As such, it is quasi-coherent by [**17**, Proposition 3.2.1], and there is a multiplicative and conditionally convergent descent spectral sequence

$$E^2_{i,j} = H^{-i}(X, \mathrm{THH}_j(\mathcal{O}_X)) \Rightarrow \mathrm{THH}_{i+j}(X).$$

We consider this spectral sequence, which we call the Hodge spectral sequence for topological Hochschild homology, in detail in Section 6 below. Since $\mathrm{THH}_j(\mathcal{O}_X)$ is zero for $j < 0$, it shows, in particular, that $\mathrm{THH}_i(X)$ is zero for $i < -\dim(X)$.

More generally, we let $r = p^{n-1}$ be a prime power and let $C_r \subset \mathbb{T}$ be the subgroup of order r. The equivariant homotopy groups

$$\mathrm{TR}^n_*(X;p) = \pi_*(\mathrm{THH}(X)^{C_r})$$

again form an anticommutative graded ring and we define $\mathrm{TR}^n_*(\mathcal{O}_X;p)$ to be the sheaf of anticommutative graded rings on the small étale site of X associated to the presheaf that to an étale morphism $f\colon U \to X$ assigns $\mathrm{TR}^n_*(U;p)$. By [**26**, Theorem F], there is a canonical natural ring isomorphism

$$W_n(\mathcal{O}_X) \xrightarrow{\ \eta\ } \mathrm{TR}^n_0(\mathcal{O}_X;p),$$

which is compatible with the restriction, Frobenius, and Verschiebung operators, and we view $\mathrm{TR}^n_*(\mathcal{O}_X;p)$ as a sheaf of anticommutative graded $W_n(\mathcal{O}_X)$-algebras via this map. As such, it is quasi-coherent. Indeed, this follows from the proof of [**28**, Proposition 6.2.4] by using [**9**, Theorem B] and [**10**, Corollary 15.4], which were proved later.

The $W_n(\mathcal{O}_X)$-modules $\mathrm{TR}^n_j(\mathcal{O}_X;p)$ together with the cup product $\cup$, the restriction R, the Frobenius F, the Verschiebung V, and Connes' operator d constitute a p-typical Witt complex over $\mathcal{O}_X$; see [**25**, Definition 4.1]. Accordingly, this p-typical Witt complex receives a unique map of p-typical Witt complexes

$$W_n\Omega^j_X \xrightarrow{\ \eta\ } \mathrm{TR}^n_j(\mathcal{O}_X;p)$$

from the p-typical de Rham-Witt complex of X, which, by definition, is the initial p-typical Witt complex over $\mathcal{O}_X$. This map is an isomorphism for $j \leqslant 1$ by [**22**].

We next recall the following fundamental periodicity theorem. The basic case $n = 1$ is Bökstedt's theorem proved in his paper [**8**] which unfortunately remains unpublished. A detailed outline of Bökstedt's proof is given in [**26**, Section 5.2], and the proof of the general case $n \geqslant 1$ is given in [**26**, Theorem 5.5].

THEOREM 3.1. *Let k be a perfect field of characteristic $p > 0$ and let n be a positive integer. The canonical map of graded $W_n(k)$-algebras*

$$S_{W_n(k)}(\mathrm{TR}^n_2(k;p)) \longrightarrow \mathrm{TR}^n_*(k;p)$$

is an isomorphism and the $W_n(k)$-module $\mathrm{TR}^n_2(k;p)$ is free of rank 1.

In Section 4, we define a preferred $W_n(k)$-module generator $\alpha_n \in \mathrm{TR}^n_2(k;p)$, the divided Bott element. It satisfies $F(\alpha_n) = \alpha_{n-1}$ and $R(\alpha_n) = p \cdot \alpha_{n-1}$, showing that, in [**26**, Theorem 5.5], we can arrange for the units λ_n to be equal to 1.

We also recall from [**21**, Theorem B] the following result, which we will use in Section 6 below to construct the Hodge spectral sequence mentioned in the Introduction.

THEOREM 3.2. *Let k be a perfect field of characteristic $p > 0$, let n be a positive integer, and let $f\colon X \to \operatorname{Spec}(k)$ be a smooth morphism. In this situation, the canonical map of graded $W_n(\mathcal{O}_X)$-algebras*

$$W_n\Omega_X^* \otimes_{f^*W_n(k)} f^* \operatorname{TR}_*^n(k;p) \longrightarrow \operatorname{TR}_*^n(\mathcal{O}_X;p)$$

is an isomorphism.

We note that, by comparison, if k is a separably closed field of any characteristic and if $f\colon X \to \operatorname{Spec}(k)$ is a smooth morphism, then the canonical map

$$K_*^M(\mathcal{O}_X) \otimes_{f^*\mathbb{Z}/p^n\mathbb{Z}} f^*K_*(k,\mathbb{Z}/p^n\mathbb{Z}) \longrightarrow K_*(\mathcal{O}_X,\mathbb{Z}/p^n\mathbb{Z})$$

is an isomorphism by Voevodsky. If $\operatorname{char}(k) \neq p$, then $K_*(k,\mathbb{Z}/p^n\mathbb{Z})$ is a symmetric algebra generated by the Bott element $\beta_{\epsilon,n} \in K_2(k,\mathbb{Z}/p^n\mathbb{Z})$ by Suslin [**33**, **34**]; and if $\operatorname{char}(k) = p$, then $K_*(k,\mathbb{Z}/p^n\mathbb{Z}) = \mathbb{Z}/p^n\mathbb{Z}$ by Geisser-Levine [**18**].

4. The divided Bott element

Let k be a perfect field of characteristic $p > 0$. We proved in [**26**, Section 5] that, as an anticommutative graded ring $\operatorname{TP}_*(k) = S_{W(k)}\{v^{\pm 1}\}$ for some $v \in \operatorname{TP}_{-2}(k)$. In this section, we make a preferred choice of the generator $v \in \operatorname{TP}_{-2}(k)$ based on the results in [**24**]. We call this generator the inverse divided Bott element.

We first define the ring isomorphism

$$W(k) \xrightarrow{\ \tau\ } \operatorname{TP}_0(k)$$

by which we consider $\operatorname{TP}_*(k)$ a graded $W(k)$-algebra. Since k is an $\mathbb{F}_p$-algebra, the Witt vector Frobenius $F\colon W_n(k) \to W_{n-1}(k)$ is equal to the common composite of the maps in the following diagram with the horizontal maps induced by the Frobenius $\varphi\colon k \to k$,

$$\begin{array}{ccc} W_n(k) & \xrightarrow{W_n(\varphi)} & W_n(k) \\ \downarrow{\scriptstyle R} & & \downarrow{\scriptstyle R} \\ W_{n-1}(k) & \xrightarrow{W_{n-1}(\varphi)} & W_{n-1}(k). \end{array}$$

Therefore, we may define a ring homomorphism

$$\lim_{n,F} W_n(k) \xrightarrow{\ \varphi^\infty\ } \lim_{n,R} W_n(k) = W(k)$$

to be the map of limits induced by the map of limit systems that, at level n, is given by the ring homomorphism $W_n(\varphi^n)\colon W_n(k) \to W_n(k)$. The map φ^∞ is an isomorphism, because k is perfect. Moreover, the composition

$$W_n(k) \xrightarrow{\ \eta\ } \operatorname{TR}_0^n(k;p) \xrightarrow{\ \gamma\ } \hat{H}^0(C_{p^n},\operatorname{THH}(k))$$

of the canonical isomorphism from [**26**, Theorem F] and the isomorphism induced by the cyclotomic structure map defines a ring isomorphism

$$\lim_{n,F} W_n(k) \xrightarrow{\ \hat\eta\ } \lim_{n,F} \hat{H}^0(C_{p^n},\operatorname{THH}(k)),$$

where the structure maps in the right-hand limit system are the restriction maps in Tate cohomology. Finally, we have an isomorphism of graded rings

$$\operatorname{TP}_*(k) \xrightarrow{\ \rho\ } \lim_{n,F} \hat{H}^{-*}(C_{p^n},\operatorname{THH}(k)),$$

which is also given by restriction maps in Tate cohomology. We now define the ring isomorphism τ to be the composite $\rho^{-1} \circ \hat{\eta} \circ (\varphi^{\infty})^{-1}$. Extension of scalars along the unique ring homomorphism $f \colon \mathbb{F}_p \to k$ defines a map of graded $W(k)$-algebras

$$W(k) \otimes_{W(\mathbb{F}_p)} \mathrm{TP}_*(\mathbb{F}_p) \longrightarrow \mathrm{TP}_*(k)$$

and that this map is an isomorphism.

We proceed to define the generator $v \in \mathrm{TP}_{-2}(\mathbb{F}_p)$, and, to this end, we recall the main results of [**24**]. Let C be the perfectoid field of p-adic complex numbers. It is defined, we recall, to be the completion of an algebraic closure of the field $\mathbb{Q}_p$ of p-adic numbers and is both complete and algebraically closed. Its valuation ring $\mathcal{O}_C$ is a complete valuation ring with value group $\mathbb{Q}$, the residue field k of which is an algebraic closure of $\mathbb{F}_p$.

We first recall the results on the algebraic K-theory of $\mathcal{O}_C$ proved by Suslin in [**33, 34**], based on the rigidity theorems of Gabber [**16**] and Gillet-Thomason [**19**]. The group $K_j(\mathcal{O}_C)$ is p-divisible if j is positive, and it is uniquely p-divisible if j is both positive and even. Hence, the Bockstein homomorphism

$$K_2(\mathcal{O}_C, \mathbb{Z}_p) \xrightarrow{\ \partial\ } T_p(K_1(\mathcal{O}_C)) = \mathrm{Hom}_{\mathbb{Z}}(\mathbb{Q}_p/\mathbb{Z}_p, K_1(\mathcal{O}_C))$$

from the p-adic K-group in degree 2 to the p-primary Tate module of the K-group in degree 1 is an isomorphism. The resulting map of graded $K_0(\mathcal{O}_C, \mathbb{Z}_p)$-algebras

$$S_{K_0(\mathcal{O}_C, \mathbb{Z}_p)}(T_p(K_1(\mathcal{O}_C))) \longrightarrow K_*(\mathcal{O}_C, \mathbb{Z}_p),$$

moreover, is an isomorphism. Here the ring $K_0(\mathcal{O}_C, \mathbb{Z}_p)$ is canonically isomorphic to $\mathbb{Z}_p$ and the p-primary Tate module $\mathbb{Z}_p(1) = T_p(K_1(\mathcal{O}_C))$ is a free $\mathbb{Z}_p$-module of rank 1. An element $\epsilon \in \mathbb{Z}_p(1)$ determines and is determined by the sequence $(\epsilon^{(v)})_{v \geqslant 0}$ of compatible p-power roots of unity in C defined by $\epsilon^{(v)} = \epsilon(p^{-v} + \mathbb{Z}_p)$ and it is a generator if and only if the pth root of unity $\epsilon^{(1)}$ is primitive. In this situation, we follow Thomason [**36**] and define the Bott element associated with ϵ to be the unique generator

$$\beta_{\epsilon} \in K_2(\mathcal{O}_C, \mathbb{Z}_p)$$

such that $\partial(\beta_{\epsilon}) = \epsilon$.

We next recall the corresponding results for TF-theory from [**24**], which, as it turns out, are completely analogous. The group $\mathrm{TF}_j(\mathcal{O}_C; p)$ is p-divisible if j is positive, and it is uniquely p-divisible if j is both positive and even. Accordingly, the Bockstein homomorphism induces an isomorphism

$$\mathrm{TF}_2(\mathcal{O}_C; p, \mathbb{Z}_p) \xrightarrow{\ \partial\ } T_p(\mathrm{TF}_1(\mathcal{O}_C; p)) = \mathrm{Hom}_{\mathbb{Z}}(\mathbb{Q}_p/\mathbb{Z}_p, \mathrm{TF}_1(\mathcal{O}_C; p)),$$

and it is proved in op. cit. that the resulting map of graded $\mathrm{TF}_0(\mathcal{O}_C; p, \mathbb{Z}_p)$-algebras

$$S_{\mathrm{TF}_0(\mathcal{O}_C; p, \mathbb{Z}_p)}(T_p(\mathrm{TF}_1(\mathcal{O}_C; p))) \longrightarrow \mathrm{TF}_*(\mathcal{O}_C; p, \mathbb{Z}_p)$$

is an isomorphism. Moreover, the ring $\mathrm{TF}_0(\mathcal{O}_C; p, \mathbb{Z}_p)$ is canonically isomorphic to Fontaine's ring of p-adic periods

$$A_{\mathrm{inf}} = \lim_{n,F} W_n(\mathcal{O}_C),$$

which plays a prominent role in p-adic Hodge theory. The p-primary Tate module

$$A_{\mathrm{inf}}\{1\} = T_p(\mathrm{TF}_1(\mathcal{O}_C; p)) = \lim_{n,F} T_p(W_n\Omega^1_{\mathcal{O}_C})$$

is a free A_{inf}-module of rank 1. However, the image by the cyclotomic trace map

$$K_*(\mathcal{O}_C, \mathbb{Z}_p) \xrightarrow{\mathrm{tr}} \mathrm{TF}_*(\mathcal{O}_C; p, \mathbb{Z}_p)$$

of the Bott element β_ϵ is not a generator of this A_{inf}-module. Instead, we proved in [**24**, Theorem B] that the image of the Bott element, which, by abuse of notation, we also denote by β_ϵ, is divisible by the non-unit

$$\mu_\epsilon = [\varepsilon] - 1 = ([\epsilon^{(n)}]_n - 1)_{n \geqslant 1} \in \lim_{n,F} W_n(\mathcal{O}_C) = A_{\mathrm{inf}}$$

and that the element $\alpha_\epsilon = \mu_\epsilon^{-1} \cdot \beta_\epsilon$ is a generator of $\mathrm{TF}_2(\mathcal{O}_C; p, \mathbb{Z}_p)$.

DEFINITION 4.1. If ϵ is a generator of the free $\mathbb{Z}_p$-module $\mathbb{Z}_p(1)$, then the associated divided Bott element is the generator

$$\alpha_\epsilon = \mu_\epsilon^{-1} \cdot \beta_\epsilon$$

of the free A_{inf}-module $\mathrm{TF}_2(\mathcal{O}_C; p, \mathbb{Z}_p) = A_{\mathrm{inf}}\{1\}$.

By functoriality, the left action of the Galois group $G = \mathrm{Gal}(C/\mathbb{Q}_p)$ on $\mathcal{O}_C$ induces an action on the graded ring $\mathrm{TF}_*(\mathcal{O}_C; p, \mathbb{Z}_p)$, and the action on the Bott element is given by

$$\sigma \cdot \beta_\epsilon = \beta_{\sigma(\epsilon)} = \chi(\sigma) \cdot \beta_\epsilon,$$

where $\chi\colon G \to \mathrm{Aut}(\mu_{p^\infty}) = \mathbb{Z}_p^*$ is the cyclotomic character. Since A_{inf} is an integral domain, we therefore conclude that

$$\sigma \cdot \alpha_\epsilon = \alpha_{\sigma(\epsilon)} = \chi(\sigma) \cdot \sigma(\mu_\epsilon)^{-1} \cdot \mu_\epsilon \cdot \alpha_\epsilon.$$

Similarly, the inverse Frobenius operator induces a homorphism of graded rings

$$\mathrm{TF}_*(\mathcal{O}_C; p, \mathbb{Z}_p) \xrightarrow{\varphi^{-1}} \mathrm{TF}_*(\mathcal{O}_C; p, \mathbb{Z}_p).$$

In degree zero, it is given, up to canonical isomorphism, by the ring automorphism

$$A_{\mathrm{inf}} \xrightarrow{\varphi^{-1}} A_{\mathrm{inf}}$$

defined as the composition

$$\lim_{n,F} W_n(\mathcal{O}_C) \xrightarrow{\mathrm{res}_S} \lim_{n,F} W_{n+1}(\mathcal{O}_C) \xrightarrow{\lim_{n,F} R} \lim_{n,F} W_n(\mathcal{O}_C)$$

of the restriction along the successor functor and the map induced by the Witt vector restriction. The cyclotomic trace map takes values in the sub-graded ring of elements fixed by the inverse Frobenius operator. In particular,

$$\varphi^{-1}(\beta_\epsilon) = \beta_\epsilon,$$

and using again that A_{inf} is an integral domain, we therefore conclude that

$$\varphi^{-1}(\alpha_\epsilon) = \varphi^{-1}(\xi_\epsilon) \cdot \alpha_\epsilon,$$

where $\xi_\epsilon \in A_{\mathrm{inf}}$ is the element defined by

$$\xi_\epsilon = \varphi(\mu_\epsilon) \cdot \mu_\epsilon^{-1} = 1 + [\epsilon] + [\epsilon]^2 + \cdots + [\epsilon]^{p-1}.$$

We remark that this element is a generator of the kernel of the ring homomorphism

$$A_{\mathrm{inf}} \xrightarrow{\theta} \mathcal{O}_C,$$

which, in our formulation, is given by the canonical projection from the limit that defines the left-hand ring to the term $n = 1$.

We recall from [**27**, Addendum 5.4.4] and [**24**, Proposition 2.1.4] that the map of graded A_{inf}-algebras induced by the cyclotomic structure map

$$\text{TF}_*(\mathcal{O}_C; p, \mathbb{Z}_p) \xrightarrow{\gamma} \text{TP}_*(\mathcal{O}_C, \mathbb{Z}_p)$$

is an isomorphism in non-negative degrees. Hence, it becomes an isomorphism after inverting the divided Bott element α_ϵ. We now define

$$v_\epsilon \in \text{TP}_{-2}(\mathcal{O}_C, \mathbb{Z}_p)$$

to be the inverse of the image of the divided Bott element α_ϵ. It is represented in the Tate spectral sequence by the element $t \in E^2_{-2,0}$ defined in Section 1. Clearly, the Galois group G acts on the inverse divided Bott element by

$$\sigma \cdot v_\epsilon = v_{\sigma(\epsilon)} = \chi(\sigma)^{-1} \cdot \sigma(\mu_\epsilon) \cdot \mu_\epsilon^{-1} \cdot v_\epsilon.$$

Moreover, the inverse Frobenius operator φ^{-1} on $\text{TF}_*(\mathcal{O}_C; p, \mathbb{Z}_p)$ gives rise to a meromorphic Frobenius operator

$$\text{TP}_*(\mathcal{O}_C, \mathbb{Z}_p) \xrightarrow{\varphi} \text{TP}_*(\mathcal{O}_C, \mathbb{Z}_p),$$

which is defined and invertible away from the divisor "$\xi_\epsilon = 0$" and satisfies

$$\varphi(v_\epsilon) = \xi_\epsilon \cdot v_\epsilon.$$

This completes our recollection of the results from [**24**].

We consider the ring homomorphism

$$A_{\text{inf}} \xrightarrow{i} W(k)$$

defined to be the composition

$$\lim_{n,F} W_n(\mathcal{O}_C) \longrightarrow \lim_{n,F} W_n(k) \xrightarrow{\varphi^\infty} \lim_{n,R} W_n(k)$$

of the map induced by the canonical projection of $\mathcal{O}_C$ onto its residue field k and the isomorphism φ^∞ above. Extension of scalars along i induces an isomorphism

$$W(k) \otimes_{A_{\text{inf}}} \text{TP}_*(\mathcal{O}_C, \mathbb{Z}_p) \xrightarrow{i^*} \text{TP}_*(k, \mathbb{Z}_p),$$

the target of which is canonically isomorphic to $\text{TP}_*(k)$.

PROPOSITION 4.2. *The class $v = i^*(v_\epsilon) \in \text{TP}_{-2}(k)$ is independent of the choice of generator $\epsilon \in \mathbb{Z}_p(1)$ and descends to a generator $v \in \text{TP}_{-2}(\mathbb{F}_p)$. Moreover, the meromorphic Frobenius operator*

$$\text{TP}_*(k) \xrightarrow{\varphi} \text{TP}_*(k)$$

is defined and invertible away from the divisor "$p = 0$" and satisfies $\varphi(v) = p \cdot v$.

PROOF. If $\epsilon \in \mathbb{Z}_p(1)$ is a fixed generator, then a general generator is of the form $\sigma(\epsilon) \in \mathbb{Z}_p(1)$, for some $\sigma \in G$. So it suffices to show that $i \colon A_{\text{inf}} \to W(k)$ maps the unit $u_\epsilon = \chi(\sigma)^{-1} \cdot \sigma(\mu_\epsilon) \cdot \mu_\epsilon^{-1} \in A_{\text{inf}}$ to $1 \in W(k)$. We may assume that $\chi(\sigma)$ is an integer, in which case, the image $u_{\epsilon,n}$ of u_ϵ by the canonical projection from A_{inf} to $W_n(\mathcal{O}_C)$ is given by

$$\begin{aligned} u_{\epsilon,n} &= \chi(\sigma)^{-1} \cdot ([\sigma(\epsilon^{(n)})]_n - 1) \cdot ([\epsilon^{(n)}]_n - 1)^{-1} \\ &= \chi(\sigma)^{-1} \cdot ([\epsilon^{(n)}]_n^{\chi(\sigma)} - 1) \cdot ([\epsilon^{(n)}]_n - 1)^{-1} \\ &= \chi(\sigma)^{-1} \cdot \textstyle\sum_{0 \leqslant j < \chi(\sigma)} [\epsilon^{(n)}]_n^j. \end{aligned}$$

But the canonical projection from $\mathcal{O}_C$ to k takes $\epsilon^{(n)}$ to 1, and hence, the induced map from $W_n(\mathcal{O}_C)$ to $W_n(k)$ takes $u_{\epsilon,n}$ to 1, as desired. This shows that the class

$$v = i^*(v_\epsilon) \in \mathrm{TP}_{-2}(k)$$

is independent of ϵ as stated. It also shows that this class is Galois fixed, and hence, that it descends to a class $v \in \mathrm{TP}_{-2}(\mathbb{F}_p)$. Similarly, we have

$$i(\xi_\epsilon) = p,$$

which proves the remaining statements. □

REMARK 4.3. It follows from Proposition 4.2 that if we define $\alpha_n \in \mathrm{TR}_2^n(\mathbb{F}_p;p)$ to be the image of $v^{-1} \in \mathrm{TF}_2(\mathbb{F}_p;p)$, then $F(\alpha_n) = \alpha_{n-1}$ and $R(\alpha_n) = p \cdot \alpha_{n-1}$. This shows that, in [**26**, Proposition 5.4], we can arrange for the units λ_n to be equal to 1.

5. The conjugate spectral sequence

We recall from the paper [**30**] the higher Cartier isomorphism and the structure of the conjugate spectral sequence converging to the crystalline cohomology of a scheme smooth and proper over a perfect field of positive characteristic.

Let k be a perfect field of characteristic $p > 0$ and let $f\colon X \to \mathrm{Spec}(k)$ be a smooth morphism. The comparison theorem [**29**, Théorème II.1.4] gives a canonical isomorphism of $W_n(k)$-modules

$$H^i_{\mathrm{crys}}(X/W_n(k)) \longrightarrow H^i(X, W_n\Omega^{\cdot}_X)$$

from the crystalline cohomology of $X/W_n(k)$ defined in [**2**] to the hypercohomology of X with coefficients in the de Rham-Witt complex. The second spectral sequence of hypercohomology gives a spectral sequence of $W_n(k)$-modules

$$E_2^{i,j} = H^i(X, \mathcal{H}^j(W_n\Omega^{\cdot}_X)) \Rightarrow H^{i+j}_{\mathrm{crys}}(X/W_n(k)).$$

Being located in the first quadrant, it converges strongly. If $f\colon X \to \mathrm{Spec}(k)$ is also proper, and hence of finite relative dimension d, then the $W_n(k)$-modules $E_2^{i,j}$ are finitely generated by [**29**, Proposition II.2.1]. Therefore, taking limits with respect to the restriction maps, we obtain a spectral sequence of $W(k)$-modules

$$E_2^{i,j} = \lim_{n,\mathcal{H}^j(R)} H^i(X, \mathcal{H}^j(W_n\Omega^{\cdot}_X)) \Rightarrow H^{i+j}_{\mathrm{crys}}(X/W(k)).$$

Indeed, by finite generation, the limit systems satisfy the Mittag-Leffler condition. To rewrite the E_2-term we recall the higher Cartier isomorphism.

We have isomorphisms of sheaves of abelian groups

$$W_n\Omega^j_X \xleftarrow{R^n} W_{2n}\Omega^j_X/(V^nW_n\Omega^j_X + dV^nW_n\Omega^{j-1}_X) \xrightarrow{F^n} \mathcal{H}^j(W_n\Omega^{\cdot}_X).$$

Indeed, the left-hand map is an isomorphism for formal reasons, and the right-hand map is an isomorphism by [**30**, Proposition III.1.4]. The inverse higher Cartier operator is defined to be the composite isomorphism

$$C^{-n} = F^n \circ R^{-n}.$$

It follows easily from the definition that the diagram of sheaves of abelian groups

$$\begin{array}{ccc} W_n\Omega^j_X & \xrightarrow{C^{-n}} & \mathcal{H}^j(W_n\Omega^{\cdot}_X) \\ \downarrow F & & \downarrow \mathcal{H}^j(R) \\ W_{n-1}\Omega^j_X & \xrightarrow{C^{-(n-1)}} & \mathcal{H}^j(W_{n-1}\Omega^{\cdot}_X) \end{array}$$

commutes, and hence, we obtain an induced isomorphism of abelian groups

$$\lim_{n,F} H^{-i}(X, W_n\Omega^j_X) \longrightarrow \lim_{n,\mathcal{H}^j(R)} H^{-i}(X, \mathcal{H}^j(W_n\Omega^{\cdot}_X)).$$

In addition, this map is an isomorphism of $W(k)$-modules, provided that we consider its domain to be a $W(k)$-module via the inverse of the isomorphism

$$\lim_{n,F} W_n(k) \xrightarrow{\varphi^\infty} \lim_{n,R} W_n(k) = W(k)$$

defined in Section 4. We record these recollections from [**29**, **30**] in the following result.

THEOREM 5.1. *If k is a perfect field of characteristic $p > 0$ and $f\colon X \to \operatorname{Spec}(k)$ a smooth and proper morphism, then there is a spectral sequence of $W(k)$-modules*

$$E_2^{i,j} = \lim_{n,F} H^i(X, W_n\Omega^j_X) \Rightarrow H^{i+j}_{\text{crys}}(X/W(k))$$

which converges strongly.

We call the spectral sequence in Theorem 5.1 the *conjugate* spectral sequence. The analysis of its E_2-term in [**30**, Théorème III.2.2] gives the following result.

COROLLARY 5.2. *Let k be a finite field of order $q = p^r$, let W be its ring of p-typical Witt vectors, let $\iota\colon W \to \mathbb{C}$ be an embedding, and let $f\colon X \to \operatorname{Spec}(k)$ be a smooth and proper morphism. In this situation, there is a strongly convergent spectral sequence of degree-wise finite dimensional $\mathbb{C}$-vector spaces*

$$E_2^{i,j} = (\lim_{n,F} H^i(X, W_n\Omega^j_X)) \otimes_{W,\iota} \mathbb{C} \Rightarrow H^{i+j}_{\text{crys}}(X/W) \otimes_{W,\iota} \mathbb{C}.$$

PROOF. The spectral sequence in the statement is obtained from the conjugate spectral sequence by extension of scalars along the ring homomorphism $\iota\colon W \to \mathbb{C}$, which is flat. Finally, by [**30**, Théorème III.2.2], each of the $W(k)$-modules $E_2^{i,j}$ in the conjugate spectral sequence is finitely generated modulo torsion. □

REMARK 5.3. In the situation of Theorem 5.1, [**30**, Théorème III.2.2] also shows that the torsion sub-$W(k)$-module of $\lim_{n,F} H^i(X, W_n\Omega^j_X)$ is annihilated by p. However, this torsion sub-$W(k)$-module may well not be finitely generated,[4] as the example of a supersingular K3 surface shows; see [**30**, IV.2.15.6] and [**29**, II.7.2].

6. The Hodge spectral sequence

In this section, we construct the Hodge spectral sequence for periodic topological cyclic homology of schemes smooth and proper over a perfect field of characteristic $p > 0$.

In general, let X be a site, let F be a presheaf of spectra on X, and let

$$F(U) \xrightarrow{\eta_U} H^{\cdot}(U, F)$$

[4] The weight spectral sequence of [**3**] does not have this pathology.

be the canonical morphism to the associated cohomology presheaf. The latter may be defined by means of the Godement construction, if the site X has enough points, and by a fibrant replacement in an appropriate model structure, in general; see [**31**, **36**]. In either case, there is a hypercohomology spectral sequence

$$E^2_{i,j} = H^{-i}(U, \pi_j(F)^{\sim}) \Rightarrow H^{-i-j}(U, F)$$

from the cohomology of U with coefficients in the sheaves assoaciated with the presheaves of homotopy groups which converges conditionally to the homotopy groups of the global sections of the cohomology presheaf. If the morphism η_U is a weak equivalence for every object U of X, then we say that the presheaf F satisfies descent. In this case, the abutment of the hypercohomlogy spectral sequence is canonically identified with $\pi_{i+j}(F(U))$.

If X is a scheme, then by [**17**, Definition 3.2.3] or [**5**, Theorem 1.3] the presheaf of spectra on the small étale site of X that to an étale morphism $h\colon U \to X$ assigns $\mathrm{THH}(U)$ satisfies descent. Hence, in this situation the hypercohomology spectral sequence takes the form

$$E^2_{i,j} = H^{-i}(U, \mathrm{THH}_j(\mathcal{O}_X)) \Rightarrow \mathrm{THH}_{i+j}(U),$$

where, as in Section 3, $\mathrm{THH}_j(\mathcal{O}_X)$ is the sheaf of $\mathcal{O}_X$-modules on the small étale site of X associated with the presheaf that to an étale morphism $h\colon U \to X$ assigns $\mathrm{THH}_j(U)$. We recall from [**17**, Proposition 3.2.1] that this $\mathcal{O}_X$-module is quasi-coherent. In particular, in the case where U is affine, the spectral sequence collapses and the edge homomorphism

$$\mathrm{THH}_j(U) \longrightarrow H^0(U, \mathrm{THH}_j(\mathcal{O}_X))$$

is an isomorphism.

We next consider the presheaf of spectra on the small étale site of X that to an étale morphism $h\colon U \to X$ assigns the Tate spectrum

$$F(U) = \hat{H}^{\cdot}(C_{p^n}, \mathrm{THH}(U)),$$

and let $\hat{H}^{-j}(C_{p^n}, \mathrm{THH}(\mathcal{O}_X))$ be the sheaf of $W_n(\mathcal{O}_X)$-modules on the small étale site of X that to an étale morphism $h\colon U \to X$ assigns $\hat{H}^{-j}(C_{p^n}, \mathrm{THH}(U))$.

LEMMA 6.1. *Let k be a perfect field of characteristic $p > 0$. If $f\colon k \to A$ is a smooth morphism of relative dimension d, then the Tate spectral sequence*

$$E^2 = \hat{H}^{-i}(C_{p^n}, \mathrm{THH}_j(A)) \Rightarrow \hat{H}^{-i-j}(C_{p^n}, \mathrm{THH}(A))$$

converges strongly and $E^{\infty}_{i,j}$ vanishes for $j \geqslant d + 2n$. In particular, the induced filtration of the abutment is of finite length.

PROOF. The spectral sequence in question takes the form

$$E^2 = \Lambda\{u_n\} \otimes S\{t^{\pm 1}, \alpha\} \otimes \Omega^*_A \Rightarrow \hat{H}^{-*}(C_{p^n}, \mathrm{THH}(A)),$$

where the classes u_n, t, and α have bidegrees $(-1, 0)$, $(-2, 0)$, and $(0, 2)$, and where Ω^*_A is located in bidegrees $(0, j)$ with $0 \leqslant j \leqslant d$. By [**26**, Lemma 5.4], the classes t and α are infinite cycles, and if $A = k$, then

$$d^{2n+1}(u_n) = \lambda \cdot t^{n+1}\alpha^n$$

for some unit $\lambda \in \mathbb{F}_p$. By comparison, the $\mathbb{T}$-Tate spectral sequence takes the form

$$E^2 = S\{t^{\pm 1}, \alpha\} \otimes \Omega^*_A \Rightarrow \hat{H}^{-*}(\mathbb{T}, \mathrm{THH}(A)),$$

and the inclusion of C_{p^n} in $\mathbb{T}$ induces a map from the latter spectral sequence to the form, which, on E^2-terms, is induced by the unit map $\eta\colon k \to \Lambda\{u_n\}$. Now, for degree reasons, all nonzero differentials in the latter spectral sequence are even. Therefore, in the former spectral sequence, the d^{2n+1}-differential is given by

$$d^{2n+1}(u_n \otimes \omega) = \lambda \cdot t^{n+1}\alpha^n \omega.$$

This shows that $E^{2n+2}_{i,j}$ vanishes for $j \geqslant d + 2n$, and hence, so does $E^{\infty}_{i,j}$. Finally, since the Tate spectral sequence always converges conditionally, we conclude that, in the situation at hand, it converges strongly; see [**6**, Theorem 7.1]. □

PROPOSITION 6.2. *Let k be a perfect field of characteristic $p > 0$, let n be a positive integer, and let $f\colon k \to A$ be a smooth morphism. If $h\colon A \to B$ is an étale morphism, then the induced map of graded $W_n(B)$-algebras*

$$W_n(B) \otimes_{W_n(A)} \hat{H}^{-*}(C_{p^n}, \mathrm{THH}(A)) \longrightarrow \hat{H}^{-*}(C_{p^n}, \mathrm{THH}(B))$$

is an isomorphism.

PROOF. The Tate spectral sequence is a spectral sequence of $W_{n+1}(A)$-modules

$$E^2_{i,j} = \hat{H}^{-i}(C_{p^n}, F^n_* \,\mathrm{THH}_j(A)) \Rightarrow R_*\hat{H}^{-i-j}(C_{p^n}, \mathrm{THH}(A)),$$

where $F^n_*(-)$ and $R_*(-)$ indicates the restriction of scalars along the Frobenius $F^n\colon W_{n+1}(A) \to A$ and restriction $R\colon W_{n+1}(A) \to W_n(A)$, respectively. We recall from [**9**, Theorem B] that the morphism $W_{n+1}(h)\colon W_{n+1}(A) \to W_{n+1}(B)$ again is étale and hence flat. Therefore, by extension of scalars along this morphism, we obtain a spectral sequence of $W_{n+1}(B)$-modules

$$\begin{aligned} E^2_{i,j} &= \hat{H}^{-i}(C_{p^n}, W_{n+1}(B) \otimes_{W_{n+1}(A)} F^n_* \,\mathrm{THH}_j(A)) \\ &\Rightarrow W_{n+1}(B) \otimes_{W_{n+1}(A)} R_*\hat{H}^{-i-j}(C_{p^n}, \mathrm{THH}(A)), \end{aligned}$$

and Lemma 6.1 shows that both this spectral sequence and the spectral sequence

$$E^2_{i,j} = \hat{H}^{-i}(C_{p^n}, F^n_* \,\mathrm{THH}_j(B)) \Rightarrow R_*\hat{H}^{-i-j}(C_{p^n}, \mathrm{THH}(B)),$$

converge strongly. The morphism $h\colon A \to B$ induces a map of spectral sequences from the former to the latter, and we claim that this map is an isomorphism. To prove this, it suffices to show that the induced map of $W_{n+1}(B)$-modules

$$W_{n+1}(B) \otimes_{W_{n+1}(A)} F^n_* \,\mathrm{THH}_j(A) \longrightarrow F^n_* \,\mathrm{THH}_j(B)$$

is an isomorphism. If $n = 0$, then this is proved in [**17**, Proposition 3.2.1], and the general case follows from this case and from the fact proved in [**10**, Corollary 15.4] that the left-hand diagram below is a cocartesian square of commutative rings.

$$\begin{array}{ccc} W_{n+1}(A) & \xrightarrow{W_{n+1}(h)} & W_{n+1}(B) \\ \downarrow{\scriptstyle F} & & \downarrow{\scriptstyle F} \\ W_n(A) & \xrightarrow{W_n(h)} & W_n(B) \end{array} \qquad \begin{array}{ccc} W_{n+1}(A) & \xrightarrow{W_{n+1}(h)} & W_{n+1}(B) \\ \downarrow{\scriptstyle R} & & \downarrow{\scriptstyle R} \\ W_n(A) & \xrightarrow{W_n(h)} & W_n(B) \end{array}$$

We therefore conclude that the map of graded $W_{n+1}(B)$-modules

$$W_{n+1}(B) \otimes_{W_{n+1}(A)} R_*\hat{H}^{-*}(C_{p^n}, \mathrm{THH}(A)) \longrightarrow R_*\hat{H}^{-*}(C_{p^n}, \mathrm{THH}(B))$$

induced by $h\colon A \to B$ is an isomorphism. Finally, by loc. cit., the right-hand diagram above also is a cocartesian square of commutative rings. This shows that the map in the statement is an isomorphism as desired. □

COROLLARY 6.3. *Let k be a perfect field of characteristic $p > 0$, let n be a positive integer, and let $f\colon X \to \operatorname{Spec}(k)$ be a smooth morphism of finite relative dimension. In this situation, the graded $W_n(\mathcal{O}_X)$-algebra $\hat{H}^{-*}(C_{p^n}, \mathrm{THH}(\mathcal{O}_X))$ is quasi-coherent.*

PROOF. As proved in [**1**, Exposé VII, Proposition 3.1], the étale morphisms $h\colon A \to B$ with $\operatorname{Spec}(A)$ open in X generate the small étale topos of X. Therefore, the statement follows from Proposition 6.2. □

PROPOSITION 6.4. *Let k be a perfect field of characteristic $p > 0$, let n be a positive integer, and let $f\colon X \to \operatorname{Spec}(k)$ is a smooth morphism of finite relative dimension. In this situation, the presheaf of spectra on the small étale site of X that to an étale morphism $h\colon U \to X$ assigns $\hat{H}^{\cdot}(C_{p^n}, \mathrm{THH}(U))$ satisfies descent.*

PROOF. Let again F denote the presheaf in question. We wish to prove that for every étale morphism $h\colon U \to X$, the canonical morphism

$$F(U) \xrightarrow{\eta_U} H^{\cdot}(U, F)$$

is a weak equivalence. Corollary 6.3 shows that, for U affine, the sheafification map

$$\pi_j(F(U)) \longrightarrow H^0(U, \pi_j(F)^{\tilde{}})$$

is an isomorphism and that the descent spectral sequence

$$E^2_{i,j} = H^{-i}(X, \pi_j(F)^{\tilde{}}) \Rightarrow H^{-i-j}(U, F)$$

collapses. But the sheafification map is equal to the composition of the map of homotopy groups induced by η_U and the edge homomorphism of the descent spectral sequence, so we conclude that η_U is a weak equivalence in this case. If $h\colon U \to X$ is a general étale morphism, then we choose a hyper-covering $\mathcal{U}$ of U by affine schemes étale over U and consider the diagram in which the two vertical morphisms are the canonical morphisms to the respective Čech cohomology spectra,

$$\begin{array}{ccc} F(U) & \xrightarrow{\quad \eta_U \quad} & H^{\cdot}(U, F) \\ \Big\downarrow{\scriptstyle \check{\eta}_{\mathcal{U}}} & & \Big\downarrow{\scriptstyle \check{\eta}_{\mathcal{U}}} \\ \check{H}^{\cdot}(\mathcal{U}, F) & \xrightarrow{\check{H}^{\cdot}(\mathcal{U}, \eta)} & \check{H}^{\cdot}(\mathcal{U}, H^{\cdot}(-, F)). \end{array}$$

The lower horizontal morphism is a weak equivalence by the case already considered, and the right-hand vertical morphism is a weak equivalence, since sheaf cohomology satisfies descent. We claim that also the left-hand vertical morphism is a weak equivalence. We write this morphism as the composition

$$\hat{H}^{\cdot}(C_{p^n}, \mathrm{THH}(U)) \longrightarrow \hat{H}^{\cdot}(C_{p^n}, \check{H}^{\cdot}(\mathcal{U}, \mathrm{THH}(-))) \longrightarrow \check{H}^{\cdot}(\mathcal{U}, \hat{H}^{\cdot}(C_{p^n}, \mathrm{THH}(-)))$$

of the morphism in Tate cohomlogy induced by the corresponding morphism in topological Hochschild homology and the canonical map from Tate cohomology of a homotopy limit to the homotopy limit of the Tate cohomology of the individual terms in the limit system. The former map is a weak equivalence, since topological Hochschild homology satisfies descent, and the latter map is a weak equivalence, since the spectra in the limit system are universally bounded below by the relative dimension of $f\colon X \to \operatorname{Spec}(k)$. □

REMARK 6.5. We do not know whether or not Corollary 6.3 and Proposition 6.4 hold for all schemes X.

We view the next result as an analogue of analytic continuation.

PROPOSITION 6.6. *Let k be a perfect field of characteristic $p > 0$, let $n \geqslant 1$ be an integer, and let $f\colon X \to \operatorname{Spec}(k)$ is a smooth morphism of relative dimension d. In this situation, the map of graded $W_n(\mathcal{O}_X)$-algebras*

$$\mathrm{TR}_i^n(\mathcal{O}_X;p) \xrightarrow{\gamma} \hat{H}^{-i}(C_{p^n}, \mathrm{THH}(\mathcal{O}_X))$$

induced by the cyclotomic structure morphism is an isomorphism for $i \geqslant d$.

PROOF. The domain and target $W_n(\mathcal{O}_X)$-modules of the map in question both are quasi-coherent. Therefore, it suffices to take $X = \mathbb{A}_k^d$ and show that the induced map of global sections is an isomorphism for $i \geqslant d$. Moreover, by [**37**, Theorem 2.4], it suffices to consider the case $n = 1$. Now, in the case $d = 0$, the result was proved in [**26**, Proposition 5.3]; see also loc. cit., Remark 5.5. In general, we have an equivalence of C_p-spectra

$$\mathrm{THH}(k) \otimes N^{\mathrm{cy}}(\langle x_1, \dots, x_d\rangle) \xrightarrow{\alpha} \mathrm{THH}(k[x_1, \dots, x_d])$$

where the second factor on the left-hand side is the cyclic bar-construction of the free commutative monoid on the indicated generators. We refer to [**23**, Proposition 3] for the definition of the map α and for the proof that it is a weak equivalence. Moreover, there is a commutative diagram of spectra

$$\begin{array}{ccc} \mathrm{THH}(k) \otimes N^{\mathrm{cy}}(\langle x_1, \dots, x_d\rangle) & \xrightarrow{\alpha} & \mathrm{THH}(k[x_1, \dots, x_d]) \\ \downarrow{\scriptstyle \gamma\otimes \mathrm{id}} & & \downarrow{\scriptstyle \gamma} \\ \hat{H}^{\cdot}(C_p, \mathrm{THH}(k)) \otimes N^{\mathrm{cy}}(\langle x_1, \dots, x_d\rangle) & \xrightarrow{\hat{\alpha}} & \hat{H}^{\cdot}(C_p, \mathrm{THH}(k[x_1, \dots, x_d])) \end{array}$$

and the lower horizontal map also is a weak equivalence; see [**26**, Corollary 9.1]. We claim that the reduced singular homology of $N^{\mathrm{cy}}(\langle x_1, \dots, x_d\rangle)$ is concentrated in the degrees $0 \leqslant i \leqslant d$. Granting this, we conclude from the case $d = 0$ that the left-hand vertical map in the diagram above induces an isomorphism of homotopy groups in degrees greater than or equal to d. But then the same holds for the right-hand vertical map, which proves the proposition. To prove the claim we use that there are canonical isomorphisms

$$\Omega^i_{\mathbb{Z}[x_1,\dots,x_d]/\mathbb{Z}} \longrightarrow \mathrm{HH}_i(\mathbb{Z}[x_1, \dots, x_d]/\mathbb{Z}) \longrightarrow \tilde{H}_i(N^{\mathrm{cy}}(\langle x_1, \dots, x_d\rangle), \mathbb{Z})$$

and the easy calculation of the left-hand side. □

THEOREM 6.7. *Let k be a perfect field of characteristic $p > 0$, let n be a positive integer, and let $f\colon X \to \operatorname{Spec}(k)$ be a smooth morphism of finite relative dimension. In this situation, the canonical map of graded $W_n(\mathcal{O}_X)$-algebras*

$$W_n\Omega_X^* \otimes_{f^*W_n(k)} f^*\hat{H}^{-*}(C_{p^n}, \mathrm{THH}(k)) \xrightarrow{\hat{\eta}} \hat{H}^{-*}(C_{p^n}, \mathrm{THH}(\mathcal{O}_X))$$

is an isomorphism.

PROOF. We consider the commutative diagram of graded $W_n(\mathcal{O}_X)$-algebras in which the horizontal maps are the canonical maps and the vertical maps are induced by the cyclotomic structure maps,

$$\begin{array}{ccc} W_n\Omega_X^* \otimes_{f^*W_n(k)} f^*\operatorname{TR}_*^n(k;p) & \xrightarrow{\quad\eta\quad} & \operatorname{TR}_*^n(\mathcal{O}_X;p) \\ \big\downarrow{\scriptstyle \operatorname{id}\otimes\gamma} & & \big\downarrow{\scriptstyle \gamma} \\ W_n\Omega_X^* \otimes_{f^*W_n(k)} f^*\hat{H}^{-*}(C_{p^n},\operatorname{THH}(k)) & \xrightarrow{\hat{\eta}} & \hat{H}^{-*}(C_{p^n},\operatorname{THH}(\mathcal{O}_X)). \end{array}$$

We recall from [**21**, Theorem B] that the top horizontal map is an isomorphism and from Proposition 6.6 that the right-hand vertical map is an isomorphism in degrees greater than or equal to the relative dimension of $f\colon X \to \operatorname{Spec}(k)$. Since the graded $W_n(\mathcal{O}_X)$-algebras in the bottom row both are 2-periodic, the theorem follows. □

THEOREM 6.8. *If X is a scheme smooth and proper morphism over a perfect field k of characteristic $p > 0$, then there is a spectral sequence of $W(k)$-modules*

$$E_{i,j}^2 = \bigoplus\nolimits_{m\in\mathbb{Z}} \lim\nolimits_{n,F} H^{-i}(X, W_n\Omega_X^{j+2m}) \Rightarrow \operatorname{TP}_{i+j}(X)$$

which converges strongly.

PROOF. We first argue that the canonical morphism

$$\operatorname{TP}(X) = \hat{H}^{\cdot}(\mathbb{T}, \operatorname{THH}(X)) \longrightarrow \operatorname{holim}_{n,F} \hat{H}^{\cdot}(C_{p^n}, \operatorname{THH}(X))$$

is a weak equivalence. It follows from [**24**, Lemma 2.1.1] that this morphism becomes a weak equivalence after p-completion, and therefore, it suffices to show that the domain and target are p-complete. This, in turn, follows from the definition of the Tate spectrum, once we shows that $\operatorname{THH}(X)$ is p-complete and bounded below. That the spectrum $\operatorname{THH}(X)$ is p-complete is an immediate consequence of its homotopy groups being k-vector spaces, and hence, annihilated by p. To see that it is bounded below, we consider the conditionally convergent descent spectral sequence of k-vector spaces

$$E_{i,j}^2 = H^{-i}(X, \operatorname{THH}_j(\mathcal{O}_X)) \Rightarrow \operatorname{THH}_{i+j}(X).$$

Since the structure morphism $f\colon X \to \operatorname{Spec}(k)$ is smooth and proper, its relative dimension d is finite. Therefore, the k-vector space $E_{i,j}^2$ is zero, unless $-d \leqslant i \leqslant 0$ and $j \geqslant 0$, which shows that the spectral sequence converges strongly and that $\operatorname{THH}_j(X)$ is zero for $j < -d$. So the canonical morphism at the beginning of the proof is a weak equivalence as desired.

Next, by Proposition 6.4, we have descent spectral sequences

$$E_{i,j}^2 = H^{-i}(X, \hat{H}^{-j}(C_{p^n}, \operatorname{THH}(\mathcal{O}_X))) \Rightarrow \hat{H}^{-i-j}(C_{p^n}, \operatorname{THH}(X)),$$

which converge strongly. Indeed, the spectral sequences converge conditionally, and $E_{i,j}^2$ is zero, unless $-d \leqslant i \leqslant 0$. Moreover, identifying

$$\hat{H}^{-*}(C_{p^n}, \operatorname{THH}(k)) = S_{W_n(k)}\{v^{\pm 1}\}$$

with $v \in \operatorname{TP}_{-2}(k)$, Theorem 6.7 gives canonical isomorphisms of $W_n(\mathcal{O}_X)$-modules

$$\bigoplus\nolimits_{m\in\mathbb{Z}} W_n\Omega_X^{j+2m} \longrightarrow \hat{H}^{-j}(C_{p^n}, \operatorname{THH}(\mathcal{O}_X)),$$

and hence, we may rewrite the descent spectral sequences above in the form

$$E^2_{i,j} = \bigoplus_{m\in\mathbb{Z}} H^{-i}(X, W_n\Omega^{j+2m}_X) \Rightarrow \hat{H}^{-i-j}(C_{p^n}, \mathrm{THH}(X)).$$

We claim that the $W_n(k)$-module $E^2_{i,j}$ is finitely generated for all values of (i,j). Indeed, since $f\colon X \to \operatorname{Spec}(k)$ is smooth and proper of relative dimension d, it follows from [**29**, Proposition II.2.1] that the $W_n(k)$-module $H^{-i}(X, W_n\Omega^{j+2m}_X)$ is finitely generated, for all integers m. But unless $0 \leqslant j+2m \leqslant d$, said $W_n(k)$-module is zero, so the claim follows. Moreover, for cohomological dimension reasons, $E^2_{i,j}$ is zero, unless $-d \leqslant i \leqslant 0$, so we conclude that also $\hat{H}^{-i}(C_{p^n}, \mathrm{THH}(X))$ is a finitely generated $W_n(k)$-module, for all integers i. Hence, taking limits with respect to the Frobenius maps, we obtain a spectral sequence of $W(k)$-modules

$$E^2_{i,j} = \bigoplus_{m\in\mathbb{Z}} \lim_{n,F} H^{-i}(X, W_n\Omega^{j+2m}_X) \Rightarrow \lim_{n,F} \hat{H}^{-i-j}(C_{p^n}, \mathrm{THH}(X)),$$

and the corresponding higher derived limits vanish. Indeed, by finite generation, the limit systems satisfy the Mittag-Leffler condition. This also implies that the canonical map

$$\mathrm{TP}_i(X) \longrightarrow \lim_{n,F} \hat{H}^{-i}(C_{p^n}, \mathrm{THH}(X))$$

is an isomorphism, so the theorem follows. □

COROLLARY 6.9. *Let k be a finite field of order $q = p^r$, let W be its ring of p-typical Witt vectors, and let $\iota\colon W \to \mathbb{C}$ be a choice of embedding. If X is a scheme smooth and proper over k, then there is a strongly convergent spectral sequence of degree-wise finite dimensional $\mathbb{C}$-vector spaces*

$$E^2_{i,j} = \bigoplus_{m\in\mathbb{Z}} (\lim_{n,F} H^{-i}(X, W_n\Omega^{j+2m}_X)) \otimes_{W,\iota} \mathbb{C} \Rightarrow \mathrm{TP}_{i+j}(X) \otimes_{W,\iota} \mathbb{C}.$$

PROOF. We have the strongly convergent spectral sequence of W-modules

$$E^2_{i,j} = \bigoplus_{m\in\mathbb{Z}} \lim_{n,F} H^{-i}(X, W_n\Omega^{j+2m}_X) \Rightarrow \mathrm{TP}_{i+j}(X)$$

from Theorem 6.8. For cohomological dimension reasons, it is concentrated in the strip $-d \leqslant i \leqslant 0$, where d is the relative dimension of X over k. Therefore, we obtain the desired spectral sequence by extension of scalars along the flat ring homomorphism $\iota\colon W \to \mathbb{C}$. Finally, by [**30**, Théorème III.2.2], we see that the $\mathbb{C}$-vector spaces $E^2_{i,j}$ are finite dimensional as in the proof of Corollary 5.2. □

7. Regularized determinants

In this section, we recall the definitions of regularized determinant and anomalous dimension following Deninger [**15**] and complete the proof of Theorem A.

The definition requires a choice of principal branch $\operatorname{Arg}(\lambda)$ of the argument of $\lambda \in \mathbb{C}^*$. We follow [**15**, Section 1] and choose $-\pi < \operatorname{Arg}(\lambda) \leqslant \pi$ and define

$$\lambda^{-s} = |\lambda|^{-s} e^{-is\operatorname{Arg}(\lambda)}$$

for $\lambda \in \mathbb{C}^*$ and $s \in \mathbb{C}$. Let V be a $\mathbb{C}$-vector space[5] of at most countably infinite dimension and let $\Theta\colon V \to V$ be a $\mathbb{C}$-linear map such that:

(1) For every $\lambda \in \mathbb{C}$, the generalized eigenspace V_λ of $\Theta\colon V \to V$ associated to the eigenvalue λ is finite dimensional.
(2) The Dirichlet series $\sum_{\lambda\in\mathbb{C}^*} \dim_{\mathbb{C}}(V_\lambda)\lambda^{-s}$ converges absolutely for $\operatorname{Re}(s) \gg 0$ and admits a meromorphic continuation $\zeta_\Theta(s)$ to the halfplane $\operatorname{Re}(s) > -\epsilon$ for some $\epsilon > 0$ which is holomorphic at $s = 0$.

[5] There is no topology on V; in particular, it is not a Hilbert space.

The anomalous dimension and regularized determinant of $\Theta\colon V \to V$ are now defined to be the complex numbers

$$\dim_\infty(\Theta \mid V) = \dim_{\mathbb{C}}(V_0) + \zeta_{\bar\Theta}(0),$$

$$\det_\infty(\Theta \mid V) = \begin{cases} e^{-\zeta'_{\bar\Theta}(0)} & \text{if } \dim_{\mathbb{C}}(V_0) = 0, \\ 0 & \text{if } \dim_{\mathbb{C}}(V_0) > 0, \end{cases}$$

where, in the top line, $\bar\Theta\colon V/V_0 \to V/V_0$ is the map induced by $\Theta\colon V \to V$. If V is finite dimensional, then $\dim_\infty(\Theta \mid V) = \dim_{\mathbb{C}}(V)$ and $\det_\infty(\Theta \mid V) = \det(\Theta \mid V)$. Moreover, if δ is a positive real number, then one has

$$\det_\infty(\delta \cdot \Theta \mid V) = \delta^{\dim_\infty(\Theta \mid V)} \det_\infty(\Theta \mid V),$$

as expected. More importantly, by [**15**, Lemma 1.2], given a commutative diagram

$$\begin{array}{ccccccccc} 0 & \longrightarrow & V' & \longrightarrow & V & \longrightarrow & V'' & \longrightarrow & 0 \\ & & \downarrow{\scriptstyle \Theta'} & & \downarrow{\scriptstyle \Theta} & & \downarrow{\scriptstyle \Theta''} & & \\ 0 & \longrightarrow & V' & \longrightarrow & V & \longrightarrow & V'' & \longrightarrow & 0 \end{array}$$

of $\mathbb{C}$-vector spaces and $\mathbb{C}$-linear maps with exact rows such that the right- and left-hand vertical maps satisfy (1)–(2), then so does the middle vertical map and

$$\dim_\infty(\Theta \mid V) = \dim_\infty(\Theta' \mid V') + \dim_\infty(\Theta'' \mid V'')$$
$$\det_\infty(\Theta \mid V) = \det_\infty(\Theta' \mid V) \cdot \det_\infty(\Theta'' \mid V'').$$

The following result is a special case of [**15**, Corollary 2.8].

PROPOSITION 7.1. *Let T be an anticommutative graded $\mathbb{C}$-algebra such that the sub-$\mathbb{C}$-vector space $T_j \subset T$ of homogeneous elements of degree j is finite dimensional for all integers j. Let $\Theta\colon T \to T$ be a graded $\mathbb{C}$-linear derivation and suppose that there exists a unit $v \in T_{-2}$ such that $\Theta(v) = \frac{2\pi i}{\log q} \cdot v$. In this situation,*

$$\det_\infty(s \cdot \mathrm{id} - \Theta \mid T_{2*+j}) = \det(\mathrm{id} - q^{-s\cdot\Theta} \mid T_j),$$
$$\dim_\infty(s \cdot \mathrm{id} - \Theta \mid T_{2*+j}) = 0.$$

for all $s \in \mathbb{C}$ and for all integers j.

REMARK 7.2. The two side of the equality in the statement of Proposition 7.1 only depend on the parity of the integer j. Indeed, we have the exact sequence

$$0 \longrightarrow \tfrac{2\pi i}{\log q}\mathbb{Z} \longrightarrow \mathbb{C} \longrightarrow \mathbb{C}^* \longrightarrow 0,$$

where the right-hand map takes $\alpha \in \mathbb{C}$ to $q^\alpha \in \mathbb{C}^*$. In its most basic form, Deninger's formula expresses the inverse non-archimedean Euler factor $1 - \lambda q^{-s}$ with $\lambda \in \mathbb{C}^*$ as the regularized product of the $s - \alpha$ with $\alpha \in \mathbb{C}$ ranging over all solutions to $q^\alpha = \lambda$.

PROOF OF THEOREM A. By applying Proposition 7.1 to the anticommutative graded $\mathbb{C}$-algebra $\mathrm{TP}_*(X) \otimes_{W,\iota} \mathbb{C}$ and a graded $\mathbb{C}$-linear derivation

$$\mathrm{TP}_*(X) \otimes_{W,\iota} \mathbb{C} \xrightarrow{\Theta} \mathrm{TP}_*(X) \otimes_{W,\iota} \mathbb{C}$$

as in the statement of the theorem, we find that for all $s \in \mathbb{C}$ and $j \in \mathbb{Z}$,

$$\det_\infty(s \cdot \mathrm{id} - \Theta \mid \mathrm{TP}_{2*+j}(X) \otimes_{W,\iota} \mathbb{C}) = \det(\mathrm{id} - q^{-s}\,\mathrm{Fr}_q^* \mid \mathrm{TP}_j(X) \otimes_{W,\iota} \mathbb{C}).$$

Moreover, by comparing the Hodge spectral sequence in Corollary 6.9 and the conjugate spectral sequence in Corollary 5.2 and using that the determinant is multiplicative on exact sequence, we furthermore conclude that

$$\det(\mathrm{id} - q^{-s}\operatorname{Fr}_q^* \mid \mathrm{TP}_j(X) \otimes_{W,\iota} \mathbb{C}) = \begin{cases} \det(\mathrm{id} - q^{-s}\operatorname{Fr}_q^* \mid H^{\mathrm{od}}_{\mathrm{crys}}(X/W) \otimes_{W,\iota} \mathbb{C}) \\ \det(\mathrm{id} - q^{-s}\operatorname{Fr}_q^* \mid H^{\mathrm{ev}}_{\mathrm{crys}}(X/W) \otimes_{W,\iota} \mathbb{C}) \end{cases}$$

respectively, as j is odd or even. Putting the two equalities together, we find that

$$\frac{\det_\infty(s \cdot \mathrm{id} - \Theta \mid \mathrm{TP}_{\mathrm{od}}(X) \otimes_{W,\iota} \mathbb{C})}{\det_\infty(s \cdot \mathrm{id} - \Theta \mid \mathrm{TP}_{\mathrm{ev}}(X) \otimes_{W,\iota} \mathbb{C})} = \frac{\det(\mathrm{id} - q^{-s}\operatorname{Fr}_q^* \mid H^{\mathrm{od}}_{\mathrm{crys}}(X/W) \otimes_{W,\iota} \mathbb{C})}{\det(\mathrm{id} - q^{-s}\operatorname{Fr}_q^* \mid H^{\mathrm{ev}}_{\mathrm{crys}}(X/W) \otimes_{W,\iota} \mathbb{C})},$$

and finally, it is proved in [**2**, Théorème VII.3.2.3] that the right-hand side is equal to the Hasse-Weil zeta function $\zeta(X, s)$. □

By the analogue of analytic continuation, Proposition 6.6, the inverse Frobenius operator φ^{-1} on $\mathrm{TF}_*(X;p)$ gives rise to a meromorphic Frobenius operator

$$\mathrm{TP}_*(X) \xrightarrow{\varphi} \mathrm{TP}_*(X),$$

which is defined and invertible after inverting p. We relate this operator to the geometric Frobenius Fr_q^* as follows. By functoriality, the Hodge spectral sequence

$$E^2_{i,j} = \textstyle\bigoplus_{m \in \mathbb{Z}} \lim_{n,F} H^{-i}(X, W_n\Omega_X^{j+2m}) \Rightarrow \mathrm{TP}_{i+j}(X)$$

is a spectral sequence of graded $\mathrm{TP}_*(k)$-modules. Moreover, on $E^2_{i,j}$, and hence, on $E^\infty_{i,j}$, the geometric Frobenius acts as

$$\operatorname{Fr}_q^* = q^w \varphi^r,$$

where $w = j + m$ is the weight and $q = p^r$.

Acknowledgments

It is a great pleasure to thank the Hausdorff Research Institute for Mathematics for its hospitality and support during the Trimester Program "Homotopy theory, manifolds, and field theories," where part of the work reported here was conducted. I am also very grateful to Alain Connes for helpful conversations and for announcing Theorem A in his paper [**13**] and to Peter Scholze for explaining the conjugate spectral sequence to me and for making me fully appreciate the importance of Bökstedt's periodicity theorem.

References

[1] *Théorie des topos et cohomologie étale des schémas. Tome 1: Théorie des topos* (French), Lecture Notes in Mathematics, Vol. 269, Springer-Verlag, Berlin-New York, 1972. Séminaire de Géométrie Algébrique du Bois-Marie 1963–1964 (SGA 4); Dirigé par M. Artin, A. Grothendieck, et J. L. Verdier. Avec la collaboration de N. Bourbaki, P. Deligne et B. Saint-Donat. MR0354652

[2] P. Berthelot, *Cohomologie cristalline des schémas de caractéristique $p > 0$* (French), Lecture Notes in Mathematics, Vol. 407, Springer-Verlag, Berlin-New York, 1974. MR0384804

[3] B. Bhatt, M. Morrow, and P. Scholze, *Topological Hochschild homology and integral p-adic Hodge theory*, arXiv:1802.03261.

[4] A. J. Blumberg and M. A. Mandell, *Algebraic K-theory and abstract homotopy theory*, Adv. Math. **226** (2011), no. 4, 3760–3812. MR2764905

[5] A. J. Blumberg and M. A. Mandell, *Localization theorems in topological Hochschild homology and topological cyclic homology*, Geom. Topol. **16** (2012), no. 2, 1053–1120. MR2928988

[6] J. M. Boardman, *Conditionally convergent spectral sequences*, Homotopy invariant algebraic structures (Baltimore, MD, 1998), Contemp. Math., vol. 239, Amer. Math. Soc., Providence, RI, 1999, pp. 49–84. MR1718076

[7] M. Bökstedt, *Topological Hochschild homology*, Preprint, Bielefeld University, 1985.

[8] M. Bökstedt, *Topological Hochschild homology of* $\mathbb{Z}$ *and* $\mathbb{Z}/p$, Bielefeld, 1985.

[9] J. Borger, *The basic geometry of Witt vectors, I: The affine case*, Algebra Number Theory **5** (2011), no. 2, 231–285. MR2833791

[10] J. Borger, *The basic geometry of Witt vectors. II: Spaces*, Math. Ann. **351** (2011), no. 4, 877–933. MR2854117

[11] A. K. Bousfield, *The localization of spectra with respect to homology*, Topology **18** (1979), no. 4, 257–281. MR551009

[12] H. Cartan, *Séminaire 1954–1955. Algébre d'Eilenberg-Mac lane et homologie*, Benjamin, New York, 1967.

[13] A. Connes, *An essay on the Riemann hypothesis*, Open problems in mathematics, Springer, [Cham], 2016, pp. 225–257. MR3526936

[14] A. Connes and C. Consani, *Cyclic homology, Serre's local factors and* λ*-operations*, J. K-Theory **14** (2014), no. 1, 1–45. MR3238256

[15] C. Deninger, *Motivic L-functions and regularized determinants*, Motives (Seattle, WA, 1991), Proc. Sympos. Pure Math., vol. 55, Amer. Math. Soc., Providence, RI, 1994, pp. 707–743. MR1265547

[16] O. Gabber, K*-theory of Henselian local rings and Henselian pairs*, Algebraic K-theory, commutative algebra, and algebraic geometry (Santa Margherita Ligure, 1989), Contemp. Math., vol. 126, Amer. Math. Soc., Providence, RI, 1992, pp. 59–70. MR1156502

[17] T. Geisser and L. Hesselholt, *Topological cyclic homology of schemes*, Algebraic K-theory (Seattle, WA, 1997), Proc. Sympos. Pure Math., vol. 67, Amer. Math. Soc., Providence, RI, 1999, pp. 41–87. MR1743237

[18] T. Geisser and M. Levine, *The* K*-theory of fields in characteristic* p, Invent. Math. **139** (2000), no. 3, 459–493. MR1738056

[19] H. A. Gillet and R. W. Thomason, *The* K*-theory of strict Hensel local rings and a theorem of Suslin*, Proceedings of the Luminy conference on algebraic K-theory (Luminy, 1983), J. Pure Appl. Algebra **34** (1984), no. 2-3, 241–254. MR772059

[20] J. P. C. Greenlees, *Representing Tate cohomology of* G*-spaces*, Proc. Edinburgh Math. Soc. (2) **30** (1987), no. 3, 435–443. MR908451

[21] L. Hesselholt, *On the p-typical curves in Quillen's* K*-theory*, Acta Math. **177** (1996), no. 1, 1–53. MR1417085

[22] L. Hesselholt, *Topological Hochschild homology and the de Rham-Witt complex for* $\mathbb{Z}_{(p)}$*-algebras*, Homotopy theory: relations with algebraic geometry, group cohomology, and algebraic K-theory, Contemp. Math., vol. 346, Amer. Math. Soc., Providence, RI, 2004, pp. 253–259. MR2066502

[23] L. Hesselholt, K*-theory of truncated polynomial algebras*, Handbook of K-theory, Springer-Verlag, New York, 2005.

[24] L. Hesselholt, *On the topological cyclic homology of the algebraic closure of a local field*, An alpine anthology of homotopy theory, Contemp. Math., vol. 399, Amer. Math. Soc., Providence, RI, 2006, pp. 133–162. MR2222509

[25] L. Hesselholt, *The big de Rham-Witt complex*, Acta Math. **214** (2015), no. 1, 135–207. MR3316757

[26] L. Hesselholt and I. Madsen, *On the* K*-theory of finite algebras over Witt vectors of perfect fields*, Topology **36** (1997), no. 1, 29–101. MR1410465

[27] L. Hesselholt and I. Madsen, *On the* K*-theory of local fields*, Ann. of Math. **158** (2003), 1–113.

[28] L. Hesselholt and I. Madsen, *On the De Rham-Witt complex in mixed characteristic* (English, with English and French summaries), Ann. Sci. École Norm. Sup. (4) **37** (2004), no. 1, 1–43. MR2050204

[29] L. Illusie, *Complexe de de Rham-Witt et cohomologie cristalline* (French), Ann. Sci. École Norm. Sup. (4) **12** (1979), no. 4, 501–661. MR565469

[30] L. Illusie and M. Raynaud, *Les suites spectrales associées au complexe de de Rham-Witt* (French), Inst. Hautes Études Sci. Publ. Math. **57** (1983), 73–212. MR699058

[31] J. F. Jardine, *Local homotopy theory*, Springer Monographs in Mathematics, Springer, New York, 2015. MR3309296

[32] T. Nikolaus and P. Scholze, *On topological cyclic homology*, arXiv:1707.01799.

[33] A. Suslin, *On the K-theory of algebraically closed fields*, Invent. Math. **73** (1983), no. 2, 241–245. MR714090

[34] A. A. Suslin, *On the K-theory of local fields*, Proceedings of the Luminy conference on algebraic K-theory (Luminy, 1983), J. Pure Appl. Algebra **34** (1984), no. 2-3, 301–318. MR772065

[35] G. Tabuada, *Finite generation of the numerical Grothendieck group*, arXiv:1704.06252.

[36] R. W. Thomason, *Algebraic K-theory and étale cohomology*, Ann. Sci. École Norm. Sup. (4) **18** (1985), no. 3, 437–552. MR826102

[37] S. Tsalidis, *Topological Hochschild homology and the homotopy descent problem*, Topology **37** (1998), no. 4, 913–934. MR1607764

Graduate School of Mathematics, Nagoya University, Japan, and Department of Mathematics, University of Copenhagen, Denmark

Email address: larsh@math.nagoya-u.ac.jp

Contemporary Mathematics
Volume **708**, 2018
http://dx.doi.org/10.1090/conm/708/14269

The stable symplectic category and a conjecture of Kontsevich

Nitu Kitchloo and Jack Morava

Abstract. We consider an oriented version of the stable symplectic category defined in previous work by the first author. We show that the group of monoidal automorphisms of this category, that fix each object, contains a natural subgroup isomorphic to the solvable quotient (or a graded-abelian quotient) of the Grothendieck–Teichmüller group. This establishes a stable version of a conjecture of Kontsevich which states that groups closely related to the Grothendieck–Teichmüller group act on the moduli space of certain field theories. The above quotient is also shown to be the motivic group of monoidal automorphisms of a canonical representation (or fiber functor) on the stable symplectic category.

Contents

1. Introduction

In his solution to the existence of deformation quantization of a Poisson manifold (M, ω), Kontsevich showed in [**18**] that there exists a quasi-equivalence of L_∞-algebras:

$$\mathrm{dQ} : \Gamma(M, \Lambda^*(M))\,[[\hbar]] \longrightarrow \mathrm{Hoch}^*(\mathcal{O}(M))\,[[\hbar]], \qquad \hbar\,\omega \leftrightarrow [\mathcal{O}_\hbar(M)],$$

where the left hand side is the DGLA of poly-vector fields (with $\mathrm{d} = 0$) that controls Poisson structures on M, and the right hand side is the DGLA of poly-differential operators on M that controls the associative deformations of the functions $\mathcal{O}(M)$ on M. Hence, up to equivalence, the Poisson structure ω yields a unique associative deformation $\mathcal{O}_\hbar(M)$. However, the map dQ is not unique. The construction of (a universal choice) of map dQ requires the construction of a formal series in two variables called a *Drinfeld associator* [**13**]. The set of all such associators is a torsor under the action of a certain pro-solvable, pro-algebraic group $\mathsf{GT}_\mathbb{Q}$ over $\mathbb{Q}$, known

2010 *Mathematics Subject Classification.* Primary 55N22.

as the Grothendieck-Teichmüller group (see §4). This group can be identified with (homotopy) automorphisms of the operad of chains on little 2-disks, which is known to act on Hochschild cohomology (via the Deligne conjecture).

This Lie algebra of $\mathsf{GT}_{\mathbb{Q}}$ given by the extension of a free Lie algebra on generators: $z_3, z_5, \ldots$, by a grading generator z_0:

$$\mathrm{Lie}\,\mathsf{GT}_{\mathbb{Q}} = \mathbb{Q}\langle z_0\rangle \ltimes \langle z_3, z_5, \ldots, z_{2k+1}, \ldots\rangle, \qquad [z_0, z_{2k+1}] = (2k+1)z_{2k+1}.$$

The above result suggests that if (M, ω) is the space of classical solutions of a given classical field theory, then the group $\mathsf{GT}_{\mathbb{Q}}$ acts on the collection of all possible rings of quantum observables: $\mathcal{O}_\hbar(M)$. Indeed, the construction of the Drinfeld associator involves the calculation of certain integrals which appear to be the same integrals that one encounters in the perturbative expansion of the Feynman path integrals. These integrals are examples of certain well-studied integrals in algebraic geometry known as *Periods* [**10**, **11**]. These periods are known to form a ring whose underlying variety is a torsor over a group known as the *Motivic Galois group*, which is very closely related to the Grothendieck-Teichmuller group.

Based on the above observations, Kontsevich conjectured [**19**] (Section 5) that groups closely related to the Grothendieck-Teichmuller group act on the Moduli space of certain Quantum Field theories.[1] We will approach Kontsevich's conjecture from the standpoint of geometric quantization. As opposed to deformation quantization, the method of geometric quantization is an attempt to describe a quantum field theory directly and functorially with respect to the space of classical solutions. Functoriality above is to be understood as asking for the existence of a "Geometric Quantization" functor with values in the category of topological vector spaces:

$$\mathrm{gQ} : \mathcal{S} \longrightarrow \mathrm{Top.\ V.S},$$

where $\mathcal{S}$ is a suitable category of correspondences between symplectic manifolds known as Weinstein's Symplectic Category [**23**, **24**]. In particular, gQ applied to the Poisson manifold of classical solutions (M, ω) in $\mathcal{S}$ should recover the quantum state space of the field theory in question.

Unfortunately, there are several reasons that the functor gQ cannot exist, the least of which is the fact that $\mathcal{S}$ is not even a category (as we shall see below). Suspending reality for the moment, let us assume that we could define the functor gQ and construct an action of $\mathsf{GT}_{\mathbb{Q}}$ on $\mathcal{S}$ that fixes each object, we would obtain an action of the Grothendieck-Teichmuller group (by pre-composition) on the set of field theories as suggested by the conjecture of Kontsevich above. It is our goal in this paper to impose some natural conditions that ensure that the above assumption is a factual statement. In order to achieve that goal, we will need to first construct an honest proxy category that replaces $\mathcal{S}$.

In [**17**] the first author defined a stabilization $h\mathbb{S}$ of the symplectic "category" $\mathcal{S}$ introduced by A. Weinstein in the 1980's [**23**, **24**]. The objects of Weinstein's category are symplectic manifolds, and the morphisms between two symplectic manifolds (M, ω) and (N, η) are lagrangian immersions into $\overline{M} \times N$, where the conjugate symplectic manifold $\overline{M}$ is defined by the pair $(M, -\omega)$. The composition $L_1 * L_2$ of two lagrangian immersions $L_1 \looparrowright \overline{M} \times N$ and $L_2 \looparrowright \overline{N} \times K$, is defined

[1] The theories in question are free massless theories.

to be the fiber product: $L_1 \times_N L_2 \longrightarrow \overline{M} \times K$. This definition does not always yield a lagrangian immersion to $\overline{M} \times K$: to do so, the pullback defining it must be transverse, so Weinstein's construction is unfortunately not a genuine category.

In [**17**], we described a way to extend the symplectic category to an honest category $h\mathbb{S}$, by introducing a moduli space of stabilized (in the sense of homotopy theory) lagrangian immersions in a symplectic manifold of the form $\overline{M} \times N$ [2]. This moduli space can be described as the infinite loop space corresponding to a certain Thom spectrum. Taking this as the space of morphisms defines a *stable symplectic (homotopy) category* $h\mathbb{S}$ that is naturally enriched over the homotopy category of spectra (under smash product) [3]. Composition in $h\mathbb{S}$ is well-defined and remains faithful to Weinstein's original definition. Geometrically, the stabilization of Weinstein's category can be seen as "inverting the symplectic manifold $\mathbb{C}$", analogous to the introduction of an inverse to the projective line in the theory of motives. In other words, we introduce a relation on the symplectic category that identifies two symplectic manifolds M and N if $M \times \mathbb{C}^k$ becomes equivalent to $N \times \mathbb{C}^k$ for some k. Note that the ring of observables corresponding to $\mathbb{C}$ is the (simple) algebra known as the Heisenberg algebra corresponding to a single free Boson. So in a suitable sense the stable symplectic category can be seen as the symplectic analog of the "derived rings of observables" studied by G. Segal in [**15**].

The stable symplectic category also has variants defined by lagrangian immersions endowed with orientations or metaplectic structures (see [**17**]). Here we study the group of monoidal automorphisms of the **oriented** stable symplectic homotopy category: $hs\mathbb{S}$. In other words, by extending coefficients to commutative algebras E over $s\Omega$, we consider a family of categories $hs\mathbb{S} \wedge_{s\Omega} \mathrm{E}$. Consequently, we have the family of groups that represent monoidal automorphisms of the categories $hs\mathbb{S} \wedge_{s\Omega} \mathrm{E}$ that fix each object. We describe this group (see corollary 3.7 and theorem 3.12), and relate it in characteristic zero to an abelian quotient of the Grothendieck–Teichmüller group. This establishes a stable version of a conjecture of Kontsevich mentioned earlier. More precisely, we will show:

THEOREM. *Define a derived group scheme* $\mathbf{G} = Spec\ \mathrm{THH}(s\Omega)$, *whose* E*-points for an arbitrary commutative* $s\Omega$*-algebra* E *is defined to be the group of homotopy classes of* $s\Omega$*-algebra maps from* $\mathrm{THH}(s\Omega)$ *to* E*:*

$$\mathbf{G}(\mathrm{E}) = \mathrm{Alg}_{s\Omega}(\mathrm{THH}(s\Omega), \mathrm{E}),$$

where $\mathrm{THH}(s\Omega)$ *denotes the commutative Hopf-algebra over* $s\Omega$ *given the by Topological Hochschild homology of the spectrum* $s\Omega$*. Then, given a commutative* $s\Omega$*-algebra* E*, the group* $\mathbf{G}(\mathrm{E})$ *acts by monoidal automorphisms on the category* $hs\mathbb{S} \wedge_{s\Omega} \mathrm{E}$ *that fix each object. Furthermore, the Lie algebra of* $\mathbf{G}$ *is rationally graded-commutative generated by classes in degree* $4k+2$ *for* $k \geq 0$*.*

The above group is also shown to be the motivic group of monoidal automorphisms of a canonical representation (or fiber functor) on a truncation of the oriented stable symplectic category. The value of this functor on an object M is the $s\Omega$-module $s\Omega(M)$ representing the space of stably immersed oriented lagrangians in M.

[2] under the assumption of monotonicity. Otherwise, one has the space of totally real immersions.

[3] In [**17**], we lifted this category to an A_∞-category $\mathbb{S}$ enriched over spectra.

This document is organized as follows: In section 2 we recall the construction of the stable symplectic homotopy category. In section 3 we study the group of monoidal automorphisms of the oriented stable symplectic homotopy category over the rational numbers which fix each object. We identify this group with a graded abelian quotient of the Grothendieck–Teichmüller group and reinterpret it as validation of a stable conjecture of Kontsevich on the action of the Grothendieck-Teichmüller group on the moduli space of field theories. In this section we also identify the above quotient as the motivic group of monoidal automorphisms of the fiber functor described above. Section 3 also contains a description of an integral model for the group of monoidal automorphisms of the oriented stable symplectic homotopy category. The Appendix summarizes some properties of the Grothendieck-Teichmüller group and related constructions. In particular, we define a Hopf-Galois analog of the Grothendieck-Teichmüller group in the category of spectra, and identify its abelianization with the group of symmetries identified in Section 3.

Before we begin, we would like to thank M. Abouzaid, V. Angelveit, A. Baker, Y. Eliashberg, D. Gepner, M. Hazewinkel, J. Lind, B. Richter and H. L. Tanaka for helpful conversations related to this project.

2. The stable symplectic homotopy category

In this section we recall the construction of the stable symplectic homotopy category [**17**]. Given a symplectic manifold (M, ω) of real dimension $2m$, we construct a spectrum $\Omega(M)$ so that the corresponding infinite loop space can be interpreted as a space whose points represent manifolds that admit totally-real immersions into $M \times \mathbb{C}^n$ for large values of n (up to an equivalence that shall be made precise later). We will say that M satisfies monotonicity if the cohomology class of the symplectic form ω is a scalar multiple of the first Chern class of M. Under the assumption of monotonicity, totally real immersions may be replaced by lagrangian immersions in the above interpretation of $\Omega(M)$.

§2.1 The basic construction

Consider the Thom spectrum $\Sigma^n \mathcal{G}(\tau \oplus \mathbb{C}^n)^{-\zeta_n}$, where the bundle ζ_n is defined by virtue of the pullback diagram:

$$\begin{array}{ccc} \mathcal{G}(\tau \oplus \mathbb{C}^n) & \xrightarrow{\zeta_n} & \mathrm{BO}(m+n) \\ \downarrow & & \downarrow \\ M & \xrightarrow{\tau \oplus \mathbb{C}^n} & \mathrm{BU}(m+n), \end{array}$$

where τ denotes (homotopy unique) complex structure on the tangent bundle of M compatible with the symplectic form ω. In [**17**], we used the work of D. Ayala [**3**] to show that for $n > 0$, the infinite loop space $\Omega^{\infty-n}(\mathcal{G}(\tau \oplus \mathbb{C}^n)^{-\zeta_n})$ can be interpreted as the moduli space of manifolds $L^{m+n} \subset \mathbb{R}^\infty \times \mathbb{R}^n$, with a proper projection onto $\mathbb{R}^n$, and endowed with a totally-real immersion $L^{m+k} \looparrowright M \times \mathbb{C}^n$ (or lagrangian immersion, under the assumption of monotonicity). More precisely, the space $\Omega^{\infty-n}(\mathcal{G}(\tau \oplus \mathbb{C}^n)^{-\zeta_n})$ is uniquely defined by the property that given a smooth manifold X, the set of homotopy classes of maps $[X, \Omega^{\infty-n}(\mathcal{G}(\tau \oplus \mathbb{C}^n)^{-\zeta_n})]$, is in bijection with concordance classes of smooth manifolds $E \subset X \times \mathbb{R}^\infty \times \mathbb{R}^n$ over X, so that the first factor projection: $\pi : E \longrightarrow X$ is a submersion, and which are endowed with a smooth map $\varphi : E \longrightarrow M \times \mathbb{C}^n$ which restricts to a totally-real immersion (resp. lagrangian) on each fiber of π. As before, we demand that the third factor projection $E \longrightarrow \mathbb{R}^n$ be fiberwise proper over X.

Now the standard inclusion $\mathbb{R}^{n_1} \subseteq \mathbb{R}^{n_2}$, induces a natural map:

$$\varphi_{n_1,n_2} : \Sigma^{n_1}\mathcal{G}(\tau \oplus \mathbb{C}^{n_1})^{-\zeta_{n_1}} \longrightarrow \Sigma^{n_2}\mathcal{G}(\tau \oplus \mathbb{C}^{n_2})^{-\zeta_{n_2}},$$

which represents the map that sends a concordance class E, to $E \times \mathbb{R}^{n_2-n_1}$, by simply taking the product with the orthogonal complement of $\mathbb{R}^{n_1}$ in $\mathbb{R}^{n_2}$.

DEFINITION 2.1. *Define the Thom spectrum $\Omega(M)$ representing the infinite loop space of stabilized totally-real (resp. lagrangian under the assumption of monotonicity) immersions in M to be the colimit:*

$$\Omega(M) = \underline{\mathcal{G}}(M)^{-\zeta} := colim_n \; \Sigma^n\mathcal{G}(\tau \oplus \mathbb{C}^n)^{-\zeta_n}.$$

Notice that by definition, we have a canonical homotopy equivalence: $\Omega(M \times \mathbb{C}) \simeq \Sigma^{-1}\Omega(M)$. Taking M to be a point, we define $\Omega = \Omega(*) = (\mathrm{U/O})^{-\zeta}$, where the bundle ζ over $\mathrm{U/O}$ is the virtual zero dimensional bundle over $(\mathrm{U/O})$ defined by the canonical inclusion $\mathrm{U/O} \longrightarrow \mathrm{BO}$.

Henceforth, we shall use the term "lagrangian immersion" to mean "totally-real immersion" if the condition of monotonicity fails to hold. We take this opportunity to also introduce the (abusive) convention of not decorating the stable vector bundle ζ by the underlying manifold M. Hopefully, the manifold M will be clear from context.

We may also describe $\Omega(M)$ as a Thom spectrum: Let the stable tangent bundle of M of virtual (complex) dimension m be given by a map $\tau : M \longrightarrow \mathbb{Z} \times \mathrm{BU}$. As the notation suggests, let $\underline{\mathcal{G}}(M)$ be defined as the pullback:

$$\begin{array}{ccc} \underline{\mathcal{G}}(M) & \xrightarrow{\zeta} & \mathbb{Z} \times \mathrm{BO} \\ \downarrow & & \downarrow \\ M & \xrightarrow{\tau} & \mathbb{Z} \times \mathrm{BU}. \end{array}$$

Then the spectrum $\Omega(M)$ is homotopy equivalent to the Thom spectrum of the stable vector bundle $-\zeta$ over $\underline{\mathcal{G}}(M)$ defined in the diagram above.

Notice that the fibration $\mathbb{Z} \times \mathrm{BO} \longrightarrow \mathbb{Z} \times \mathrm{BU}$ is a principal bundle up to homotopy, with fiber being the infinite loop space $\mathrm{U/O}$. Hence, the spectrum $\Omega(M)$ is homotopy equivalent to a Ω-module spectrum. Now, observe that we have the equivalence, up to homotopy, of $\mathrm{U/O}$-spaces:

$$\underline{\mathcal{G}}(M) \times_{\mathrm{U/O}} \underline{\mathcal{G}}(N) \simeq \underline{\mathcal{G}}(M \times N).$$

This translates to a canonical homotopy equivalence:

$$\mu : \Omega(M) \wedge_{\Omega} \Omega(N) \simeq \Omega(M \times N).$$

Let us now describe the stable symplectic homotopy category $h\mathbb{S}$. The objects of this category will be symplectic manifolds (M, ω) (see remark 2.3), endowed with a compatible almost complex structure.

DEFINITION 2.2. *The spectrum $\Omega(M, N)$ of morphisms in $h\mathbb{S}$ from M to N is the Ω-module spectrum:*

$$\Omega(M, N) := \Omega(\overline{M} \times N).$$

REMARK 2.3. *Notice that objects in $h\mathbb{S}$ need not be compact. The price we pay for this, familiar from other contexts, is that we simply lose the identity morphisms for non-compact objects.*

The next step is to define composition. The simplest case

$$\Omega(M,*) \wedge_\Omega \Omega(*,N) \longrightarrow \Omega(M,N),$$

is the map μ constructed earlier. For the general case, consider $k+1$ objects objects M_i with $0 \le i \le k$, and let the space $\underline{\mathcal{G}}(\Delta)$ be defined by the pullback:

$$\begin{array}{ccc} \underline{\mathcal{G}}(\Delta) & \longrightarrow & \underline{\mathcal{G}}(\overline{M}_0 \times M_1 \times \cdots \times \overline{M}_{k-1} \times M_k) \\ \downarrow \xi & & \downarrow \\ \overline{M}_0 \times (M_1 \times \cdots \times M_{k-1}) \times M_k & \xrightarrow{\Delta} & \overline{M}_0 \times (M_1 \times \overline{M}_1) \times \cdots \times (M_{k-1} \times \overline{M}_{k-1}) \times M_k \end{array}$$

where Δ denotes the product to the diagonals $\Delta : M_i \longrightarrow M_i \times \overline{M}_i$, for $0 < i < k$.

Now notice that the fibrations defining the pullback above are direct limits of smooth fibrations with compact fiber. Furthermore, the map Δ is a proper map for any choice of $k+1$-objects (even if they are non-compact). In particular, we may construct the Pontrjagin–Thom collapse map along the top horizontal map by defining it as a direct limit of Pontrjagin–Thom collapses for each stage.

Let ζ_i denote the individual structure maps $\underline{\mathcal{G}}(\overline{M}_{i-1} \times M_i) \longrightarrow \mathbb{Z} \times \mathrm{BO}$, and let $\eta(\Delta)$ denote the normal bundle of Δ. Performing the Pontrjagin–Thom construction along the top horizontal map in the above diagram yields a morphism of spectra:

$$\varphi : \Omega(M_0, M_1) \wedge_\Omega \cdots \wedge_\Omega \Omega(M_{k-1}, M_k) \simeq \Omega(\overline{M}_0 \times M_1 \times \cdots \times \overline{M}_{k-1} \times M_k) \longrightarrow \underline{\mathcal{G}}(\Delta)^{-\lambda}$$

where $\lambda : \underline{\mathcal{G}}(\Delta) \longrightarrow \mathbb{Z} \times \mathrm{BO}$ is the formal difference of the bundle $\bigoplus \zeta_i$ and the pullback bundle $\xi^* \eta(\Delta)$.

The next step in defining composition is to show that $\underline{\mathcal{G}}(\Delta)^{-\lambda}$ is canonically homotopy equivalent to $\Omega(M_0, M_k) \wedge (M_1 \times \cdots \times M_{k-1})_+$, where $(M_1 \times \cdots \times M_{k-1})_+$ denotes the manifold $M_1 \times \cdots \times M_{k-1}$ with a disjoint basepoint. To achieve this, it is sufficient to construct a U/O-equivariant map over $\overline{M}_0 \times (M_1 \times \cdots \times M_{k-1}) \times M_k$:

$$\psi : \underline{\mathcal{G}}(\overline{M}_0 \times M_k) \times (M_1 \times \cdots \times M_{k-1}) \longrightarrow \underline{\mathcal{G}}(\Delta),$$

that pulls λ back to the bundle $\zeta \times 0$. The construction of ψ is straightforward. We define:

$$\psi(\lambda, m_1, \ldots, m_{k-1}) = \lambda \oplus \Delta(T_{m_1}(M_1)) \oplus \cdots \oplus \Delta(T_{m_{k-1}}(M_{k-1})),$$

where $\Delta(T_m(M)) \subset T_{(m,m)}(M \times \overline{M})$ denotes the diagonal lagrangian subspace. Now let $\pi : \underline{\mathcal{G}}(\Delta)^{-\lambda} \longrightarrow \Omega(M_0, M_k)$ be the projection map that collapses $M_1 \times \cdots \times M_{k-1}$ to a point.

DEFINITION 2.4. *We define the composition map to be the induced composite:*

$$\pi\varphi : \Omega(M_0, M_1) \wedge_\Omega \cdots \wedge_\Omega \Omega(M_{k-1}, M_k) \longrightarrow \underline{\mathcal{G}}(\Delta)^{-\lambda} \longrightarrow \Omega(M_0, M_k).$$

We leave it to the reader to check that composition as defined above is homotopy associative.

§2.2 The identity morphism

We now show that a compact manifold (M, ω) has an identity morphism:

PROPOSITION 2.5. *Let M be a compact manifold, and let $[id] : \mathrm{S} \longrightarrow \Omega(M, M)$ be a representative of homotopy class of the diagonal (lagrangian) embedding $\Delta : M \longrightarrow \overline{M} \times M$. Then $[id]$ is indeed the identity for the composition defined above.*

PROOF. Given two manifolds M, N, let $\Delta(M) \subset \overline{M} \times M$ be a diagonal representative of $[id]$ as above. Observe that $\overline{N} \times \Delta(M) \times M$ is transverse to $\overline{N} \times M \times \Delta(M)$ inside $\overline{N} \times M \times \overline{M} \times M$. They intersect along $\overline{N} \times \Delta_3(M)$, where $\Delta_3(M) \subset M \times \overline{M} \times M$ is the triple (thin) diagonal. Hence we get a diagram

$$\begin{array}{ccccc} \Omega(N,M) \wedge \mathrm{S} & \longrightarrow & \Omega(N,M) \wedge \Delta(M)^{-\tau} & \xrightarrow{\Delta^{-\tau}} & \Omega(N,M) \wedge_{\Omega} \Omega(M,M) \\ & \searrow & \downarrow & & \downarrow \\ & & \Omega(N,M) & \xrightarrow{=} & \Omega(N,M) \end{array}$$

commutative up to homotopy, where the right vertical map is composition, and the left vertical map is the Pontrjagin–Thom collapse over the inclusion map

$$\overline{N} \times M = \overline{N} \times \Delta_3(M) \longrightarrow \overline{N} \times M \times \Delta(M).$$

Now consider the following factorization of the identity map:

$$\overline{N} \times M = \overline{N} \times \Delta_3(M) \longrightarrow \overline{N} \times M \times \Delta(M) \longrightarrow \overline{N} \times M$$

where the last map is the projection onto the first two factors. Performing the Pontrjagin–Thom collapse over this composite shows that

$$\Omega(N, M) \wedge \mathrm{S} \longrightarrow \Omega(N, M) \wedge \Delta(M)^{-\tau} \longrightarrow \Omega(N, M).$$

is the identity. It follows that right multiplication by $[id] : \mathrm{S} \to \Omega(M, M)$ induces the identity map on $\Omega(N, M)$, up to homotopy. A similar argument works for left multiplication. □

REMARK 2.6. *Recall that given arbitrary symplectic manifolds M and N, the composition map:*

$$\Omega(M, *) \wedge_{\Omega} \Omega(*, N) \simeq \Omega(M, N)$$

induces a natural decomposition of $\Omega(M, N)$. In particular, arbitrary compositions can be canonically factored using this decomposition, and can be computed by applying the "inner product"

$$\Omega(*, N) \wedge_{\Omega} \Omega(N, *) \longrightarrow \Omega.$$

to the factors. When M is compact, this pairing defines an equivalence

$$\Omega(\overline{M}) \longrightarrow \mathrm{Hom}_{\Omega}(\Omega(M), \Omega).$$

It will also be important below that $h\mathbb{S}$ is a symmetric-monoidal category, with a product given by the cartesian product of symplectic manifolds.

DEFINITION 2.7. *The construction of the* **oriented** *stable symplectic homotopy category $s\mathbb{S}$ is completely analogous, but with O replaced by SO; the commutative ring spectrum $s\Omega = (\mathrm{U}/\mathrm{SO})^{-\zeta}$ defines its coefficients.*

Now Ω is an (Eilenberg–MacLane) $\mathrm{H}(\mathbb{Z}/2)$-algebra, so we can regard $\mathbb{S}$ as a category with morphism objects enriched over a classical differential graded algebra. This is not the case for $s\Omega$, but its rationalization $s\Omega \otimes \mathbb{Q}$ is again a generalized Eilenberg–MacLane spectrum, with

$$s\Omega_* \otimes \mathbb{Q} = \Lambda_{\mathbb{Q}}[y_{4i+1},\ i \geq 0]$$

an exterior algebra on certain odd - degree generators. Moreover, the category $s\mathbb{S}$ simplifies considerably when rationalized. In particular, the Thom isomorphism

$$s\Omega(M)_* \otimes \mathbb{Q} \cong \mathrm{H}_*(M, s\Omega_* \otimes \mathbb{Q})$$

identifies $s\mathbb{S} \otimes \mathbb{Q}$ with an (Arnol'd-Hörmander-Maslov…) category of symplectic manifolds, whose morphisms are classical cohomological correspondences with compact support, but with coefficients in the graded ring $s\Omega_* \otimes \mathbb{Q} \cong \mathrm{H}_*(\mathrm{U}/\mathrm{SO}_+, \mathbb{Q})$.

3. A stable version of Kontsevich's conjecture

As mentioned in the introduction, based on explicit computations, Kontsevich conjectures in [**19**] (section 5) that the motivic Galois group acts on the "moduli space" of free massless field theories, and that in even dimensions, this action factors through the Grothendieck–Teichmüller group. Kontsevich was led to this conjecture by his work on the question of uniqueness of the deformation quantization functor. However, we plan to approach his conjecture from the standpoint of geometric quantization. Note that the moduli space of field theories is not a well defined object and so the above conjecture can only be formulated once we have an appropriate definition in place.

Now the $(n+1)$-dimensional classical field theories that occur in physics are typically described by an action functional. The (conjectural) framework of quantization proceeds as follows: By taking solutions of the Euler-Lagrange equations about the germ of an n-manifold, one obtains a functor from the $(n+1)$-dimensional cobordism category to a variant of the symplectic category. [4] Then the corresponding quantum field theory is constructed as the composite of the above functor with the geometric quantization functor, the latter having the symplectic category as its domain and taking values in the category of topological vector spaces. An automorphism of the symplectic category that fixes every object would therefore induce a deformation of any such field theory.

In this section, we will provide a rigorous framework for Kontsevich's conjecture: we will compute a canonical subgroup of the group of monoidal automorphisms of the oriented *stable* symplectic (homotopy) category, and show that it is indeed an abelian quotient of the Grothendieck–Teichmüller group, thereby showing that a stable version of Kontsevich's conjecture is indeed true.

§3.1 Monoidal automorphisms of the stable symplectic category

We begin this section by defining the group of monoidal automorphisms of the stable symplectic homotopy category. Recall that this category $h\mathbb{S}$ is enriched over

[4] The infinite dimensional nature of symmetries force us to also consider infinite dimensional symplectic manifolds.

the category of Ω-module spectra. Given a commutative Ω-algebra E, let $h\mathbb{S} \wedge_{\Omega} \mathrm{E}$ denote the category enriched over E-modules constructed by extending coefficients.

DEFINITION 3.1. *Define the group of automorphisms of the stable symplectic category as a derived group G_{Ω}, whose* E*-points: $G_{\Omega}(\mathrm{E})$, for any commutative Ω-algebra* E*, are defined as the group of monoidal automorphisms of the category $h\mathbb{S} \wedge_{\Omega} \mathrm{E}$ that fix each object. We similarly define the group $G_{s\Omega}$ as the group of automorphisms of the oriented stable symplectic homotopy category.*

Our first order of business will be to consider certain endomorphisms of $hs\mathbb{S} \wedge_{s\Omega} \mathrm{E}$ that will eventually turn out to be the elements in the Lie algebra of $G_{s\Omega}(\mathrm{E})$. We begin with some preliminary constructions that don't require orientability:

Given a symplectic manifold M, recall that $\Omega(M)$ was a defined as a Thom spectrum: $\underline{\mathcal{G}}(M)^{-\zeta}$. Here $\pi : \underline{\mathcal{G}}(M) \longrightarrow M$ was a principal U/O-bundle, supporting a stable real vector bundle $\zeta : \underline{\mathcal{G}}(M) \longrightarrow \mathrm{BO}$ of virtual dimension m. The bundle π is classified by the map:

$$\tau(\Omega, M) : M \longrightarrow \mathrm{B(U/O)}\,.$$

THEOREM 3.2. *Given an object of $h\mathbb{S}$ represented by a symplectic manifold (M, ω), then $\tau(M, \Omega)$ factors through the map $\tau(M) : M \longrightarrow \mathrm{BU}$ that classfies the tangent bundle of M, followed by the projection map* $\mathrm{BU} \longrightarrow \mathrm{B(U/O)}$. *Furthermore, the restriction of $\tau(\Omega, M \times \overline{M})$ along the diagonal: $\Delta : M \longrightarrow M \times \overline{M}$ is trivial.*

PROOF. Notice that the bundle $\pi : \underline{\mathcal{G}}(M) \longrightarrow M$ was induced from the frame bundle of M, with structure group U, along the left action of U on U/O, so we can lift the map $\tau(\Omega, M)$, to BU, to get the factorization through $\tau(M)$. This proves the first part of the claim. Now notice that the restriction of the tangent bundle of $M \times \overline{M}$ along the diagonal Δ is canonically isomorphic to the complexification of the tangent bundle of M. This may be restated as saying that one has a unique lift $\tau(\Delta)$ that makes the following commute:

$$\begin{array}{ccc} M & \xrightarrow{\tau(\Delta)} & \mathrm{BO} \\ \downarrow{\scriptstyle \Delta} & & \downarrow \\ M \times \overline{M} & \xrightarrow{\tau} & \mathrm{BU}\,. \end{array}$$

It follows that $\tau(\Omega, M \times \overline{M})$ is trivial when restricted along Δ. □

REMARK 3.3. *The projection map* $\mathrm{BU} \longrightarrow \mathrm{B(U/O)}$ *is injective in cohomology away from two. It is easy to see that its image is generated by classes d_{2i+1} in degree $4i + 2$, whose generating function can be expressed in terms of the Chern classes as:*

$$\sum_i d_i = \frac{\sum_i (-1)^i c_i}{\sum_i c_i} \equiv 1 - \sum_i 2c_{2i+1} + \textit{decomposables}.$$

Similarly, if one considers the inclusion $\mathrm{BO} \longrightarrow \mathrm{BU}$, *then this map is injective in homology away from two. If we let $\sum_i b_i$ denote the homogeneous generators in the* *image of* $\mathrm{H}_*(\mathbb{CP}^\infty) \subset \mathrm{H}_*(\mathrm{BU})$, *then the image of* $\mathrm{H}_*(\mathrm{BO})$ *is the sub (polynomial) algebra generated by classes* a_i *(away from two), given by:*

$$\sum_i a_i = (\sum_i b_i)(\sum_i (-1)^i b_i) \equiv 1 + \sum_{i>0} 2b_{2i} + \textit{decomposables}.$$

DEFINITION 3.4. *Henceforth, we work in the oriented category* $hs\mathbb{S}$, *and we assume that* $\pi_* \mathrm{E}$ *is a* $\mathbb{Q}$*-vector space. Define* $\mathcal{P}_{\mathrm{E}}(\mathrm{B(U/SO)})$ *to be the graded* E^**-submodule of* $\tilde{\mathrm{E}}^*(\mathrm{B(U/SO)})$ *consisting of primitive elements in the (commutative) Hopf algebra* $\mathrm{E}^*(\mathrm{B(U/SO)})$.

THEOREM 3.5. $\mathcal{P}_{\mathrm{E}}(\mathrm{B(U/SO)})$ *induces primitive graded endomorphisms of the category* $hs\mathbb{S} \wedge_{s\Omega} \mathrm{E}$ *that fix each object. In other words, there is a natural map of graded* E^**-modules:*

$$\mathcal{P}(\mathrm{E}) : \mathcal{P}_{\mathrm{E}}(\mathrm{B(U/SO)}) \longrightarrow \mathrm{End}(hs\mathbb{S} \wedge_{s\Omega} \mathrm{E}).$$

Furthermore, the image of $\mathcal{P}(\mathrm{E})$ *is contained in the subgroup of primitive functors, defined as functors* φ *that fix each object, and are additive with respect to the monoidal structure on the morphisms:*

$$\varphi(X \wedge_{\mathrm{E}} Y) = \varphi(X) \wedge_{\mathrm{E}} Y + X \wedge_{\mathrm{E}} \varphi(Y),$$

for all morphisms X *and* Y.

PROOF. Fix objects (M, ω) and (N, η) of $hs\mathbb{S}$. Given an element $\alpha \in \mathcal{P}_{\mathrm{E}}(\mathrm{B(U/SO)})$, we define the action of $\mathcal{P}(\mathrm{E})(\alpha)$ on

$$s\Omega(M, N) \wedge_{s\Omega} \mathrm{E} = s\Omega(\overline{M} \times N) \wedge_{s\Omega} \mathrm{E} := s\Omega(\overline{M} \times N)_{\mathrm{E}}$$

as the cap product with α, described as the composite map:

$$\alpha_* : s\Omega(\overline{M} \times N)_{\mathrm{E}} \longrightarrow s\Omega(\overline{M} \times M)_{\mathrm{E}} \wedge (\overline{M} \times N)_+ \\ \longrightarrow s\Omega(\overline{M} \times N)_{\mathrm{E}} \wedge \mathrm{B(U/SO)}_+ \longrightarrow s\Omega(\overline{M} \times N)_{\mathrm{E}},$$

where the first map is induced by the Thom diagonal map:

$$s\underline{\mathcal{G}}(\overline{M} \times N)^{-\zeta} \longrightarrow s\underline{\mathcal{G}}(\overline{M} \times N)^{-\zeta} \wedge (\overline{M} \times N)_+.$$

The second map is induced by $\tau(s\Omega, \overline{M} \times N)$, and the third map above is given by capping with the class α. Now recall that $s\underline{\mathcal{G}}(\overline{M} \times N)$ is equivalent to the external product bundle $s\underline{\mathcal{G}}(\overline{M}) \times_{\mathrm{U/SO}} s\underline{\mathcal{G}}(N)$. In particular, the element $\tau(s\Omega, \overline{M} \times N)$ decomposes as the composite:

$$\overline{M} \times N \longrightarrow \mathrm{B(U/SO)} \times \mathrm{B(U/SO)} \longrightarrow \mathrm{B(U/SO)}.$$

Since α is a primitive class, the pullback of α along $\tau(s\Omega, \overline{M} \times N)$ is given by $\alpha_* \wedge 1 + 1 \wedge \alpha_*$ under the decomposition $s\Omega(\overline{M} \times N)_{\mathrm{E}} = s\Omega(\overline{M})_{\mathrm{E}} \wedge_{\mathrm{E}} s\Omega(N)_{\mathrm{E}}$. This argument shows that the elements α_* are primitive with respect to the monoidal structure.

Next we will show that the construction $\alpha \mapsto \alpha_*$ is functorial. This will yield a map of E^*-modules that we seek:

$$\mathcal{P}(\mathrm{E}) : \mathcal{P}_{\mathrm{E}}(\mathrm{B(U/SO)}) \longrightarrow \mathrm{End}(hs\mathbb{S} \wedge_{s\Omega} \mathrm{E}).$$

To show functoriality, we need to show that the diagram:

$$\begin{array}{ccc} s\Omega(L,M)_{\mathrm{E}} \wedge_{\mathrm{E}} s\Omega(M,N)_{\mathrm{E}} & \xrightarrow{\alpha_* \wedge \alpha_*} & s\Omega(L,M)_{\mathrm{E}} \wedge_{\mathrm{E}} s\Omega(M,N)_{\mathrm{E}} \\ \downarrow & & \downarrow \\ s\Omega(L,N)_{\mathrm{E}} & \xrightarrow{\quad\alpha_*\quad} & s\Omega(L,N)_{\mathrm{E}}, \end{array}$$

commutes; where the vertical maps are induced by composition in $sh\mathbb{S}$, and the top horizontal map: $\alpha_* \wedge \alpha_* : s\Omega(L,M)_{\mathrm{E}} \wedge_{\mathrm{E}} s\Omega(M,N)_{\mathrm{E}} \longrightarrow s\Omega(L,M)_{\mathrm{E}} \wedge_{\mathrm{E}} s\Omega(M,N)_{\mathrm{E}}$ denotes the external smash product of the two maps $\alpha(\overline{L} \times M)_*$ and $\alpha(\overline{M} \times N)_*$, where we write $\alpha(X)_*$ to indicate that it is an operator on $s\Omega(X)_{\mathrm{E}}$.

By the primitivity of α_*, we write $\alpha(X \times Y)_*$ as the sum $\alpha(X)_* \wedge 1 + 1 \wedge \alpha(Y)_*$. This decomposition allows us to reduce the general case to the special case when L and N are a point. In other words, we would like to show that the following special case of the above diagram commutes:

$$\begin{array}{ccc} s\Omega(*,M)_{\mathrm{E}} \wedge_{\mathrm{E}} s\Omega(M,*)_{\mathrm{E}} & \xrightarrow{\alpha_* \wedge \alpha_*} & s\Omega(*,M)_{\mathrm{E}} \wedge_{\mathrm{E}} s\Omega(M,*)_{\mathrm{E}} \\ \downarrow & & \downarrow \\ \mathrm{E} & \xrightarrow{\quad 0\quad} & \mathrm{E}\,. \end{array}$$

To show this, recall that the composition $s\Omega(*,M)_{\mathrm{E}} \wedge_{\mathrm{E}} s\Omega(M,*)_{\mathrm{E}} \longrightarrow \mathrm{E}$ is obtained by restricting along the diagonal $\Delta : M \longrightarrow M \times \overline{M}$. By claim 3.2, we see that the restriction of $\tau(s\Omega, M \times \overline{M})$ along Δ is trivial. It follows that the $\alpha_* \wedge \alpha_*$ followed by composition is trivial. □

§3.2 The structure of primitives

Let $\mathcal{P}_{\mathrm{E}}(\mathrm{BU}) \subset \tilde{\mathrm{E}}^*(\mathrm{BU})$ denote the submodule of primitives. For complex oriented theories it is a standard fact that this is a free (completed) E^*-module generated by the Newton polynomials $N_k(c_1, \ldots, c_k)$, in the Chern classes. These Newton polynomials $N_i(\sigma_1, \cdots, \sigma_i)$ are defined by writing the power symmetric functions $x_1^i + x_2^i + \cdots + x_i^i$ in terms of the elementary symmetric functions $\sigma_1, \sigma_2, \cdots, \sigma_i$.

Rationally, the classes N_k can be expressed in terms of the Chern classes c_k, or the classes d_k of remark 3.3, by comparing the homogeneous terms in the formal graded equalities (see [**20**] Ch.1):

$$\sum_{k\geq 0} c_k = \prod_{i\geq 0} \exp \frac{(-1)^i N_{i+1}}{i+1}, \qquad \sum_{k\geq 0} d_k = \prod_{i\geq 0} \exp \frac{-2N_{2i+1}}{2i+1}.$$

Up to a scaling factor of $k!$, $N_k(c_1, \ldots, c_k)$ is the homogeneous degree $2k$ term in the Chern character ch_k for the universal virtual vector bundle over BU. Notice that any theory E such that E_* is a $\mathbb{Q}$-vector space, is complex orientable.

THEOREM 3.6. *Assume that* E_* *is a* $\mathbb{Q}$*-vector space. Fix a complex orientation on* E *(see remark* 3.9 *below). Then, in cohomology, the map induced by the projection:*

$$\mathcal{P}_{\mathrm{E}}(\mathrm{B(U/SO)}) \longrightarrow \mathcal{P}_{\mathrm{E}}(\mathrm{BU})$$

is injective onto the free (completed) E^**-module generated by the primitives* $N_{2k+1}(c_1, \ldots, c_{2k+1})$ *in degree* $4k+2$*, with* $k \geq 0$*. In particular, the image of*

$\mathcal{P}(\mathrm{E})$ is generated by operators acting on $s\Omega(M)$ via multiplication with the classes $ch_{2k+1}(\tau)$ (compare with [**19**] *Theorem 9).*

PROOF. Since E_* is a $\mathbb{Q}$-vector space, we may assume E is a generalized Eilenberg–MacLane spectrum. Notice that the projection map $\mathrm{BU} \longrightarrow \mathrm{B(U/SO)}$ is a map of H-spaces, and in homology, it maps the indecomposable elements in degrees $4k+2$ isomorphically onto the indecomposables in the Hopf-algebra $\mathrm{E}_*(\mathrm{B(U/SO)})$ (see remark 3.3). Dually, it follows that $\mathcal{P}_E(\mathrm{B(U/SO)})$ maps isomorphically to the completed subspace generated by primitives in degree $4k+2$ in $\mathcal{P}_{\mathrm{E}}(\mathrm{BU})$. By the above discussion, this is the completed subspace generated by the odd Newton polynomials in the Chern classes. □

COROLLARY 3.7. *Working in characteristic zero, let $G(\mathrm{E})$ denote the (pro) abelian group generated by the formal exponentials of the form $\mathrm{Exp}(t\, ch_{2k+1}(\tau))$, with t being any homogeneous element of degree $4k+2$ in the $\mathbb{Q}$-vector space E_*, where $k \geq 0$. Then $G(\mathrm{E})$ acts by degree-preserving monoidal automorphisms on the category $sh\mathbb{S} \wedge_{s\Omega} \mathrm{E}$ that fix each object. In particular, $G(\mathrm{E})$ is a subgroup of the Galois group of automorphisms $G_{s\Omega}(\mathrm{E})$. Presently, we will describe a canonical integral form for this group.*

REMARK 3.8. *Notice that the space of primitives we have identified in theorem 3.6 contain a class N_1 which has no corresponding partner in the Lie algebra of $\mathsf{GT}_{\mathbb{Q}}$. If we had considered the stable symplectic category enriched over the spectrum $(\mathrm{SU/SO})^{-\zeta}$ instead, then the space of primitives would indeed begin with N_3 as expected.*

REMARK 3.9. *The reader may wish to verify that the completed subspace generated by the odd Newton polynomials in the Chern classes is independent of the choice of complex orientation. In particular, the same holds for the group $G(\mathrm{E})$. The reader may also verify that the action of $G(\mathrm{E})$ restricts to the identity on the subgroup: $\pi_0(s\Omega(M,N)) \subset \pi_*(s\Omega(M,N) \wedge_{s\Omega} \mathrm{E})$.*

REMARK 3.10. *Given a Calabi-Yau manifold M, homological mirror symmetry gives rise to an action of the Grothendieck–Teichmüller group on $H^*(M,\mathbb{C})$ under the identification of $H^*(M,\mathbb{C})$ with the cohomology of poly-vector fields on M (see* [**19**], *Theorem 9). Remarkably, this action is exactly the same as that of $G(\mathrm{H}(\mathbb{C}))$ on $s\Omega(M)_{\mathrm{H}(\mathbb{C})}$, once we use the Thom isomorphism to identify $s\Omega(M)_{\mathrm{H}(\mathbb{C})}$ with $H^*(M,\mathbb{C})$. It would be very interesting to give a geometric description of this identification. Also see remark 4.1.*

§3.3 An integral candidate for the (abelianized) Grothendieck–Teichmüller group

In theorems 3.5 and 3.6 we described the Lie algebra of the group $G(\mathrm{E})$ as being the primitives in $\tilde{\mathrm{E}}^*(\mathrm{B(U/SO)})$. This implies that the cotangent space of $G(\mathrm{E})$ at the identity element should be interpreted as the vector space dual to these primitives. This dual space can be canonically identified with a subspace of the indecomposables: $Q(\mathrm{E}_*(\mathrm{B(U/SO)})) = \mathrm{I}/\mathrm{I}^2$, where I is the augmentation ideal in $\mathrm{E}_*(\mathrm{B(U/SO)})$. This suggests that one must think of the commutative ring spectrum $s\Omega \wedge \mathrm{B(U/SO)}_+$ as "functions on a derived avatar" of the abelianized Grothendieck–Teichmuller group.

The spectrum $s\Omega \wedge \mathrm{B(U/SO)}_+$ can be constructed functorially from $s\Omega$: Indeed, we know by [**5**] (Prop. 7.3), that $s\Omega \wedge \mathrm{B(U/SO)}_+$ is equivalent to $\mathrm{THH}(s\Omega)$ as commutative algebras. Notice in fact, that $s\Omega \wedge \mathrm{B(U/SO)}_+$ is a commutative Hopf-algebra spectrum in the category of $s\Omega$-module spectra [**2**]. The above discussion leads naturally to the:

DEFINITION 3.11. *Define a derived group scheme* $\mathbf{G} = Spec\ \mathrm{THH}(s\Omega)$, *whose* E*-points for an arbitrary commutative* $s\Omega$*-algebra* E *is defined to be the group of homotopy classes of* $s\Omega$*-algebra maps from* $\mathrm{THH}(s\Omega)$ *to* E*:*

$$\mathbf{G}(\mathrm{E}) = \mathrm{Alg}_{s\Omega}(\mathrm{THH}(s\Omega), \mathrm{E}) = \mathrm{Alg}_{\mathbb{S}}(\mathrm{B(U/SO)}_+, \mathrm{E}).$$

THEOREM 3.12. *Given a commutative* $s\Omega$*-algebra* E, *the group* $\mathbf{G}(\mathrm{E})$ *acts by degree preserving monoidal automorphisms on the category* $hs\mathbb{S} \wedge_{s\Omega} \mathrm{E}$ *that fix each object.*

PROOF. The proof of this theorem is similar to the proof of theorem 3.5. As before, given $\beta \in \mathbf{G}(\mathrm{E})$, we get an automorphism of $s\Omega(\overline{M} \times N)_{\mathrm{E}}$ given by the cap product with β:

$$\beta_* : s\Omega(\overline{M} \times N)_{\mathrm{E}} \longrightarrow s\Omega(\overline{M} \times N)_{\mathrm{E}} \wedge (\overline{M} \times N)_+ \\ \longrightarrow s\Omega(\overline{M} \times N)_{\mathrm{E}} \wedge \mathrm{B(U/SO)}_+ \longrightarrow s\Omega(\overline{M} \times N)_{\mathrm{E}}.$$

Since β is a map of algebras, we see that β_* preserves the monoidal structure. So the only thing left to check is that β_* is functorial. As in the proof of theorem 3.5, we may reduce this question to showing that the following diagram commutes:

$$\begin{array}{ccc} s\Omega(*, M)_{\mathrm{E}} \wedge_{\mathrm{E}} s\Omega(M, *)_{\mathrm{E}} & \xrightarrow{\beta_* \wedge \beta_*} & s\Omega(*, M)_{\mathrm{E}} \wedge_{\mathrm{E}} s\Omega(M, *)_{\mathrm{E}} \\ \downarrow & & \downarrow \\ \mathrm{E} & \xrightarrow{=} & \mathrm{E}. \end{array}$$

Again, as in the proof of 3.5, this requires showing that the restriction of $\tau(s\Omega, M \times \overline{M})$ along Δ below, factors through the unit of E:

$$M_+ \longrightarrow M_+ \wedge \overline{M}_+ \longrightarrow \mathrm{B(U/SO)}_+ \longrightarrow \mathrm{E}.$$

But this follows from theorem 3.2, and the fact that β is a ring map. □

§3.4 The (abelianized) Grothendieck–Teichmüller group as a motivic group

Given a symplectic manifold (M, ω) recall that the morphism spectrum $\Omega(*, M)$ in $h\mathbb{S}$ can be identified with the Ω-module spectrum $\Omega(M)$. Let $\pi_0(\mathbb{S})$ denote the truncation of $\mathbb{S}$, so that the objects of $\pi_0(\mathbb{S})$ are the same as those of $h\mathbb{S}$, and morphisms given by $\pi_0(\Omega(M, N))$. Given a commutative Ω-algebra E, right composition in $h\mathbb{S}$ gives rise to a functor $\mathcal{F}_{\mathrm{E}}$ with values in E-module spectra:

$$\mathcal{F}_{\mathrm{E}} : \pi_0(\mathbb{S}) \longrightarrow \mathrm{E}\,\mathcal{S}, \qquad \mathcal{F}_{\mathrm{E}}(M) = \Omega(M)_{\mathrm{E}} := \Omega(M) \wedge_{\Omega} \mathrm{E},$$

where $\mathrm{E}\,\mathcal{S}$ denotes the homotopy category of E-module spectra. Recall that the category $h\mathbb{S}$ is a symmetric-monoidal category, with the monoidal structure given by the cartesian product of symplectic manifolds. Since $\Omega(M \times N)$ is equivalent to $\Omega(M) \wedge_{\Omega} \Omega(N)$, the functor $\mathcal{F}_{\mathrm{E}}$ is monoidal.

Constructions analogous to the ones described above can be made in oriented stable symplectic homotopy category: $hs\mathbb{S}$. Furthermore, these constructions remain nontrivial when tensored with $\mathbb{Q}$. Henceforth, we will work over $\mathbb{Q}$. Now, given a commutative $s\Omega$-algebra E, and an element $\beta \in G(\mathrm{E})$ as defined in corollary 3.7, recall that we have an E-module automorphism:

$$\beta_*(M) : s\Omega(M)_{\mathrm{E}} \longrightarrow s\Omega(M)_{\mathrm{E}}.$$

Furthermore, β_* is monoidal, i.e. $\beta_*(M \times N) = \beta_*(M) \wedge \beta_*(N)$ under the monoidal structure of $hs\mathbb{S}$. In addition, from remark 3.9, we know that composition induces a map:

$$\mathcal{F}_{\mathrm{E}} : \pi_0(s\Omega(M,N)) \longrightarrow [s\Omega(M)_{\mathrm{E}}, s\Omega(N)_{\mathrm{E}}],$$

lands inside operators that commute with β_*. In other words, we see that β_* is a natural automorphism of the functor $\mathcal{F}_{\mathrm{E}}$. In particular, we see that the abelianized Grothendieck–Teichmüller groupscheme is a natural sub groupscheme of the motivic groupscheme of monoidal automorphisms of the functor $\mathrm{E} \longmapsto \mathcal{F}_{\mathrm{E}}$.

REMARK 3.13. *Recall that the map:*

$$\mathcal{F}_{\mathrm{E}} : \pi_0(s\Omega(M,N)) \longrightarrow [s\Omega(M)_{\mathrm{E}}, s\Omega(N)_{\mathrm{E}}]$$

is an equivalence if M is a compact manifold. In particular, $\mathcal{F}_{\mathrm{E}}$ is a rigid monoidal functor on the subcategory of $\pi_0(\mathbb{S})$ generated by compact symplectic manifolds (M,ω).

REMARK 3.14. *Consider the map* $\mathrm{BU} \longrightarrow \mathrm{THH}(s\Omega)$ *induced by the inclusion of the unit* $\mathrm{S} \longrightarrow s\Omega$*:*

$$\mathrm{BU} \longrightarrow \mathrm{B(U/SO)}_+ \longrightarrow s\Omega \wedge \mathrm{B(U/SO)}_+ = \mathrm{THH}(s\Omega).$$

This map may be shown to factor through $\mathrm{BU} \longrightarrow K(s\Omega)$ *lifting the Dennis trace map. This suggests a close relation between the Waldhausen K-theory of $s\Omega$ and the pro-abelian group scheme* **G**. *This is strikingly reminiscent of Kato's higher classfield theory* [**16**]*[Theorem 2.1], which relates the algebraic K-theory of higher local fields to the Galois groups of their abelian extensions.*

§3.5 Final remarks and speculation

In the sections that follow, we borrow notation from [**22**], where $\mathrm{S}[\mathrm{G}_+]$ denotes the suspension spectrum of G, viewed as a kind of group ring, for an H-space G.

Kontsevich's 1999 paper suggests that an action of $\mathsf{GT}_{\mathbb{Q}}$ defines a deformation of the complexified $\hat{A}$-genus, which can be interpreted as associated to the formal group law with $\Gamma(x)^{-1}$ as its exponential (where $\Gamma(x)$ denotes the Gamma-function, [**22**][§2.3.1]): We seek to generalize Kontsevich's result to our context. Let $\mathrm{MU} \longrightarrow \mathrm{MSO}$ denote the forgetful map, and let $\mathrm{MU} \longrightarrow \mathrm{S}[\mathrm{BU}_+] \wedge \mathrm{MU}$ denote the diagonal map. Then given a commutative ring spectrum E with an SO-orientation $\rho : \mathrm{MSO} \longrightarrow \mathrm{E}$, we have a map of ring spectra Γ_ρ:

$$\Gamma_\rho : \mathrm{MU} \longrightarrow \mathrm{S}[\mathrm{BU}_+] \wedge \mathrm{MU} \longrightarrow \mathrm{S}[\mathrm{B(U/SO)}_+] \wedge \mathrm{MSO} \longrightarrow \mathrm{S}[\mathrm{B(U/SO)}_+] \wedge \mathrm{E},$$

Note that the map Γ_ρ above can be expressed as a morphism:

$$\Gamma_\rho : \mathrm{MU} \longrightarrow \mathrm{THH}(s\Omega) \wedge_{s\Omega} \mathrm{E}\,.$$

Using this description, one may generalize Kontsevich's construction in our framework: namely, one may define a torsor of deformations of the ρ-orientation under the action of the group $\mathbf{G}(\mathrm{E})$ as follows: Given an element in $\mathbf{G}(\mathrm{E})$ represented by a ring map: $\beta : \mathrm{B(U/SO)}_+ \longrightarrow \mathrm{E}$, the corresponding deformation of ρ is given by capping β with Γ_ρ:

$$\rho_\beta = \beta \cap \Gamma_\rho : \mathrm{MU} \longrightarrow \mathrm{S}[\mathrm{B(U/SO)}_+] \wedge \mathrm{E} \longrightarrow \mathrm{E}\,.$$

There is a metaplectic analog of this whole picture. Recall ([**17**], §8), that the unitary group U admits a natural double cover $\tilde{\mathrm{U}}$ that supports the square-root of the determinant homomorphism. The forgetful map $\mathrm{U} \longrightarrow \mathrm{SO}$ lifts to a unique map $\tilde{\mathrm{U}} \longrightarrow \mathrm{Spin}$. Therefore, given a spectrum E with a Spin-orientation $\rho : \mathrm{MSpin} \longrightarrow \mathrm{E}$, we have the corresponding:

$$\tilde{\Gamma}_\rho : \mathrm{M}\tilde{\mathrm{U}} \longrightarrow \mathrm{THH}(s\tilde{\Omega}) \wedge_{s\tilde{\Omega}} \mathrm{E},$$

leading to a $\mathbf{G}(\mathrm{E})$-torsor of deformations of ρ as before.

REMARK 3.15. *Notice that the maps:*

$$\mathrm{M}\tilde{\mathrm{U}} \longrightarrow \mathrm{S}[\mathrm{B}(\tilde{\mathrm{U}}/\mathrm{Spin})_+] \wedge \mathrm{MSpin}, \qquad \mathrm{MU} \longrightarrow \mathrm{S}[\mathrm{B(U/SO)}_+] \wedge \mathrm{MSO}$$

are equivalences away from the prime two. In addition, the spaces $\mathrm{B}(\tilde{\mathrm{U}}/\mathrm{Spin})$, $\mathrm{B(U/SO)}$ *and* $(\mathrm{Sp/U})$ *are also equivalent away from two.*

An example of such a deformation is the one suggested by Kontsevich above: Let E denote complex K-theory with complex coefficients: $\mathrm{KU} \otimes \mathbb{C}$ supporting the $\hat{A}$-orientation (or rather, its scalar extension by $\mathbb{C}$): $\hat{A} : \mathrm{MSpin} \longrightarrow \mathrm{KU} \otimes \mathbb{C}$. In [**22**][§2.3.1]), we construct a deformation $\hat{A}_\zeta$ by specializing the generators of the polynomial algebra $\mathrm{H}_*(\mathrm{Sp/U}, \mathbb{Q})$ to odd zeta-values in $\mathrm{KU} \otimes \mathbb{C}$, graded using suitable powers of the Bott class. The genus associated to $\hat{A}_\zeta$ is therefore the composite:

$$\hat{A}_\zeta : \mathrm{MU}_* \longrightarrow \mathrm{H}_*(\mathrm{Sp/U}_+, \mathrm{KU} \otimes \mathbb{Q}) \longrightarrow \mathrm{KU}_* \otimes \mathbb{C}$$

which restricts to the $\hat{A}$-genus on MSp (which is MSO, or MSpin away from two), and sends manifolds of dimension $\equiv 2 \bmod 4$ to $i\mathbb{R} \subset \mathbb{C}$; more precisely, the primitives of $\mathrm{H}_*(\mathrm{Sp/U}_+, \mathbb{Q})$, interpreted as symmetric functions, are sent (as explained in the appendix below) to odd zeta-values, graded using the Bott class.

Let us now describe the geometry behind this deformation induced by the group $\mathbf{G}(\mathrm{E})$. First consider the spectrum $\mathrm{B(U/SO)}_+ \wedge \mathrm{MSO}$. Using the language used in Section §2, we may identify an element in

$$\pi_k(\mathrm{B(U/SO)}_+ \wedge \mathrm{MSO}) = \pi_k(\mathrm{THH}(s\Omega) \wedge_{s\Omega} \mathrm{MSO})$$

as the stabilization (with respect to the integer n) of the data given by a concordance class of oriented manifolds embedded in euclidean space: $M^{k+n} \subset \mathbb{R}^\infty \times \mathbb{R}^n$, that are proper over $\mathbb{R}^n$ and are endowed with a principal bundle of oriented lagrangian grassmannians which is classified by a map $\theta : M^{k+n} \longrightarrow \mathrm{B(U/SO)}$. In this language, the map $\Gamma_\rho : \pi_k(\mathrm{MU}) \longrightarrow \pi_k(\mathrm{THH}(s\Omega) \wedge_{s\Omega} \mathrm{MSO})$ described above identifies a stably almost complex manifold M^k with the underlying oriented manifold, endowed with the formal negative of the bundle ξ, where $\xi : s\underline{\mathcal{G}}(M) \longrightarrow M$ is the bundle of oriented lagrangians in the tangent bundle (see §2). Given an E-valued genus ρ, the action of $\mathbf{G}(\mathrm{E})$, in the language used above, corresponds to deforming ρ along different choices of multiplicative E-theory characteristic classes of the bundle θ. The analogous statements hold in the metaplectic case.

REMARK 3.16. *It is a compelling question to ask about the meaning of these (lagrangian-bundle) deformations in terms of elliptic differential operators (like the Dirac operator) that define genera. This framework bears a striking resemblance to the construction of the analytic torsion classes* [**6**] *and one would like to know if there is any relation.*

4. Appendix: Grothendieck-Teichmüller groups

We recall some of the complex history [**1**](§25.9) of this subject, which has deep connections to homotopy theory but which may be unfamiliar.

For topologists, the operad defined by Artin's braid groups is a good place to start: it has few automorphisms, but its completions are less rigid. Ihara has studied its system of profinite completions, but for our purposes Drinfeld's work [**13**] on the automorphism group of its system of Malcev $\mathbb{Q}$-completions will be more relevant. Kontsevich compared the latter object to the group of homotopy automorphisms of the rational chains on the little disk operad [**14**], and suggested [**19**][§4.4] that both these objects are isomorphic to the motivic Galois group of a certain [**12**] Tannakian category of mixed Tate motives.

We follow Kontsevich's lead here, and refer to all these conjecturally equivalent pro-algebraic $\mathbb{Q}$-groupschemes as **the** Grothendieck-Teichmüller group $\mathsf{GT}_{\mathbb{Q}}$; we take it to be an extension of the form:

$$1 \to \mathfrak{F} \longrightarrow \mathsf{GT}_{\mathbb{Q}} \longrightarrow \mathbb{G}_m \to 1$$

with $\mathfrak{F}$ pro-unipotent, defined by a graded Lie algebra $\mathfrak{f}$ free on generators z_{2i+1}, $i > 0$, associated to the generators of the rank one abelian groups $\mathrm{K}^{\mathrm{alg}}_{4n+1}(\mathbb{Z})$ through the manifestation of $\mathsf{GT}_{\mathbb{Q}}$ as the motivic group of a $\mathbb{Q}$-linear category $\mathsf{MTM}_{\mathbb{Z}}$ of mixed Tate motives, generated by certain cell-like objects $\mathbb{Z}(n)$ satisfying

$$\mathrm{Ext}^*_{\mathsf{MTM}}(\mathbb{Z}(0), \mathbb{Z}(n)) = \mathbb{Q} \text{ if } * = 0 \text{ and } n = 0, \quad = \mathrm{K}^{\mathrm{alg}}_{2n+1}(\mathbb{Z}) \otimes \mathbb{Q} \text{ if } * = 1,$$

these groups being zero otherwise. The derived group $\mathfrak{F}$ of $\mathsf{GT}_{\mathbb{Q}}$ has, as its abelianization, a groupscheme represented by a commutative and cocommutative Hopf algebra over $\mathbb{Q}$, with primitives in topological degree 4k+2, corresponding to $\mathrm{H}_*(\mathrm{Sp}/\mathrm{U}; \mathbb{Q})$. Borel's work on regulators thus relates the generators z_{2i+1} to the (conjecturally transcendental) zeta-values $\zeta(2i+1)$.

The affine groupscheme $\mathfrak{F}$ has a $\mathbb{G}_m$-action, and the associated commutative Hopf algebra of functions $\mathsf{QSymm} \otimes \mathbb{Q}$ is dual to the universal enveloping algebra of its Lie algebra $\mathfrak{f}$. That is the free associative $\mathbb{Q}$-algebra:

$$U(\mathfrak{f}) = \mathbb{Q}\langle\langle z_{2i+1} \mid i > 0 \rangle\rangle = \mathsf{NSymm} \otimes \mathbb{Q}$$

of rational noncommutative symmetric functions [**4**, **9**], with diagonal defined by the juxtaposition coproduct. The (self-dual) Hopf algebra Symm of classical symmetric functions is dual to the abelianization of NSymm: the abelianization $\mathfrak{f} \to \mathfrak{f}_{\mathrm{ab}}$ defines a homomorphism $U(\mathfrak{f}) \to U(\mathfrak{f}_{\mathrm{ab}})$ dual to the inclusion of the symmetric functions in the quasisymmetric ones, defining a quotient

$$1 \to \mathfrak{F}_{\mathrm{ab}} \longrightarrow \widetilde{\mathsf{GT}}_{\mathbb{Q}} \longrightarrow \mathbb{G}_m \to 1$$

of $\mathsf{GT}_{\mathbb{Q}}$ with $\mathfrak{f}_{\mathrm{ab}}$ as its (graded abelian) Lie algebra.

One consequence of deep work of Hatcher, Waldhausen, Bökstedt, Rognes and others is a rational equivalence [**21**](§3)

$$(\mathrm{S} \vee \Sigma\, \mathrm{kO})_{\mathbb{Q}} \longrightarrow \mathrm{K}(\mathrm{S})_{\mathbb{Q}}$$

of spectra, which yields a canonical identification

$$\mathrm{S}_{\mathbb{Q}} \wedge^{L}_{K(\mathrm{S})} \mathrm{S}_{\mathbb{Q}} \;\cong\; \mathsf{QSymm}_{\mathbb{Q}}$$

of the rational covariant Koszul dual of Waldhausen's K-theory of the sphere spectrum with a model for the Hopf algebra of functions on $\mathsf{GT}_{\mathbb{Q}}$. The analogous Koszul dual of the stabilization morphism

$$\mathrm{S}[\mathrm{SU}/\mathrm{SO}] \longrightarrow \mathrm{S} \wedge \Sigma\, \mathrm{kO}$$

of ringspectra defines a morphism

$$\mathrm{S}_{\mathbb{Q}} \wedge^{L}_{\mathrm{S}[\mathrm{SU}/\mathrm{SO}]} \mathrm{S}_{\mathbb{Q}} \;\cong\; \mathsf{Symm}_{\mathbb{Q}} \to \mathsf{QSymm}_{\mathbb{Q}}$$

representing the quotient $\mathsf{GT}_{\mathbb{Q}} \to \widetilde{\mathsf{GT}}_{\mathbb{Q}}$.

The map that sends the Newton's power sums

$$N_n \;:=\; \sum_{k \geq 1} x_k^n \in \mathsf{Symm} \subset \mathsf{QSymm}$$

to the real number given by the Riemann-zeta value $\zeta(n)$, under the specialization that sends $x_k \mapsto 1/k$ can, with some care, be extended from Symm to a ring homomomorphism:

$$\boldsymbol{\zeta} : \mathsf{QSymm} \longrightarrow \mathrm{MZV} \subset \mathbb{R}$$

with values in a certain graded algebra of real multizeta values[5].

Remarkably enough, these multizeta values play an important role in Connes, Kreimer, and Marcolli's Galois-theoretic reinterpretation [**10**, **11**] of the classical BPS renormalization theory of Feynman integrals, in which MZVs appear ubiquitously in explicit computations. Kontsevich found an action [**19**][§4.6 Th 9] of the abelianization of $\mathsf{GT}_{\mathbb{Q}}$ on a moduli space for deformation quantizations of Poisson manifolds, through an action of the little disk operad on Hochschild homology. These developments led Cartier [**8**] to suggest that $\mathsf{GT}_{\mathbb{Q}}$ is in some sense a *cosmic* Galois group.

REMARK 4.1. *As noted in Remark* 3.10, *Kontsevich* [**18**]*[§4.6.2, 8.4] shows that the graded cohomology algebra*

$$\mathrm{H}^*(M, \Lambda^* T)$$

of polyvector fields on a complex manifold is isomorphic to a kind of Hochschild cohomology

$$\mathrm{HH}^*(M) \;=\; \mathrm{Ext}^*_{\mathcal{O}_{M \times M}}(\mathcal{O}_M, \mathcal{O}_M)$$

defined in terms of the coherent sheaf of holomorphic functions on $M \times M$ and its diagonal. If M is Calabi-Yau, we can identify its tangent and cotangent bundle, obtaining an isomorphism of both cohomologies with the Hodge cohomology of M, defining an action of (some version of) the Grothendieck-Teichmüller group on the de Rham cohomology of M. The homotopy-theoretic point of view suggests the Calabi-Yau hypothesis may be unnecessarily strong.

[5]The case $n = 1$, ie $i = 0$, which needs special treatment [**7**], is interestingly absent from the constructions considered above.

References

[1] Y. André, *Une introduction aux motifs (motifs purs, motifs mixtes, périodes)* (French, with English and French summaries), Panoramas et Synthèses [Panoramas and Syntheses], vol. 17, Société Mathématique de France, Paris, 2004. MR2115000

[2] C. Ausoni and J. Rognes, *Rational algebraic K-theory of topological K-theory*, Geom. Topol. **16** (2012), no. 4, 2037–2065. MR2975299

[3] D. Ayala, *Geometric cobordism categories*, ProQuest LLC, Ann Arbor, MI, 2009. Thesis (Ph.D.)–Stanford University. MR2713365

[4] A. Baker and B. Richter, *Quasisymmetric functions from a topological point of view*, Math. Scand. **103** (2008), no. 2, 208–242. MR2484353

[5] A. J. Blumberg, *Topological Hochschild homology of Thom spectra which are E_∞ ring spectra*, J. Topol. **3** (2010), no. 3, 535–560. MR2684512

[6] J.-M. Bismut and J. Lott, *Flat vector bundles, direct images and higher real analytic torsion* (English, with English and French summaries), J. Amer. Math. Soc. **8** (1995), no. 2, 291–363. MR1303026

[7] P. Cartier, *Fonctions polylogarithmes, nombres polyzêtas et groupes pro-unipotents* (French, with French summary), Astérisque **282** (2002), Exp. No. 885, viii, 137–173. Séminaire Bourbaki, Vol. 2000/2001. MR1975178

[8] P. Cartier, *A mad day's work: from Grothendieck to Connes and Kontsevich. The evolution of concepts of space and symmetry [in Les relations entre les mathématiques et la physique théorique, 23–42, Inst. Hautes Études Sci., Bures-sur-Yvette, 1998; MR1667896 (2000c:01028)]*, Bull. Amer. Math. Soc. (N.S.) **38** (2001), no. 4, 389–408. Translated from the French by Roger Cooke. MR1848254

[9] P. Cartier, *A primer of Hopf algebras*, in Frontiers in number theory, physics, and geometry. II, 537 - 615, Springer (2007).

[10] A. Connes and M. Marcolli, *Renormalization, the Riemann-Hilbert correspondence, and motivic Galois theory*, Frontiers in number theory, physics, and geometry. II, Springer, Berlin, 2007, pp. 617–713. MR2290770

[11] Y. André, *Galois theory, motives and transcendental numbers*, Renormalization and Galois theories, IRMA Lect. Math. Theor. Phys., vol. 15, Eur. Math. Soc., Zürich, 2009, pp. 165–177. MR2588609

[12] P. Deligne and A. B. Goncharov, *Groupes fondamentaux motiviques de Tate mixte* (French, with English and French summaries), Ann. Sci. École Norm. Sup. (4) **38** (2005), no. 1, 1–56. MR2136480

[13] V. G. Drinfel′d, *On quasitriangular quasi-Hopf algebras and on a group that is closely connected with* $\mathrm{Gal}(\overline{\mathbf{Q}}/\mathbf{Q})$ (Russian), Algebra i Analiz **2** (1990), no. 4, 149–181; English transl., Leningrad Math. J. **2** (1991), no. 4, 829–860. MR1080203

[14] B. Fresse, *Homotopy of operads and Grothendieck-Teichmüller groups*, available at `http://math.univ-lille1.fr/~fresse/OperadHomotopyBook/`

[15] G. Segal, *A geometric perspective on quantum field theory*, Algebraic topology: applications and new directions, Contemp. Math., vol. 620, Amer. Math. Soc., Providence, RI, 2014, pp. 281–294. MR3290096

[16] K. Kato, Generalized class-field theory, available at `http://www.mathunion.org/ICM/ICM1990.1/Main/icm1990.1.0419.0428.ocr.pdf`

[17] N. Kitchloo, *The stable symplectic category and quantization*, Algebraic topology: applications and new directions, Contemp. Math., vol. 620, Amer. Math. Soc., Providence, RI, 2014, pp. 251–279. MR3290095

[18] M. Kontsevich, *Deformation quantization of Poisson manifolds*, Lett. Math. Phys. **66** (2003), no. 3, 157–216. MR2062626

[19] M. Kontsevich, *Operads and motives in deformation quantization*, Lett. Math. Phys. **48** (1999), no. 1, 35–72. Moshé Flato (1937–1998). MR1718044

[20] I. G. Macdonald, *Symmetric functions and Hall polynomials*, 2nd ed., Oxford Mathematical Monographs, The Clarendon Press, Oxford University Press, New York, 1995. With contributions by A. Zelevinsky; Oxford Science Publications. MR1354144

[21] J. Morava, *Homotopy-theoretically enriched categories of noncommutative motives*, Res. Math. Sci. **2** (2015), Art. 8, 16. MR3355473

[22] J. Morava, *An integral lift of the Γ-genus*, available at `arXiv:1101.1647`.

[23] A. Weinstein, *Symplectic categories*, Port. Math. **67** (2010), no. 2, 261–278. MR2662868
[24] A. Weinstein, *Symplectic geometry*, Bull. Amer. Math. Soc. (N.S.) **5** (1981), no. 1, 1–13. MR614310

Department of Mathematics, Johns Hopkins University, Baltimore, Maryland 21218
Email address: nitu@math.jhu.edu

Department of Mathematics, Johns Hopkins University, Baltimore, Maryland 21218
Email address: jack@math.jhu.edu

Contemporary Mathematics
Volume **708**, 2018
http://dx.doi.org/10.1090/conm/708/14267

Universal Gysin formulas for the universal Hall-Littlewood functions

Masaki Nakagawa and Hiroshi Naruse

ABSTRACT. It is known that the usual Schur S- and P-polynomials can be described via the Gysin homomorphisms for flag bundles in the ordinary cohomology theory. Recently, P. Pragacz generalized these *Gysin formulas* to the Hall-Littlewood polynomials. In this paper, we introduce a *universal* analogue of the Hall-Littlewood polynomials, which we call the *universal Hall-Littlewood functions*, and give Gysin formulas for various flag bundles in the complex cobordism theory. Furthermore, we give two kinds of the *universal* analogue of the Schur polynomials, and some Gysin formulas for these functions are established.

CONTENTS

1. Introduction

1.1. Gysin formulas. Any continuous map $f : X \longrightarrow Y$ between topological spaces defines *pull-back* homomorphism $f^* : H^i(Y) \longrightarrow H^i(X)$ in cohomology, and *push-forward* homomorphism $f_* : H_i(X) \longrightarrow H_i(Y)$ in homology for all $i \in \mathbb{Z}$. If X is a compact oriented smooth manifold of dimension m, the m-th homology group $H_m(X)$ is isomorphic to $\mathbb{Z}$, with a generator $[X]$ the *fundamental class* of X, and the following *Poincaré duality map*

$$\mathcal{P}_X : H^i(X) \longrightarrow H_{m-i}(X),\ \alpha \longmapsto \alpha \cap [X]$$

2010 *Mathematics Subject Classification.* Primary 05E05, 55N20; Secondary 14M15, 14N15, 55N22, 57R77.

Key words and phrases. Schur S-, P-, and Q-functions, Hall-Littlewood function, complex-oriented generalized cohomology theory, Gysin map.

The first author is partially supported by the Grant-in-Aid for Scientific Research (C) 15K04876, Japan Society for the Promotion of Science.

The second author is partially supported by the Grant-in-Aid for Scientific Research (B) 16H03921, Japan Society for the Promotion of Science.

is an isomorphism. If $f : X \longrightarrow Y$ is a smooth map, with X and Y compact oriented smooth manifolds of dimensions m and n respectively, then we get a push-forward homomorphism in cohomology:

$$f_* : H^i(X) \xrightarrow{\sim} H_{m-i}(X) \xrightarrow{f_*} H_{m-i}(Y) \xleftarrow{\sim} H^{i-(m-n)}(Y).$$

This map is called a *Gysin map* (*homomorphism*), *push-forward*, or *Umkehr map*. Intuitively the intersection of cycles in homology is turned into the product of cohomology classes by means of Poincaré duality. This conversion, together with the computation of the Gysin maps, enables us to make some geometric problems into algebraic computations. For example, the computation of the Gysin map for various flag bundles is used to determine the cohomology class corresponding to a Schubert variety (see e.g., Akyildiz [**2**], Damon [**17**, Theorem 3 (Chern's formula)]). Furthermore, the similar computation is applied to determining the cohomology class of *degeneracy locus* (see e.g., Damon [**17**, Corollary 3], Fulton [**23**, Chapter 14], Porteous [**58**, p.298]). Thus the computation of various Gysin maps has a lot of applications in geometry, and there are many formulas describing Gysin maps. These formulas are called *Gysin formulas* or *push-forward formulas* in general. Although we do not intend to survey these formulas thoroughly here, we shall quote some results related to our work:

- Gysin formulas for flag bundles, Grassmann bundles, or projective bundles are described in Borel-Hirzebruch [**7**, §8], [**8**, §20], Buch [**14**, §7], Damon [**16**], [**17**], Darondeau-Pragacz [**18**], Fel'dman [**22**, §4], Fulton [**23**, §14.2], Fulton-Pragacz [**25**, Chapter IV, Appendices E, F], Harris-Tu [**27**, §2], Ilori [**36**], Kajimoto-Sugawara [**41**], Pragacz [**59**, §2], [**61**, §4], [**62**], Quillen [**63**], Sugawara [**66**], Tu [**69**], [**70**], Vishik [**71**, §5.7].
- For a connected complex (semi-simple) Lie group $G_{\mathbb{C}}$ with a Borel subgroup B and a parabolic subgroup P containing B, Gysin formulas for the natural projection $G_{\mathbb{C}}/B \longrightarrow G_{\mathbb{C}}/P$ are described in Akyildiz [**2**], Akyildiz-Carrell [**3**], [**4**], Fulton-Pragacz [**25**, Appendix E], Brion [**12**], Kajimoto [**40**].

As we mentioned above, most of these formulas are formulated in the ordinary cohomolgy (or Chow) theory, and many different approaches such as the *residue symbol* (Damon [**16**]), the *zeros of holomorphic vector fields on flag varieties* (Akyildiz-Carrell [**3**], [**4**]), *representation theory* (Brion [**12**]), and the *equivariant localization formula of Atiyah-Bott-Berline-Vergne for a torus action* (Tu [**69**], [**70**]) have been used to prove them.

On the other hand, it is known that the ordinary cohomology theory is a special case of *complex-oriented* generalized cohomology theories which corresponds to an *additive* formal group law (see Example 2.2 (1)). Therefore it is natural to ask if the above Gysin formulas in the ordinary cohomology theory can be generalized to any complex-oriented generalized cohomology theory. For example, Buch [**14**, §7] proved a K-theoretic analogue of Pragacz's Gysin formula [**59**]. Quillen [**63**] stated (without proof) the Gysin formula for a projective bundle in the complex cobordism theory. As shown by Quillen [**63**], the complex cobordism theory $MU^*(-)$ is *universal* among complex-oriented generalized cohomology theories, and therefore it is desirable to formulate these Gysin formulas in the complex cobordism theory. The first main purpose of this paper is to establish the Gysin formulas for general flag bundles in the complex cobordism theory.[1] One of our main results, Theorem 4.10, is a generalization of Pragacz's result [**62**, Proposition 5] to the complex cobordism theory. Note that Pragacz derived his formula from Brion's result [**12**, Proposition 2.1] which is proved by the Weyl character formula and the Grothendieck-Riemann-Roch theorem. Here we adopt a topological approach which was developed by Bressler-Evens [**10**, §1]. Main tools are the *Becker-Gottlieb transfer* [**6**, Theorem 4.3] and the *Brumfiel-Madsen formula* [**13**, Theorem 3.5]. It should be remarked that most of our results in this paper seems to be valid in the *algebraic cobordism theory* due to Levine-Morel [**48**].

1.2. Gysin formulas for Schur functions. In the previous subsection §1.1, we collected various Gysin formulas related to our work. Some formulas involve symmetric functions such as *Schur S-* and *P-functions* (see e.g., Fulton-Pragacz [**25**, Chapter IV], Pragacz [**59**, §2], [**61**, §4]). In order to clarify what we are considering, we shall give one typical example: Let $E \overset{p}{\longrightarrow} X$ be a complex vector bundle of rank n over a variety X. Let $\pi : G^1(E) \longrightarrow X$ be the associated Grassmann bundle of hyperplanes in E.[2] On $G^1(E)$, we have the *tautological exact sequence* of vector bundles:

$$0 \longrightarrow S \hookrightarrow \pi^*(E) \twoheadrightarrow Q \longrightarrow 0.$$

Let $\xi := c_1(Q) \in H^2(G^1(E))$ be the first Chern class of the line bundle Q, and $x_1 := \xi, x_2, \ldots, x_n$ be the *Chern roots* of E.[3] Then as for the Gysin homomorphism $\pi_* : H^*(G^1(E)) \longrightarrow H^*(X)$, it is well-known that

$$\pi_*(\xi^k) = s_{k-n+1}(E) \quad (k \geq 0), \tag{1.1}$$

where $s_i(E)$ is the i-th *Segre class* of E (see e.g., Fulton-Pragacz [**25**, §4.1]). Since the Chern classes $c_i(E)$ can be identified with the i-th elementary symmetric polynomial $e_i(\mathbf{x}_n)$ in $\mathbf{x}_n = (x_1, \ldots, x_n)$,[4] the Segre class $s_j(E)$ can be identified with the j-th homogeneous complete symmetric polynomial $h_j(\mathbf{x}_n)$, which is nothing but the Schur S-polynomial $s_{(j)}(\mathbf{x}_n)$ corresponding to the "one-row" (j). Therefore the formula (1.1) can be interpreted as[5]

$$\pi_*(x_1^k) = s_{(k-n+1)}(\mathbf{x}_n). \tag{1.2}$$

The formula (1.2) can be generalized to the full flag bundle $\tau : \mathcal{F}\ell(E) \longrightarrow X$ as follows: Let $\lambda = (\lambda_1, \lambda_2, \ldots, \lambda_n)$ $(\lambda_1 \geq \lambda_2 \geq \cdots \geq \lambda_n \geq 0)$ be a partition of length $\leq n$. Then Fulton-Pragacz [**25**, (4.1)] (see also Pragacz [**59**, Lemma 2.3], [**61**, Proposition 4.4], [**62**, Example 8]) showed the following formula[6]:

$$\tau_*(x_1^{\lambda_1+n-1} x_2^{\lambda_2+n-2} \cdots x_n^{\lambda_n}) = s_\lambda(\mathbf{x}_n). \tag{1.3}$$

The formula (1.3) gives the usual Schur S-polynomial as the push-forward image of the Gysin map from the full flag bundle, and is called the *Jacobi-Trudi identity*

[1] *Universal Gysin formulas* in the title of this paper signifies Gysin formulas in the complex cobordism theory which is *universal* among all complex-oriented generalized cohomology theories.

[2] One can naturally identify $G^1(E)$ with the associated projective bundle $P(E^\vee)$ of lines in the dual bundle $E^\vee$.

[3] By the *splitting principle*, the vector bundle E splits into the sum of line bundles when pulled-back to the *full flag bundle* $\mathcal{F}\ell(E)$ via the projection $\tau : \mathcal{F}\ell(E) \longrightarrow X$. The Chern roots of E are the first Chern classes of these line bundles on $\mathcal{F}\ell(E)$.

[4] by means of $\tau^* : H^*(X) \longrightarrow H^*(\mathcal{F}\ell(E))$, which is known to be injective.

[5] Strictly speaking, this formula should be considered in $H^*(\mathcal{F}\ell(E))$.

[6] This formula is also considered in $H^*(\mathcal{F}\ell(E))$.

in Fulton-Pragacz [**25**, §4.1, p.42]. The essentially identical formula has been obtained by many authors (Damon [**17**, Corollary 2], Harris-Tu [**27**, Proposition 2.3], Manivel [**51**, Exercise 3.8.3], Pragacz [**61**, Proposition 4.4], [**62**, Example 8], Sugawara [**66**, Theorem A]). An analogous formula for Schur P-polynomial is also given by Pragacz [**59**, Corollary 2.7], [**62**, Example 11]. Recently, Pragacz [**62**] succeeded in generalizing the above Gysin formulas for Schur S- and P-polynomials to the *Hall-Littlewood polynomials*, which interpolate between Schur S- and P-polynomials (see Macdonald [**49**, Chapter III, §2]).

On the other hand, it is well-known that the usual Schur S-polynomials (resp. P-polynomials) *represent* the *Schubert classes* of the complex Grassmannian (resp. Lagrangian Grassmannian) in the ordinary cohomology (see e.g., Fulton [**24**, §9.4], Pragacz [**60**, §6]). In order to generalize the above facts to other cohomology theories such as K-theory, complex cobordism theory (or algebraic cobordism theory) and their torus equivariant versions, various generalizations, analogues, and deformations of Schur functions have been introduced by many authors. We shall quote some of them for convenience of the readers (in the following, λ (resp. ν) is understood to be a partition (resp. strict partition) of length $\leq n$, and $\mathbf{x}_n = (x_1, \ldots, x_n)$ is a sequence of n independent variables, and $\mathbf{b} = (b_1, b_2, \ldots)$ is a sequence of "deformation parameters"):

- The *factorial Schur polynomials* $s_\lambda(\mathbf{x}_n|\mathbf{b})$ (see Ikeda-Naruse [**33**, §5.1], Macdonald [**49**, Chapter 1, Example 20], [**50**, 6th variation], Molev-Sagan [**55**]) represent the Schubert classes of the torus equivariant cohomology of the complex Grassmannian (see Knutson-Tao [**43**, §6], Ikeda-Naruse [**33**, Theorem 5.4]).
- The *factorial Schur* P- and Q-*polynomials* $P_\nu(\mathbf{x}_n|\mathbf{b})$, $Q_\nu(\mathbf{x}_n|\mathbf{b})$ introduced by Ivanov [**37**, Definitions 2.10 and 2.13] (see also Ikeda-Mihalcea-Naruse [**34**, §4.2]) represent the Schubert classes of the torus equivariant cohomology of the orthogonal and Lagrangian Grassmannians (see Ikeda [**32**, Theorem 6.2], Ikeda-Naruse [**33**, Theorem 8.7]).
- The *factorial Grothendieck polynomials* $G_\lambda(\mathbf{x}_n|\mathbf{b})$ introduced by McNamara [**52**, Definition 4.1] (see also Ikeda-Naruse [**35**, (2.13), (2.14)]) represent the Schubert classes of the torus equivariant K-theory of the complex Grassmannian (see Ikeda-Naruse [**35**]).
- The K-*theoretic factorial* P- and Q-*polynomials* $GP_\nu(\mathbf{x}_n|\mathbf{b})$, $GQ_\nu(\mathbf{x}_n|\mathbf{b})$ introduced by Ikeda-Naruse [**35**, Definition 2.1] represent the Schubert classes of the torus equivariant K-theory of the orthogonal and Lagrangian Grassmannians (see Ikeda-Naruse [**35**, Theorem 8.3]).
- The *universal factorial Schur* (S-) *functions*[7] $s^{\mathbb{L}}_\lambda(\mathbf{x}_n|\mathbf{b})$, P- and Q-*functions* $P^{\mathbb{L}}_\nu(\mathbf{x}_n|\mathbf{b})$, $Q^{\mathbb{L}}_\nu(\mathbf{x}_n|\mathbf{b})$ were introduced by the authors (see [**56**, Definitions 4.1 and 4.10] or (3.1) and (3.4) in this paper). These functions are *universal* analogue of the above polynomials.

Thus our second main purpose of this paper is the introduction of the *universal* analogue of the Hall-Littlewood polynomials, and to establish the Gysin formulas for various Schur functions in the complex cobordism theory. The universal analogue of the Hall-Littlewood polynomials denoted by $H^{\mathbb{L}}_\lambda(\mathbf{x}_n; t)$ are defined in Definition 3.2, and we call them the *universal Hall-Littlewood functions*. Then our main result

[7]Notice that $s_\lambda(\mathbf{x}_n|\mathbf{b})$ and $G_\lambda(\mathbf{x}_n|\mathbf{b})$ are polynomials in $\mathbf{x}_n$, whereas $s^{\mathbb{L}}_\lambda(\mathbf{x}_n|\mathbf{b})$ is a formal power series in $\mathbf{x}_n$.

in this direction is Corollary 4.11 which gives the universal Hall-Littlewood function as the push-forward image of the Gysin map from a partial flag bundle, thus generalizing Pragacz's result (see Corollary 4.4). The Jacobi-Trudi identity (1.3) can also be formulated in the complex cobordism theory (see Corollary 4.8). With regard to the universal Schur functions $s_\lambda^{\mathbb{L}}(\mathbf{x}_n)$ and the universal Hall-Littlewood functions $H_\lambda^{\mathbb{L}}(\mathbf{x}_n; t)$, a comment is in order: The usual Hall-Littlewood polynomial denoted by $P_\lambda(x_1, \ldots, x_n; t)$ in Macdonald's book [**49**, Chapter III, §2] reduces to the usual Schur polynomial $s_\lambda(x_1, \ldots, x_n)$ under the specialization $t = 0$. However, we found that our $H_\lambda^{\mathbb{L}}(\mathbf{x}_n; t)$ does not necessarily reduce to the universal Schur function $s_\lambda^{\mathbb{L}}(\mathbf{x}_n)$ when $t = 0$. Thus the specialization $H_\lambda^{\mathbb{L}}(\mathbf{x}_n; 0)$ gives another universal analogue of Schur functions, which we call the *new universal Schur functions.*[8] We shall discuss these new functions denoted $\mathbb{S}_\lambda^{\mathbb{L}}(\mathbf{x}_n)$ (and their factorial version denoted $\mathbb{S}_\lambda^{\mathbb{L}}(\mathbf{x}|\mathbf{b})$) separately in the final section §5. Among our results in this section are Theorem 5.6 which reveals the difference between the "old" and the "new" universal factorial Schur functions. As an application of the Gysin formulas for the new universal factorial Schur functions, we formulate the *Thom-Porteous formula* in the complex cobordism theory (see Theorem 5.11). Moreover, we shall show that the new universal factorial Schur functions *represent* the Schubert classes of the complex Grassmannian in the complex cobordism theory (see Theorem 5.12).

1.3. Organization of the paper. The paper is organized as follows: In Section 2, we shall give topological preliminaries needed to develop our work. The key concepts are *complex-oriented generalized cohomology theory*, *Gysin maps*, *Becker-Gottlieb transfer*, *Brumfiel-Madsen formula*, *Bressler-Evens formula.* Various types of *Gysin formulas* are reviewed at the end of this section. Section 3 is devoted to the introduction of the *universal* analogue of the usual Schur S-, P-, Q-, and Hall-Littlewood polynomials. Especially, the *universal Hall-Littlewood functions*, which are the central theme of this paper, are introduced. In Section 4, after reviewing the Gysin formulas for various Schur functions, we shall give the *universal* analogues of these formulas. In order to establish these formulas, the Bressler-Evens formula plays the crucial role. In Section 5, we introduce the *new universal factorial Schur functions.* If we set all the deformation parameters to be 0, the *new universal Schur functions* can be obtained, and these functions coincide with the universal Hall-Littlewood functions under the specialization $t = 0$. We also give some Gysin formulas for these new universal Schur functions.

Acknowledgments. We would like to thank Takeshi Ikeda, Thomas Hudson, Tomoo Matsumura for helpful comments and valuable conversations. Especially, Tomoo Matsumura kindly explained his recent work (Hudson-Matsumura [**31**]) to us, that is closely related to our current work.

2. Topological preliminaries

2.1. Complex-oriented generalized cohomology theory. A *generalized cohomology theory* $h^*(-) = \bigoplus_{n\in\mathbb{Z}} h^n(-)$ is a contravariant functor from the category $\mathcal{CW}$ of CW complexes to the category of graded abelian groups which satisfies all the axioms of Eilenberg-Steenrod [**21**, Chapter I, 3c] except the *dimension axiom.* Thus $h^n(\mathrm{pt})$ is not necessarily zero even if $n \neq 0$. Here pt means a space

[8] It is desirable to give these functions more specific name which characterize them (a proposal will be given in §5.5).

consisting of a single point. The cohomology $h^* := h^*(\mathrm{pt})$ of a point is called the *coefficient group*. In what follows, we assume that the theory $h^*(-)$ is *multiplicative*, that is, for CW pairs (X, A) and (Y, B), there exists an *external product*

$$h^k(X, A) \otimes h^l(Y, B) \longrightarrow h^{k+l}((X, A) \times (Y, B)) \quad k, l \in \mathbb{Z},$$

that satisfies certain axioms (see e.g., Dold [**20**, §4]). Under this assumption, h^* becomes a graded-commutative ring, and for a space X, the cohomology ring $h^*(X)$ has an h^*-module structure. Furthermore, if we consider the *infinite* CW complexes, we need suitable axioms about limits such as the *additivity axiom* or *wedge axiom* due to Milnor [**53**]. In what follows, when we refer to a *generalized cohomology theory*, it means a multiplicative generalized cohomology theory defined on $\mathcal{CW}$ satisfying the additivity axiom unless otherwise stated. Let $\tilde{h}^*(-)$ denote the corresponding *reduced* cohomology theory,[9] and let $j : \mathbb{C}P^1 \hookrightarrow \mathbb{C}P^\infty$ be the canonical inclusion of $\mathbb{C}P^1 \approx S^2$ into the infinite complex projective space.

DEFINITION 2.1 (Adams [**1**], Part II, p.37; Switzer [**67**], §16.27). A generalized cohomology theory $h^*(-)$ is called complex-orientable if there exists an element $x^h \in \tilde{h}^2(\mathbb{C}P^\infty)$ such that $j^*(x^h)$ is a generator of $\tilde{h}^2(\mathbb{C}P^1) \cong \tilde{h}^2(S^2) \cong \tilde{h}^0(S^0) \cong h^0(\mathrm{pt}) = h^0$.

If this element x^h is specified, then $h^*(-)$ is said to be *complex-oriented*, and x^h is called the *orientation class*. Then it is known that the cohomology ring of the infinite projective space $\mathbb{C}P^\infty$ is $h^*(\mathbb{C}P^\infty) = h^*[[x^h]]$, a formal power series ring with the given generator $x^h \in \tilde{h}^2(\mathbb{C}P^\infty)$ (see Adams [**1**, Part II, Lemma 2.5]).

Complex-orientability implies a lot of useful properties. We recall here some of them.

2.1.1. *Chern classes.* For a complex vector bundle $E \xrightarrow{p} X$, one can define the *h^*-theory Chern classes* $c_i^h(E) \in h^{2i}(X)$ $(i = 0, 1, 2, \ldots, n = \operatorname{rank} E)$ by the usual Grothendieck's method (see Conner-Floyd [**15**, Theorem 7.6], Grothendieck [**26**, §3], Switzer [**67**, Theorem 16.2]). The total Chern class of E is given by $c^h(E) := \sum_{i=0}^n c_i^h(E)$. Then the usual Whitney product formula is given by $c^h(E \oplus F) = c^h(E) \cdot c^h(F)$ for two complex vector bundles E, F.

2.1.2. *Formal group law.* If L, M are complex line bundles over X, then

$$c_1^h(L \otimes M) = F_h(c_1^h(L), c_1^h(M)),$$

where

$$F_h(X, Y) = X + Y + \sum_{i,j \geq 1} a_{i,j}^h X^i Y^j \in h^*[[X, Y]] \quad (a_{i,j}^h \in h^{2(1-i-j)})$$

is a (one dimensional commutative) formal group law over the graded ring h^* associated with the cohomology theory $h^*(-)$. Then the formal power series $F_h(X, Y)$ satisfies the conditions

(i) $F_h(X, 0) = X$, $F_h(0, Y) = Y$,
(ii) $F_h(X, Y) = F_h(Y, X)$,
(iii) $F_h(X, F_h(Y, Z)) = F_h(F_h(X, Y), Z)$.

[9] For a CW-complex X with base point x_0, the reduced cohomology $\tilde{h}^*(X)$ is defined to be $h^*(X, x_0)$.

We shall use this formal group law to define the *formal sum*, *formal inverse*, and *formal subtraction*. For two indeterminates X, Y, the formal sum $X +_F Y$ is defined as

$$X +_F Y := F_h(X, Y) = X + Y + \sum_{i,j \geq 1} a^h_{i,j} X^i Y^j \in h^*[[X, Y]].$$

Denote by

$$[-1]_F(X) = \iota_F(X) = \overline{X} = -X + \sum_{j \geq 2} c^h_j X^j \in h^*[[X]]$$

the formal inverse series. To be precise, $[-1]_F(X)$ is the unique formal power series satisfying the condition $F_h(X, [-1]_F(X)) \equiv 0$, or equivalently $X +_F [-1]_F(X) = 0$. This formal inverse allows us to define the formal subtraction:

$$X -_F Y := X +_F [-1]_F(Y) = X +_F \overline{Y}.$$

Finally, we define $[0]_F(X) := 0$, and inductively,

$$[n]_F(X) := [n-1]_F(X) +_F X = F_h([n-1]_F(X), X) = \underbrace{X +_F X +_F \cdots +_F X}_{n}$$

for $n \geq 1$. We also define $[-n]_F(X) := [n]_F([-1]_F(X)) = [-1]_F([n]_F(X))$ for $n \geq 1$. We call $[n]_F(X)$ the *n-series* in the following.

EXAMPLE 2.2.

(1) For the ordinary cohomology theory (with integer coefficients) $h = H$, the coefficient ring is $H^* = H^*(\mathrm{pt}) = \mathbb{Z}$ ($H^0 = \mathbb{Z}$, $H^k = 0$ $(k \neq 0)$). We choose the standard orientation, namely the class of a hyperplane $x^H \in \tilde{H}^2(\mathbb{C}P^\infty)$. Then the associated formal group law is the *additive* formal group law $F_H(X, Y) = F_a(X, Y) = X + Y$, and the formal inverse is given by $[-1]_H(X) = -X$.

(2) For the (topological) K-theory $h = K$, the coefficient ring is $K^* = K^*(\mathrm{pt}) = \mathbb{Z}[\beta, \beta^{-1}]$, with $\beta := 1 - \eta_1^\vee \in K^{-2}(\mathrm{pt}) \cong \tilde{K}(S^2)$, where η_1 stands for the *tautological* (or Hopf) line bundle over $\mathbb{C}P^1 \cong S^2$, and $\eta_1^\vee$ its dual. We choose the standard orientation $x^K := \beta^{-1}(1 - \eta_\infty^\vee) \in \tilde{K}^2(\mathbb{C}P^\infty)$, where η_∞ stands for the tautological line bundle over $\mathbb{C}P^\infty$.[10] Then the associated formal group law is the *multiplicative* formal group law $F_K(X, Y) = F_m(X, Y) = X + Y - \beta XY$, and the formal inverse is given by

$$[-1]_K(X) = -\frac{X}{1 - \beta X} = -X - \beta X^2 - \beta^2 X^3 - \beta^3 X^4 - \cdots.$$

(3) For the complex cobordism theory $h = MU$, the coefficient ring $MU^* = MU^*(\mathrm{pt})$ is a polynomial algebra over $\mathbb{Z}$ on generators of degrees $-2, -4, \ldots$ (see e.g., Adams [**1**, Part II, Theorem 8.1]). As in Adams [**1**, Part II, Examples (2.4)], Ravenel [**65**, Example 4.1.3], we take the orientation class $x^{MU} \in \tilde{MU}^2(\mathbb{C}P^\infty)$ to be the (stable) homotopy class of the map $\mathbb{C}P^\infty \simeq BU(1) \xrightarrow{\sim} MU(1)$, where $MU(1)$ denotes the Thom

---

[10] We adopt the convention due to Bott [**9**, Theorem 7.1], Levine-Morel [**48**, Example 1.1.5] so that the K-theory first Chern class of a line bundle L (over a space X) is given by $c_1^K(L) = \beta^{-1}(1 - L^\vee)$, where $L^\vee$ denotes the dual bundle of L. In this convention, the orientation class x^K is equal to the K-theory first Chern class of the bundle η_∞, namely $c_1^K(\eta_\infty) = \beta^{-1}(1 - \eta_\infty^\vee)$.

space of the universal line bundle over $BU(1)$. Then the associated formal group law

$$F_{MU}(X,Y) = X + Y + \sum_{i,j\geq 1} a_{i,j}^{MU} X^i Y^j, \quad a_{i,j}^{MU} \in MU^{2(1-i-j)},$$

is a *universal formal group law* first shown by Quillen [**63**, Theorem 2]. To be precise, for any formal group law F over a commutative ring R with unit, there exists a unique ring homomorphism $\theta : MU^* \longrightarrow R$ such that $F(X,Y) = (\theta_* F_{MU})(X,Y) := X + Y + \sum_{i,j\geq 1} \theta(a_{i,j}^{MU}) X^i Y^j$. Quillen also showed that the coefficient ring MU^* is isomorphic to the *Lazard ring* $\mathbb{L}$ (see §3.1 in this paper).

2.2. Gysin maps. For a certain kind of map $f : X \longrightarrow Y$ between spaces, the so-called *Gysin map*, *push-forward*, or *Umkehr map*, usually denoted by $f_* : h^*(X) \longrightarrow h^*(Y)$ can be defined. Here are some examples of Gysin maps:

- (Classical Gysin map in the ordinary cohomology theory): For a smooth map

$$f : M \longrightarrow N,$$

between compact oriented smooth manifolds, the *Gysin map*

$$f_* : H^q(M) \longrightarrow H^{q-(\dim M - \dim N)}(N)$$

can be defined by $f_* := \mathcal{P}_N^{-1} \circ f_* \circ \mathcal{P}_M$, where $\mathcal{P}_M$ (resp. $\mathcal{P}_N$) denotes the Poincaré duality isomorphism from cohomology to homology.

- (Integration along (over) the fiber): For a fibration $F \hookrightarrow E \xrightarrow{\pi} B$ with the base B simply-connected and the fiber F a compact connected manifold, Borel-Hirzebruch [**7**, §8] defined a push-forward map called the *integration along the fiber*:

$$\pi_* = \natural : H^q(E) \longrightarrow H^{q-\dim F}(B).$$

- (Gysin map in the K-theory $K(-)$): For a proper morphism $f : X \longrightarrow Y$ of non-singular varieties, Grothendieck constructed an additive map $f_! : K_\circ(X) \longrightarrow K_\circ(Y)$ defined by

$$f_!([\mathcal{F}]) := \sum_{i\geq 0} (-1)^i [R^i f_*(\mathcal{F})],$$

for $\mathcal{F}$ a coherent sheaf. Here $R^i f_*(\mathcal{F})$ is *Grothendieck's higher direct image sheaf* (see e.g., Fulton [**23**, §15.1]).

- (Gysin map in the complex cobordism theory $MU^*(-)$): In [**64**], Quillen gave a geometric interpretation of the complex cobordism theory $MU^*(X)$, where X is assumed to be a manifold. In his interpretation, an element of $MU^*(X)$ is given by a *cobordism class* of a *proper* and *complex-oriented* map $f : Z \longrightarrow X$. A proper complex-oriented map $g : X \longrightarrow Y$ of dimension d induces a map

$$g_* : MU^q(X) \longrightarrow MU^{q-d}(Y)$$

which sends the cobordism class of $f : Z \longrightarrow X$ into the cobordism class of $g \circ f : Z \longrightarrow Y$.

All these Gysin maps have the common properties which can be axiomatized as follows (see Bressler-Evens [**10**, p.801, (E)], Levine-Morel [**48**, Definition 1.1.2], Quillen [**64**, §1]): Let $h^*(-)$ be a complex-oriented generalized cohomology theory defined on a suitable category of spaces. For a morphism $f : X \longrightarrow Y$, one has a Gysin map $f_* : h^*(X) \longrightarrow h^*(Y)$ having the following basic properties:

(1) (Naturality): For a composite $g \circ f : X \xrightarrow{f} Y \xrightarrow{g} Z$, one has $(g \circ f)_* = g_* \circ f_*$.
(2) (Projection formula): For $x \in h^*(X)$ and $y \in h^*(Y)$, one has $f_*(f^*(y) \cdot x) = y \cdot f_*(x)$.
(3) (Base-change): For the following commutative diagram

$$\begin{array}{ccc} X \times_Z Y & \xrightarrow{\tilde{f}} & Y \\ {\scriptstyle \tilde{g}}\downarrow & & \downarrow{\scriptstyle g} \\ X & \xrightarrow{f} & Z, \end{array}$$

one has $f^* \circ g_* = \tilde{g}_* \circ \tilde{f}^*$. Here $X \times_Z Y$ denotes the fiber product of X and Y over Z, namely $X \times_Z Y = \{(x, y) \in X \times Y \mid f(x) = g(y)\}$.

2.3. Becker-Gottlieb transfer. Let $F \overset{\iota}{\hookrightarrow} E \xrightarrow{\pi} B$ be a fiber bundle whose fiber F is a compact smooth manifold, whose structure group G is a compact Lie group acting smoothly on F, and whose base space B is a finite complex. In this setting, Becker-Gottlieb [**6**, §3] constructed a *stable* map[11]

$$\tau(\pi) : B^+ \longrightarrow E^+,$$

called the *Becker-Gottlieb transfer*. Here B^+ means the union of B with a point. For any generalized cohomology theory $h^*(-)$, the map $\tau(\pi) : B^+ \longrightarrow E^+$ induces a "wrong-way" degree-preserving homomorphism

$$\tau(\pi)^* : h^*(E) \longrightarrow h^*(B),$$

which is also called the Becker-Gottlieb transfer. $\tau(\pi)^*$ is *not* a ring homomorphism, but is an $h^*(B)$-module homomorphism (Becker-Gottlieb [**6**, (5.3)]), that is, the following formula holds:

$$\tau(\pi)^*(\pi^*(x) \cdot y) = x \cdot \tau(\pi)^*(y), \quad x \in h^*(B),\ y \in h^*(E).$$

Furthermore the Gysin map and the Becker-Gottlieb transfer are related as follows (see Becker-Gottlieb [**6**, Theorem 4.3]):

$$\tau(\pi)^*(x) = \pi_*(\chi^h(T_\pi) \cdot x) \quad (x \in h^*(E)). \tag{2.1}$$

Here T_π is the *(tangent) bundle along the fibers (of* π*)* (see e.g., Borel-Hirzebruch [**7**, §7.4]),[12] and $\chi^h(T_\pi)$ denotes the h^*-theory Euler class of T_π. Here T_π is regarded as a real vector bundle, and assumed to be h^*-*oriented* (For the notion of h^*-orientablility, see Atiyah-Bott-Shapiro [**5**, §12], Dold [**20**, §4]). The h^*-theory Euler

[11] For a suitable positive integer m, we have a map

$$\tau(\pi) : S^m(B^+) \longrightarrow S^m(E^+),$$

where S denotes the reduced suspension.

[12] If we assume that E, B are also smooth manifolds, T_π is a subbundle of the tangent bundle $T(E)$ of E. The fiber of T_π over a point $y \in E$ consists of all tangent vectors at the point y which are tangent to the fiber ($\cong F$) through y. If we denote by $\iota : F \hookrightarrow E$ the fiber inclusion, then $\iota^*(T_\pi) \cong T(F)$, the tangent bundle of F.

class $\chi^h(T_\pi)$ is defined with respect to this orientation. In practice (see Bressler-Evens [**10**, §1] and §2.5), we require that the fiber F is smooth and *almost complex*, and the structure group of F preserves the almost complex structure. Hence the tangent bundle $T(F)$ has a complex vector bundle structure, and so does the bundle along the fibers T_π. If the cohomology theory $h^*(-)$ is complex-oriented, then the h^*-theory Euler class $\chi^h(T_\pi)$ is nothing but the h^*-theory *top* Chern class $c_f^h(T_\pi)$, where $2f$ is the *real* dimension of F.

2.4. Brumfiel-Madsen formula. Let G be a compact connected (semi-simple) Lie group (of rank ℓ) with a maximal torus T ($\cong (U(1))^\ell$). Let H be a closed connected subgroup of G of maximal rank, i.e., $T \subset H$. Denote by W_G and W_H the Weyl group of G and H respectively. We then have a natural inclusion $W_H \subset W_G$. Suppose that $G \hookrightarrow P \longrightarrow B$ is a principal G-bundle, and consider the following associated bundles:

$$\begin{array}{lll} G/T \hookrightarrow & E_1 := P \times_G (G/T) \xrightarrow{\pi_1} & B, \\ G/H \hookrightarrow & E_2 := P \times_G (G/H) \xrightarrow{\pi_2} & B. \end{array}$$

Then there is a fiber bundle $H/T \hookrightarrow E_1 \xrightarrow{\pi} E_2$, where the projection π is induced from the natural projection (also denoted by the same symbol) $\pi : G/T \longrightarrow G/H$, and we have the following commutative diagram:

$$\begin{array}{ccc} E_1 = P \times_G (G/T) & \xrightarrow{\pi} & E_2 = P \times_G (G/H) \\ \pi_1 \big\downarrow & & \big\downarrow \pi_2 \\ B & \xrightarrow{=} & B. \end{array} \tag{2.2}$$

The usual right action of the Weyl group W_G on G/T induces a right action on $E_1 = P \times_G (G/T)$ over B, i.e., a bundle map over B. As a subgroup of W_G, the Weyl group W_H of H also acts on E_1, which is a bundle map over E_2. Therefore the coset $\overline{w} = wW_H \in W_G/W_H$ defines a well-defined map $\pi \circ w : E_1 \xrightarrow{w} E_1 \xrightarrow{\pi} E_2$, which induces a homomorphism in cohomology: $w \circ \pi^* : h^*(E_2) \xrightarrow{\pi^*} h^*(E_1) \xrightarrow{w} h^*(E_1)$. Then Brumfiel-Madsen established the following useful formula:

THEOREM 2.3 (Brumfiel-Madsen [**13**], Theorem 3.5; Bressler-Evens [**10**], Theorem 1.3). *In the above setting, we have*

$$\pi_1^* \circ \tau(\pi_2)^* = \sum_{\overline{w} \in W_G/W_H} w \circ \pi^*.$$

As a special case where $H = T$, we have

COROLLARY 2.4 (Bressler-Evens [**10**], Corollary 1.4).

$$\pi_1^* \circ \tau(\pi_1)^* = \sum_{w \in W_G} w.$$

We apply Corollary 2.4 to the case where $P = EG$, the universal space[13] for G, so that $B = BG$, the classifying space of G, and $E_1 = EG \times_G (G/T) \simeq BT$. In this case, the fibration $G/T \hookrightarrow E_1 \xrightarrow{\pi_1} B$ becomes the following classical Borel fibration:

$$G/T \overset{\iota}{\hookrightarrow} BT \xrightarrow{\rho = \rho(T,G)} BG.$$

[13] EG is a contractible space on which G acts freely.

Thus we have the following:

$$\rho^* \circ \tau(\rho)^* = \sum_{w \in W_G} w. \tag{2.3}$$

Combining (2.1) and (2.3), we have

$$\rho^* \circ \rho_*(\chi^h(T_\rho) \cdot f) = \sum_{w \in W_G} w \cdot f \quad \text{for } f \in h^*(BT). \tag{2.4}$$

From this formula, Bressler-Evens [**10**] derived a useful formula which is explained briefly in the next subsection.

2.5. Bressler-Evens formula. Before stating their result, we shall recall some facts from Lie theory. Let $T \subset G$ be as above and $G/T \overset{\iota}{\hookrightarrow} BT \overset{\rho}{\longrightarrow} BG$ the Borel fibration. Let $G_{\mathbb{C}}$ and $T_{\mathbb{C}}$ be the complexification of G and T respectively. Thus $G_{\mathbb{C}}$ is a connected complex (semi-simple) Lie group with maximal compact subgroup G, and $T_{\mathbb{C}} \cong (\mathbb{C}^*)^\ell$. Denote by B a Borel subgroup of $G_{\mathbb{C}}$ containing $T_{\mathbb{C}}$. Then the natural inclusion $G \hookrightarrow G_{\mathbb{C}}$ induces a diffeomorphism $G/T \overset{\sim}{\to} G_{\mathbb{C}}/B$. By this identification, the *full flag manifold* G/T is equipped with a complex structure. Let $\mathfrak{g}$ and $\mathfrak{t}$ be the Lie algebras of G and T, and $\mathfrak{g}_{\mathbb{C}} = \mathfrak{g} \otimes_{\mathbb{R}} \mathbb{C}$ and $\mathfrak{t}_{\mathbb{C}} = \mathfrak{t} \otimes_{\mathbb{R}} \mathbb{C}$ their complexification. Then $\mathfrak{g}_{\mathbb{C}}$ and $\mathfrak{t}_{\mathbb{C}}$ are the Lie algebras of $G_{\mathbb{C}}$ and $T_{\mathbb{C}}$ respectively. Then we have the root space decomposition

$$\mathfrak{g}_{\mathbb{C}} = \mathfrak{t}_{\mathbb{C}} \oplus \bigoplus_{\alpha \in \Delta^+} (\mathfrak{g}_\alpha \oplus \mathfrak{g}_{-\alpha}), \tag{2.5}$$

where $\mathfrak{g}_\alpha = \{x \in \mathfrak{g}_{\mathbb{C}} \,|\, [h, x] = \alpha(h)x \ (\forall h \in \mathfrak{t}_{\mathbb{C}})\}$, and the system of *positive* roots $\Delta^+ \subset \mathrm{Hom}_{\mathbb{C}}(\mathfrak{t}_{\mathbb{C}}, \mathbb{C})$ corresponds to the Lie algebra $\mathfrak{b}$ of B. Thus $\mathfrak{b} = \mathfrak{t}_{\mathbb{C}} \oplus \bigoplus_{\alpha \in \Delta^+} \mathfrak{g}_\alpha$. We set $\Delta^- := -\Delta^+$, the system of *negative* roots, and $\Delta := \Delta^+ \sqcup \Delta^-$, the system of roots. It is well-known that the tangent bundle $T(G/T)$ to G/T is isomorphic to the vector bundle $G \times_T (\mathfrak{g}/\mathfrak{t})$ associated with the principal T-bundle $T \hookrightarrow G \longrightarrow G/T$ and T-module $\mathfrak{g}/\mathfrak{t}$. Since we have the natural identification $\mathfrak{g}/\mathfrak{t} \cong \mathfrak{g}_{\mathbb{C}}/\mathfrak{b}$, we have the following isomorphism as *complex* vector bundles:

$$T(G/T) \cong G \times_T (\mathfrak{g}_{\mathbb{C}}/\mathfrak{b}).$$

Hence the tangent bundle along the fibers T_ρ is isomorphic to the *complex* vector bundle $ET \times_T (\mathfrak{g}_{\mathbb{C}}/\mathfrak{b})$ associated with the universal T-bundle $T \hookrightarrow ET \to BT$. Thus

$$T_\rho \cong ET \times_T (\mathfrak{g}_{\mathbb{C}}/\mathfrak{b}). \tag{2.6}$$

For each character $\chi \in \mathrm{Hom}\,(T, U(1)) \cong \mathrm{Hom}\,(T_{\mathbb{C}}, \mathbb{C}^*) = \widehat{T_{\mathbb{C}}}$, we have the associated complex line bundle L_χ over BT defined by $L_\chi := ET \times_T \mathbb{C} = (ET \times \mathbb{C})/(y, v) \sim (y \cdot t, \chi(t)^{-1}v)$. Each root $\alpha \in \Delta \subset \mathrm{Hom}_{\mathbb{C}}(\mathfrak{t}_{\mathbb{C}}, \mathbb{C})$ defines a character $\chi_\alpha \in \mathrm{Hom}\,(T_{\mathbb{C}}, \mathbb{C}^*)$, and we have the associated complex line bundle L_{χ_α}, which is also denoted by L_α for simplicity. By (2.5) and (2.6), we have

$$T_\rho \cong ET \times_T \bigoplus_{\alpha \in \Delta^+} \mathfrak{g}_{-\alpha} \cong \bigoplus_{\alpha \in \Delta^+} L_{-\alpha}.$$

Therefore the h^*-theory top Chern class (Euler class) of T_ρ is given by

$$c^h_{\mathrm{top}}(T_\rho) = c^h_{\mathrm{top}}\left(\bigoplus_{\alpha \in \Delta^+} L_{-\alpha}\right) = \prod_{\alpha \in \Delta^+} c^h_1(L_{-\alpha}). \tag{2.7}$$

From (2.7) and the formula (2.4), Bressler-Evens deduced the following:[14]

THEOREM 2.5 (Bressler-Evens [**10**], Theorem 1.8). *Let $h^*(-)$ be a complex-oriented generalized cohomology theory. We assume that the coefficient ring h^* is torsion-free.*[15] *Then for $f \in h^*(BT)$, we have*

$$\rho^* \circ \rho_*(f) = \sum_{w \in W_G} w \cdot \left[\frac{f}{\prod_{\alpha \in \Delta^+} c_1^h(L_{-\alpha})} \right].$$

One can easily extend Theorem 2.5 to the case of *partial flag manifolds.* Thus let $\Theta \subset \Pi := \{\text{simple roots}\} \subset \Delta^+$ be a subset of the set of simple roots, and $P = P_\Theta$ be the corresponding *parabolic subgroup*, and put $H = H_\Theta := G \cap P$. Then it is known that H is the centralizer of the toral subgroup defined by "$\alpha = 0$" for $\forall \alpha \in \Theta$, and hence a closed connected subgroup of G of maximal rank, i.e., $T \subset H$. The homogeneous manifold G/H has a complex structure (see e.g., Borel-Hirzebruch [**7**, §13.5]), and we have the following Borel fibration

$$G/H \overset{\iota_H}{\hookrightarrow} BH \overset{\sigma = \rho(H,G)}{\longrightarrow} BG.$$

There exists a natural identification $G/H \overset{\sim}{\to} G_{\mathbb{C}}/P$, and this homogeneous manifold is called a partial flag manifold (see e.g., Borel-Hirzebruch [**7**, §14.3]). Denote by $\Delta_H = \Delta_H^+ \sqcup \Delta_H^-$ the system of roots of H with respect to T. The Lie algebra $\mathfrak{p} = \mathfrak{p}_\Theta$ of P is given by

$$\mathfrak{p} = \mathfrak{b} \oplus \bigoplus_{\alpha \in \Delta_H^+} \mathfrak{g}_{-\alpha} = \mathfrak{t}_{\mathbb{C}} \oplus \bigoplus_{\alpha \in \Delta^+} \mathfrak{g}_\alpha \oplus \bigoplus_{\alpha \in \Delta_H^+} \mathfrak{g}_{-\alpha}. \tag{2.8}$$

The tangent bundle of $G/H \overset{\sim}{\to} G_{\mathbb{C}}/P$ is given by

$$T(G/H) \cong G \times_H (\mathfrak{g}_{\mathbb{C}}/\mathfrak{p}).$$

Hence the tangent bundle along the fibers T_σ is isomorphic to the complex vector bundle $EH \times_H (\mathfrak{g}_{\mathbb{C}}/\mathfrak{p})$ associated with the universal H-bundle $H \hookrightarrow EH \longrightarrow BH$. Thus

$$T_\sigma \cong EH \times_H (\mathfrak{g}_{\mathbb{C}}/\mathfrak{p}). \tag{2.9}$$

We apply Theorem 2.3 to the case where $P = EG$ and $H = H_\Theta$, so that $B = BG$, and $E_2 = EG \times_G (G/H) \simeq BH$. Then the fibration $G/H \hookrightarrow E_2 \overset{\pi_2}{\longrightarrow} B$ becomes the Borel fibration $G/H \overset{\iota_H}{\hookrightarrow} BH \overset{\sigma}{\longrightarrow} BG$, and the commutative diagram (2.2) yields the following commutative diagram:

$$\begin{array}{ccc} BT & \xrightarrow{\pi = \rho(T,H)} & BH \\ {\scriptstyle \rho = \rho(T,G)} \downarrow & & \downarrow {\scriptstyle \sigma = \rho(H,G)} \\ BG & \xrightarrow{=} & BG. \end{array} \tag{2.10}$$

Then by Theorem 2.3, we have

$$\rho^* \circ \tau(\sigma)^* = \sum_{\overline{w} \in W_G/W_H} w \circ \pi^*. \tag{2.11}$$

[14] For the ordinary cohomology theory $h = H$, this formula was already proved by Borel-Hirzebruch [**8**, Theorem 20.3] (see also Tu [**70**, §11.1]).

[15] We made this assumption for simplicity. See Bressler-Evens [**11**, Remark 1.10] for less restrictive assumptions.

Combining (2.1) and (2.11), we have

$$\rho^* \circ \sigma_*(c^h_{\text{top}}(T_\sigma) \cdot f) = \sum_{\overline{w} \in W_G/W_H} w \cdot \pi^*(f) \quad \text{for } f \in h^*(BH). \tag{2.12}$$

On the other hand, pulling back the tangent bundle along the fibers T_σ to BT via the map $\pi = \rho(T, H)$, we have from (2.9), (2.5), and (2.8),

$$\pi^*(T_\sigma) \cong ET \times_T (\mathfrak{g}_{\mathbb{C}}/\mathfrak{p}) \cong ET \times_T \bigoplus_{\alpha \in \Delta^+ \setminus \Delta^+_H} \mathfrak{g}_{-\alpha} \cong \bigoplus_{\alpha \in \Delta^+ \setminus \Delta^+_H} L_{-\alpha}.$$

Therefore the h^*-theory top Chern class (Euler class) of $\pi^*(T_\sigma)$ is given by

$$\pi^* c^h_{\text{top}}(T_\sigma) = c^h_{\text{top}}\left(\bigoplus_{\alpha \in \Delta^+ \setminus \Delta^+_H} L_{-\alpha} \right) = \prod_{\alpha \in \Delta^+ \setminus \Delta^+_H} c^h_1(L_{-\alpha}). \tag{2.13}$$

Then by the analogous argument to that of Bressler-Evens [**10**, Theorem 1.8], (2.13), and the formula (2.12), one obtains the following (Note that by the commutativity of the diagram (2.10), one has $\rho^* \circ \sigma_* = \pi^* \circ \sigma^* \circ \sigma_*$):

COROLLARY 2.6. *For $f \in h^*(BH)$, we have*

$$\pi^* \circ \sigma^* \circ \sigma_*(f) = \sum_{\overline{w} \in W_G/W_H} w \cdot \left[\frac{\pi^*(f)}{\prod_{\alpha \in \Delta^+ \setminus \Delta^+_H} c^h_1(L_{-\alpha})} \right]. \tag{2.14}$$

2.6. Various Gysin formulas. As mentioned in the introduction, various types of *Gysin formulas* related to the Gysin maps are known (see e.g., Akyildiz [**2**], Akyildiz-Carrell [**4**], Buch [**14**], Damon [**16**], [**17**], Darondeau-Pragacz [**18**], Fel'dman [**22**], Fulton [**23**], Fulton-Pragacz [**25**], Harris-Tu [**27**], Ilori [**36**], Jozefiak-Lascoux-Pragacz [**38**], Kajimoto [**40**], Quillen [**63**], Pragacz [**59**], [**61**], [**62**], Sugawara [**66**], Tu [**69**], [**70**]). In this subsection, we shall take up typical examples of these formulas.

2.6.1. *Gysin formulas of type "$G_{\mathbb{C}}/B \longrightarrow G_{\mathbb{C}}/P$".* First recall the result due to Akyildiz-Carrell [**4**]. In order to state their result, we shall use the same notation as in §2.5 with a slightly minor change. So Let $G_{\mathbb{C}} \supset B \supset T_{\mathbb{C}}$ be as in §2.5. Consider the parabolic subgroup $P = P_\Theta$ corresponding to a subset $\Theta \subset \Pi = \{\text{simple roots}\} \subset \Delta^+$. Thus the homogeneous variety $G_{\mathbb{C}}/P$ is a partial flag variety. Denote by W_Θ (resp. Δ_Θ) the Weyl group (resp. root system) corresponding to P_Θ. Let $\chi \in \widehat{T}_{\mathbb{C}} = \text{Hom}\,(T_{\mathbb{C}}, \mathbb{C}^*)$ be a character. By composing the natural projection[16] $B \longrightarrow\!\!\!\!\!\rightarrow T_{\mathbb{C}}$ with $\chi : T_{\mathbb{C}} \longrightarrow \mathbb{C}^*$, we have a character $\chi_B = \chi : B \longrightarrow \mathbb{C}^*$. Then one can define a complex line bundle M_χ over $G_{\mathbb{C}}/B$ in the usual manner. By assigning each character $\chi \in \widehat{T}_{\mathbb{C}}$ the first Chern class $c_1(M_\chi) \in H^2(G_{\mathbb{C}}/B; \mathbb{C})$, the *characteristic homomorphism*[17]

$$c : R := \text{Sym}\,(\widehat{T}_{\mathbb{C}}) \to H^*(G_{\mathbb{C}}/B; \mathbb{C})$$

[16]Recall that B is the semi-direct product of $T_{\mathbb{C}}$ and its unipotent part.

[17]In topology, it is customary that the character group $\widehat{T}_{\mathbb{C}} \cong \text{Hom}\,(T, U(1))$ is identified with $H^1(T) \cong H^2(BT)$ (this latter identification is given by the *negative transgression* (see Borel-Hirzebruch [**7**, §10.1])). Under this identification, one has the isomorphism $\text{Sym}\,(\widehat{T}_{\mathbb{C}}) \cong H^*(BT; \mathbb{C})$ as algebras, and the characteristic homomorphism c can be identified with the induced homomorphism $\iota^* : H^*(BT; \mathbb{C}) \longrightarrow H^*(G/T; \mathbb{C})$.

is defined. Here $\mathrm{Sym}(\widehat{T}_{\mathbb{C}})$ means the symmetric algebra of $\widehat{T}_{\mathbb{C}}$ over $\mathbb{C}$. Let $\pi : G_{\mathbb{C}}/B \longrightarrow G_{\mathbb{C}}/P$ be the natural projection. Then Akyildiz-Carrell showed the following formula (see also Brion [**12**]):

THEOREM 2.7 (Akyildiz-Carrell [**4**], Theorem 1; Brion [**12**], Proposition 1.1). *The Gysin homomorphism* $\pi_* : H^*(G_{\mathbb{C}}/B;\mathbb{C}) \longrightarrow H^*(G_{\mathbb{C}}/P;\mathbb{C})$ *is given by*

$$\pi^* \circ \pi_* (c(f)) = c\left(\sum_{w \in W_\Theta} \frac{\det(w)\, w \cdot f}{\prod_{\alpha \in \Delta_\Theta^+} \alpha} \right) \quad \textit{for } f \in R. \tag{2.15}$$

Here $\det(w)$ means $(-1)^{\ell(w)}$, where $\ell(w)$ denotes the length of the Weyl group element w. Since $w \cdot \prod_{\alpha \in \Delta_\Theta^+} \alpha = (-1)^{\ell(w)} \prod_{\alpha \in \Delta_\Theta^+} \alpha$ for any $w \in W_\Theta$, the above formula (2.15) can also be written as follows:

$$\pi^* \circ \pi_* (c(f)) = c\left(\sum_{w \in W_\Theta} w \cdot \left[\frac{f}{\prod_{\alpha \in \Delta_\Theta^+} \alpha} \right] \right) \quad \text{for } f \in R.$$

Akyildiz-Carrell proved this formula by the method based on the zeros of holomorphic vector fields on relevant flag varieties. Brion proved this formula by the Weyl character formula and the Grothendieck-Riemann-Roch theorem.

2.6.2. *Gysin formulas of type "$\mathcal{F}\ell(E) \longrightarrow X$".* Next recall the result due to Fulton-Pragacz [**25**, Chapter IV]. Let $E \xrightarrow{p} X$ be a complex vector bundle of rank n over a *variety*.[18] Denote by $\tau = \tau_E : \mathcal{F}\ell(E) \longrightarrow X$ the associated flag bundle parametrizing successive flags of *quotients* of E of ranks $n-1, \ldots, 2, 1$. Thus we have the *tautological sequence of flag of quotient bundles*

$$\tau^* E = Q^n \twoheadrightarrow Q^{n-1} \twoheadrightarrow \cdots \twoheadrightarrow Q^2 \twoheadrightarrow Q^1 \twoheadrightarrow Q^0 = 0,$$

where $\mathrm{rank}\,(Q^i) = i$ $(i = 0, 1, 2, \ldots, n)$.[19] Define the line bundles $L^i := \mathrm{Ker}\,(Q^i \twoheadrightarrow Q^{i-1})$ $(i = 1, 2, \ldots, n)$ over $\mathcal{F}\ell(E)$. Put $x_i := c_1(L^i) \in H^2(\mathcal{F}\ell(E))$ $(i = 1, 2, \ldots, n)$ (the *Chern roots* of E). Then Fulton-Pragacz showed the following formula:

THEOREM 2.8 (Pragacz [**59**], Lemma 2.4; [**61**], Proposition 4.3 (ii); Fulton-Pragacz [**25**], p.41). *For a polynomial* $f(X_1, \ldots, X_n) \in H^*(X)[X_1, \ldots, X_n]$, *we have*

$$\tau^* \circ \tau_*(f(x_1, \ldots, x_n)) = \sum_{w \in S_n} w \cdot \left[\frac{f(x_1, \ldots, x_n)}{\prod_{1 \le i < j \le n} (x_i - x_j)} \right].$$

Thus the Gysin map τ_* is given by a certain *symmetrizing operator* called the *Jacobi symmetrizer* in Fulton-Pragacz [**25**, §4.1]. As is well-known, the flag bundle $\mathcal{F}\ell(E)$ can be constructed as a sequence of projective bundles, and Fulton-Pragacz proved this formula by the induction on the rank of E.

---

[18] Actually it is enough to assume that the base space is some *nice space*, say, a *paracompact space*, so that the classification theorem of vector bundles holds.

[19] Here we followed the convention as in e.g., Pragacz [**59**, §2]. On $\mathcal{F}\ell(E)$, we also have the *tautological sequence of flag of subbundles*

$$0 = S_0 \subset S_1 \subset S_2 \subset \cdots \subset S_{n-1} \subset S_n = \tau^* E,$$

where $\mathrm{rank}\,(S_i) = i$ $(i = 0, 1, 2, \ldots, n)$. These two tautological sequences are related by $Q^i = \tau^* E / S_{n-i}$. Therefore if we define the line bundles $L_i := S_i / S_{i-1}$ $(i = 1, 2, \ldots, n)$, then we have $L^i = L_{n+1-i}$ $(i = 1, 2, \ldots, n)$.

2.6.3. *Application of the Bressler-Evens formula.* Most of these Gysin formulas are formulated in the ordinary cohomology rings or Chow rings, and proved by many different ways. We remark that the Bressler-Evens formulas (Theorem 2.5 and Corollary 2.6) enable us to show these Gysin formulas by a unified manner.

For Theorem 2.7, one can argue as follows: Put $H = G \cap P$ as in §2.5. Then by the *classification theorem of principal bundles*, we have a *classifying map* $h : G/H \longrightarrow BH$ and its lift $\tilde{h} : G/T \longrightarrow BT$, and the following diagram is commutative:

$$\begin{array}{ccc} G_{\mathbb{C}}/B \cong G/T & \xrightarrow{\tilde{h} \simeq \iota} & BT \\ \pi \downarrow & & \downarrow \rho = \rho(T,H) \\ G_{\mathbb{C}}/P \cong G/H & \xrightarrow{h \simeq \iota_H} & BH. \end{array}$$

Note that the above classifying map h (resp. $\tilde{h}$) coincides with the fiber inclusion $\iota_H : G/H \hookrightarrow BH$ (resp. $\iota : G/T \hookrightarrow BT$) up to homotopy. Then by Theorem 2.5 and the base-change property of Gysin maps, we compute

$$\begin{aligned} \pi^* \circ \pi_* (c(f)) &= \pi^* \circ \pi_* \circ \iota^*(f) = \pi^* \circ \iota_H^* \circ \rho_*(f) = \iota^* \circ \rho^* \circ \rho_*(f) \\ &= \iota^* \left(\sum_{w \in W_\Theta} w \cdot \left[\frac{f}{\prod_{\alpha \in \Delta_\Theta^+} c_1(L_{-\alpha})} \right] \right) \\ &= c \left(\sum_{w \in W_\Theta} w \cdot \left[\frac{f}{\prod_{\alpha \in \Delta_\Theta^+} \alpha} \right] \right), \end{aligned}$$

as required. Here we used the convention that $c_1(L_\alpha) = -\alpha$ for a root $\alpha \in \Delta$.

For Theorem 2.8, one can argue as follows: By the *classification theorem of complex vector bundles*, we have the *classifying map* $h : X \longrightarrow BU(n)$, and its lift $\tilde{h} : \mathcal{F}\ell(E) \longrightarrow BT^n$, and the following diagram is commutative:

$$\begin{array}{ccc} \mathcal{F}\ell(E) & \xrightarrow{\tilde{h}} & BT^n \\ \tau \downarrow & & \downarrow \rho \\ X & \xrightarrow{h} & BU(n). \end{array}$$

Let $\chi_i : T^n \longrightarrow U(1)$ be the character which takes an element $t = \text{diag}\,(t_1, \dots, t_n) \in T^n$ to the i-th entry $t_i \in U(1)$ $(i = 1, 2, \dots, n)$. The line bundles L_{χ_i} over BT^n can be constructed as in §2.5. Put $y_i := -c_1(L_{\chi_i}) = c_1((L_{\chi_i})^\vee) \in H^2(BT^n)$ $(i = 1, 2, \dots, n)$ (notice our convention). Then the positive root system for $G = U(n)$ is given by $\Delta^+ = \{y_i - y_j \ (1 \le i < j \le n)\}$ as a subset of $H^2(BT^n)$. The Weyl group $W_{U(n)}$ of $U(n)$ can be identified with the symmetric group S_n by the usual manner. Let $\gamma^n \longrightarrow BU(n)$ be the *universal* or *canonical* vector bundle over $BU(n)$ (see Milnor-Stasheff [**54**, §14]). Then the associated flag bundle $\mathcal{F}\ell(\gamma^n) \longrightarrow BU(n)$ can be identified with the Borel fibration $BT^n \xrightarrow{\rho} BU(n)$. As noted in §2.6.2, there is the tautological sequence of flag of subbundles $S_0 \subset S_1 \subset S_2 \subset \cdots \subset S_{n-1} \subset S_n = \rho^*(\gamma^n)$ over BT^n. The usual line bundles L_i $(i = 1, 2, \dots, n)$ over BT^n are defined by $L_i := S_i/S_{i-1}$ $(i = 1, 2, \dots, n)$. Then it is easily verified that the line bundle L_i can be identified with the line bundle L_{χ_i}. Therefore as for the Chern roots of E, we have

$$x_i = c_1(L^i) = c_1(L_{n+1-i}) = c_1(\tilde{h}^*(L_{n+1-i})) = \tilde{h}^*(c_1(L_{n+1-i})) = \tilde{h}^*(-y_{n+1-i}).$$

Then by Theorem 2.5 and the base-change property of Gysin maps, we compute

$$\begin{aligned}\tau^* \circ \tau_*(f(x_1, \dots, x_n)) &= \tau^* \circ \tau_* \circ \tilde{h}^*(f(-y_n, \dots, -y_1)) \\ &= \tilde{h}^* \circ \rho^* \circ \rho_*(f(-y_n, \dots, -y_1)) \\ &= \tilde{h}^* \left(\sum_{w \in W_{U(n)}} w \cdot \left[\frac{f(-y_n, \dots, -y_1)}{\prod_{\alpha \in \Delta^+} c_1(L_{-\alpha})} \right] \right) \\ &= \tilde{h}^* \left(\sum_{w \in S_n} w \cdot \left[\frac{f(-y_n, \dots, -y_1)}{\prod_{1 \le i < j \le n}(y_i - y_j)} \right] \right) \\ &= \sum_{w \in S_n} w \cdot \left[\frac{f(x_1, \dots, x_n)}{\prod_{1 \le i < j \le n}(x_i - x_j)} \right],\end{aligned}$$

as required.

2.6.4. *Thom-Porteous formula.* Finally, we briefly review the *Thom-Porteous formula* (see Porteous [**58**, p.298, Proposition 1.3]) as an application of Gysin formulas. Here we adopt the formulation as in Fulton [**23**, §14.4], Fulton-Pragacz [**25**, §2.1], Pragacz [**59**], [**61**] which is slightly different from Porteous' original one. Let $E \xrightarrow{p_E} X$ and $F \xrightarrow{p_F} X$ be complex vector bundles of ranks e and f on a variety X. Let $\varphi : E \longrightarrow F$ be a vector bundle homomorphism. For each point x, denote by $\varphi_x : E_x = p_E^{-1}(x) \longrightarrow F_x = p_F^{-1}(x)$ the linear map on the fiber. Then we set

$$D_r(\varphi) := \{x \in X \mid \operatorname{rank} \varphi_x \le r\} \subset X,$$

which is called the rth *degeneracy locus* of φ ($r = 0, 1, \dots, \min(e, f)$). It is known that if the map φ is *sufficiently generic*, the subvariety $D_r(\varphi)$ has codimension $(e-r)(f-r)$, and defines a cohomology class $[D_r(\varphi)] \in H^{2(e-r)(f-r)}(X)$. Then Thom [**68**] observed that there must be a polynomial in the Chern classes of E and F which is equal to $[D_r(\varphi)]$. Thom posed a problem to find such a polynomial, and later Porteous gave the answer (see Fulton-Pragacz [**25**, Chapter II, (2.1), (2.5)], Pragacz [**59**, p.414]):

$$[D_r(\varphi)] = \det (c_{f-r-i+j}(F - E))_{1 \le i,j \le e-r}. \tag{2.16}$$

Here $c(F-E)$ is defined to be $c(F)/c(E)$. Notice that the right-hand side of (2.16) is equal to the *relative* version of the Schur polynomial $s_{((e-r)^{(f-r)})}(F - E)$, where $((e-r)^{(f-r)})$ is a *rectangular partition* with $(f-r)$ rows and $(e-r)$ columns (for the notation, see Fulton-Pragacz [**25**, §3.2]), and the above formula becomes as follows:

$$[D_r(\varphi)] = s_{((e-r)^{(f-r)})}(F - E). \tag{2.17}$$

We shall give an outline of the proof of the above formula for reader's convenience: Let $\pi_F : G^{f-r}(F) \longrightarrow X$ be the Grassmann bundle parametrizing rank $(f-r)$ quotient bundles of F. On $G^{f-r}(F)$, we have the tautological exact sequence of vector bundles:

$$0 \longrightarrow S_F \hookrightarrow \pi_F^*(F) \twoheadrightarrow Q_F \longrightarrow 0.$$

Then the vector bundle homomorphism $\pi_F^*(E) \xrightarrow{\pi_F^*\varphi} \pi_F^*(F) \twoheadrightarrow Q_F$ over $G^{f-r}(F)$ gives a cross-section $s_\varphi \in \Gamma(\operatorname{Hom}(\pi_F^*(E), Q_F)) \cong \Gamma(\pi_F^*(E)^\vee \otimes Q_F)$. Denote by $Z(s_\varphi) \subset G^{f-r}(F)$ the zero locus of φ. Then for an element $W \in G^{f-r}(F)$ with $\pi_F(W) = x \in X$, one sees immediately that $W \in Z(s_\varphi)$ implies $\operatorname{Im} \varphi_x \subset W$, and hence $\operatorname{rank} \varphi_x \le \dim W = r$. Thus we have $x \in D_r(\varphi)$. From this, the set $Z(s_\varphi)$

maps onto $D_r(\varphi)$. Then under appropriate conditions, the class $[Z(s_\varphi)]$ is given by the top Chern class $c_{e(f-r)}(\pi_F^*(E)^\vee \otimes Q_F)$. Therefore we have the following formula:

$$\pi_{F*}(c_{e(f-r)}(\pi_F^*(E)^\vee \otimes Q_F)) = [D_r(\varphi)],$$

and we have to compute the left-hand side of the above equation. This can be done by making use of Gysin formulas. Let $x_1, \ldots, x_f$ (resp. $a_1, \ldots, a_e$) be the Chern roots of F (resp. E) as in §2.6.2. The Chern roots of Q_F are $x_1, \ldots, x_{f-r}$. By the splitting principle, the top Chern class $c_{e(f-r)}(\pi_F^*(E)^\vee \otimes Q_F)$ is given by the product $\prod_{i=1}^{f-r}\prod_{j=1}^{e}(x_i - a_j)$. On the other hand, by a similar argument as in the previous subsection §2.6.2, the Gysin map $\pi_{F*} : H^*(G^{f-r}(F)) \longrightarrow H^*(X)$ is described by the following symmetrizing operator (see also Pragacz [**59**, Lemma 2.5], [**61**, Proposition 4.2]):

$$\pi_{F*}(g(x_1, \ldots, x_f)) = \sum_{\overline{w} \in S_f/S_{f-r}\times S_r} w \cdot \left[\frac{g(x_1, \ldots, x_f)}{\prod_{1\le i\le f-r}\prod_{f-r+1\le j\le f}(x_i - x_j)}\right]$$

for a polynomial $g(X_1, \ldots, X_f) \in H^*(X)[X_1, \ldots, X_f]^{S_{f-r}\times S_r}$. From this description, one can compute $\pi_{F*}(\prod_{i=1}^{f-r}\prod_{j=1}^{e}(x_i - a_j))$, and obtain the formula (2.17).[20] We remark that the K-theoretic analogue of this formula is also given by Buch [**14**, Theorem 2.3]. In §5.4, we shall generalize the Thom-Porteous formula for cohomology to the complex cobordism theory.

3. Universal Hall-Littlewood functions

As mentioned in the introduction (see also Example 2.2), Quillen [**63**] showed that the complex cobordism theory $MU^*(-)$ (with the associated formal group law F_{MU}) has the following *universal* property: for any complex-oriented cohomology theory $h^*(-)$ (with the associated formal group law F_h), there exists a homomorphism of rings $\theta : MU^* \longrightarrow h^*$ such that $F_h(X,Y) = (\theta_* F_{MU})(X,Y) = X + Y + \sum_{i,j\ge 1}\theta(a_{i,j}^{MU})X^iY^j$. Thus it will be sufficient to consider the case when $h = MU$, for general case follows immediately from the universal one by the specialization $a_{i,j}^{MU} \longmapsto \theta(a_{i,j}^{MU})$ $(i, j \ge 1)$. Recall that, by Quillen again, the coefficient ring $MU_* = MU^{-*}$ is isomorphic to the Lazard ring $\mathbb{L}$. In our previous paper [**56**], we introduced the *universal Schur (S-) functions* $s_\lambda^{\mathbb{L}}(\mathbf{x}_n)$ for λ partitions, and the *universal Schur P- and Q-functions* $P_\nu^{\mathbb{L}}(\mathbf{x}_n)$, $Q_\nu^{\mathbb{L}}(\mathbf{x}_n)$ for ν strict partitions. In this section, we introduce the *universal Hall-Littlewood functions* $H_\lambda^{\mathbb{L}}(\mathbf{x}_n; t)$ which will

---

[20] As in Ikeda-Naruse [**33**, §5.1], Molev-Sagan [**55**, §2], let us introduce the following notation (cf. §3.2): Set

$$(t|a)^k := \prod_{i=1}^{k}(t - a_i) = (t - a_1)(t - a_2)\cdots(t - a_k)$$

for any integer $k \ge 0$ (Here $a = (a_1, \ldots, a_e) = (a_1, \ldots, a_e, 0, 0, \ldots)$ is a sequence of the Chern roots of E). Then one can rewrite

$$\prod_{i=1}^{f-r}\prod_{j=1}^{e}(x_i - a_j) = \prod_{i=1}^{f-r}(x_i|a)^e.$$

Then one computes $\pi_{F*}(\prod_{i=1}^{f-r}(x_i|a)^e)$ by the above symmetrizing operator description of π_{F*}, and obtains the *factorial Schur polynomial* $s_{((e-r)^{(f-r)})}(\mathbf{x}_f|a)$, which is equal to $s_{((e-r)^{(f-r)})}(F-E)$.

be expected to interpolate between the universal Schur S-functions and the universal Schur P-functions. Since these functions will be of independent interest in terms of, e.g., algebraic combinatorics, so apart from geometry, we shall deal with these functions purely algebraically, and slightly changes the notation concerning the formal group law in this section.

3.1. Lazard ring $\mathbb{L}$ and the universal formal group law. We begin with collecting the basic facts about the Lazard ring. We use the convention as in Levine-Morel's book [**48**]. In [**45**], Lazard considered a universal commutative formal group law of rank one $(\mathbb{L}, F_{\mathbb{L}})$, where the ring $\mathbb{L}$, called the *Lazard ring*, is isomorphic to the polynomial ring in countably infinite number of variables with integer coefficients, and $F_{\mathbb{L}} = F_{\mathbb{L}}(u, v)$ is the *universal formal group law* (for a construction and basic properties of $\mathbb{L}$, see Levine-Morel [**48**, §1.1]):

$$F_{\mathbb{L}}(u, v) = u + v + \sum_{i,j\geq 1} a^{\mathbb{L}}_{i,j} u^i v^j \in \mathbb{L}[[u, v]].$$

This is a formal power series in u, v with coefficients $a^{\mathbb{L}}_{i,j}$ of formal variables which satisfies the axiom of the formal group law (see §2.1). For the universal formal group law, we shall use the notation (see Levine-Morel [**48**, §2.3.2])

$$\begin{aligned} u +_{\mathbb{L}} v &= F_{\mathbb{L}}(u, v) && \text{(formal sum)},\\ \overline{u} &= [-1]_{\mathbb{L}}(u) = \chi_{\mathbb{L}}(u) && \text{(formal inverse of } u).\end{aligned}$$

Note that $\overline{u} \in \mathbb{L}[[u]]$ is a formal power series in u with initial term $-u$, and first few terms appear in Levine-Morel [**48**, §2.5.1, p.41]. The n-series $[n]_{F_{\mathbb{L}}}(u)$ introduced in §2.1.2 shall be denoted simply by $[n]_{\mathbb{L}}(u)$ in the sequel. In what follows, we regard $\mathbb{L}$ as a graded algebra over $\mathbb{Z}$, and the grading of $\mathbb{L}$ is given by $\deg(a^{\mathbb{L}}_{i,j}) = 1 - i - j$ $(i, j \geq 1)$ (see Levine-Morel [**48**, §1.1, p.5]). Be aware that in topology, it is customary to give $a^{\mathbb{L}}_{i,j}$ the *cohomological* degree $2(1 - i - j)$.

3.2. Universal factorial Schur S-, P-, and Q-functions. In this subsection, we recall the definitions of the *universal factorial Schur S-, P-, and Q-functions* following Nakagawa-Naruse [**56**, §4]. Besides the variables $\mathbf{x} = (x_1, x_2, \ldots)$, we prepare another set of variables $\mathbf{b} = (b_1, b_2, \ldots)$. We provide the variables $\mathbf{x} = (x_1, x_2, \ldots)$ and $\mathbf{b} = (b_1, b_2, \ldots)$ with degree $\deg(x_i) = \deg(b_i) = 1$ for $i = 1, 2, \ldots$. In what follows, when considering polynomials or formal power series $f(x_1, x_2, \ldots)$ with coefficients in $\mathbb{L}$ (or $\mathbb{L}[[\mathbf{b}]]$), we shall call the degree with respect to $x_1, x_2, \ldots, b_1, b_2, \ldots$, and $a^{\mathbb{L}}_{i,j}$ $(i, j \geq 1)$ the *total* degree of $f(x_1, x_2, \ldots)$.

For an integer $k \geq 1$, we define a generalization of the ordinary k-th power t^k by

$$[t|\mathbf{b}]^k := \prod_{i=1}^{k} (t +_{\mathbb{L}} b_i) = (t +_{\mathbb{L}} b_1)(t +_{\mathbb{L}} b_2) \cdots (t +_{\mathbb{L}} b_k)$$

and its variant by

$$[[t|\mathbf{b}]]^k := (t +_{\mathbb{L}} t)[t|\mathbf{b}]^{k-1} = (t +_{\mathbb{L}} t)(t +_{\mathbb{L}} b_1)(t +_{\mathbb{L}} b_2) \cdots (t +_{\mathbb{L}} b_{k-1}),$$

where we set $[t|\mathbf{b}]^0 = [[t|\mathbf{b}]]^0 := 1$. For a partition,[21] i.e., a non-increasing sequence of non-negative integers $\lambda = (\lambda_1, \dots, \lambda_r)$ $(\lambda_1 \geq \lambda_2 \geq \cdots \geq \lambda_r \geq 0)$,[22] we set

$$[\mathbf{x}|\mathbf{b}]^\lambda := \prod_{i=1}^r [x_i|\mathbf{b}]^{\lambda_i} \quad \text{and} \quad [[\mathbf{x}|\mathbf{b}]]^\lambda := \prod_{i=1}^r [[x_i|\mathbf{b}]]^{\lambda_i}.$$

Let $\mathcal{P}_n$ denote the set of all partitions of length $\leq n$. For a positive integer k, we set $\rho(k) := (k, k-1, \dots, 2, 1)$. For partitions $\lambda, \mu \in \mathcal{P}_n$, $\lambda + \mu$ is a partition of length $\leq n$ defined by $(\lambda + \mu)_i := \lambda_i + \mu_i$ $(1 \leq i \leq n)$. With this notation, the *universal factorial Schur (S-) function* $s^{\mathbb{L}}_\lambda(\mathbf{x}_n|\mathbf{b}) = s^{\mathbb{L}}_\lambda(x_1, \dots, x_n|\mathbf{b})$ corresponding to a partition $\lambda = (\lambda_1, \dots, \lambda_n) \in \mathcal{P}_n$ is defined to be

$$s^{\mathbb{L}}_\lambda(\mathbf{x}_n|\mathbf{b}) = s^{\mathbb{L}}_\lambda(x_1, \dots, x_n|\mathbf{b}) := \sum_{w \in S_n} w \cdot \left[\frac{[\mathbf{x}|\mathbf{b}]^{\lambda + \rho(n-1)}}{\prod_{1 \leq i < j \leq n} (x_i +_{\mathbb{L}} \overline{x}_j)} \right]. \tag{3.1}$$

We also define

$$s^{\mathbb{L}}_\lambda(\mathbf{x}_n) = s^{\mathbb{L}}_\lambda(x_1, \dots, x_n) := s^{\mathbb{L}}_\lambda(x_1, \dots, x_n|\mathbf{0}). \tag{3.2}$$

The function $s^{\mathbb{L}}_\lambda(\mathbf{x}_n)$ will be called just the *universal Schur function.*

REMARK 3.1. The non-equivariant version $s^{\mathbb{L}}_\lambda(\mathbf{x}_n)$ is already defined by Fel'dman [**22**, Definition 4.2]. These are called the *generalized Schur polynomials* there. In that paper, the author also established a Gysin formula for these generalized Schur polynomials (see [**22**, Theorem 4.5]).[23]

Since

$$x_i +_{\mathbb{L}} \overline{x}_j = (x_i - x_j)(1 + \text{higher degree terms in } x_i \text{ and } x_j \text{ with coefficients in } \mathbb{L}),$$

the function $s^{\mathbb{L}}_\lambda(\mathbf{x}_n|\mathbf{b})$ is a formal power series with coefficients in $\mathbb{L}$ in the variables $x_1, \dots, x_n$ and $b_1, b_2, \dots, b_{\lambda_1+n-1}$. It is also a homogeneous formal power series of total degree $|\lambda| = \sum_{i=1}^n \lambda_i$, the size of λ. In (3.1), if we put $a^{\mathbb{L}}_{i,j} = 0$ for all $i, j \geq 1$ and $b_i = -a_i$ $(i = 1, 2, \dots)$, where $a = (a_1, a_2, \dots,)$ is another sequence of parameters, the functions $s^{\mathbb{L}}_\lambda(\mathbf{x}_n|\mathbf{b})$ reduce to the factorial Schur polynomials usually denoted by $s_\lambda(\mathbf{x}_n|a)$ (for its definition, see Ikeda-Naruse [**33**, §5.1], Macdonald [**49**, I, §3, Example 20], Molev-Sagan [**55**, p.4431]). If we put $a^{\mathbb{L}}_{1,1} = \beta$ and $a^{\mathbb{L}}_{i,j} = 0$ for all $(i,j) \neq (1,1)$,[24] then $s^{\mathbb{L}}_\lambda(\mathbf{x}_n|\mathbf{b})$ reduce to the factorial Grothendieck polynomials $G_\lambda(\mathbf{x}_n|\mathbf{b})$ (for its definition, see Ikeda-Naruse [**35**, (2.12), (2.13)], McNamara [**52**, Definition 4.1]). Thus our functions $s^{\mathbb{L}}_\lambda(\mathbf{x}_n|\mathbf{b})$ are generalizations of these polynomials and hence *universal* in this sense. Note that unlike the usual factorial Schur

---

[21] For notation and terminology on partitions which will be used throughout this paper, we mainly follow those in Macdonald's book [**49**, Chapter I].

[22] It is customary not to distinguish two partitions which differ only by a string of zeros at the end.

[23] Fel'dman's formula is a generalization of a formula of Fulton-Pragacz [**25**, (4.2)] (see also (4.2) in this paper) to the complex cobordism theory as well as to general partial flag bundles. However, we think that his formula should be modified correctly. In the later section §4.2.2, we shall establish a Gysin formula for the universal Schur functions, thus correcting Fel'dman's formula.

[24] Notice that the sign convention of β is opposite from the one given in Example 2.2. In the rest of this paper, we shall use this sign convention that fits in with the listed references here.

and Grothendieck polynomials, the function $s^{\mathbb{L}}_{\emptyset}(\mathbf{x}_n|\mathbf{b})$ corresponding to the empty partition $\emptyset = (0^n)$ is not equal to 1. For instance, we have

$$s^{\mathbb{L}}_{\emptyset}(\mathbf{x}_2|\mathbf{b}) = \frac{x_1 +_{\mathbb{L}} b_1}{x_1 +_{\mathbb{L}} \overline{x}_2} + \frac{x_2 +_{\mathbb{L}} b_1}{x_2 +_{\mathbb{L}} \overline{x}_1} = 1 + a^{\mathbb{L}}_{1,2}x_1x_2 + a^{\mathbb{L}}_{1,1}a^{\mathbb{L}}_{1,2}b_1x_1x_2 + \cdots \neq 1.$$

As mentioned in Remark 3.1, in order to formulate Fel'dman's Gysin formula for the universal Schur functions, we need to extend the above definition to arbitrary sequences of non-negative integers. Thus, for a sequence $I = (I_1, I_2, \ldots, I_n)$ of non-negative integers, we define $s^{\mathbb{L}}_I(\mathbf{x}_n)$ to be

$$s^{\mathbb{L}}_I(\mathbf{x}_n) := \sum_{w \in S_n} w \cdot \left[\frac{\mathbf{x}^{I+\rho(n-1)}}{\prod_{1 \le i < j \le n}(x_i +_{\mathbb{L}} \overline{x}_j)} \right]. \tag{3.3}$$

Next let us recall the definition of the universal factorial Schur P- and Q-functions. Denote by $\mathcal{SP}_n$ the set of all strict partitions of length $\le n$, i.e., a sequence of positive integers $\nu = (\nu_1, \ldots, \nu_k)$ with $k \le n$ such that $\nu_1 > \cdots > \nu_k > 0$. Then, for a strict partition $\nu = (\nu_1, \ldots, \nu_k) \in \mathcal{SP}_n$, the *universal factorial Schur P- and Q-functions* are defined to be

(3.4)

$$P^{\mathbb{L}}_{\nu}(\mathbf{x}_n|\mathbf{b}) = P^{\mathbb{L}}_{\nu}(x_1, \ldots, x_n|\mathbf{b}) := \frac{1}{(n-k)!} \sum_{w \in S_n} w \cdot \left[[\mathbf{x}|\mathbf{b}]^{\nu} \prod_{i=1}^{k} \prod_{j=i+1}^{n} \frac{x_i +_{\mathbb{L}} x_j}{x_i +_{\mathbb{L}} \overline{x}_j} \right],$$

$$Q^{\mathbb{L}}_{\nu}(\mathbf{x}_n|\mathbf{b}) = Q^{\mathbb{L}}_{\nu}(x_1, \ldots, x_n|\mathbf{b}) := \frac{1}{(n-k)!} \sum_{w \in S_n} w \cdot \left[[[\mathbf{x}|\mathbf{b}]]^{\nu} \prod_{i=1}^{k} \prod_{j=i+1}^{n} \frac{x_i +_{\mathbb{L}} x_j}{x_i +_{\mathbb{L}} \overline{x}_j} \right],$$

where the symmetric group S_n acts only on the x-variables $x_1, \ldots, x_n$ by permutations. We also define

$$\begin{aligned} P^{\mathbb{L}}_{\nu}(\mathbf{x}_n) &= P^{\mathbb{L}}_{\nu}(x_1, \ldots, x_n) := P^{\mathbb{L}}_{\nu}(x_1, \ldots, x_n|\mathbf{0}), \\ Q^{\mathbb{L}}_{\nu}(\mathbf{x}_n) &= Q^{\mathbb{L}}_{\nu}(x_1, \ldots, x_n) := Q^{\mathbb{L}}_{\nu}(x_1, \ldots, x_n|\mathbf{0}). \end{aligned}$$

The functions $P^{\mathbb{L}}_{\nu}(\mathbf{x}_n|\mathbf{b})$ and $Q^{\mathbb{L}}_{\nu}(\mathbf{x}_n|\mathbf{b})$ are formal power series with coefficients in $\mathbb{L}$ in the variables $x_1, \ldots, x_n$ and $b_1, b_2, \ldots, b_{\nu_1}$ for $P^{\mathbb{L}}_{\nu}$ (resp. $b_1, b_2 \ldots, b_{\nu_1 - 1}$ for $Q^{\mathbb{L}}_{\nu}$). These are homogeneous formal power series of total degree $|\nu|$. In (3.4), if we put $a^{\mathbb{L}}_{i,j} = 0$ for all $i, j \ge 1$ and $b_i = -a_i$ $(i = 1, 2, \ldots)$, the functions $P^{\mathbb{L}}_{\nu}(\mathbf{x}_n|\mathbf{b})$, $Q^{\mathbb{L}}_{\nu}(\mathbf{x}_n|\mathbf{b})$ reduce to the *factorial Schur P- and Q-polynomials* $P_{\nu}(\mathbf{x}_n|a)$, $Q_{\nu}(\mathbf{x}_n|a)$ (for their definitions, see Ikeda-Mihalcea-Naruse [**34**, §4.2]), Ikeda-Naruse [**33**, Definition 8.1], Ivanov [**37**, Definitions 2.10 and 2.13]). If we put $a^{\mathbb{L}}_{1,1} = \beta$ and $a^{\mathbb{L}}_{i,j} = 0$ for all $(i,j) \neq (1,1)$, then $P^{\mathbb{L}}_{\nu}(\mathbf{x}_n|\mathbf{b})$, $Q^{\mathbb{L}}_{\nu}(\mathbf{x}_n|\mathbf{b})$ reduce to the *K-theoretic factorial Schur P- and Q-polynomials* $GP_{\nu}(\mathbf{x}_n|\mathbf{b})$, $GQ_{\nu}(\mathbf{x}_n|\mathbf{b})$ due to Ikeda-Naruse [**35**, Definition 2.1]. Thus our functions $P^{\mathbb{L}}_{\nu}(\mathbf{x}_n|\mathbf{b})$, $Q^{\mathbb{L}}_{\nu}(\mathbf{x}_n|\mathbf{b})$ are generalizations of these polynomials and hence *universal* in this sense.

3.3. Universal Hall-Littlewood functions. In this subsection, we introduce the *universal Hall-Littlewood functions* which interpolate between the S-functions and the P-functions. We use the notation as in Pragacz [**62**]. Let $\lambda = (\lambda_1, \ldots, \lambda_n) \in \mathcal{P}_n$ be a partition of length $\ell(\lambda) \le n$. Consider the maximal "intervals" $I_1, I_2, \ldots, I_d$ in $[n] := \{1, 2, \ldots, n\}$, where the sequence λ is "constant".

Thus we have a decomposition

$$[n] = I_1 \sqcup I_2 \sqcup \cdots \sqcup I_d \quad \text{(disjoint union)}.$$

Here and in what follows, we keep the following convention: when we refer to such a decomposition, we always arrange "intervals" $I_1, I_2, \ldots$ in increasing order, that is, $\max I_r < \min I_{r+1}$ for each r. Let m_r be the "length" of the interval I_r for $r = 1, 2, \ldots, d$, namely the cardinality of I_r so that $\sum_{r=1}^{d} m_r = n$. We write $n(i)$ the number of the interval containing i for $i \in [n]$, namely if i is in I_r, then $n(i) = r$. Notice that, since $\lambda = (\lambda_1, \ldots, \lambda_n)$ is a partition, $n(i) < n(j)$ is equivalent to $\lambda_i > \lambda_j$ for $i, j \in [n]$. We define a subgroup S_n^{λ} of S_n as the stabilizer of λ. Thus

$$S_n^{\lambda} = \prod_{i=1}^{d} S_{m_i} = S_{m_1} \times S_{m_2} \times \cdots \times S_{m_d}.$$

Denote by $\ell_{\mathbb{L}}(x) \in \mathbb{L} \otimes \mathbb{Q}[[x]]$ the *logarithm*[25] of $F_{\mathbb{L}}$, i.e., a unique formal power series with leading term x such that

$$\ell_{\mathbb{L}}(a +_{\mathbb{L}} b) = \ell_{\mathbb{L}}(a) + \ell_{\mathbb{L}}(b).$$

Using the logarithm $\ell_{\mathbb{L}}(x)$, one can rewrite the n-series $[n]_{\mathbb{L}}(x)$ for a non-negative integer n, aforementioned in §3.1, as

$$\ell_{\mathbb{L}}([n]_{\mathbb{L}}(x)) = \ell_{\mathbb{L}}(\underbrace{x +_{\mathbb{L}} + \cdots +_{\mathbb{L}} x}_{n}) = \underbrace{\ell_{\mathbb{L}}(x) + \cdots + \ell_{\mathbb{L}}(x)}_{n} = n \cdot \ell_{\mathbb{L}}(x),$$

in other words, $[n]_{\mathbb{L}}(x) = \ell_{\mathbb{L}}^{-1}(n \cdot \ell_{\mathbb{L}}(x))$. This formula allows us to define

$$[t]_{\mathbb{L}}(x) := \ell_{\mathbb{L}}^{-1}(t \cdot \ell_{\mathbb{L}}(x))$$

for an indeterminate t. This is a natural extension of $t \cdot x$ as well as the n-series $[n]_{\mathbb{L}}(x)$.[26]

DEFINITION 3.2 (Universal Hall-Littlewood function). With the above notation, for a partition $\lambda \in \mathcal{P}_n$, we define

$$H_{\lambda}^{\mathbb{L}}(\mathbf{x}_n; t) := \sum_{\overline{w} \in S_n / S_n^{\lambda}} w \cdot \left[\mathbf{x}^{\lambda} \prod_{1 \le i < j \le n,\; n(i) < n(j)} \frac{x_i +_{\mathbb{L}} [t]_{\mathbb{L}}(\overline{x}_j)}{x_i +_{\mathbb{L}} \overline{x}_j} \right]. \tag{3.5}$$

If we put $a_{i,j}^{\mathbb{L}} = 0$ for all $i, j \ge 1$, the functions $H_{\lambda}^{\mathbb{L}}(\mathbf{x}_n; t)$ reduce to the usual Hall-Littlewood polynomials denoted by $P_{\lambda}(x_1, \ldots, x_n; t)$ in Macdonald's book [**49**, Chapter III, §2, (2.2)]. For the usual Hall-Littlewood polynomial $P_{\lambda}(x_1, \ldots, x_n; t)$, it is known that

$$P_{\lambda}(x_1, \ldots, x_n; 0) = s_{\lambda}(x_1, \ldots, x_n)$$

[25] See e.g., Kono-Tamaki [**44**, Lemma 6.27], Levine-Morel [**48**, Lemma 4.1.29], Quillen [**63**], Ravenel [**65**, Appendix A2]. If we put $a_{i,j}^{\mathbb{L}} = 0$ for all $i, j \ge 1$, then $F_{\mathbb{L}}(u, v)$ reduces to the additive formal group law $F_a(u, v) = u + v$, and both $\ell_{\mathbb{L}}(x)$ and $\ell_{\mathbb{L}}^{-1}(x)$ reduce to x. If we put $a_{1,1}^{\mathbb{L}} = \beta$ and $a_{i,j}^{\mathbb{L}} = 0$ for all $(i, j) \ne (1, 1)$, then $F_{\mathbb{L}}(u, v)$ reduces to the multiplicative formal group law $F_m(u, v) = u + v + \beta uv$. In this case, $\ell_{\mathbb{L}}(x)$ reduces to $\beta^{-1} \log(1 + \beta x) = \sum_{i=0}^{\infty} \frac{(-\beta)^i}{i+1} x^{i+1}$, and $\ell_{\mathbb{L}}^{-1}(x)$ reduces to $\beta^{-1}(e^{\beta x} - 1) = \sum_{i=0}^{\infty} \frac{\beta^i}{(i+1)!} x^{i+1}$.

[26] $[t]_{\mathbb{L}}(x)$ reduces to just $t \cdot x$ for the additive formal group law $F_a(u, v) = u + v$, and $\beta^{-1}\{(1 + \beta x)^t - 1\} = \sum_{i=1}^{\infty} \frac{(t)_i}{i!} \beta^{i-1} x^i$ for the multiplicative formal group law $F_m(u, v) = u + v + \beta uv$. Here $(t)_i$ means $t(t-1)(t-2) \cdots (t-i+1)$.

under the specialization $t = 0$ (see Macdonald [**49**, Chapter III, §2, (2.3)]), and

$$P_\lambda(x_1, \ldots, x_n; 1) = m_\lambda(x_1, \ldots, x_n),$$

under the specialization $t = 1$ (see Macdonald [**49**, Chapter III, §2, (2.4)]). Here m_λ denotes the monomial symmetric polynomial corresponding to λ. Moreover, for a strict partition ν of length $\ell(\nu) \leq n$, one obtains that

$$P_\nu(x_1, \ldots, x_n; -1) = P_\nu(x_1, \ldots, x_n) \tag{3.6}$$

under the specialization $t = -1$ (see Macdonald [**49**, Chapter III, §8, Example 1]).

For the universal Hall-Littlewood functions $H^{\mathbb{L}}_\lambda(\mathbf{x}_n; t)$, it follows immediately from (3.5) that $H^{\mathbb{L}}_\lambda(\mathbf{x}_n; 1) = m_\lambda(\mathbf{x}_n)$ under the specialization $t = 1$. Let us next consider the specialization $t = -1$. Let $\nu = (\nu_1, \ldots, \nu_k) \in \mathcal{SP}_n$ be a strict partition with length $\ell(\nu) = k \leq n$. Then we have a decomposition $[n] = I_1 \sqcup \cdots \sqcup I_k \sqcup I_{k+1}$, where $I_r = \{r\}$ $(r = 1, \ldots, k)$ and $I_{k+1} = \{k+1, \ldots, n\}$. Therefore we have $m_r = 1$ $(r = 1, \ldots, k)$, $m_{k+1} = n - k$ and $n(i) = i$ $(i = 1, \ldots, k)$, $n(i) = k + 1$ $(i = k+1, \ldots, n)$. The stabilizer of ν is given by $S_n^\nu = (S_1)^k \times S_{n-k}$. Therefore it follows from Definition 3.2 and (3.4), we have

$$\begin{aligned} H^{\mathbb{L}}_\nu(\mathbf{x}_n; -1) &= \sum_{\overline{w} \in S_n/(S_1)^k \times S_{n-k}} w \cdot \left[\mathbf{x}^\nu \prod_{1 \leq i < j \leq n,\ 1 \leq i \leq k} \frac{x_i +_{\mathbb{L}} x_j}{x_i +_{\mathbb{L}} \overline{x}_j} \right] \\ &= \frac{1}{(n-k)!} \sum_{w \in S_n} w \cdot \left[\mathbf{x}^\nu \prod_{i=1}^{k} \prod_{j=i+1}^{n} \frac{x_i +_{\mathbb{L}} x_j}{x_i +_{\mathbb{L}} \overline{x}_j} \right] = P^{\mathbb{L}}_\nu(\mathbf{x}_n). \end{aligned}$$

We now consider the specialization $t = 0$. Let $\lambda = (\lambda_1, \ldots, \lambda_n) \in \mathcal{P}_n$ be a partition with length $\ell(\lambda) \leq n$. Then we have a decomposition $[n] = I_1 \sqcup I_2 \sqcup \cdots \sqcup I_d$ as above. Letting m_r be the cardinality of I_r for $r = 1, \ldots, d$, one can rewrite $\lambda = (n_1^{m_1}\ n_2^{m_2}\ \ldots\ n_d^{m_d})$ $(n_1 > n_2 > \cdots > n_d \geq 0)$. We put $\nu(r) := m_1 + \cdots + m_r$ for $r = 1, 2, \ldots, d$ and $\nu(0) := 0$. Then the specialization $t = 0$ gives

$$\begin{aligned} H^{\mathbb{L}}_\lambda(\mathbf{x}_n; 0) &= \sum_{\overline{w} \in S_n/S_n^\lambda} w \cdot \left[\mathbf{x}^\lambda \prod_{1 \leq i < j \leq n,\ n(i) < n(j)} \frac{x_i}{x_i +_{\mathbb{L}} \overline{x}_j} \right] \\ &= \sum_{\overline{w} \in S_n/S_n^\lambda} w \cdot \left[\prod_{i=1}^{n} x_i^{\lambda_i} \prod_{1 \leq i < j \leq n,\ n(i) < n(j)} \frac{x_i}{x_i +_{\mathbb{L}} \overline{x}_j} \right] \\ &= \sum_{\overline{w} \in S_n/S_n^\lambda} w \cdot \left[\frac{\prod_{r=1}^{d} \left(\prod_{m_1 + \cdots + m_{r-1} < i \leq m_1 + \cdots + m_r} x_i^{n_r + n - (m_1 + \cdots + m_r)} \right)}{\prod_{1 \leq i < j \leq n,\ n(i) < n(j)} (x_i +_{\mathbb{L}} \overline{x}_j)} \right] \\ &= \sum_{\overline{w} \in S_n/S_n^\lambda} w \cdot \left[\frac{\prod_{r=1}^{d} \left(\prod_{\nu(r-1) < i \leq \nu(r)} x_i^{n_r + n - \nu(r)} \right)}{\prod_{1 \leq i < j \leq n,\ n(i) < n(j)} (x_i +_{\mathbb{L}} \overline{x}_j)} \right]. \end{aligned} \tag{3.7}$$

We shall consider this function in later section §5.

4. Applications of Gysin formulas to the Schur functions

In §2.6, we reviewed various Gysin formulas in the ordinary cohomology (or Chow) theory, and in the previous section §3, we introduced *universal* analogues of the ordinary Schur S-, P-, Q-, and Hall-Littlewood functions. In this section, we pursue the Gysin formulas in the complex cobordism theory which relate Gysin maps for flag bundles to these universal Schur functions. The following §4.1 is devoted to the recollection of various known Gysin formulas in the ordinary cohomology theory, and their generalization to the complex cobordism theory will be treated in §4.2. Our main tool for establishing Gysin formulas in the complex cobordism theory is the Bressler-Evens formula reviewed in §2.5.

4.1. Gysin formulas for various Schur functions. We use the same notation as in §2.6.2. Let $E \xrightarrow{p} X$ be a complex vector bundle of rank n (over a variety). Denote by $\tau = \tau_E : \mathcal{F}\ell(E) \longrightarrow X$ the associated flag bundle parametrizing successive flags of quotient bundles of E of ranks $n-1, \dots, 2, 1$. The usual Schur polynomial $s_\lambda(X_1, \dots, X_n)$ corresponding to a partition $\lambda \in \mathcal{P}_n$ is a symmetric polynomial in the n-variables $X_1, \dots, X_n$, and therefore it can be written as a polynomial in the elementary symmetric polynomials $e_i(X_1, \dots, X_n)$'s. Let $x_1, \dots, x_n$ be the Chern roots of E as in §2.6.2. Then the Chern classes $c_i(E)$'s can be identified with $e_i(x_1, \dots, x_n)$'s as usual, and hence $s_\lambda(x_1, \dots, x_n)$ can be expressed as a polynomial in $c_i(E)$'s. Let us define a cohomology class $s_\lambda(E) \in H^{2|\lambda|}(X)$ to be $\tau^*(s_\lambda(E)) := s_\lambda(x_1, \dots, x_n) \in H^{2|\lambda|}(\mathcal{F}\ell(E))$.[27] Then the following formula is known:

PROPOSITION 4.1 (Pragacz [**59**], Lemma 2.3; Fulton-Pragacz [**25**], (4.1)). *The image of the monomial $\mathbf{x}^{\lambda+\rho(n-1)} = x_1^{\lambda_1+n-1}x_2^{\lambda_2+n-2}\cdots x_n^{\lambda_n}$ under the Gysin homomorphism $\tau_* : H^*(\mathcal{F}\ell(E)) \longrightarrow H^*(X)$ is given by*

$$\tau_*(\mathbf{x}^{\lambda+\rho(n-1)}) = s_\lambda(E). \tag{4.1}$$

The formula (4.1) is called the *Jacobi-Trudi identity*, from which Fulton-Pragacz [**25**] derived some useful formulas for Grassmann bundles, which we recall a bit later.

Furthermore, the analogous Gysin formulas which relate the Hall-Littlewood polynomials and more general flag bundles are considered in Pragacz [**62**]. Let us recall these formulas. We use the same notation as in §3.3 (see also Pragacz [**62**]). Let $\lambda = (\lambda_1, \dots, \lambda_n) \in \mathcal{P}_n$ be a partition of length $\le n$. Then we have a decomposition

$$[n] = \{1, 2, \dots, n\} = I_1 \sqcup I_2 \sqcup \cdots \sqcup I_d,$$

where λ is "constant" on each I_r $(r = 1, 2, \dots, d)$. Denote by m_r the length of the interval I_r for $r = 1, 2, \dots, d$ so that $\sum_{r=1}^d m_r = n$, and $n(i)$ the number of the interval containing i for $i \in [n]$. We put $\nu(p) = \sum_{r=1}^p m_r$ for $p = 1, \dots, d$ and $\nu(0) = 0$. The stabilizer of λ is denoted by S_n^λ. Associated to a complex vector bundle $E \longrightarrow X$, one can define a "$(d-1)$-step flag bundle" with steps of lengths m_r

$$\eta_\lambda : \mathcal{F}\ell^\lambda(E) \longrightarrow X,$$

[27] It is well-known that the induced homomorphism $\tau^* : H^*(X) \hookrightarrow H^*(\mathcal{F}\ell(E))$ is injective, and hence the cohomology class $s_\lambda(E)$ is well-defined.

parametrizing flags of quotient bundles of E of ranks

$$n-m_d=\nu(d-1), n-m_d-m_{d-1}=\nu(d-2),\dots,n-m_d-m_{d-1}-\dots-m_2=\nu(1).$$

If $\lambda=\emptyset$, the empty partition, then $\mathcal{F}\ell^{\emptyset}(E)$ is understood to be the base space X. Here, for later discussion, we shall fix the notation about *partial flag bundles* associated to a complex vector bundle $E \xrightarrow{p} X$ of rank n (see Fulton's book [**24**, §9.1, 10.6]): For a sequence of integers $0<r_1<r_2<\cdots<r_k<n$, let us denote by $\mathcal{F}\ell_{r_1,r_2,\dots,r_k}(E)$ a partial flag bundle consisting of flags of subbundles $0\subsetneq S_1\subsetneq S_2\subsetneq\cdots\subsetneq S_k\subsetneq E$ with $\operatorname{rank} S_i=r_i$ $(i=1,\dots,k)$. Since giving a flag of subbundles of E as above is equivalent to giving a flag of quotient bundles $E\twoheadrightarrow Q^k\twoheadrightarrow Q^{k-1}\twoheadrightarrow\cdots\twoheadrightarrow Q^1\twoheadrightarrow 0$ with $\operatorname{rank} Q^i=n-r_{k+1-i}$ $(i=1,\dots,k)$, this partial flag bundle is also denoted by $\mathcal{F}\ell^{n-r_1,n-r_2,\dots,n-r_k}(E)$. Moreover, dualizing the sequence of vector bundles

$$0\hookrightarrow S_1\hookrightarrow S_2\hookrightarrow\cdots\hookrightarrow S_k\hookrightarrow E\twoheadrightarrow Q^k\twoheadrightarrow Q^{k-1}\twoheadrightarrow\cdots\twoheadrightarrow Q^1\twoheadrightarrow 0$$

gives the sequence of vector bundles

$$\begin{aligned}0\hookrightarrow (Q^1)^\vee\hookrightarrow (Q^2)^\vee\hookrightarrow\cdots\quad &\hookrightarrow (Q^k)^\vee\hookrightarrow E^\vee\\ &\twoheadrightarrow (S_k)^\vee\twoheadrightarrow (S_{k-1})^\vee\twoheadrightarrow\cdots\twoheadrightarrow (S_1)^\vee\twoheadrightarrow 0.\end{aligned}$$

Therefore we have the canonical isomorphism

$$\begin{aligned}\mathcal{F}\ell_{r_1,r_2,\dots,r_k}(E)=\mathcal{F}\ell^{n-r_1,n-r_2,\dots,n-r_k}(E) &\cong \mathcal{F}\ell_{n-r_k,n-r_{k-1},\dots,n-r_1}(E^\vee)\\ &=\mathcal{F}\ell^{r_k,r_{k-1},\dots,r_1}(E^\vee).\end{aligned}$$

Using this notation, we can write

$$\begin{aligned}\mathcal{F}\ell^\lambda(E) &=\mathcal{F}\ell^{\nu(d-1),\nu(d-2),\dots,\nu(1)}(E)=\mathcal{F}\ell_{n-\nu(d-1),n-\nu(d-2),\dots,n-\nu(1)}(E)\\ &\cong\mathcal{F}\ell_{\nu(1),\nu(2),\dots,\nu(d-1)}(E^\vee)=\mathcal{F}\ell^{n-\nu(1),n-\nu(2),\dots,n-\nu(d-1)}(E^\vee).\end{aligned}$$

EXAMPLE 4.2 (See Pragacz [**62**], Example 2).

(1) Let $\nu=(\nu_1,\dots,\nu_k)\in\mathcal{SP}_n$ be a strict partition with length $\ell(\nu)=k\le n$. Then as we saw at the end of §3.3, we have $d=k+1$, and

$$(m_1,\dots,m_k,m_{k+1})=(\underbrace{1,1,\dots,1}_{k},n-k),\quad S_n^\nu=(S_1)^k\times S_{n-k}.$$

Then the corresponding flag bundle $\eta_\nu:\mathcal{F}\ell^\nu(E)\longrightarrow X$ is often denoted by $\tau^k=\tau_E^k:\mathcal{F}\ell^{k,k-1,\dots,2,1}(E)\longrightarrow X$, and parametrizes flags of successive quotient bundles of E of ranks $k,k-1,\dots,2,1$. As a special case of this example, the flag bundle corresponding to the partition $\rho(n-1)=(n-1,n-2,\dots,2,1,0)$ is the full flag bundle $\tau:\mathcal{F}\ell^{n-1,n-2,\dots,2,1}(E)=\mathcal{F}\ell(E)\longrightarrow X$.

(2) Let $\lambda=(\underbrace{a,\dots,a}_{q},\underbrace{b,\dots,b}_{n-q})=(a^q\,b^{n-q})\in\mathcal{P}_n$ be a partition with $a>b\ge 0$. Then we have a decomposition $[n]=I_1\sqcup I_2$, where $I_1=[1,q]=\{1,\dots,q\}$ and $I_2=[q+1,n]=\{q+1,\dots,n\}$. Therefore we have

$$(m_1,m_2)=(q,n-q),\quad\text{and}\quad S_n^\lambda=S_q\times S_{n-q}.$$

Thus the corresponding flag bundle is the Grassmann bundle $\pi:G^q(E)\longrightarrow X$ parametrizing rank q quotient bundles of E.

Now we recall the useful formula for Grassmann bundle derived from the Jacobi-Trudi identity mentioned above. Let $\pi : G^q(E) \longrightarrow X$ be the Grassmann bundle parametrizing rank q quotient bundles of E as in Example 4.2 (2). On $G^q(E)$, we have the tautological exact sequence of vector bundles:

$$0 \longrightarrow S \hookrightarrow \pi^*(E) \twoheadrightarrow Q \longrightarrow 0,$$

where rank $S = n - q$ and rank $Q = q$. Set $r := n - q$. Then by the repeated applications of the Jacobi-Trudi identity (4.1), Fulton-Pragacz [**25**, (4.2)] showed the following formula:[28] For any partitions $\lambda = (\lambda_1, \ldots, \lambda_q)$, $\mu = (\mu_1, \ldots, \mu_r)$,

$$\pi_*(s_\lambda(Q) \cdot s_\mu(S)) = s_{\lambda_1 - r, \ldots, \lambda_q - r,\, \mu_1, \ldots, \mu_r}(E). \tag{4.2}$$

In §4.2.2 below, we shall give a generalization of the above formula (4.2) as a special case of the corrected version of Fel'dman's Gysin formula.

Let us recall Gysin formulas for more general flag bundles. The following proposition is a generalization of Theorem 2.8 to the partial flag bundle $\eta_\lambda : \mathcal{F}\ell^\lambda(E) \longrightarrow X$, which was proved by Pragacz [**62**] as a particular case of Brion [**12**, Proposition 2.1]:

PROPOSITION 4.3 (Pragacz [**62**], Proposition 5). *For an S_n^λ-invariant polynomial $f(X_1, \ldots, X_n) \in H^*(X)[X_1, \ldots, X_n]^{S_n^\lambda}$, we have*

$$(\eta_\lambda)_*(f(x_1, \ldots, x_n)) = \sum_{\overline{w} \in S_n/S_n^\lambda} w \cdot \left[\frac{f(x_1, \ldots, x_n)}{\prod_{1 \le i < j \le n,\, n(i) < n(j)} (x_i - x_j)} \right].$$

Here the element $f(x_1, \ldots, x_n)$ is regarded as an element in $H^*(\mathcal{F}\ell^\lambda(E)) \hookrightarrow H^*(\mathcal{F}\ell(E))$.[29]

From this, Pragacz showed (implicitly) the following formula:

COROLLARY 4.4. *For the Gysin homomorphism $(\eta_\lambda)_* : H^*(\mathcal{F}\ell^\lambda(E)) \longrightarrow H^*(X)$, the following formula holds:*

$$(\eta_\lambda)_* \left(\mathbf{x}^\lambda \prod_{1 \le i < j \le n,\, n(i) < n(j)} (x_i - t x_j) \right) = P_\lambda(E; t),$$

where the cohomology class $P_\lambda(E; t)$ is defined from the Hall-Littlewood polynomial $P_\lambda(x_1, \ldots, x_n; t)$ in the same way as $s_\lambda(E)$ at the beginning of this subsection.

By Corollary 4.4, one can deduce Gysin formula for Schur P-polynomials (see Pragacz [**62**, Examples 2 and 11, Corollary 6]):

COROLLARY 4.5 (Pragacz [**62**], Corollary 6). *In the same setting as in Example 4.2 (1), the following formula holds:*

$$(\tau^k)_* \left(\mathbf{x}^\nu \prod_{1 \le i \le k,\, 1 \le i < j \le n} (x_i - t x_j) \right) = P_\nu(E; t).$$

---

[28] Note that in case the sequence $(\lambda_1 - r, \ldots, \lambda_q - r,\, \mu_1, \ldots, \mu_r)$ is *not* a partition, the right-hand side is either 0 or $\pm s_\nu(E)$ for some partition ν (see Fulton-Pragacz [**25**, p.42, Footnote]).

[29] Strictly speaking, this formula should be considered in $H^*(\mathcal{F}\ell(E))$ via the pull-back $H^*(X) \overset{\eta_\lambda^*}{\hookrightarrow} H^*(\mathcal{F}\ell^\lambda(E)) \hookrightarrow H^*(\mathcal{F}\ell(E))$ (cf. Corollary 2.6). In what follows, we often use such abbreviation to simplify the presentation.

Since we know that $P_\nu(\mathbf{x}_n; -1) = P_\nu(\mathbf{x}_n)$ (see (3.6)), we obtain the following:

COROLLARY 4.6. *With the above notation, the following formula holds:*

$$(\tau^k)_*\left(\mathbf{x}^\nu \prod_{1\le i\le k,\ 1\le i<j\le n}(x_i+x_j)\right) = P_\nu(E).$$

4.2. Universal Gysin formulas for the universal Schur functions.

4.2.1. *Application of the Bressler-Evens formula.* Since the Bressler-Evens formula (Theorem 2.5) is formulated in complex-oriented generalized cohomology theories, it can be applied especially to the complex cobordism theory $MU^*(-)$. We use the same notation as in §2.6.2. For a complex vector bundle $E \xrightarrow{p} X$ of rank n, one can associate the MU^*-theory Chern classes[30] $c_i^{MU}(E) \in MU^{2i}(X)$ $(i = 1, 2, \ldots, n)$ and $c_0^{MU}(E) := 1$. Put $x_i = x_i^{MU} := c_1^{MU}(L^i) \in MU^2(\mathcal{F}\ell(E))$ $(i = 1, 2, \ldots, n)$ (the MU^*-theory Chern roots of E). Then the MU^*-cohomology of $\mathcal{F}\ell(E)$ is given as follows (see e.g., Hornbostel-Kiritchenko [**28**, Theorem 2.6]):

$$MU^*(\mathcal{F}\ell(E)) = MU^*(X)[x_1, \ldots, x_n]/\left(\prod_{i=1}^n (1+x_i) = c^{MU}(E)\right),$$

where $c^{MU}(E) = \sum_{i=0}^n c_i^{MU}(E)$ is the total Chern class of E. Then we obtain the following result, which is a *universal* analogue of Theorem 2.8:

THEOREM 4.7. *With the same notation as in Theorem 2.8, we have*

$$\tau^* \circ \tau_*(f(x_1, \ldots, x_n)) = \sum_{w\in S_n} w \cdot \left[\frac{f(x_1, \ldots, x_n)}{\prod_{1\le i<j\le n}(x_i +_{\mathbb{L}} \overline{x}_j)}\right]$$

for a polynomial $f(X_1, \ldots, X_n) \in MU^*(X)[X_1, \ldots, X_n]$.

Thus the Gysin map τ_* in the complex cobordism theory is also given by a certain symmetrizing operator which is a universal analogue of the Jacobi symmetrizer (see Theorem 2.8).

By Theorem 4.7 and (3.1), we obtain the following corollary, which is a universal analogue of Proposition 4.1:

COROLLARY 4.8 (Characterization of the universal Schur functions). *The image of the monomial* $\mathbf{x}^{\lambda+\rho(n-1)} = x_1^{\lambda_1+n-1}x_2^{\lambda_2+n-2}\cdots x_n^{\lambda_n}$ *under the Gysin homomorphism* $\tau_* : MU^*(\mathcal{F}\ell(E)) \longrightarrow MU^*(X)$ *is given by*

$$\tau_*(\mathbf{x}^{\lambda+\rho(n-1)}) = s_\lambda^{\mathbb{L}}(E).$$

Here the characteristic class $s_\lambda^{\mathbb{L}}(E) \in MU^{2|\lambda|}(X)$ *is defined by the same manner as* $s_\lambda(E) \in H^{2|\lambda|}(X)$.

REMARK 4.9. Since $\tau^*(s_\lambda^{\mathbb{L}}(E)) = s_\lambda^{\mathbb{L}}(\mathbf{x}_n) \in MU^*(\mathcal{F}\ell(E))$, and τ^* is injective, the above formula can be written as

$$\tau_*(\mathbf{x}^{\lambda+\rho(n-1)}) = s_\lambda^{\mathbb{L}}(\mathbf{x}_n).$$

More generally, the symmetrizing operator description of τ_* (Theorem 4.7) yields formally the following formula:

$$\tau_*([\mathbf{x}|\mathbf{b}]^{\lambda+\rho(n-1)}) = s_\lambda^{\mathbb{L}}(\mathbf{x}_n|\mathbf{b}). \tag{4.3}$$

[30] For the historical reason, they are also called *Conner-Floyd Chern classes* (see Adams [**1**, Part I, §4], Conner-Floyd [**15**, Corollary 8.3]).

Here $\mathbf{b} = (b_1, b_2, \ldots)$ is a sequence of certain elements in $MU^*(X)$, which behaves as scalars with respect to τ_*.

The universal analogue of Proposition 4.3 can also be obtained. Under the same setting as in Proposition 4.3, we have a classifying map $h : X \longrightarrow BU(n)$ and its lift $\tilde{h} : \mathcal{F}\ell^\lambda(E) \longrightarrow B(U(m_1) \times U(m_2) \times \cdots \times U(m_d))$, and the following diagram is commutative:

$$\begin{array}{ccc} \mathcal{F}\ell^\lambda(E) & \xrightarrow{\tilde{h}} & B(U(m_1) \times U(m_2) \times \cdots \times U(m_d)) \\ \eta_\lambda \downarrow & & \downarrow \sigma \\ X & \xrightarrow{h} & BU(n). \end{array}$$

By Corollary 2.6 and the base-change property of Gysin maps, we obtain immediately the following generalization of Proposition 4.3:

THEOREM 4.10. *For an S_n^λ-invariant polynomial $f(X_1, \ldots, X_n) \in MU^*(X)[X_1, \ldots, X_n]^{S_n^\lambda}$, we have*

$$(\eta_\lambda)_*(f(x_1, \ldots, x_n)) = \sum_{\overline{w} \in S_n/S_n^\lambda} w \cdot \left[\frac{f(x_1, \ldots, x_n)}{\prod_{1 \le i < j \le n,\ n(i) < n(j)} (x_i +_{\mathbb{L}} \overline{x}_j)} \right].$$

From this, we have the following corollary which is a universal analogue of Corollary 4.4:

COROLLARY 4.11 (Characterization of the universal Hall-Littlewood functions). *For the Gysin homomorphism $(\eta_\lambda)_* : MU^*(\mathcal{F}\ell^\lambda(E)) \longrightarrow MU^*(X)$, the following formula holds*:

$$(\eta_\lambda)_* \left(\mathbf{x}^\lambda \prod_{1 \le i < j \le n,\ n(i) < n(j)} (x_i +_{\mathbb{L}} [t]_{\mathbb{L}}(\overline{x}_j)) \right) = H_\lambda^{\mathbb{L}}(E; t).$$

For a strict partition $\nu \in \mathcal{SP}_n$, we saw that $H_\nu^{\mathbb{L}}(\mathbf{x}_n; -1) = P_\nu^{\mathbb{L}}(\mathbf{x}_n)$ at the end of §3.3. From this and Corollary 4.11, we obtain the following corollary, which is a universal analogue of Corollary 4.6:

COROLLARY 4.12. *With the same notation as in Corollary 4.6, we have*

$$(\tau^k)_* \left(\mathbf{x}^\nu \prod_{1 \le i \le k,\ 1 \le i < j \le n} (x_i +_{\mathbb{L}} x_j) \right) = P_\nu^{\mathbb{L}}(E).$$

4.2.2. *Fel'dman's Gysin formula.* As promised in the footnote after Remark 3.1, we shall reformulate Fel'dman's Gysin formula [**22**, Theorem 4.5] in the complex cobordism theory, and prove it. To this end, we use the universal Schur function $s_I^{\mathbb{L}}(\mathbf{x}_n)$ for a sequence of non-negative integers I defined in §3.2. For two sequences $I = (I_1, I_2, \ldots, I_q)$, $J = (J_1, J_2, \ldots, J_r)$ of non-negative integers, denote by IJ their *juxtaposition*, i.e.,

$$IJ := (I_1, I_2, \ldots, I_q, J_1, J_2, \ldots, J_r).$$

Let $\lambda = (\lambda_1, \ldots, \lambda_n) \in \mathcal{P}_n$ be a partition of length $\le n$, and rewrite it as $\lambda = (n_1^{m_1}\, n_2^{m_2} \cdots n_d^{m_d})$, $n_1 > n_2 > \cdots > n_d \ge 0$, according to the decomposition $[n] = I_1 \sqcup I_2 \sqcup \cdots \sqcup I_d$. Then consider the $(d-1)$-step partial flag bundle

$$\eta_\lambda : \mathcal{F}\ell^\lambda(E) \longrightarrow X$$

associated with a complex vector bundle $E \longrightarrow X$ (for the notation, see §4.1). We are concerned with the induced Gysin map in the complex cobordism theory:

$$(\eta_\lambda)_* : MU^*(\mathcal{F}\ell^\lambda(E)) \longrightarrow MU^*(X).$$

On $\mathcal{F}\ell^\lambda(E)$, we have the tautological sequence of flag of sub and quotient bundles:

$$S_0 = 0 \hookrightarrow S_1 \hookrightarrow \cdots \hookrightarrow S_r \hookrightarrow \cdots \hookrightarrow S_{d-1} \hookrightarrow S_d = \eta_\lambda^*(E),$$
$$\eta_\lambda^*(E) = Q^d \twoheadrightarrow Q^{d-1} \twoheadrightarrow \cdots \twoheadrightarrow Q^r \twoheadrightarrow \cdots \twoheadrightarrow Q^1 \twoheadrightarrow Q^0 = 0,$$

where, $Q^r := \eta_\lambda^*(E)/S_{d-r}$ $(r = 0, 1, 2, \dots, d)$. Define the vector bundles over $\mathcal{F}\ell^\lambda(E)$ by

$$E_r := \operatorname{Ker}(Q^r \twoheadrightarrow Q^{r-1}) \quad (r = 1, 2, \dots, d).$$

Then we have $\eta_\lambda^*(E) \cong \bigoplus_{r=1}^d E_r$, and the MU^*-theory Chern roots of E_r are given by $x_{\nu(r-1)+1}, \dots, x_{\nu(r)}$ $(r = 1, 2, \dots, d)$. Here $x_1, \dots, x_n$ are the MU^*-theory Chern roots of E (see the beginning of §4.2.1), and $\nu(r) := \sum_{i=1}^r m_i$ for $r = 1, 2, \dots, d$ and $\nu(0) := 0$ (see the end of §3.3). With this notation, our version of Fel'dman's Gysin formula is stated as follows:

THEOREM 4.13 (cf. Fel'dman [**22**], Theorem 4.5). *For sequences of non-negative integers $I^{(r)} = (I_1^{(r)}, I_2^{(r)}, \dots, I_{m_r}^{(r)})$ $(r = 1, 2, \dots, d)$, we have*

$$(\eta_\lambda)_* \left(\prod_{r=1}^d s^{\mathbb{L}}_{I^{(r)} + ((n-\nu(r))^{m_r})}(E_r) \right) = s^{\mathbb{L}}_{I^{(1)} I^{(2)} \cdots I^{(d)}}(E).$$

Here the partition $((n-\nu(r))^{m_r})$ means $(\underbrace{n-\nu(r), n-\nu(r), \dots, n-\nu(r)}_{m_r})$.

PROOF. As in Fel'dman's proof, it suffices to prove the assertion for the case of a Grassmann bundle. Therefore we may assume that the partition λ is of the form $\lambda = (a^q\, b^{n-q})$ with $a > b \geq 0$, and consider the associated Grassmann bundle $\pi : G^q(E) \longrightarrow X$ (see Example 4.2 (2)). On $G^q(E)$, we have the tautological exact sequence of vector bundles:

$$0 \longrightarrow S \hookrightarrow \pi^*(E) \twoheadrightarrow Q \longrightarrow 0.$$

Note that $E_1 = Q$, and $E_2 = \operatorname{Ker}(\pi^*(E) \twoheadrightarrow Q) \cong S$. For given two sequences of non-negative integers $I = (I_1, I_2, \dots, I_q)$, $J = (J_1, J_2, \dots, J_{n-q})$, consider the universal Schur function corresponding to the juxtaposition $IJ = (I_1, \dots, I_q, J_1, \dots, J_{n-q})$:

$$s^{\mathbb{L}}_{IJ}(x_1, \dots, x_q, x_{q+1}, \dots, x_n) = \sum_{w \in S_n} w \cdot \left[\frac{\mathbf{x}^{IJ+\rho(n-1)}}{\prod_{1 \leq i < j \leq n}(x_i +_{\mathbb{L}} \overline{x}_j)} \right].$$

We decompose the denominator and the numerator inside the bracket in the right-hand side as

$$\prod_{1 \leq i < j \leq n} (x_i +_{\mathbb{L}} \overline{x}_j) = \prod_{1 \leq i < j \leq q} (x_i +_{\mathbb{L}} \overline{x}_j) \times \prod_{q+1 \leq i < j \leq n} (x_i +_{\mathbb{L}} \overline{x}_j) \times \prod_{\substack{1 \leq i \leq q \\ q+1 \leq j \leq n}} (x_i +_{\mathbb{L}} \overline{x}_j),$$

$$\begin{aligned}
\mathbf{x}^{IJ+\rho(n-1)} &= \prod_{i=1}^q x_i^{I_i+n-i} \times \prod_{i=q+1}^n x_i^{J_{i-q}+n-i} \\
&= (x_1 \cdots x_q)^{n-q} \times \prod_{i=1}^q x_i^{I_i+q-i} \times \prod_{i=q+1}^n x_i^{J_{i-q}+n-i}.
\end{aligned}$$

Since $\prod_{1\le i\le q,\, q+1\le j\le n}(x_i +_{\mathbb{L}} \overline{x}_j)$ and $(x_1\cdots x_q)^{n-q}$ are both $S_q \times S_{n-q}$-invariant, we compute

$$\begin{aligned}
s^{\mathbb{L}}_{IJ}(x_1,\dots,x_q,x_{q+1},\dots,x_n) &= \sum_{w\in S_n} w\cdot\left[\frac{\mathbf{x}^{IJ+\rho(n-1)}}{\prod_{1\le i<j\le n}(x_i +_{\mathbb{L}} \overline{x}_j)}\right]\\
&= \sum_{\overline{w}\in S_n/S_q\times S_{n-q}} w\cdot\left[\frac{(x_1\cdots x_q)^{n-q}}{\prod_{1\le i\le q,\, q+1\le j\le n}(x_i +_{\mathbb{L}} \overline{x}_j)}\right.\\
&\quad\times \left.\sum_{(u,v)\in S_q\times S_{n-q}} u\cdot\left[\frac{\prod_{i=1}^{q} x_i^{I_i+q-i}}{\prod_{1\le i<j\le q}(x_i +_{\mathbb{L}} \overline{x}_j)}\right]\times v\cdot\left[\frac{\prod_{i=q+1}^{n} x_i^{J_{i-q}+n-i}}{\prod_{q+1\le i<j\le n}(x_i +_{\mathbb{L}} \overline{x}_j)}\right]\right]\\
&= \sum_{\overline{w}\in S_n/S_q\times S_{n-q}} w\cdot\left[\frac{(x_1\cdots x_q)^{n-q} s^{\mathbb{L}}_I(x_1,\dots,x_q) s^{\mathbb{L}}_J(x_{q+1},\dots,x_n)}{\prod_{1\le i\le q,\, q+1\le j\le n}(x_i +_{\mathbb{L}} \overline{x}_j)}\right]\\
&= \pi_*((x_1\cdots x_q)^{n-q} s^{\mathbb{L}}_I(x_1,\dots,x_q)\, s^{\mathbb{L}}_J(x_{q+1},\dots,x_n))\\
&= \pi_*(s^{\mathbb{L}}_{I+((n-q)^q)}(x_1,\dots,x_q)\, s^{\mathbb{L}}_J(x_{q+1},\dots,x_n)).
\end{aligned}$$

In the final step, we used the formula "$(x_1\cdots x_n)^c \times s^{\mathbb{L}}_I(\mathbf{x}_n) = s^{\mathbb{L}}_{I+(c^n)}(\mathbf{x}_n)$" for any non-negative integer c, which follows immediately from the definition of $s^{\mathbb{L}}_I(\mathbf{x}_n)$. Thus we have

$$\pi_*(s^{\mathbb{L}}_{I+((n-q)^q)}(x_1,\dots,x_q) s^{\mathbb{L}}_J(x_{q+1},\dots,x_n)) = s^{\mathbb{L}}_{IJ}(x_1,\dots,x_n).$$

Rewriting this expression in terms of vector bundles, we have the desired formula:

$$\pi_*(s^{\mathbb{L}}_{I+((n-q)^q)}(E_1)\, s^{\mathbb{L}}_J(E_2)) = s^{\mathbb{L}}_{IJ}(E).$$

□

5. New universal factorial Schur functions

5.1. Definition of the new universal factorial Schur functions. In §3.3, we defined the universal Hall-Littlewood function $H^{\mathbb{L}}_\lambda(\mathbf{x}_n;t)$ for a partition $\lambda\in\mathcal{P}_n$. We saw that for a strict partition $\nu\in\mathcal{SP}_n$, this function reduces to the universal Schur P-function $P^{\mathbb{L}}_\nu(\mathbf{x}_n)$ under the specialization $t=-1$. As announced in §3.3, we shall consider the specialization $t=0$ in this subsection. In fact, unlike the usual Hall-Littlewood polynomial $P_\lambda(x_1,\dots,x_n;t)$, the universal Hall-Littlewood function $H^{\mathbb{L}}_\lambda(\mathbf{x}_n;t)$ need not coincide with the universal Schur function $s^{\mathbb{L}}_\lambda(\mathbf{x}_n)$ under the specialization $t=0$. Thus one obtains another *universal* analogue of the Schur polynomial, which we call the "new universal (factorial) Schur function". We shall use the same notation as in §3.3. Let $\lambda=(\lambda_1,\dots,\lambda_n)\in\mathcal{P}_n$ be a partition of length $\le n$. Then we rewrite $\lambda=(n_1^{m_1}\, n_2^{m_2}\,\dots\, n_d^{m_d})$, $n_1>n_2>\cdots>n_d\ge 0$ as before. Put $\nu(r)=\sum_{i=1}^r m_i$ for $r=1,2,\dots,d$ and $\nu(0)=0$. Define

$$(5.1)\qquad (\mathbf{x}|\mathbf{b})^{[\lambda]} := \prod_{r=1}^{d}\left(\prod_{\nu(r-1)<i\le\nu(r)} [x_i|\mathbf{b}]^{n_r+n-\nu(r)}\right).$$

When the parameters $\mathbf{b}=\mathbf{0}$, then we write simply $\mathbf{x}^{[\lambda]}$ for $(\mathbf{x}|\mathbf{0})^{[\lambda]}$. With the above notation, we make the following definition:

DEFINITION 5.1 (New universal factorial Schur functions). For a partition $\lambda = (\lambda_1, \ldots, \lambda_n) \in \mathcal{P}_n$, we define

$$\mathbb{S}^{\mathbb{L}}_{\lambda}(\mathbf{x}_n|\mathbf{b}) = \mathbb{S}^{\mathbb{L}}_{\lambda}(x_1, \ldots, x_n|\mathbf{b}) := \sum_{\overline{w} \in S_n/S_n^{\lambda}} w \cdot \left[\frac{(\mathbf{x}|\mathbf{b})^{[\lambda]}}{\prod_{1 \le i < j \le n,\ n(i) < n(j)} (x_i +_{\mathbb{L}} \overline{x}_j)} \right].$$

We also define

$$\mathbb{S}^{\mathbb{L}}_{\lambda}(\mathbf{x}_n) = \mathbb{S}^{\mathbb{L}}_{\lambda}(x_1, \ldots, x_n) := \mathbb{S}^{\mathbb{L}}_{\lambda}(x_1, \ldots, x_n|\mathbf{0}).$$

It follows immediately from Definition 5.1 and (3.7) that when $t = 0$, the universal Hall-Littlewood function $H^{\mathbb{L}}_{\lambda}(\mathbf{x}_n; t)$ reduces to the new universal Schur function $\mathbb{S}^{\mathbb{L}}_{\lambda}(\mathbf{x}_n)$. Moreover, one sees directly from the definition that if we specialize the universal formal group law $F_{\mathbb{L}}(u, v)$ to the additive one $F_a(u, v) = u + v$ (and $b_i = -a_i$ $(i = 1, 2, \ldots)$), then the new universal factorial Schur function $\mathbb{S}^{\mathbb{L}}_{\lambda}(\mathbf{x}_n|\mathbf{b})$ reduces to the factorial Schur function $s_{\lambda}(\mathbf{x}_n|a)$ (see the proof of Macdonald [**49**, Chapter III, §1, (1.5)]). On the other hand, it is not obvious from Definition 5.1 that $\mathbb{S}^{\mathbb{L}}_{\lambda}(\mathbf{x}_n|\mathbf{b})$ reduces to the factorial Grothendieck polynomial $G_{\lambda}(\mathbf{x}_n|\mathbf{b})$ under the specialization from $F_{\mathbb{L}}(u, v)$ to the multiplicative one $F_m(u, v) = u + v + \beta uv$. We shall show that this is the case after establishing some Gysin formulas for the new universal factorial Schur functions (see §5.2). Thus the new universal factorial Schur functions are also *universal* analogues of the usual Schur functions. Note that, by definition, $\mathbb{S}^{\mathbb{L}}_{\emptyset}(\mathbf{x}_n|\mathbf{b}) = 1$ for the empty partition $\emptyset$, whereas $s^{\mathbb{L}}_{\emptyset}(\mathbf{x}_n|\mathbf{b}) \neq 1$ (see §3.2).

EXAMPLE 5.2.

(1) Let $\lambda \in \mathcal{P}_n$ be a partition such that $\lambda_1 > \lambda_2 > \cdots > \lambda_{n-1} > \lambda_n \ge 0$. Then we have a decomposition $[n] = I_1 \sqcup I_2 \sqcup \cdots \sqcup I_n$, where $I_i = \{i\}$. Therefore we have

$$(m_1, m_2, \ldots, m_n) = (\underbrace{1, 1, \ldots, 1}_{n}) \quad \text{and} \quad S_n^{\lambda} = \underbrace{S_1 \times \cdots \times S_1}_{n} = \{1\}.$$

In this case, one sees immediately from (5.1) that $(\mathbf{x}|\mathbf{b})^{[\lambda]} = [\mathbf{x}|\mathbf{b}]^{\lambda + \rho(n-1)}$, and hence

$$s^{\mathbb{L}}_{\lambda}(\mathbf{x}_n|\mathbf{b}) = \mathbb{S}^{\mathbb{L}}_{\lambda}(\mathbf{x}_n|\mathbf{b}).$$

Thus for such a strict partition λ, the function $s^{\mathbb{L}}_{\lambda}(\mathbf{x}_n|\mathbf{b})$ coincides with the function $\mathbb{S}^{\mathbb{L}}_{\lambda}(\mathbf{x}_n|\mathbf{b})$. However, for a partition with equal parts, the difference does occur (see the end of §5.2).

(2) Let $k \ge 1$ be a positive integer, and consider the case where λ is "one-row", namely $\lambda = (k) = (k\ 0^{n-1})$. Then we have a decomposition $[n] = I_1 \sqcup I_2$, where $I_1 = \{1\}$, $I_2 = [2, n] = \{2, \ldots, n\}$. Thus we have

$$(m_1, m_2) = (1, n-1) \quad \text{and} \quad S_n^{\lambda} = S_1 \times S_{n-1},$$

and one sees immediately that $\mathbf{x}^{[\lambda]} = x_1^{k+n-1}$. Therefore the function $\mathbb{S}^{\mathbb{L}}_k(\mathbf{x}_n) := \mathbb{S}^{\mathbb{L}}_{(k)}(\mathbf{x}_n)$ corresponding to the one-row (k) is given by

$$\mathbb{S}^{\mathbb{L}}_k(\mathbf{x}_n) = \sum_{\overline{w} \in S_n/S_1 \times S_{n-1}} w \cdot \left[\frac{x_1^{k+n-1}}{\prod_{j=2}^{n} (x_1 +_{\mathbb{L}} \overline{x}_j)} \right] = \sum_{i=1}^{n} \frac{x_i^{k+n-1}}{\prod_{j \neq i} (x_i +_{\mathbb{L}} \overline{x}_j)}. \tag{5.2}$$

Here we assumed that $k \ge 1$ is a positive integer. However, the right-hand side of (5.2) makes sense for $k \ge 1 - n$, which could be non-positive, or actually negative if $n \ge 2$. Therefore one can formally define $\mathbb{S}^{\mathbb{L}}_k(\mathbf{x}_n)$ for each integer $k \geq 1-n$ by the right-hand side of the above expression. For instance, one has

$$\mathbb{S}_0^{\mathbb{L}}(\mathbf{x}_n) = \sum_{i=1}^{n} \frac{x_i^{n-1}}{\prod_{j \neq i}(x_i +_{\mathbb{L}} \overline{x}_j)},$$

which differs from 1.[31] Moreover, as we will see in the next subsection §5.3, one can define $\mathbb{S}_k^{\mathbb{L}}(\mathbf{x}_n)$ for any integer $k \in \mathbb{Z}$.

REMARK 5.3. In our previous paper [**56**, §4.5], we investigated various properties of the universal factorial Schur functions $s_\lambda^{\mathbb{L}}(\mathbf{x}_n|\mathbf{b})$ such as "vanishing property", "basis theorem". The new universal factorial Schur functions $\mathbb{S}_\lambda^{\mathbb{L}}(\mathbf{x}_n|\mathbf{b})$ also have the similar properties. We shall discuss this problem elsewhere.

5.2. Gysin formulas for the new universal Schur functions. In this subsection, we shall establish Gysin formulas for the new universal factorial Schur functions. Using these formulas, we are able to compare the new universal factorial Schur functions ("new" functions for short) $\mathbb{S}_\lambda^{\mathbb{L}}(\mathbf{x}_n|\mathbf{b})$ with the universal factorial Schur functions ("old" functions for short) $s_\lambda^{\mathbb{L}}(\mathbf{x}_n|\mathbf{b})$ introduced in §3.2.

5.2.1. *Gysin formulas for the new universal Schur functions.* We begin with the following theorem: From Theorem 4.10, we have immediately

THEOREM 5.4 (Characterization of the new universal Schur functions). *For the Gysin homomorphism $(\eta_\lambda)_* : MU^*(\mathcal{F}\ell^\lambda(E)) \longrightarrow MU^*(X)$, the following formula holds*:

$$(\eta_\lambda)_*(\mathbf{x}^{[\lambda]}) = \mathbb{S}_\lambda^{\mathbb{L}}(\mathbf{x}_n),$$

where $x_1, \ldots, x_n$ are the MU^*-theory Chern roots of E as before. More generally, for a sequence $\mathbf{b} = (b_1, b_2, \ldots)$ of certain elements in $MU^*(X)$, the following version holds:

$$(\eta_\lambda)_*((\mathbf{x}|\mathbf{b})^{[\lambda]}) = \mathbb{S}_\lambda^{\mathbb{L}}(\mathbf{x}_n|\mathbf{b}). \tag{5.3}$$

EXAMPLE 5.5.

(1) Let us consider the "one-row" case $\lambda = (k)$ as in Example 5.2 (2). Thus the corresponding flag bundle is the projective bundle $\pi_1 : G^1(E) \longrightarrow X$, and Theorem 5.4 implies the following formula:
$$\pi_{1*}(x_1^{k+n-1}) = \mathbb{S}_k^{\mathbb{L}}(\mathbf{x}_n).$$
(2) Let $\lambda = (\underbrace{a, \ldots, a}_{q}, \underbrace{b, \ldots, b}_{n-q}) = (a^q\, b^{n-q}) \in \mathcal{P}_n$ be a partition with $a > b \geq 0$ as in Example 4.2 (2). Then the corresponding flag bundle is the Grassmann bundle $\pi : G^q(E) \longrightarrow X$, and one sees directly that $\mathbf{x}^{[\lambda]} = x_1^{a+n-q} \cdots x_q^{a+n-q} x_{q+1}^b \cdots x_n^b$, and Theorem 5.4 implies
$$\pi_*(x_1^{a+n-q} \cdots x_q^{a+n-q} x_{q+1}^b \cdots x_n^b) = \mathbb{S}_{(a^q\, b^{n-q})}^{\mathbb{L}}(\mathbf{x}_n).$$
(3) Let $\lambda \in \mathcal{P}_n$ be a strict partition such that $\lambda_1 > \lambda_2 > \cdots > \lambda_{n-1} > \lambda_n \geq 0$ as in Example 5.2 (1). Then the corresponding partial flag bundle is the full flag bundle $\tau : \mathcal{F}\ell(E) \longrightarrow X$. In this case, one obtains that $\mathbf{x}^{[\lambda]} = \mathbf{x}^{\lambda+\rho(n-1)}$, and Corollary 4.8 and Theorem 5.4 imply
$$s_\lambda^{\mathbb{L}}(\mathbf{x}_n) = \tau_*(\mathbf{x}^{\lambda+\rho(n-1)}) = \mathbb{S}_\lambda^{\mathbb{L}}(\mathbf{x}_n),$$
which was already seen in Example 5.2 (1) (with $\mathbf{b} = \mathbf{0}$).

[31] Note that $\mathbb{S}_\emptyset^{\mathbb{L}}(\mathbf{x}_n) = 1$ by definition.

5.2.2. *Comparison of "new" functions $\mathbb{S}^{\mathbb{L}}_{\lambda}(\mathbf{x}_n|\mathbf{b})$ with "old" functions $s^{\mathbb{L}}_{\lambda}(\mathbf{x}_n|\mathbf{b})$.* More generally, the above Gysin formulas (Theorem 5.4) can be formulated as those between two partial flag bundles of the form "$\pi^{\mu}_{\lambda} : \mathcal{F}\ell^{\mu}(E) \twoheadrightarrow \mathcal{F}\ell^{\lambda}(E)$", where the partition μ is a "refinement" of λ. Let us explain what the word "refinement" means: We use the notation in the beginning of this subsection (see also §3.3 and §4.1). For two positive integers a, b with $a < b$, denote by $[a, b]$ the set of integers i such that $a \le i \le b$. Let $\lambda, \mu \in \mathcal{P}_n$ be two partitions of length $\le n$. Then as explained in §3.3, one obtains two decompositions

$$[1, n] = I_1 \sqcup I_2 \sqcup \cdots \sqcup I_d, \quad [1, n] = J_1 \sqcup J_2 \sqcup \cdots \sqcup J_e$$

of the interval $[1, n]$ corresponding to λ, μ respectively. Suppose that the decomposition $[1, n] = J_1 \sqcup \cdots \sqcup J_e$ is a refinement of the decomposition $[1, n] = I_1 \sqcup \cdots \sqcup I_d$ in the usual sense, i.e., $I_1 = J_1 \sqcup \cdots \sqcup J_{k_1}$, $I_2 = J_{k_1+1} \sqcup \cdots \sqcup J_{k_2}$, and so on, for some positive integers $1 \le k_1 < k_2 < \cdots \le e$. Then we say that μ is a "refinement" of λ.[32] By the construction of the associated partial flag bundle $\mathcal{F}\ell^{\lambda}(E)$ (see §4.1), we have a natural projection from $\mathcal{F}\ell^{\mu}(E)$ to $\mathcal{F}\ell^{\lambda}(E)$, denoted by π^{μ}_{λ}. Furthermore, for three partitions λ, μ, and $\nu \in \mathcal{P}_n$, if ν is a refinement of μ, and μ is a refinement of λ, then ν is also a refinement of λ, and the relation $\pi^{\mu}_{\lambda} \circ \pi^{\nu}_{\mu} = \pi^{\nu}_{\lambda}$ holds. For example, the partition $\rho(n-1) = (n-1, n-2, \ldots, 1, 0) \in \mathcal{P}_n$ is a refinement of any partition $\lambda \in \mathcal{P}_n$, and any partition $\lambda \in \mathcal{P}_n$ is a refinement of the empty partition $\emptyset = (0^n)$. In particular, we are concerned with the following three projections and the induced Gysin homomorphisms in complex cobordism:

$$\begin{aligned} &\pi^{\rho(n-1)}_{\emptyset} = \tau : \mathcal{F}\ell^{\rho(n-1)}(E) = \mathcal{F}\ell(E) \twoheadrightarrow \mathcal{F}\ell^{\emptyset}(E) = X, \\ &\pi^{\lambda}_{\emptyset} = \eta_{\lambda} : \mathcal{F}\ell^{\lambda}(E) \twoheadrightarrow \mathcal{F}\ell^{\emptyset}(E) = X, \\ &\pi^{\rho(n-1)}_{\lambda} : \mathcal{F}\ell^{\rho(n-1)}(E) = \mathcal{F}\ell(E) \twoheadrightarrow \mathcal{F}\ell^{\lambda}(E). \end{aligned} \tag{5.4}$$

The theorem below (Theorem 5.6) will be useful in describing the difference between the "new" function $\mathbb{S}^{\mathbb{L}}_{\lambda}(\mathbf{x}_n|\mathbf{b})$ and the "old" one $s^{\mathbb{L}}_{\lambda}(\mathbf{x}_n|\mathbf{b})$. Note that τ is written as a composite $\eta_{\lambda} \circ \pi^{\rho(n-1)}_{\lambda}$, and hence $\tau_* = (\eta_{\lambda})_* \circ (\pi^{\rho(n-1)}_{\lambda})_*$.

THEOREM 5.6.

(1) *For the Gysin map $(\pi^{\rho(n-1)}_{\emptyset})_* = \tau_* : MU^*(\mathcal{F}\ell(E)) \longrightarrow MU^*(X)$, the following formula holds:*

$$(\pi^{\rho(n-1)}_{\emptyset})_*([\mathbf{x}|\mathbf{b}]^{\lambda+\rho(n-1)}) = s^{\mathbb{L}}_{\lambda}(\mathbf{x}_n|\mathbf{b}).$$

(2) *For the Gysin map $(\pi^{\lambda}_{\emptyset})_* = (\eta_{\lambda})_* : MU^*(\mathcal{F}\ell^{\lambda}(E)) \longrightarrow MU^*(X)$, the following formula holds:*

$$(\pi^{\lambda}_{\emptyset})_*((\mathbf{x}|\mathbf{b})^{[\lambda]}) = \mathbb{S}^{\mathbb{L}}_{\lambda}(\mathbf{x}_n|\mathbf{b}).$$

(3) *For the Gysin map $(\pi^{\rho(n-1)}_{\lambda})_* : MU^*(\mathcal{F}\ell(E)) \longrightarrow MU^*(\mathcal{F}\ell^{\lambda}(E))$, the following formula holds:*

$$(\pi^{\rho(n-1)}_{\lambda})_*([\mathbf{x}|\mathbf{b}]^{\lambda+\rho(n-1)}) = (\mathbf{x}|\mathbf{b})^{[\lambda]} \cdot \prod_{r=1}^{d} s^{\mathbb{L}}_{\emptyset}(x_{\nu(r-1)+1}, \ldots, x_{\nu(r)}|\mathbf{b}[+[n_r+n-\nu(r)]]).$$

Here $\mathbf{b} = (b_1, b_2, \ldots)$ is a sequence of elements in $MU^*(X)$, and $\mathbf{b}[+m] := (b_{m+1}, b_{m+2}, \ldots)$ for a positive integer m.

[32] Note that even though μ is a "refinement" of λ, this does not necessarily mean $\mu \subset \lambda$.

PROOF. (1) and (2): The Gysin maps $(\pi_{\emptyset}^{\rho(n-1)})_* = \tau_*$ and $(\pi_{\emptyset}^{\lambda})_* = (\eta_\lambda)_*$ were already considered in (4.3) and (5.3) respectively.

(3) We shall show the assertion when the partition $\lambda \in \mathcal{P}_n$ is of the form $\lambda = (a^q\, b^{n-q})$ (see Example 5.5 (2)). In this case, one sees

$$\begin{aligned}
[\mathbf{x}|\mathbf{b}]^{\lambda+\rho(n-1)} &= \prod_{i=1}^{q}[x_i|\mathbf{b}]^{a+n-i} \times \prod_{i=q+1}^{n}[x_i|\mathbf{b}]^{b+n-i}\\
&= \prod_{i=1}^{q}[x_i|\mathbf{b}]^{a+n-q} \times \prod_{i=q+1}^{n}[x_i|\mathbf{b}]^{b} \times \prod_{i=1}^{q}[x_i|\mathbf{b}[+(a+n-q)]]^{q-i} \times \prod_{i=q+1}^{n}[x_i|\mathbf{b}[+b]]^{n-i}\\
&= (\mathbf{x}|\mathbf{b})^{[\lambda]} \times \prod_{i=1}^{q}[x_i|\mathbf{b}[+(a+n-q)]]^{q-i} \times \prod_{i=q+1}^{n}[x_i|\mathbf{b}[+b]]^{n-i}.
\end{aligned}$$

Since $(\mathbf{x}|\mathbf{b})^{[\lambda]}$ is $S_q \times S_{n-q}$-invariant, we have by the Bressler-Evens formula (Theorem 2.5),

$$\begin{aligned}
&(\pi_{\lambda}^{\rho(n-1)})_*([\mathbf{x}|\mathbf{b}]^{\lambda+\rho(n-1)})\\
&= \sum_{w\in S_q\times S_{n-q}} w\cdot\left[\frac{[\mathbf{x}|\mathbf{b}]^{\lambda+\rho(n-1)}}{\prod_{1\le i<j\le q}(x_i +_{\mathbb{L}} \overline{x}_j) \times \prod_{q+1\le i<j\le n}(x_i +_{\mathbb{L}} \overline{x}_j)}\right]\\
&= (\mathbf{x}|\mathbf{b})^{[\lambda]} \times \sum_{w\in S_q\times S_{n-q}} w\cdot\left[\frac{\prod_{i=1}^{q}[x_i|\mathbf{b}[+(a+n-q)]]^{q-i} \times \prod_{i=q+1}^{n}[x_i|\mathbf{b}[+b]]^{n-i}}{\prod_{1\le i<j\le q}(x_i +_{\mathbb{L}} \overline{x}_j) \times \prod_{q+1\le i<j\le n}(x_i +_{\mathbb{L}} \overline{x}_j)}\right]\\
&= (\mathbf{x}|\mathbf{b})^{[\lambda]} \times s_{\emptyset}^{\mathbb{L}}(x_1,\dots,x_q|\mathbf{b}[+(a+n-q)]) \times s_{\emptyset}^{\mathbb{L}}(x_{q+1},\dots,x_n|\mathbf{b}[+b]),
\end{aligned}$$

and the theorem holds.

For an arbitrary partition $\lambda \in \mathcal{P}_n$, write λ as the following form: $\lambda = (n_1^{m_1} \cdots n_d^{m_d})$, $n_1 > \cdots > n_d \ge 0$. Here $m_i > 0$ for $i = 1,2,\dots,d$, and $\sum_{i=1}^{d} m_i = n$. Put $\nu(r) = \sum_{i=1}^{r} m_i$ for $r = 1,2,\dots,d$ and $\nu(0) = 0$. Then analogous computation to the above case leads to the result of $(\pi_{\lambda}^{\rho(n-1)})_*([\mathbf{x}|\mathbf{b}]^{\lambda+\rho(n-1)})$. Note that in the case where $\lambda = (a^q\, b^{n-q})$, the "parameter shift" is given by $a+n-q = n_1+n-\nu(1)$ and $b = n_2+n-\nu(2)$. □

With the aide of Theorem 5.6, we observe that

(5.5)
$$\begin{aligned}
s_{\lambda}^{\mathbb{L}}(\mathbf{x}_n|\mathbf{b}) &= \tau_*([\mathbf{x}|\mathbf{b}]^{\lambda+\rho(n-1)}) = (\eta_\lambda)_* \circ (\pi_{\lambda}^{\rho(n-1)})_*([\mathbf{x}|\mathbf{b}]^{\lambda+\rho(n-1)})\\
&= (\eta_\lambda)_*\left((\mathbf{x}|\mathbf{b})^{[\lambda]} \cdot \prod_{r=1}^{d} s_{\emptyset}^{\mathbb{L}}(x_{\nu(r-1)+1},\dots,x_{\nu(r)}|\mathbf{b}[+[n_r+n-\nu(r)]])\right),\\
\mathbb{S}_{\lambda}^{\mathbb{L}}(\mathbf{x}_n|\mathbf{b}) &= (\eta_\lambda)_*((\mathbf{x}|\mathbf{b})^{[\lambda]}).
\end{aligned}$$

Thus the difference between the "new" function $\mathbb{S}_{\lambda}^{\mathbb{L}}(\mathbf{x}_n|\mathbf{b})$ and the "old" one $s_{\lambda}^{\mathbb{L}}(\mathbf{x}_n|\mathbf{b})$ is given by the product of "old" functions corresponding to the empty partition $\emptyset$. From this, "new" function $\mathbb{S}_{\lambda}^{\mathbb{L}}(\mathbf{x}_n|\mathbf{b})$ reduce to $G_\lambda(\mathbf{x}_n|\mathbf{b})$ under the specialization from $F_{\mathbb{L}}(u,v)$ to $F_m(u,v)$ because of the fact $G_{\emptyset}(\mathbf{x}_n|\mathbf{b}) = 1$ (see e.g., Ikeda-Naruse [**35**, §2.4]).

5.3. Generating function for $\mathbb{S}_k^{\mathbb{L}}(\mathbf{x}_n)$. As we mentioned in Example 5.2 (2), the new universal Schur functions $\mathbb{S}_k^{\mathbb{L}}(\mathbf{x}_n) := \mathbb{S}_{(k)}^{\mathbb{L}}(\mathbf{x}_n)$ for $k \geq 1$ can be extended to non-positive integers. Thus one has the functions $\mathbb{S}_k^{\mathbb{L}}(\mathbf{x}_n)$ for all integers $k \in \mathbb{Z}$. In this subsection, we shall give the generating function for these. For this purpose, we make use of Quillen's result. Recall from Quillen [**63**] that the *normalized invariant differential form*[33] $\omega_{\mathbb{L}}(t)$ associated with the universal formal group law $F_{\mathbb{L}}(u, v) = u + v + \sum_{i,j\geq 1} a_{i,j}^{\mathbb{L}} u^i v^j$ is defined by

$$\omega_{\mathbb{L}}(t) := \frac{dt}{\dfrac{\partial F_{\mathbb{L}}}{\partial v}(t, 0)} = \frac{dt}{1 + \sum_{i\geq 1} a_{i,1}^{\mathbb{L}} t^i}.$$

The logarithm $\ell_{\mathbb{L}}(x)$ of $F_{\mathbb{L}}$ is then determined by the equations

$$\ell'_{\mathbb{L}}(t)\, dt = \omega_{\mathbb{L}}(t), \quad \ell_{\mathbb{L}}(0) = 0.$$

Then Quillen gave the following formula (see also Damon [**16**, p.650, Proposition]):

THEOREM 5.7 (Quillen [**63**], Theorem 1). *Let $E \longrightarrow X$ be a complex vector bundle of rank n, let $\pi_1 : G^1(E) \cong P(E^\vee) \longrightarrow X$ be the associated projective bundle of lines in the dual $E^\vee$ of E, and let $Q \cong \mathcal{O}(1)$ be the canonical quotient line bundle on $G^1(E) \cong P(E^\vee)$. Then the Gysin homomorphism $\pi_{1*} : MU^*(G^1(E)) \longrightarrow MU^*(X)$ is given by the residue formula*

$$\pi_{1*}(f(\xi)) = \mathrm{Res}_{t=0} \frac{f(t)\, \omega_{\mathbb{L}}(t)}{\prod_{i=1}^n (t +_{\mathbb{L}} \overline{x}_i)} = \mathrm{Res}_{t=0} \frac{f(t)\, dt}{(1 + \sum_{i\geq 1} a_{i,1}^{\mathbb{L}} t^i) \prod_{i=1}^n (t +_{\mathbb{L}} \overline{x}_i)}. \tag{5.6}$$

Thus $\pi_{1*}(f(\xi))$ is given by the coefficient of t^{-1} in the Laurent series

$$\frac{f(t)}{(1 + \sum_{i\geq 1} a_{i,1}^{\mathbb{L}} t^i) \prod_{i=1}^n (t +_{\mathbb{L}} \overline{x}_i)}.$$

Here $f(t) \in MU^*(X)[t]$, $\xi := c_1^{MU}(Q) \in MU^2(G^1(E))$, and $x_1 = \xi, x_2, \ldots, x_n$ are the MU^*-theory Chern roots of E.

On the other hand, in the same setting as in Examples 5.2 and 5.5, together with Theorem 4.10, we obtain

$$\pi_{1*}(f(\xi)) = \sum_{i=1}^n \frac{f(x_i)}{\prod_{j\neq i} (x_i +_{\mathbb{L}} \overline{x}_j)}. \tag{5.7}$$

The equivalence of (5.6) and (5.7) is shown by Vishik [**71**, Lemma 5.36]. Therefore if we take $f(t) = t^{k+n-1}$ ($k \geq 1$), we have by Example 5.2 (2),

$$\begin{aligned} \mathbb{S}_k^{\mathbb{L}}(\mathbf{x}_n) &= \sum_{i=1}^n \frac{x_i^{k+n-1}}{\prod_{j\neq i}(x_i +_{\mathbb{L}} \overline{x}_j)} = \pi_{1*}(\xi^{k+n-1}) \\ &= \mathrm{Res}_{t=0} \frac{t^{k+n-1}\, dt}{(1 + \sum_{i\geq 1} a_{i,1}^{\mathbb{L}} t^i) \prod_{i=1}^n (t +_{\mathbb{L}} \overline{x}_i)}. \end{aligned} \tag{5.8}$$

---

[33] If we put $a_{i,j}^{\mathbb{L}} = 0$ for all $i, j \geq 1$, then $\omega_{\mathbb{L}}(t)$ reduces to dt. If we put $a_{1,1}^{\mathbb{L}} = \beta$ and $a_{i,j}^{\mathbb{L}} = 0$ for all $(i, j) \neq (1, 1)$, then $\omega_{\mathbb{L}}(t)$ reduces to $\dfrac{dt}{1 + \beta t}$.

Thus the coefficient of t^{-1} in the formal Laurent series

$$\frac{t^{k+n-1}}{(1+\sum_{i\geq 1} a^{\mathbb{L}}_{i,1} t^i)\prod_{i=1}^n (t +_{\mathbb{L}} \overline{x}_i)}$$

is equal to $\mathbb{S}^{\mathbb{L}}_k(\mathbf{x}_n)$.

REMARK 5.8. It should be remarked that the above residue formula is closely related to the recent work of Darondeau-Pragacz [**18**]. The following comment might be useful when one wants to relate the residue formula to their work. We borrow the notation from [**18**, §0]. For a Laurent series $f(t)$ in one indeterminate t, we shall denote by $[t^k](f(t))$ the coefficient of t^k in $f(t)$. Then the above residue formula means

$$\begin{aligned}\mathbb{S}^{\mathbb{L}}_k(\mathbf{x}_n) &= [t^{-1}]\left(\frac{t^{k+n-1}}{(1+\sum_{i\geq 1} a^{\mathbb{L}}_{i,1} t^i)\prod_{i=1}^n (t +_{\mathbb{L}} \overline{x}_i)}\right)\\ &= [t^{n-1}]\left(t^{k+n-1}\cdot\frac{t^n}{(1+\sum_{i\geq 1} a^{\mathbb{L}}_{i,1} t^i)\prod_{i=1}^n (t +_{\mathbb{L}} \overline{x}_i)}\right).\end{aligned}$$

If one wants to consider the ordinary cohomology theory, we put $a^{\mathbb{L}}_{i,j} = 0$ for all $i, j \geq 1$. Then the above formula leads to

$$s_k(\mathbf{x}_n) = [t^{n-1}]\left(t^{k+n-1}\cdot\frac{t^n}{\prod_{i=1}^n (t - x_i)}\right).$$

Here the left-hand side is the usual Schur polynomial corresponding to the one row (k), namely, the k-th homogeneous complete symmetric polynomial. We see easily that the rational function $\dfrac{t^n}{\prod_{i=1}^n (t - x_i)}$ in the right-hand side is the "reversed Segre polynomial" $s_{1/t}(E)$ of E if we think the variables $x_1, \ldots, x_n$ the Chern roots of a complex vector bundle E. This is the fundamental formula given in [**18**, §0, p.2].[34]

From the above interpretation, we can obtain the generating function for $\mathbb{S}^{\mathbb{L}}_k(\mathbf{x}_n)$ $(k \in \mathbb{Z})$. We argue as follows: Set

$$F_n(t) := \frac{t^n}{(1+\sum_{i\geq 1} a^{\mathbb{L}}_{i,1} t^i)\prod_{i=1}^n (t +_{\mathbb{L}} \overline{x}_i)}.$$

Then it follows from the above interpretation that the coefficient of t^{-k} of $F_n(t)$ is equal to $\mathbb{S}^{\mathbb{L}}_k(\mathbf{x}_n)$ for each positive integer $k \geq 1$. By defining $\mathbb{S}^{\mathbb{L}}_k(\mathbf{x}_n)$ for $k \leq 0$ by the same procedure, one obtains

$$F_n(t) = \sum_{k\in\mathbb{Z}} \mathbb{S}^{\mathbb{L}}_k(\mathbf{x}_n) t^{-k}.$$

Thus we have the following:

THEOREM 5.9. *The generating function for* $\mathbb{S}^{\mathbb{L}}_k(\mathbf{x}_n)$ $(k \in \mathbb{Z})$ *is given by*

$$\sum_{k\in\mathbb{Z}} \mathbb{S}^{\mathbb{L}}_k(\mathbf{x}_n) u^k = F_n(t)|_{t=u^{-1}} = \frac{u^{-n}}{(1+\sum_{i\geq 1} a^{\mathbb{L}}_{i,1} u^{-i})\prod_{i=1}^n (u^{-1} +_{\mathbb{L}} \overline{x}_i)}. \tag{5.9}$$

[34]Since Darondeau-Pragacz uses the projective bundle of lines $G_1(E) = P(E) \longrightarrow X$, not $G^1(E) \cong P(E^\vee)$, in their formulation, one has to change E in their result to its dual $E^\vee$.

REMARK 5.10.

(1) In Hudson-Matsumura [**31**, Definition 3.1], the Segre class in the algebraic cobordism theory $\mathscr{S}_m(E)$ $(m \in \mathbb{Z})$ of a complex vector bundle E is defined by using the push-forward image from the projective bundle $G^1(E) \cong P(E^\vee)$.[35] By definition, these classes coincide with our $\mathbb{S}^{\mathbb{L}}_m(E)$. Furthermore, they obtained the generating function $\mathscr{S}(E;u) := \sum_{m\in\mathbb{Z}} \mathscr{S}_m(E)u^m$ of the Segre classes [**31**, Theorem 3.6] by a different method from ours. One can check directly that their result coincides with our Theorem 5.9. As for their argument, readers are recommended to consult Hudson-Ikeda-Matsumura-Naruse [**29**, §3.1], [**30**, §2.2]. In that paper, the K-theoretic Segre classes are introduced by the same way as above, and the generating function of the stable Grothendieck polynomials is given [**29**, Theorems 3.1 and 8.1], [**30**, Theorem 2.8]. Their result can also be obtained from Theorem 5.9 by the specialization $a^{\mathbb{L}}_{1,1} = \beta$ and $a^{\mathbb{L}}_{i,j} = 0$ for all $(i,j) \neq (1,1)$.

(2) Since our formula (5.9) is universal, one can obtain the generating function of the "h^*-theory Segre classes" of vector bundle by specializing the universal formal group law $F_{\mathbb{L}}(u,v)$ to the formal group law $F_h(u,v)$ corresponding to a given complex-oriented cohomology theory $h^*(-)$. For instance, we recently obtained a concrete expression of the "elliptic Schur functions" corresponding to a cohomology theory, denoted $\mathcal{SE}^*(-)$, whose formal group law is that of a singular cubic curve in Weierstrass form, called hyperbolic (see Lenart-Zainoulline [**46**], [**47**]).

5.4. Application of the Gysin formulas for the new universal Schur functions: Thom-Porteous formula for the complex cobordism theory. Using the new universal Schur functions, one can formulate the Thom-Porteous formula (2.17) in the universal setting. We use the same notation as in §2.6.4. As explained in that subsection, in order to obtain the class determined by $D_r(\varphi)$, we have to compute the image $\pi_{F*}(c^{MU}_{e(f-r)}(\pi^*_F(E)^\vee \otimes Q_F))$. Let $x_1, \ldots, x_f$ (resp. $b_1, \ldots, b_e$) be the MU^*-theory Chern roots of F (resp. E). The Chern roots of Q_F are $x_1, \ldots, x_{f-r}$. Let $\lambda = ((e-r)^{(f-r)})$ be the rectangular partition with $(f-r)$ rows and $(e-r)$ columns as in §2.6.4. Then by the splitting principle, the top Chern class is given by

$$c^{MU}_{e(f-r)}(\pi^*_F(E)^\vee \otimes Q_F) = \prod_{i=1}^{f-r}\prod_{j=1}^{e}(x_i +_{\mathbb{L}} \overline{b}_j) = \prod_{i=1}^{f-r}[x_i|\overline{\mathbf{b}}_e]^e = (\mathbf{x}|\overline{\mathbf{b}}_e)^{[\lambda]}.$$

Here $\overline{\mathbf{b}}_e = (\overline{b}_1, \ldots, \overline{b}_e, 0, 0, \ldots)$. Therefore by (5.3) (see also Theorem 5.4 and Example 5.5 (2)), one obtains[36]

$$\pi_{F*}(c^{MU}_{e(f-r)}(\pi^*_F(E)^\vee \otimes Q_F)) = \pi_{F*}((\mathbf{x}|\overline{\mathbf{b}}_e)^{[\lambda]}) = \mathbb{S}^{\mathbb{L}}_\lambda(\mathbf{x}_f|\overline{\mathbf{b}}_e).$$

---

[35] For the K-theoretic analogue of the Segre classes of a vector bundle, see Buch [**14**, Lemma 7.1].

[36] Since we have $\mathbb{S}^{\mathbb{L}}_{(e^{(f-r)})}(\mathbf{x}_{f-r}|\overline{\mathbf{b}}_e) = \prod_{i=1}^{f-r}[x_i|\overline{\mathbf{b}}_e]^e$, this formula can be written as

$$\pi_{F*}(\mathbb{S}^{\mathbb{L}}_{(e^{(f-r)})}(\mathbf{x}_{f-r}|\overline{\mathbf{b}}_e)) = \mathbb{S}^{\mathbb{L}}_{((e-r)^{(f-r)})}(\mathbf{x}_f|\overline{\mathbf{b}}_e).$$

THEOREM 5.11 (Thom-Porteous formula for the complex cobordism theory). *If the codimension of $D_r(\varphi)$ is $(e-r)(f-r)$, then the class determined by $D_r(\varphi)$ is given by the new universal factorial Schur function $\mathbb{S}^{\mathbb{L}}_{((e-r)^{(f-r)})}(\mathbf{x}_f|\overline{\mathbf{b}}_e)$.*

5.5. Application of the Gysin formulas for the new universal Schur functions: Class of Damon's resolution. As we mentioned in the introduction §1.2, it is well-known that the usual Schur polynomials $s_\lambda(\mathbf{x}_d)$, with λ contained in the rectangular partition $((n-d)^d)$, represent the Schubert classes in the ordinary cohomology ring $H^*(G_d(\mathbb{C}^n))$ of the complex Grassmannian $G_d(\mathbb{C}^n)$ of d-dimensional linear subspaces in $\mathbb{C}^n$ (see, e.g., Fulton [**24**, §9.4]). In this subsection, by making use of the Damon's resolution of the Schubert varieties (Damon [**17**, §3, p.258]), together with our Gysin formula (5.3), we shall show that the new universal Schur functions $\mathbb{S}^{\mathbb{L}}_\lambda(\mathbf{y}_d)$, here $y_1,\dots,y_d$ are the MU^*-theory Chern roots of the dual bundle $S^\vee$ of the tautological vector bundle S on $G_d(\mathbb{C}^n)$, give the "correct" representatives of the "Schubert classes" in the complex cobordism ring $MU^*(G_d(\mathbb{C}^n))$.

5.5.1. *Bundle of Schubert varieties.* In fact, we can formulate our assertion in more general situation: Let $E \xrightarrow{p} X$ be a complex vector bundle of rank n over a variety X. For a positive integer $1 \le d \le n-1$, consider the associated Grassmann bundle $\pi : G_d(E) \longrightarrow X$ of d-dimensional subspaces in the fibers of E. On $G_d(E)$, one has the tautological exact sequence of vector bundles

$$0 \longrightarrow S \hookrightarrow \pi^*(E) \twoheadrightarrow Q \longrightarrow 0,$$

where $\operatorname{rank} S = d$ and $\operatorname{rank} Q = n-d$. Suppose that we are given a complete flag of subbundles

$$F_\bullet : 0 = F_0 \subset F_1 \subset \cdots \subset F_i \subset \cdots \subset F_{n-1} \subset F_n = E,$$

where $\operatorname{rank} F_i = i$ $(i = 0,1,2,\dots,n)$. For any partition $\lambda = (\lambda_1,\dots,\lambda_d) \subset ((n-d)^d)$, the *bundle of Schubert varieties*[37] $\Omega_\lambda(F_\bullet) \subset G_d(E)$ is defined by the so-called "Schubert conditions", namely

$$\Omega_\lambda(F_\bullet) := \{W \in G_d(E) \mid W \subset E_x \ (x \in X),\ \dim(W \cap (F_{n-d+i-\lambda_i})_x) \ge i \ (i=1,\dots,d)\}.$$

5.5.2. *Damon's resolution of singularities of* $\Omega_\lambda(F_\bullet)$. Following Damon [**17**, §3, p.258], we will construct a resolution of singularities (or desingularization) of $\Omega_\lambda(F_\bullet)$ (with a slight modification). For a partition $\lambda = (\lambda_1,\dots,\lambda_d) \subset ((n-d)^d)$, we have a decomposition $[d] = I_1 \sqcup I_2 \sqcup \cdots \sqcup I_k$ as in §3.3. Putting $m_i := \sharp I_i$ $(i = 1,2,\dots,k)$, we can rewrite

$$\lambda = (n_1^{m_1}\, n_2^{m_2} \cdots n_k^{m_k}) \quad (n-d \ge n_1 > n_2 > \cdots > n_k \ge 0).$$

We put $\nu(p) := \sum_{i=1}^p m_i$ $(p = 1,2,\dots,k)$ and $\nu(0) := 0$. Since the variety $\Omega_\lambda(F_\bullet)$ is determined by the Schubert conditions corresponding to "outside corners" of the Young diagram of λ (see Fulton [**24**, §9.4, Exercise 18]), one can also define $\Omega_\lambda(F_\bullet)$ by

(5.10)
$$\Omega_\lambda(F_\bullet) = \{W \in G_d(E) \mid W \subset E_x \ (x \in X),\\ \dim(W \cap (F_{n-d+\nu(p)-n_p})_x) \ge \nu(p) \ (p = 1,2,\dots,k)\}.$$

[37] cf. Damon [**17**, §1, p.251], Hudson-Ikeda-Matsumura-Naruse [**29**, §4.1], [**30**, §3.1], Darondeau-Pragacz [**19**, §1.1].

Now consider the partial flag bundle $\varpi : \mathcal{F}\ell_{\nu(1),\nu(2),\dots,\nu(k-1)}(S) \longrightarrow G_d(E)$ associated with the tautological subbundle S over $G_d(E)$. The fiber over a point $W \in G_d(E)$ consists of partial flags in a d-dimensional linear subspace $W \subset E_x$ for some point $x \in X$, namely a nested sequence of linear subspaces of the form

$$W_1 \subset W_2 \subset \cdots \subset W_{k-1} \subset W_k = W,$$

where $\dim W_p = \nu(p)$ $(p = 1, \dots, k)$. On $\mathcal{F}\ell_{\nu(1),\nu(2),\dots,\nu(k-1)}(S)$, one has the tautological sequence of flag of sub and quotient bundles

$$D_1 \hookrightarrow D_2 \hookrightarrow \cdots \hookrightarrow D_p \hookrightarrow \cdots \hookrightarrow D_{k-1} \hookrightarrow D_k = \varpi^*(S),$$

$$\varpi^*(S) \twoheadrightarrow Q_1 \twoheadrightarrow Q_2 \twoheadrightarrow \cdots \twoheadrightarrow Q_p \twoheadrightarrow \cdots \twoheadrightarrow Q_{k-1} \twoheadrightarrow Q_k = 0,$$

where $\operatorname{rank} D_p = \nu(p)$, and Q_p is defined by $Q_p := \varpi^*(S)/D_p$ $(p = 1, 2, \dots, k)$. Then the partial flag bundle $\mathcal{F}\ell_{\nu(1),\nu(2),\dots,\nu(k-1)}(S)$ is constructed as a tower of Grassmann bundles (here we omit the pull-back notation of vector bundles as is customary):

$$\begin{aligned}\mathcal{F}\ell_{\nu(1),\nu(2),\dots,\nu(k-1)}(S) = G_{m_k}(Q_{k-1}) \underset{=}{\xrightarrow{\varpi_k}} G_{m_{k-1}}(Q_{k-2}) \xrightarrow{\varpi_{k-1}} \cdots \\ \xrightarrow{\varpi_3} G_{m_2}(Q_1) \xrightarrow{\varpi_2} G_{m_1}(S) \xrightarrow{\varpi_1} G_d(E).\end{aligned}$$

The natural projection $\varpi = \varpi_1 \circ \varpi_2 \circ \cdots \circ \varpi_k : \mathcal{F}\ell_{\nu(1),\nu(2),\dots,\nu(k-1)}(S) \longrightarrow G_d(E)$ sends a point $(0 \subset W_1 \subset \cdots \subset W_{k-1} \subset W_k = W) \in \mathcal{F}\ell_{\nu(1),\nu(2),\dots,\nu(k-1)}(S)$ to the point $W \in G_d(E)$. Note that the tautological exact sequence of vector bundles over the Grassmann bundle $G_{m_p}(Q_{p-1})$ is regarded as

$$0 \longrightarrow D_p/D_{p-1} \hookrightarrow Q_{p-1} \twoheadrightarrow Q_p \longrightarrow 0.$$

We then define a subvariety X_k of $\mathcal{F}\ell_{\nu(1),\nu(2),\dots,\nu(k-1)}(S)$ by

$$\begin{aligned}X_k \quad := \{&(0 \subset W_1 \subset \cdots \subset W_{k-1} \subset W_k = W) \in \mathcal{F}\ell_{\nu(1),\nu(2),\dots,\nu(k-1)}(S) \\ &\mid W \subset E_x \ (x \in X),\ W_p \subset (F_{n-d+\nu(p)-n_p})_x \ (p = 1, 2, \dots, k)\}.\end{aligned}$$

By definition, a point $(0 \subset W_1 \subset \cdots \subset W_{k-1} \subset W_k = W) \in X_k$ satisfies the conditions $W_p \subset (F_{n-d+\nu(p)-n_p})_x$, and hence $W \cap (F_{n-d+\nu(p)-n_p})_x \supset W_p$ for $p = 1, 2, \dots, k$. Therefore its image $W \in G_d(E)$ under the projection ϖ satisfies the conditions $\dim (W \cap (F_{n-d+\nu(p)-n_p})_x) \geq \nu(p)$ $(p = 1, 2, \dots, k)$. Thus W is in $\Omega_\lambda(F_\bullet)$ by the definition (5.10) of $\Omega_\lambda(F_\bullet)$. The map $\varpi|_{X_k} : X_k \longrightarrow \Omega_\lambda(F_\bullet)$ is a resolution of singularities of $\Omega_\lambda(F_\bullet)$ constructed by Damon.[38]

5.5.3. *Class of Damon's resolution.* Having constructed the resolution X_k of $\Omega_\lambda(F_\bullet)$, we then define the *Damon class* $\delta_\lambda \in MU^*(G_d(E))$ associated to a partition $\lambda \subset ((n-d)^d)$ as the push-forward image $\varpi_*([X_k])$ of the class $[X_k] \in MU^*(\mathcal{F}\ell_{\nu(1),\nu(2),\dots\nu(k)}(S))$. One can compute the class $[X_k]$ explicitly by the standard fact about the top Chern class of a vector bundle (see e.g., Quillen [**64**, §2], Levine-Morel [**48**, Lemma 6.6.7]): The vector bundle homomorphism $D_p/D_{p-1} \hookrightarrow E \twoheadrightarrow E/F_{n-d+\nu(p)-n_p}$ over $\mathcal{F}\ell_{\nu(1),\nu(2),\dots,\nu(k-1)}(S)$ defines a section s_p of the vector bundle $\operatorname{Hom}(D_p/D_{p-1}, E/F_{n-d+\nu(p)-n_p}) \cong (D_p/D_{p-1})^\vee \otimes E/F_{n-d+\nu(p)-n_p}$ for $p = 1, 2, \dots, k$, and the conditions $W_p \subset (F_{n-d+\nu(p)-n_p})_x$ for $p = 1, 2, \dots, k$ means that this section s_p vanishes. Thus the variety $X_k \subset \mathcal{F}\ell_{\nu(1),\nu(2),\dots,\nu(k)}(S)$ is given by the zero locus of the section $s = \bigoplus_{p=1}^k s_p$ of the vector bundle $\bigoplus_{p=1}^k (D_p/D_{p-1})^\vee \otimes$

[38]Damon's resolution is one of the *small resolution* of singularities of a Schubert variety constructed by Zelevinsky [**72**].

$E/F_{n-d+\nu(p)-n_p}$. Therefore the class $[X_k]$ is given by the top Chern class of this vector bundle, that is,

$$[X_k] = \prod_{p=1}^{k} c^{MU}_{m_p(n_p+d-\nu(p))}((D_p/D_{p-1})^\vee \otimes E/F_{n-d+\nu(p)-n_p}).$$

In order to proceed with the computation, we consider the full flag bundle $\mathcal{F}\ell_{1,\dots,d-1}(S) = \mathcal{F}\ell(S) \longrightarrow G_d(E)$. On $\mathcal{F}\ell(S)$, one has the tautological sequence of flag of subbundles

$$S_1 \hookrightarrow S_2 \hookrightarrow \cdots \hookrightarrow S_i \hookrightarrow \cdots \hookrightarrow S_{d-1} \hookrightarrow S_d = S,$$

where rank $S_i = i$ $(i=1,\dots,d)$. By the natural projection $\mathcal{F}\ell(S) \twoheadrightarrow \mathcal{F}\ell_{\nu(1),\dots,\nu(k)}(S)$, the bundle D_p over $\mathcal{F}\ell_{\nu(1),\nu(2),\dots,\nu(k)}(S)$ is pulled-back to $\bigoplus_{i=1}^{\nu(p)} S_i$. Now we set

$$y_i := c_1^{MU}((S_i/S_{i-1})^\vee) \quad (i = 1, 2, \dots, d).$$

These are the MU^*-theory Chern roots of the dual bundle $S^\vee$. Also we set

$$b_i := c_1^{MU}(F_{n+1-i}/F_{n-i}) \in MU^*(X) \ (i = 1, 2, \dots, n).$$

These are the MU^*-theory Chern roots of E. Then by the splitting principle, we have

$$\begin{aligned}
[X_k] &= \prod_{p=1}^{k} c^{MU}_{m_p(n_p+d-\nu(p))}((D_p/D_{p-1})^\vee \otimes E/F_{n-d+\nu(p)-n_p}) \\
&= \prod_{p=1}^{k} \left(\prod_{\substack{\nu(p-1)<i\le\nu(p) \\ 1\le j\le n_p+d-\nu(p)}} (y_i +_{\mathbb{L}} b_j) \right) \\
&= \prod_{p=1}^{k} \left(\prod_{\nu(p-1)<i\le\nu(p)} [y_i|\mathbf{b}_n]^{n_p+d-\nu(p)} \right) = (\mathbf{y}|\mathbf{b}_n)^{[\lambda]}.
\end{aligned}$$

Here $\mathbf{b}_n = (b_1, \dots, b_n, 0, 0, \dots)$. Thus we have

$$\delta_\lambda = \varpi_*((\mathbf{y}|\mathbf{b}_n)^{[\lambda]}).$$

Now, for a partition

$$\lambda = (n_1^{m_1}\, n_2^{m_2} \,\cdots\, n_k^{m_k}), \quad n-d \ge n_1 > n_2 > \cdots > n_k \ge 0,$$

consider the partial flag bundle $\eta_\lambda : \mathcal{F}\ell^\lambda(S^\vee) \longrightarrow G_d(E)$ associated with the vector bundle $S^\vee \longrightarrow G_d(E)$. Since we have

$$\begin{aligned}
\mathcal{F}\ell^\lambda(S^\vee) &= \mathcal{F}\ell^{d-m_k, d-m_k-m_{k-1}, \dots, d-m_k-m_{k-1}-\cdots-m_2}(S^\vee) \\
&= \mathcal{F}\ell_{m_k, m_k+m_{k-1}, \dots, m_k+m_{k-1}+\cdots+m_2}(S^\vee) \\
&\cong \mathcal{F}\ell_{m_1, m_1+m_2, \dots, m_1+m_2+\cdots+m_{k-1}}(S) = \mathcal{F}\ell_{\nu(1),\nu(2),\dots,\nu(k-1)}(S),
\end{aligned}$$

the partial flag bundle $\varpi : \mathcal{F}\ell_{\nu(1),\nu(2),\dots,\nu(k-1)}(S) \longrightarrow G_d(E)$ is identified with the partial flag bundle $\eta_\lambda : \mathcal{F}\ell^\lambda(S^\vee) \longrightarrow G_d(E)$. Therefore computing $\varpi_*((\mathbf{y}|\mathbf{b}_n)^{[\lambda]})$ is equivalent to computing $(\eta_\lambda)_*((\mathbf{y}|\mathbf{b}_n)^{[\lambda]})$. Our Gysin formula (5.3) then yields the following:

THEOREM 5.12. *The Damon class δ_λ is represented by the new universal factorial Schur function. More precisely, we have*

$$\delta_\lambda = \mathbb{S}^{\mathbb{L}}_\lambda(\mathbf{y}_d|\mathbf{b}_n) \in MU^*(G_d(E)).$$

By Theorem 5.12, one may call the new universal factorial Schur function $\mathbb{S}^{\mathbb{L}}_\lambda(\mathbf{x}_n|\mathbf{b})$ the universal factorial Schur function of *Damon type*.

5.6. Concluding remarks and related work. So far, we have considered two types of universal analogues of usual Schur polynomials, namely, the *universal factorial Schur functions* $s^{\mathbb{L}}_\lambda(\mathbf{x}_n|\mathbf{b})$ and the *new universal factorial Schur functions* or the *universal factorial Schur functions of Damon type* $\mathbb{S}^{\mathbb{L}}_\lambda(\mathbf{x}_n|\mathbf{b})$. In closing this paper, we shall introduce another type of universal analogue of Schur polynomials, which was essentially introduced by Hudson-Matsumura [**31**, Definition 4.1], and state briefly our new results. Details will be discussed in our forthcoming paper [**57**]. Hudson-Matsumura introduced the class κ_λ (in the algebraic cobordism ring $\Omega^*(G_d(E))$) using the *Kempf-Laksov resolution* (see op. cit. [**31**, §4.2], Kempf-Laksov [**42**]) of the Schubert varieties. They call the class κ_λ the *Kempf-Laksov class*. Using the Gysin formula (5.3) and Theorem 4.10, we are able to show that the Kempf-Laksov class κ_λ is given explicitly by (here r is the length of λ)

$$\begin{aligned}\kappa_\lambda &= (\eta_{\rho(r)})_*([\mathbf{y}|\mathbf{b}_n]^{\lambda+\rho(r-1)+(d-r)^r}) \\ &= \sum_{\overline{w}\in S_d/(S_1)^r\times S_{d-r}} w\cdot\left[\frac{[\mathbf{y}|\mathbf{b}_n]^{\lambda+\rho(r-1)+(d-r)^r}}{\prod_{1\le i\le r}\prod_{i<j\le d}(y_i +_{\mathbb{L}} \overline{y}_j)}\right]. \end{aligned} \tag{5.11}$$

Motivated by this formula, we make the following definition:

DEFINITION 5.13 (Universal factorial Schur functions of Kempf-Laksov type[39]). For a partition $\lambda \in \mathcal{P}_n$ with length $\ell(\lambda) = r \le n$, we define

$$\begin{aligned} s^{\mathbb{L},KL}_\lambda(\mathbf{x}_n|\mathbf{b}) &:= \sum_{\overline{w}\in S_n/(S_1)^r\times S_{n-r}} w\cdot\left[\frac{[\mathbf{x}|\mathbf{b}]^{\lambda+\rho(r-1)+(n-r)^r}}{\prod_{1\le i\le r}\prod_{i<j\le n}(x_i +_{\mathbb{L}} \overline{x}_j)}\right] \\ &= \sum_{\overline{w}\in S_n/(S_1)^r\times S_{n-r}} w\cdot\left[\frac{\prod_{i=1}^r [x_i|\mathbf{b}]^{\lambda_i-i+n}}{\prod_{1\le i\le r}\prod_{i<j\le n}(x_i +_{\mathbb{L}} \overline{x}_j)}\right]. \end{aligned}$$

We also define

$$s^{\mathbb{L},KL}_\lambda(\mathbf{x}_n) := s^{\mathbb{L},KL}_\lambda(\mathbf{x}_n|\mathbf{0}).$$

From Theorem 4.10, we obtain the following theorem:

THEOREM 5.14 (Characterization of the universal Schur functions of K-L type). *For the Gysin homomorphism $(\tau^r)_* : MU^*(\mathcal{F}\ell^{r,r-1,\dots,2,1}(E)) \longrightarrow MU^*(X)$, the following formula holds:*

$$(\tau^r)_*(\mathbf{x}^{\lambda+\rho(r-1)+(n-r)^r}) = s^{\mathbb{L},KL}_\lambda(E).$$

Here $x_1, \dots, x_n$ are the MU^*-theory Chern roots of E as in §4.2.1.

On the other hand, following Hudson-Matsumura [**31**, Definition 3.1], we define the *MU^*-theory k-th Segre class* $\mathscr{S}^{\mathbb{L}}_k(E)$ of a complex vector bundle E to be

$$\mathscr{S}^{\mathbb{L}}_k(E) := \pi_{1*}(x_1^{k+n-1}) \tag{5.12}$$

[39] "K-L type" for short.

in the same setting as in Theorem 5.7. Let

$$\mathscr{S}_t^{\mathbb{L}}(E) = \mathscr{S}^{\mathbb{L}}(E;t) = \sum_{k\in\mathbb{Z}} \mathscr{S}_k^{\mathbb{L}}(E)t^k$$

be the *Segre series* of E in the complex cobordism theory. Then by Example 5.5 (1) and (5.12), we see immediately that $\mathbb{S}_k^{\mathbb{L}}(E) = \mathscr{S}_k^{\mathbb{L}}(E)$, and therefore the Segre series $\mathscr{S}_t^{\mathbb{L}}(E)$ is given by Theorem 5.9. Using the Segre series $\mathscr{S}_t^{\mathbb{L}}(E)$, the *universal push-forward formula for full flag bundle of type A* established by Darondeau-Pragacz [**18**, pp.6–7][40] can be generalized to the complex cobordism theory.

THEOREM 5.15 (Darondeau-Pragacz formula in complex cobordism). *In the same setting as in Theorem* 5.14, *one has*

$$(\tau^r)_*(f(x_1,\dots,x_r)) = [t_1^{n-1}\cdots t_r^{n-1}]\left(f(t_1,\dots,t_r)\prod_{1\le i<j\le r}(t_j +_{\mathbb{L}} \overline{t}_i)\prod_{1\le i\le r}\mathscr{S}_{1/t_i}^{\mathbb{L}}(E)\right)$$

for a polynomial $f(X_1,\dots,X_r)\in MU^*(X)[X_1,\dots,X_r]$.

This formula, together with Theorem 5.14, enables us to obtain the generating function of the universal Schur functions of K-L type (see Nakagawa-Naruse [**57**]).

References

[1] J. F. Adams, *Stable homotopy and generalised homology*, University of Chicago Press, Chicago, Ill.-London, 1974. Chicago Lectures in Mathematics. MR0402720

[2] E. Akyıldız, *Gysin homomorphism and Schubert calculus*, Pacific J. Math. **115** (1984), no. 2, 257–266. MR765186

[3] E. Akyıldız and J. B. Carrell, *Zeros of holomorphic vector fields and the Gysin homomorphism*, Singularities, Part 1 (Arcata, Calif., 1981), Proc. Sympos. Pure Math., vol. 40, Amer. Math. Soc., Providence, RI, 1983, pp. 47–54. MR713044

[4] E. Akyıldız and J. B. Carrell, *An algebraic formula for the Gysin homomorphism from G/B to G/P*, Illinois J. Math. **31** (1987), no. 2, 312–320. MR882116

[5] M. F. Atiyah, R. Bott, and A. Shapiro, *Clifford modules*, Topology **3** (1964), no. suppl. 1, 3–38. MR0167985

[6] J. C. Becker and D. H. Gottlieb, *The transfer map and fiber bundles*, Topology **14** (1975), 1–12. MR0377873

[7] A. Borel and F. Hirzebruch, *Characteristic classes and homogeneous spaces. I*, Amer. J. Math. **80** (1958), 458–538. MR0102800

[8] A. Borel and F. Hirzebruch, *Characteristic classes and homogeneous spaces. II*, Amer. J. Math. **81** (1959), 315–382. MR0110105

[9] R. Bott, *Lectures on $K(X)$*, Mathematics Lecture Note Series, W. A. Benjamin, Inc., New York-Amsterdam, 1969. MR0258020

[10] P. Bressler and S. Evens, *The Schubert calculus, braid relations, and generalized cohomology*, Trans. Amer. Math. Soc. **317** (1990), no. 2, 799–811. MR968883

[11] P. Bressler and S. Evens, *Schubert calculus in complex cobordism*, Trans. Amer. Math. Soc. **331** (1992), no. 2, 799–813. MR1044959

[12] M. Brion, *The push-forward and Todd class of flag bundles*, Parameter spaces (Warsaw, 1994), Banach Center Publ., vol. 36, Polish Acad. Sci. Inst. Math., Warsaw, 1996, pp. 45–50. MR1481478

[13] G. Brumfiel and I. Madsen, *Evaluation of the transfer and the universal surgery classes*, Invent. Math. **32** (1976), no. 2, 133–169. MR0413099

[14] A. S. Buch, *Grothendieck classes of quiver varieties*, Duke Math. J. **115** (2002), no. 1, 75–103. MR1932326

[40]Note that this type of Gysin formula is also obtained by Ilori [**36**, p.623, Theorem], Kaji-Terasoma [**39**, Theorem 0.4 (Push-Forward Formula)].

[15] P. E. Conner and E. E. Floyd, *The relation of cobordism to K-theories*, Lecture Notes in Mathematics, No. 28, Springer-Verlag, Berlin-New York, 1966. MR0216511

[16] J. Damon, *The Gysin homomorphism for flag bundles*, Amer. J. Math. **95** (1973), 643–659. MR0348760

[17] J. Damon, *The Gysin homomorphism for flag bundles: applications*, Amer. J. Math. **96** (1974), 248–260. MR0358798

[18] L. Darondeau and P. Pragacz, *Universal Gysin formulas for flag bundles*, Internat. J. Math. **28** (2017), no. 11, 1750077, 23. MR3714353

[19] L. Darondeau and P. Pragacz, *Gysin maps, duality and Schubert classes*, arXiv:1602.01983.

[20] A. Dold, *Relations between ordinary and extraordinary homology*, Colloquium on Algebraic Topology, 1-10 August, 1962.

[21] S. Eilenberg and N. Steenrod, *Foundations of algebraic topology*, Princeton University Press, Princeton, New Jersey, 1952. MR0050886

[22] K. È. Fel′dman, *An equivariant analogue of the Poincaré-Hopf theorem* (Russian, with English and Russian summaries), Zap. Nauchn. Sem. S.-Peterburg. Otdel. Mat. Inst. Steklov. (POMI) **267** (2000), no. Geom. i Topol. 5, 303–318, 332–333; English transl., J. Math. Sci. (N. Y.) **113** (2003), no. 6, 906–914. MR1809835

[23] W. Fulton, *Intersection theory*, 2nd ed., Ergebnisse der Mathematik und ihrer Grenzgebiete. 3. Folge. A Series of Modern Surveys in Mathematics [Results in Mathematics and Related Areas. 3rd Series. A Series of Modern Surveys in Mathematics], vol. 2, Springer-Verlag, Berlin, 1998. MR1644323

[24] W. Fulton, *Young tableaux*, London Mathematical Society Student Texts, vol. 35, Cambridge University Press, Cambridge, 1997. With applications to representation theory and geometry. MR1464693

[25] W. Fulton and P. Pragacz, *Schubert varieties and degeneracy loci*, Lecture Notes in Mathematics, vol. 1689, Springer-Verlag, Berlin, 1998. Appendix J by the authors in collaboration with I. Ciocan-Fontanine. MR1639468

[26] A. Grothendieck, *La théorie des classes de Chern* (French), Bull. Soc. Math. France **86** (1958), 137–154. MR0116023

[27] J. Harris and L. Tu, *Chern numbers of kernel and cokernel bundles*, Invent. Math. **75** (1984), no. 3, 467–475. MR735336

[28] J. Hornbostel and V. Kiritchenko, *Schubert calculus for algebraic cobordism*, J. Reine Angew. Math. **656** (2011), 59–85. MR2818856

[29] T. Hudson, T. Ikeda, T. Matsumura, and H. Naruse, *Determinantal and Pfaffian formulas of K-theoretic Schubert calculus*, arXiv:1504.02828v2.

[30] T. Hudson, T. Ikeda, T. Matsumura, and H. Naruse, *Degeneracy loci classes in K-theory — determinantal and Pfaffian formula*, Adv. Math. **320** (2017), 115–156. MR3709102

[31] T. Hudson and T. Matsumura, *Segre classes and Kempf-Laksov formula in algebraic cobordism*, arXiv:1602.05704.

[32] T. Ikeda, *Schubert classes in the equivariant cohomology of the Lagrangian Grassmannian*, Adv. Math. **215** (2007), no. 1, 1–23. MR2354984

[33] T. Ikeda and H. Naruse, *Excited Young diagrams and equivariant Schubert calculus*, Trans. Amer. Math. Soc. **361** (2009), no. 10, 5193–5221. MR2515809

[34] T. Ikeda, L. C. Mihalcea, and H. Naruse, *Double Schubert polynomials for the classical groups*, Adv. Math. **226** (2011), no. 1, 840–886. MR2735777

[35] T. Ikeda and H. Naruse, *K-theoretic analogues of factorial Schur P- and Q-functions*, Adv. Math. **243** (2013), 22–66. MR3062739

[36] S. A. Ilori, *A generalization of the Gysin homomorphism for flag bundles*, Amer. J. Math. **100** (1978), no. 3, 621–630. MR501229

[37] V. N. Ivanov, *Interpolation analogues of Schur Q-functions*, Zap. Nauchn. Sem. S.-Peterburg. Otdel. Mat. Inst. Steklov. (POMI) **307** (2004), no. Teor. Predst. Din. Sist. Komb. i Algoritm. Metody. 10, 99–119, 281–282; English transl., J. Math. Sci. (N.Y.) **131** (2005), no. 2, 5495–5507. MR2050689

[38] T. Józefiak, A. Lascoux, and P. Pragacz, *Classes of determinantal varieties associated with symmetric and skew-symmetric matrices*, Math. USSR Izv. **18** (1982), 575–586.

[39] H. Kaji and T. Terasoma, *Degree formulae for Grassmann bundles, II*, arXiv:1508.01663.

[40] H. Kajimoto, *The Poincaré duality and the Gysin homomorphism for flag manifolds*, Hiroshima Math. J. **27** (1997), no. 2, 189–207. MR1461171

[41] T. Sugawara and H. Kajimoto, *The Gysin homomorphisms for flag bundles of the symplectic groups*, Kyushu J. Math. **52** (1998), no. 2, 287–297. MR1645447

[42] G. Kempf and D. Laksov, *The determinantal formula of Schubert calculus*, Acta Math. **132** (1974), 153–162. MR0338006

[43] A. Knutson and T. Tao, *Puzzles and (equivariant) cohomology of Grassmannians*, Duke Math. J. **119** (2003), no. 2, 221–260. MR1997946

[44] A. Kono and D. Tamaki, *Generalized cohomology*, Translations of Mathematical Monographs, vol. 230, American Mathematical Society, Providence, RI, 2006. Translated from the 2002 Japanese edition by Tamaki; Iwanami Series in Modern Mathematics. MR2225848

[45] M. Lazard, *Sur les groupes de Lie formels à un paramètre* (French), Bull. Soc. Math. France **83** (1955), 251–274. MR0073925

[46] C. Lenart and K. Zainoulline, *Towards generalized cohomology Schubert calculus via formal root polynomials*, arXiv:1408.5952.

[47] C. Lenart and K. Zainoulline, *A Schubert basis in equivariant elliptic cohomology*, New York J. Math. **23** (2017), 711–737. MR3665585

[48] M. Levine and F. Morel, *Algebraic cobordism*, Springer Monographs in Mathematics, Springer, Berlin, 2007. MR2286826

[49] I. G. Macdonald, *Symmetric functions and Hall polynomials*, 2nd ed., Oxford Mathematical Monographs, The Clarendon Press, Oxford University Press, New York, 1995. With contributions by A. Zelevinsky; Oxford Science Publications. MR1354144

[50] I. G. Macdonald, *Schur functions: theme and variations*, Séminaire Lotharingien de Combinatoire (Saint-Nabor, 1992), Publ. Inst. Rech. Math. Av., vol. 498, Univ. Louis Pasteur, Strasbourg, 1992, pp. 5–39. MR1308728

[51] L. Manivel, *Symmetric functions, Schubert polynomials and degeneracy loci*, SMF/AMS Texts and Monographs, vol. 6, American Mathematical Society, Providence, RI; Société Mathématique de France, Paris, 2001. Translated from the 1998 French original by John R. Swallow; Cours Spécialisés [Specialized Courses], 3. MR1852463

[52] P. J. McNamara, *Factorial Grothendieck polynomials*, Electron. J. Combin. **13** (2006), no. 1, Research Paper 71, 40. MR2240776

[53] J. Milnor, *On axiomatic homology theory*, Pacific J. Math. **12** (1962), 337–341. MR0159327

[54] J. W. Milnor and J. D. Stasheff, *Characteristic classes*, Princeton University Press, Princeton, N. J.; University of Tokyo Press, Tokyo, 1974. Annals of Mathematics Studies, No. 76. MR0440554

[55] A. I. Molev and B. E. Sagan, *A Littlewood-Richardson rule for factorial Schur functions*, Trans. Amer. Math. Soc. **351** (1999), no. 11, 4429–4443. MR1621694

[56] M. Nakagawa and H. Naruse, *Generalized (co)homology of the loop spaces of classical groups and the universal factorial Schur P- and Q-functions*, Schubert calculus—Osaka 2012, Adv. Stud. Pure Math., vol. 71, Math. Soc. Japan, [Tokyo], 2016, pp. 337–417. MR3644829

[57] M. Nakagawa and H. Naruse, *Generating functions for the universal Hall-Littlewood P- and Q-functions*, arXiv:1705.04791.

[58] I. R. Porteous, *Simple singularities of maps*, Proceedings of Liverpool Singularities Symposium, I (1969/70), Springer, Berlin, 1971, pp. 286–307. Lecture Notes in Math., Vol. 192. MR0293646

[59] P. Pragacz, *Enumerative geometry of degeneracy loci*, Ann. Sci. École Norm. Sup. (4) **21** (1988), no. 3, 413–454. MR974411

[60] P. Pragacz, *Algebro-geometric applications of Schur S- and Q-polynomials*, Topics in invariant theory (Paris, 1989/1990), Lecture Notes in Math., vol. 1478, Springer, Berlin, 1991, pp. 130–191. MR1180989

[61] P. Pragacz, *Symmetric polynomials and divided differences in formulas of intersection theory*, Parameter spaces (Warsaw, 1994), Banach Center Publ., vol. 36, Polish Acad. Sci. Inst. Math., Warsaw, 1996, pp. 125–177. MR1481485

[62] P. Pragacz, *A Gysin formula for Hall-Littlewood polynomials*, Proc. Amer. Math. Soc. **143** (2015), no. 11, 4705–4711. MR3391029

[63] D. Quillen, *On the formal group laws of unoriented and complex cobordism theory*, Bull. Amer. Math. Soc. **75** (1969), 1293–1298. MR0253350

[64] D. Quillen, *Elementary proofs of some results of cobordism theory using Steenrod operations*, Advances in Math. **7** (1971), 29–56 (1971). MR0290382

[65] D. C. Ravenel, *Complex cobordism and stable homotopy groups of spheres*, 2nd ed. AMS Chelsea Publishing, Amer. Math. Soc., Providence, Rhode Island, 2004.

[66] T. Sugawara, *The Gysin homomorphism for generalized flag bundles*, Mem. Fac. Sci. Kyushu Univ. Ser. A **42** (1988), no. 2, 131–144. MR963202

[67] R. M. Switzer, *Algebraic topology—homotopy and homology*, Classics in Mathematics, Springer-Verlag, Berlin, 2002. Reprint of the 1975 original [Springer, New York; MR0385836 (52 #6695)]. MR1886843

[68] R. Thom, *Les ensembles singuliers d'une application différentiable et leurs propriétés homologiques*, Séminaire de Topologie de Strasbourg, (December 1957).

[69] L. W. Tu, *Computing characteristic numbers using fixed points*, A celebration of the mathematical legacy of Raoul Bott, CRM Proc. Lecture Notes, vol. 50, Amer. Math. Soc., Providence, RI, 2010, pp. 185–206. MR2648896

[70] L. W. Tu, *Computing the Gysin map using fixed points*, Algebraic geometry and number theory, Progr. Math., vol. 321, Birkhäuser/Springer, Cham, 2017, pp. 135–160. MR3675380

[71] A. Vishik, *Symmetric operations in algebraic cobordism*, Adv. Math. **213** (2007), no. 2, 489–552. MR2332601

[72] A. V. Zelevinskiĭ, *Small resolutions of singularities of Schubert varieties* (Russian), Funktsional. Anal. i Prilozhen. **17** (1983), no. 2, 75–77. MR705051

Graduate School of Education, Okayama University, Okayama 700-8530, Japan
Email address: nakagawa@okayama-u.ac.jp

Graduate School of Education, University of Yamanashi, Kofu 400-8510, Japan
Email address: hnaruse@yamanashi.ac.jp

Contemporary Mathematics
Volume **708**, 2018
http://dx.doi.org/10.1090/conm/708/14266

Graded multiplications on iterated bar constructions

Bruno Stonek

ABSTRACT. We define a bar construction endofunctor on the category of commutative augmented monoids A of a symmetric monoidal category $\mathcal{V}$ endowed with a left adjoint monoidal functor $F : s\mathbf{Set} \to \mathcal{V}$. To do this, we need to carefully examine the monoidal properties of the well-known (reduced) simplicial bar construction $B_\bullet(\mathbb{1}, A, \mathbb{1})$. We define a geometric realization $|-|$ with respect to the image under F of the canonical cosimplicial simplicial set. This guarantees good monoidal properties of $|-|$: it is monoidal, and given a left adjoint monoidal functor $G : \mathcal{V} \to \mathcal{W}$, there is a monoidal transformation $|G-| \Rightarrow G|-|$. We can then consider $BA = |B_\bullet A|$ and the iterations $B^n A$. We establish the existence of a graded multiplication on these objects, provided the category $\mathcal{V}$ is cartesian and A is a ring object. The examples studied include simplicial sets and modules, topological spaces, chain complexes and spectra.

Introduction

The *bar construction* is an algebraic machine showing up in different settings: the flavor that we consider allows for the construction, for example, of classifying spaces BG of topological commutative monoids G, of reduced Hochschild homology $HH^k(A, k)$ where A is a commutative augmented algebra over a commutative ring k, and of reduced topological Hochschild homology $THH^R(A, R)$ where A is a commutative augmented algebra over a commutative ring spectrum R.

A very general incarnation of its *two-sided* version appears in [**EKMM97**, XII.1.1]: given a category $\mathcal{C}$, a monad T, a T-algebra C and a right T-functor F, we can build a simplicial object $B_\bullet(F, T, C)$ in $\mathcal{C}$. A particular case is given as follows: if $\mathcal{V}$ is a symmetric monoidal category, A is a monoid in $\mathcal{V}$, N is a right A-module and M is a left A-module, then $B_\bullet(M, A, N) \in s\mathcal{V}$, where $s\mathcal{V}$ denotes the category of simplicial objects of $\mathcal{V}$.

We shall focus in the following particular case, which we call the *simplicial (reduced) bar construction*: it is the object of study of Section 1. Denote by $\mathbb{1}$ the unit of $\mathcal{V}$. If A is an *augmented* commutative monoid in $\mathcal{V}$, meaning that it comes with a commutative monoid morphism $A \to \mathbb{1}$, then $\mathbb{1}$ becomes a left and right A-module, and we can consider $B_\bullet(\mathbb{1}, A, \mathbb{1})$. This is the object that we will denote $B_\bullet A$.

2010 *Mathematics Subject Classification.* Primary 55P43; Secondary 55P20, 55N20, 18D10.
Key words and phrases. Bar construction, geometric realization.
The author acknowledges support from the project ANR-16-CE40-0003 ChroK.

This construction is a strong symmetric monoidal functor from augmented commutative monoids in $\mathcal{V}$ to simplicial augmented commutative monoids in $\mathcal{V}$, as is well-known. A fact which to our knowledge has not appeared in the literature is the 2-categorical aspect of it of **Corollary 1.6**: we explain how a strong symmetric monoidal functor $F : \mathcal{V} \to \mathcal{W}$ begets a natural isomorphism $B_\bullet F \Rightarrow FB_\bullet$.

Our primary interest lies, however, in the *iteration* of the bar construction. For this we need a notion of geometric realization, from simplicial objects in $\mathcal{V}$ down to objects in $\mathcal{V}$, which we study in Section 2. This is not hard to understand abstractly. Given a fixed cosimplicial object D in $\mathcal{V}$, the geometric realization $|X|$ of $X \in s\mathcal{V}$ could be defined as the tensor product of functors $X \otimes_{\mathbb{\Delta}} D$.

However, since the simplicial bar construction is a simplicial augmented commutative monoid, we need the geometric realization to take such an object to an augmented commutative monoid, in order to be able to iterate the composition $B = |B_\bullet|$. The definition just given has no reason to satisfy that. For this reason, and because all the examples that we want to consider follow this pattern, we consider only cosimplicial objects which are induced from the canonical cosimplicial simplicial set, namely the Yoneda embedding (see Section 2.1 for what we mean by "induced"). The geometric realization functor defined by such a cosimplicial object does indeed preserve commutative monoids: this is **Theorem 2.9**.

But the advantages of having chosen such a cosimplicial object do not end there. Again, we obtain a 2-categorical result: if $G : \mathcal{V} \to \mathcal{W}$ is a strong symmetric monoidal functor which is a left adjoint, then **Theorem 2.11** provides us with a *monoidal* natural isomorphism $|G - | \Rightarrow G| - |$ of functors $s\mathcal{V} \to \mathcal{W}$. The fact that this natural isomorphism is monoidal is doubly important. First, it is an interesting fact per se. As a particular instance of this theorem, we obtain for example that the isomorphism of spectra

$$\Sigma^\infty_+ |X_\bullet| \cong |\Sigma^\infty_+ X_\bullet|$$

where X is a simplicial topological space, is actually monoidal (Example 4.5.1). Second, the monoidality of this natural isomorphism implies that it begets a natural isomorphism $|GA| \cong G|A|$ of *augmented commutative monoids*, if A is a simplicial augmented commutative monoid, and this is crucial to our desires.

Having set this up, given an augmented commutative monoid A we can define the *bar construction* BA as the geometric realization of the simplicial bar construction of A, and in Section 3 we set out to iterate it. We get a graded augmented commutative monoid $(B^n A)_{n \geq 0}$, i.e. a sequence of augmented commutative monoids. Our quest is now to find a graded multiplication

$$B^n A \otimes B^m A \to B^{n+m} A.$$

To achieve this, we will need the monoid A itself to have an additional multiplication. We need to specialize the context: instead of considering general symmetric monoidal categories, we will need to consider *cartesian* ones. Such a category allows for *ring objects*. Note that a ring object is in particular an augmented commutative monoid with a trivial augmentation. It should also be noted that from any symmetric monoidal category we can define an interesting cartesian category, the one formed by its cocommutative comonoids.

If A is a ring object in a cartesian category, then the iterated bar constructions $B^*A = (B^nA)_{n\geq 0}$ is a graded ring object. This is the content of **Theorem 3.5**. Moreover, if G is a left adjoint, cartesian functor between cartesian categories, then we get a natural isomorphism of graded ring objects $B^*GA \cong GB^*A$.

In the fourth and final section we unveil the examples. When we take the category of topological spaces as our cartesian category (Example 4.2), then given a topological abelian group A, BA is the bar construction introduced by Milgram [**Mil67**]. When A is discrete, B^nA serves as a model for an Eilenberg-Mac Lane space $K(A, n)$. Starting from a ring R, the graded multiplication

$$K(R, n) \times K(R, m) \to K(R, n + m)$$

is the one found by Ravenel and Wilson in [**RW80**], which gives the cup product in singular cohomology with coefficients in R. We can also run this machine in the category of simplicial sets (Example 4.1). If A is a simplicial abelian group, then B^nA gives a simplicial model for a $K(A, n)$; this is well-known. But we also get that starting from a simplicial ring R, B^*R is a graded simplicial ring, which under the geometric realization functor corresponds to the topological construction of Ravenel and Wilson. This was to be expected.

We can also work in some symmetric monoidal category of spectra (Example 4.5). Fix R a commutative ring spectrum. Then if A is an augmented commutative R-algebra, BA is the topological Hochschild homology $THH^R(A, R)$ over R relative to R. The iterations B^nA model the *higher* topological Hochschild homology

$$THH^{R,[n]}(A, R) = S^n \odot A,$$

where $\odot$ denotes the tensoring of augmented commutative R-algebras over pointed topological spaces. Now, if A is a ring object in cocommutative R-coalgebras, then we get a graded multiplication in higher THH (4.15). Moreover, if A is of the form $R[S]$ where S is a ring, then

$$THH^{R,[*]}(R[S], R) \cong R[K(S, *)],$$

a natural isomorphism of graded ring objects of cocommutative R-coalgebras (4.16). Here $R[-]$ denotes $R \wedge_{\mathbb{S}} \Sigma^\infty_+(-)$.

Analogously, we can work in the symmetric monoidal category of simplicial k-modules, for k a commutative ring (Example 4.3). Here the iterated bar constructions will rather yield Pirashvili's [**Pir00**] higher reduced Hochschild homology. On the other hand, Theorem 3.5 is not applicable in the context of the symmetric monoidal category of non-negatively graded chain complexes of k-modules, the problem being that the normalized Moore functor from simplicial modules into it is not *strong* monoidal (Example 4.4). Note that if A is an augmented commutative differential graded algebra, then BA is one as well: this construction of BA coincides with the classical one introduced by Eilenberg and Mac Lane [**EML53**, Theorem 11.1].

We will make some use of the theory of monoidal categories. We have chosen to defer this material to the appendix.

We let **Top** denote the cartesian closed category of compactly generated weakly Hausdorff spaces, and we let k be a commutative ring.

In this article, a *spectrum* will be understood to be an S-module in the sense of [**EKMM97**]. A *(commutative) ring spectrum* will mean a (commutative) S-algebra.

Acknowledgments. The author would like to thank Ignacio López Franco for the useful discussions on category theory. The contents of this article will be a part of the author's PhD dissertation at Université Paris 13.

1. Simplicial bar construction

Let $\mathcal{V}$ be a symmetric monoidal category with unit object $\mathbb{1}$. The symmetric monoidal category $\mathbf{Mon}(\mathcal{V})^{\mathrm{aug}}$ has as objects the augmented monoids of $\mathcal{V}$, i.e. monoids M with a monoid map $M \to \mathbb{1}$ (see A.4). We refer to A.5 for some background on simplicial objects.

Recall that $s\mathcal{V}$, the category of simplicial objects of $\mathcal{V}$, is symmetric monoidal with pointwise tensor product and unit $c\mathbb{1}$, the constant simplicial object at $\mathbb{1}$.

DEFINITION 1.1. The *simplicial (reduced) bar construction* in $\mathbf{Mon}(\mathcal{V})^{\mathrm{aug}}$ is the functor

$$B_\bullet : \mathbf{Mon}(\mathcal{V})^{\mathrm{aug}} \to s\mathcal{V}$$

defined as follows. If $A \in \mathbf{Mon}(\mathcal{V})^{\mathrm{aug}}$ with multiplication $\mu : A \otimes A \to A$, unit $\eta : \mathbb{1} \to A$ and augmentation $\varepsilon : A \to \mathbb{1}$, then $B_n(A) = A^{\otimes n}$, where $A^{\otimes 0}$ means $\mathbb{1}$. The faces $d_i : A^{\otimes n} \to A^{\otimes n-1}$, $i = 0, \ldots, n$ are defined as

$$d_0 = \varepsilon \otimes \mathrm{id}^{\otimes n-1},$$

$$d_i = \mathrm{id}^{\otimes i-1} \otimes \mu \otimes \mathrm{id}^{\otimes n-i-1} \qquad \text{if } i = 1, \ldots, n-1,$$

$$d_n = \mathrm{id}^{\otimes n-1} \otimes \varepsilon,$$

and the degeneracies $s_i : A^{\otimes n} \to A^{\otimes n+1}$ are

$$s_i = \mathrm{id}^{\otimes i} \otimes \eta \otimes \mathrm{id}^{\otimes n-i} \qquad \text{for all } i = 0, \ldots, n.$$

This bar construction is a particular case of the two-sided bar construction [**May72**, Chapter 10]: it is $B_\bullet(\mathbb{1}, -, \mathbb{1})$ where if $A \in \mathbf{Mon}(\mathcal{V})^{\mathrm{aug}}$ then $\mathbb{1}$ is viewed as a left and right A-module via the augmentation $\varepsilon : A \to \mathbb{1}$.

The proof of the following proposition is elementary.

PROPOSITION 1.2. *The functor* $B_\bullet : \mathbf{Mon}(\mathcal{V})^{\mathrm{aug}} \to s\mathcal{V}$ *is strong symmetric monoidal.*

REMARK 1.3. The symmetric monoidal category $s\mathcal{V}$ itself admits a monoidal bar construction

$$B_\bullet : s\mathbf{Mon}(\mathcal{V})^{\mathrm{aug}} \to s^2\mathcal{V}$$

which is levelwise $B_\bullet : \mathbf{Mon}(\mathcal{V})^{\mathrm{aug}} \to s\mathcal{V}$.

1.1. Behavior over monoidal functors.

We will now provide a comparison theorem for the bar constructions in two symmetric monoidal categories $\mathcal{V}, \mathcal{W}$ related by a functor $F : \mathcal{V} \to \mathcal{W}$. We first prove a lemma that isolates the part relating purely to the monoidal structure, and then deal with the simplicial structure.

LEMMA 1.4. *Let $F : \mathcal{V} \to \mathcal{W}$ be a lax symmetric monoidal functor between symmetric monoidal categories, with structure morphisms*

$$\nabla : FA \otimes FB \to F(A \otimes B).$$

There are lax symmetric monoidal functors and a monoidal transformation

$$\begin{array}{ccccc} & \overset{F^n}{\nearrow} & \mathcal{W}^n & \overset{\otimes_n}{\searrow} & \\ \mathcal{V}^n & & \Downarrow \nabla_n & & \mathcal{W} \\ & \underset{\otimes_n}{\searrow} & \mathcal{V} & \underset{F}{\nearrow} & \end{array}$$

for every $n \geq 1$. The functors are strong if F is strong.

In particular, there are monoidal transformations

$$\mathcal{V} \overset{F(-)^{\otimes n}}{\underset{F((-)^{\otimes n})}{\Downarrow \nabla_n}} \mathcal{W}.$$

between lax symmetric monoidal functors, which are strong if F is strong.

PROOF. The first part of this lemma can be proven by elementary means, using Mac Lane's coherence theorems [**ML98**, VII.2, XI.2]. We give an alternative, more sophisticated proof, essentially due to Mike Shulman [**Shu**].

The statement that we can define $\otimes_n$ and that it is strong symmetric monoidal, and the statement that we can define ∇_n and that it is symmetric monoidal are two instances of the same result. Lack [**Lac00**, 3.6] proved a coherence theorem for pseudomonoids X in monoidal bicategories $\mathcal{C}$. By means of it, one obtains a 1-cell $\mu_n : X^n \to X$. Moreover, when $\mathcal{C}$ and X are symmetric, then the multiplication $\mu : X \otimes X \to X$ is strong symmetric monoidal (i.e. a strong symmetric morphism of pseudomonoids), therefore μ_n is strong symmetric, too. This sole result encompasses both of the above results, by taking $\mathcal{C}$ to be the 2-category with products **Cat** for the first case, and by taking $\mathcal{C}$ to be **Oplax**(**2**, **Cat**), the 2-category with products consisting of oplax functors from the interval category to **Cat** for the second case.

The particular case is obtained by prewhiskering with the iterated diagonal functor $\mathcal{V} \to \mathcal{V}^n$, which is strong symmetric monoidal. Indeed, whiskering preserves monoidality of natural transformations [**AM10**, 3.21, 3.24]. □

PROPOSITION 1.5. *Let $F : \mathcal{V} \to \mathcal{W}$ be a normal lax symmetric monoidal functor between symmetric monoidal categories. Then there is a monoidal transformation*

$$\begin{array}{ccc} \mathbf{Mon}(\mathcal{V})^{\mathrm{aug}} & \xrightarrow{B_\bullet} & s\mathcal{V} \\ {\scriptstyle F}\downarrow & \nearrow\!\!\!\nearrow & \downarrow{\scriptstyle F} \\ \mathbf{Mon}(\mathcal{W})^{\mathrm{aug}} & \xrightarrow[B_\bullet]{} & s\mathcal{W} \end{array}$$

which is a natural isomorphism when F is strong.

PROOF. First of all, the fact that F is *normal* lax symmetric implies the existence of such an induced normal lax symmetric F on *augmented* monoids (A.4.2) and on simplicial objects (A.5).

Gathering the monoidal transformations of Lemma 1.4 together for all n, we obtain a monoidal transformation

$$\mathcal{V} \underset{\coprod_{n\geq 0} F((-)^{\otimes n})}{\overset{\coprod_{n\geq 0} F(-)^{\otimes n}}{\overset{\Downarrow \nabla}{\rightrightarrows}}} \mathrm{Fun}(\mathbb{N}, \mathcal{W})$$

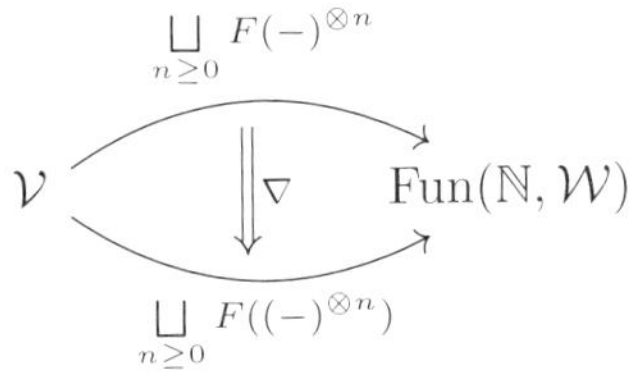

where $\mathbb{N}$ is the discrete category on the natural numbers.

All there is left is to prove is that, when we start from $\mathbf{Mon}(\mathcal{V})^{\mathrm{aug}}$, the components of ∇ are really morphisms in $s\mathcal{W} = \mathrm{Fun}(\mathbb{\Delta}^{\mathrm{op}}, \mathcal{W})$, i.e. that they are compatible with the faces and degeneracies of the simplicial bar construction.

Let $A \in \mathbf{Mon}(\mathcal{V})^{\mathrm{aug}}$ with multiplication $\mu : A \otimes A \to A$. The face map $d_1 : B_3(A) \to B_2(A)$ is

$$\mu \otimes \mathrm{id} : A \otimes A \otimes A \to A \otimes A,$$

and the face map $d_1 : B_3(FA) \to B_2(FA)$ is the composition of the two vertical maps on the left of the following diagram, whose commutativity we need to check.

$$\begin{array}{ccc}
FA \otimes FA \otimes FA & \xrightarrow{\nabla_3} & F(A \otimes A \otimes A) \\
{\scriptstyle \nabla\otimes\mathrm{id}}\big\downarrow & & \\
F(A \otimes A) \otimes FA & & \big\downarrow{\scriptstyle F(\mu\otimes\mathrm{id})} \\
{\scriptstyle F\mu\otimes\mathrm{id}}\big\downarrow & & \\
FA \otimes FA & \xrightarrow[\nabla]{} & F(A \otimes A)
\end{array}$$

But the proof of Lemma 1.4 guarantees that we can take $\nabla_3 = \nabla \circ (\nabla \otimes \mathrm{id})$, therefore this commutativity is an application of naturality of ∇. All the other differentials $d_i : A^{\otimes n} \to A^{\otimes(n-1)}$ for $i \neq 0, n$ at each level are built in the same fashion, so the proof adapts. For the extremal differentials d_0 and d_n which use the augmentation $\varepsilon : A \to \mathbb{1}$ of A, there is another diagram chase. For example, for $d_0 : A \otimes A \otimes A \to A \otimes A$, the verification is the commutativity of the following outer diagram. The inner diagrams commute by naturality and unitality of ∇.

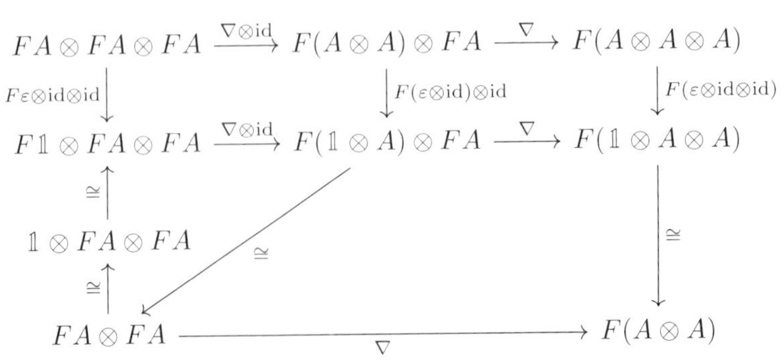

$$\begin{array}{ccccc}
FA \otimes FA \otimes FA & \xrightarrow{\nabla\otimes\mathrm{id}} & F(A \otimes A) \otimes FA & \xrightarrow{\nabla} & F(A \otimes A \otimes A) \\
{\scriptstyle F\varepsilon\otimes\mathrm{id}\otimes\mathrm{id}}\big\downarrow & & \big\downarrow{\scriptstyle F(\varepsilon\otimes\mathrm{id})\otimes\mathrm{id}} & & \big\downarrow{\scriptstyle F(\varepsilon\otimes\mathrm{id}\otimes\mathrm{id})} \\
F\mathbb{1} \otimes FA \otimes FA & \xrightarrow{\nabla\otimes\mathrm{id}} & F(\mathbb{1} \otimes A) \otimes FA & \xrightarrow{\nabla} & F(\mathbb{1} \otimes A \otimes A) \\
{\scriptstyle \cong}\big\uparrow & \swarrow{\scriptstyle \cong} & & & \\
\mathbb{1} \otimes FA \otimes FA & & & & \big\downarrow{\scriptstyle \cong} \\
{\scriptstyle \cong}\big\uparrow & & & & \\
FA \otimes FA & & \xrightarrow[\nabla]{} & & F(A \otimes A)
\end{array}$$

(the diagonal arrow $\cong$ goes from $F(\mathbb{1} \otimes A) \otimes FA$ to $FA \otimes FA$.)

Compatibility with the degeneracies follows from a similar diagram chase. $\square$

1.2. Commutative augmented monoids. We denote by $\mathbf{CMon}(\mathcal{V})$ the symmetric monoidal category of commutative monoids in $\mathcal{V}$ (see A.1), and by $\mathbf{Ab}(\mathcal{C})$ the cartesian category of abelian group objects in a cartesian category $\mathcal{C}$ (see A.2). Combining Propositions 1.2 and 1.5 we obtain the following result.

COROLLARY 1.6. *There is an induced strong symmetric monoidal functor*

$$B_\bullet : \mathbf{CMon}(\mathcal{V})^{\mathrm{aug}} \to s(\mathbf{CMon}(\mathcal{V})^{\mathrm{aug}}).$$

Let $F : \mathcal{V} \to \mathcal{W}$ be a normal lax symmetric monoidal functor between symmetric monoidal categories. There is a natural transformation displayed in the following diagram, which is a natural isomorphism when F is strong.

$$\begin{array}{ccc} \mathbf{CMon}(\mathcal{V})^{\mathrm{aug}} & \xrightarrow{B_\bullet} & s(\mathbf{CMon}(\mathcal{V})^{\mathrm{aug}}) \\ {\scriptstyle F}\downarrow & \nearrow\!\!\!\!\nearrow & \downarrow{\scriptstyle F} \\ \mathbf{CMon}(\mathcal{W})^{\mathrm{aug}} & \xrightarrow[B_\bullet]{} & s(\mathbf{CMon}(\mathcal{W})^{\mathrm{aug}}) \end{array}$$

If $\mathcal{C}$ is a cartesian category, there is an induced cartesian functor

$$B_\bullet : \mathbf{Ab}(\mathcal{C}) \to s\mathbf{Ab}(\mathcal{C}),$$

and if $F : \mathcal{C} \to \mathcal{D}$ is a cartesian functor between cartesian categories, then there is a natural isomorphism

$$\begin{array}{ccc} \mathbf{Ab}(\mathcal{C}) & \xrightarrow{B_\bullet} & s\mathbf{Ab}(\mathcal{C}) \\ {\scriptstyle F}\downarrow & \nearrow\!\!\!\!\nearrow & \downarrow{\scriptstyle F} \\ \mathbf{Ab}(\mathcal{D}) & \xrightarrow[B_\bullet]{} & s\mathbf{Ab}(\mathcal{D}). \end{array}$$

PROOF. The first claim follows from Proposition 1.2 by passing to commutative augmented monoids, and applying the Eckmann-Hilton argument (A.1) and the arguments in the appendix which allow us to permute the order of taking simplicial objects and augmented objects.

The second claim follows from applying $\mathbf{CMon}(-)^{\mathrm{aug}}$ to the diagram in Proposition 1.5 and using the result from A.4.2 which states that a monoidal transformation between normal lax symmetric monoidal functors indeed induces a natural transformation in the respective categories of augmented commutative monoids.

If $\mathcal{C}$ is cartesian, then $\mathbf{Ab}(\mathcal{C})$ and $s\mathcal{C}$ are also cartesian, and Proposition 1.2 says that $B_\bullet : \mathbf{Mon}(\mathcal{C})^{\mathrm{aug}} \to s\mathcal{C}$ is cartesian. The proof is now as above, but passing to abelian group objects instead of augmented commutative objects. Indeed, there is an equivalence of categories $\mathbf{Ab}(\mathcal{C})^{\mathrm{aug}} \cong \mathbf{Ab}(\mathcal{C})$ (A.4.2). □

2. Geometric realization

We now need a notion of "intrinsic" geometric realization of a simplicial object in a category $\mathcal{V}$ to an object in $\mathcal{V}$. By the quoted term we mean a functor $s\mathcal{V} \to \mathcal{V}$, in contrast with what happens with the standard geometric realization of a simplicial set into a topological space (see Example 4.2), which we call "extrinsic" and which we analyze in Remark 2.5.2. Recall that the geometric realization of a simplicial space can be described as a tensor product of functors [**ML98**, IX.6]: if $X_\bullet$ is a simplicial space, then

$$|X_\bullet| = X_\bullet \otimes_\Delta \Delta^\bullet_{\mathrm{top}} \in \mathbf{Top}$$

where $\Delta^\bullet_{\text{top}} : \mathbb{\Delta} \to \mathbf{Top}$ is the standard cosimplicial space that takes $[n]$ to the topological n-simplex. This defines a functor $|-| : s\mathbf{Top} \to \mathbf{Top}$.

More generally, we have the following definition.

DEFINITION 2.1. Let $\mathcal{V}$ be a cocomplete symmetric monoidal category with a given cosimplicial object $D^\bullet : \mathbb{\Delta} \to \mathcal{V}$. We define the *geometric realization (with respect to $D^\bullet$)* as the functor

$$|-|_{D^\bullet} := - \otimes_{\mathbb{\Delta}} D^\bullet : s\mathcal{V} \to \mathcal{V}. \tag{2.2}$$

Explicitly, if $X_\bullet \in s\mathcal{V}$, then $|X_\bullet|_{D^\bullet}$ can be expressed as a coend, or even more explicitly as a coequalizer, as follows:

$$|X_\bullet|_{D^\bullet} = \int^n X_n \otimes D^n = \operatorname{coeq}\left(\coprod_{[n]\xrightarrow{f}[m]\in \operatorname{Arr}\mathbb{\Delta}} X_m \otimes D^n \rightrightarrows \coprod_{[p]\in\mathbb{\Delta}} X_p \otimes D^p \right)$$

where the parallel arrows are induced by the maps $X_m \otimes D^n \to X_n \otimes D^n$ and $X_m \otimes D^n \to X_m \otimes D^m$ given by the action of f.

LEMMA 2.3. *If $\mathcal{V}$ is moreover* closed, *then the functor $|-|_{D^\bullet} : s\mathcal{V} \to \mathcal{V}$ is a left adjoint. The right adjoint is given by the functor*

$$\underline{\mathcal{V}}(D^\bullet, -) : \mathcal{V} \to s\mathcal{V}, \qquad V \mapsto ([n] \mapsto \underline{\mathcal{V}}(D^n, V))$$

where $\underline{\mathcal{V}}(-,-) \in \mathcal{V}$ denotes the internal hom object of $\mathcal{V}$.

PROOF. The proof is an elementary manipulation of adjunctions and coends, which uses that the set $\operatorname{Nat}(F, G)$ of natural transformations from F to G is equal to the end $\int_X \mathcal{V}(FX, GX)$ for functors F and G out of some small category. □

COROLLARY 2.4. *The geometric realization of a simplicial object $cX \in s\mathcal{V}$ constant at $X \in \mathcal{V}$ is isomorphic to $X \otimes D^0$.*

PROOF. Indeed,

$$\begin{aligned}\mathcal{V}(|cX|_{D^\bullet}, V) &\cong \operatorname{Nat}(cX, \underline{\mathcal{V}}(D^\bullet, V)) \cong \mathcal{V}(X, \lim \underline{\mathcal{V}}(D^\bullet, V)) \cong \lim \mathcal{V}(X, \underline{\mathcal{V}}(D^\bullet, V)) \cong \\ &\cong \lim \mathcal{V}(X \otimes D^\bullet, V) \cong \mathcal{V}(\operatorname{colim}(X \otimes D^\bullet), V) \cong \mathcal{V}(X \otimes D^0, V)\end{aligned}$$

naturally in $V \in \mathcal{V}$. Yoneda's lemma implies $|cX|_{D^\bullet} \cong X \otimes D^0$. Here we have used the fact that the colimit of a cosimplicial object is its zeroth component, since $\mathbb{\Delta}$ has $[0]$ as its final object. □

REMARK 2.5. It is interesting to note that geometric realization can be extended to an enriched context, in two different ways. Let $\mathcal{C}$ be a cocomplete category enriched, tensored and cotensored over a closed symmetric monoidal category $\mathcal{V}$.

(1) If we are given a cosimplicial object $D^\bullet : \mathbb{\Delta} \to \mathcal{V}$, then we can define $|-|_{D^\bullet} : s\mathcal{C} \to \mathcal{C}$ as $|X_\bullet|_{D^\bullet} = \int^n X_n \odot D^n$ where $\odot$ denotes the tensoring of $\mathcal{C}$ over $\mathcal{V}$. The proof of Lemma 2.3 adapts to prove that $|-|_{D^\bullet} : s\mathcal{C} \to \mathcal{C}$ is a left adjoint. If we take $\mathcal{C} = \mathcal{V}$, we recover the construction above.

(2) If instead we are given a cosimplicial object $D^\bullet : \mathbb{\Delta} \to \mathcal{C}$, then we can define $|-|^e_{D^\bullet} : s\mathcal{V} \to \mathcal{C}$ as $|X_\bullet|^e_{D^\bullet} = \int^n D^n \odot X_n$. Again, $|-|^e_{D^\bullet}$ is a left adjoint, and taking $\mathcal{C} = \mathcal{V}$ recovers the construction above. Since this functor takes simplicial objects in one category and begets objects in a different category, we call it *extrinsic geometric realization.*

2.1. Cosimplicial objects induced by the Yoneda embedding. We will concentrate on closed symmetric monoidal categories with a cosimplicial object induced by the Yoneda embedding on simplicial sets. More precisely,

> *From here on we let $\mathcal{V}$ be a cocomplete, closed symmetric monoidal category together with a lax symmetric monoidal functor $F : s\mathbf{Set} \to \mathcal{V}$ which is a left adjoint. Let $\Delta^\bullet : \mathbb{\Delta} \to s\mathbf{Set}$ be the Yoneda embedding: we consider the cosimplicial object on $\mathcal{V}$ given by $F\Delta^\bullet : \mathbb{\Delta} \to \mathcal{V}$. We denote by $|-|$ the geometric realization functor $|-|_{F\Delta^\bullet} : s\mathcal{V} \to \mathcal{V}$.*

An additional hypothesis of normality for F will be specified when necessary; notably, whenever we need F to induce a functor on *augmented* objects.

The particularities of the Yoneda embedding will allow us to do things we could not do with the geometric realization with respect to an abstract cosimplicial object (see e.g. formula (2.10)).

REMARK 2.6. Let $X \in \mathcal{V}$. Since Δ^0 is the unit of the cartesian category $s\mathbf{Set}$, the unit of F has the form $\nabla_0 : \mathbb{1} \to F\Delta^0$. Therefore we get an arrow

$$X \xrightarrow[\cong]{\rho^{-1}} X \otimes \mathbb{1} \xrightarrow{\mathrm{id}\otimes\nabla_0} X \otimes F\Delta^0 \cong |cX| \tag{2.7}$$

which is an isomorphism if F is normal. The last isomorphism is provided by Corollary 2.4.

We recall the *density theorem* [**Rie14**, 1.4.6]: if $H : I^{\mathrm{op}} \to \mathcal{C}$ is a functor from the opposite of a small category I into a cocomplete category $\mathcal{C}$, then

$$Hj \cong \int^i I(j,i) \cdot Hi. \tag{2.8}$$

Here and henceforth $\cdot$ denotes the tensoring of $\mathcal{C}$ over $\mathbf{Set}$. Explicitly, if $A \in \mathbf{Set}$ and $C \in \mathcal{C}$, then $A \cdot C$ is the coproduct $\coprod_{i\in A} C$.

Note that since $F : s\mathbf{Set} \to \mathcal{V}$ is a left adjoint it preserves coproducts, and therefore $F(A \cdot X) \cong A \cdot FX$.

THEOREM 2.9. *The geometric realization functor $|-| : s\mathcal{V} \to \mathcal{V}$ has a lax symmetric monoidal structure, which is strong (resp. normal) if F is strong (resp. normal).*

PROOF. Let us express the simplicial set $\Delta^n \times \Delta^m$ as a coend, using the density theorem.

$$\begin{aligned}(\Delta^n \times \Delta^m)(j) &\cong \int^i \mathbb{\Delta}(j,i) \times (\Delta^n \times \Delta^m)(i) = \int^i \Delta^i(j) \times (\mathbb{\Delta} \times \mathbb{\Delta})((i,i),(n,m)) \\ &= \left(\int^i (\mathbb{\Delta} \times \mathbb{\Delta})((i,i),(n,m)) \cdot \Delta^i\right)(j)\end{aligned} \tag{2.10}$$

Let $X_\bullet, Y_\bullet \in s\mathcal{V}$. We will now repeatedly use Fubini's theorem for coends [**ML98**, IX.8], the fact that $- \otimes -$ commutes with colimits (hence with coends) separately in each variable, and the fact that F commutes with coends. Finally, we use the density theorem again.

$$\begin{aligned}
|X_\bullet| \otimes |Y_\bullet| &= \int^n X_n \otimes F\Delta^n \otimes \int^m Y_m \otimes F\Delta^m \\
&\cong \int^{n,m} X_n \otimes Y_m \otimes F\Delta^n \otimes F\Delta^m \\
&\to \int^{n,m} X_n \otimes Y_m \otimes F(\Delta^n \times \Delta^m) \\
&\cong \int^{n,m} X_n \otimes Y_m \otimes F\left(\int^i (\mathbb{\Delta} \times \mathbb{\Delta})((i,i),(n,m)) \cdot \Delta^i\right) \\
&\cong \int^{n,m} X_n \otimes Y_m \otimes \int^i (\mathbb{\Delta} \times \mathbb{\Delta})((i,i),(n,m)) \cdot F\Delta^i \\
&\cong \int^i \left(\int^{n,m} (\mathbb{\Delta} \times \mathbb{\Delta})((i,i),(n,m)) \cdot (X_n \otimes Y_m)\right) \otimes F\Delta^i \\
&\cong \int^i X_i \otimes Y_i \otimes F\Delta^i = |X_\bullet \otimes Y_\bullet|
\end{aligned}$$

The unit of $|-|$ is furnished by (2.7) applied to $X = \mathbb{1}$.

It is a long, if tedious verification that these morphisms endow $|-|$ with the structure of a lax symmetric monoidal functor. □

2.2. Behavior under monoidal functors.

From here on we let $\mathcal{W}$ be a cocomplete, closed symmetric monoidal category together with a lax symmetric monoidal functor $G : \mathcal{V} \to \mathcal{W}$ which is a left adjoint. We endow $\mathcal{W}$ with the cosimplicial object $GF\Delta^\bullet : \mathbb{\Delta} \to \mathcal{W}$, and we denote by $|-|$ the geometric realization functor $|-|_{GF\Delta^\bullet} : s\mathcal{W} \to \mathcal{W}$.

Of course, Theorem 2.9 applies to $\mathcal{W}$ as well: $|-| : s\mathcal{W} \to \mathcal{W}$ has a lax symmetric monoidal structure which is strong (resp. normal) if F and G are strong (resp. normal). Note that we will be using $|-|$ to denote both $|-|_{F\Delta^\bullet}$ and $|-|_{GF\Delta^\bullet}$: which one we mean should be clear from context.

Theorem 2.11. *There is a monoidal transformation between lax symmetric monoidal functors*

$$s\mathcal{V} \underset{G|-|}{\overset{|G-|}{\rightrightarrows}} \mathcal{W}, \qquad \Downarrow \tau \tag{2.12}$$

which is an isomorphism if G is strong.

In particular, this holds when $\mathcal{V} = s\mathbf{Set}$ and $F = \mathrm{id}_{s\mathbf{Set}}$.

Proof. Since G is a left adjoint, it preserves coends, and thus we get

$$\int^n GX_n \otimes GF\Delta^n \to \int^n G(X_n \otimes F\Delta^n) \cong G \int^n X_n \otimes F\Delta^n$$

for $X_\bullet \in s\mathcal{V}$, defining the desired natural transformation. We need to check it is monoidal, i.e. that the following diagram commutes, for $X_\bullet, Y_\bullet \in s\mathcal{V}$.

$$\begin{array}{ccccc} |GX_\bullet| \otimes |GY_\bullet| & \longrightarrow & |GX_\bullet \otimes GY_\bullet| & \longrightarrow & |G(X_\bullet \otimes Y_\bullet)| \\ \downarrow{\scriptstyle \tau\otimes\tau} & & & & \downarrow{\scriptstyle \tau} \\ G|X_\bullet| \otimes G|Y_\bullet| & \longrightarrow & G(|X_\bullet| \otimes |Y_\bullet|) & \longrightarrow & G|X_\bullet \otimes Y_\bullet| \end{array}$$

The horizontal arrows intertwine the monoidal structure of G and of its induced functor $G : s\mathcal{V} \to s\mathcal{W}$ with the monoidal structure of the geometric realizations $|-|$ in $s\mathcal{V}$ and in $s\mathcal{W}$ obtained in Theorem 2.9. These structures are defined via a fairly long string of isomorphisms, whence the difficulty of reproducing the necessary diagram proof. One can expand the diagram into one big rectangle filled with coends, and juggle around with naturality and monoidality properties of F and G. The gist of the proof is that the geometric realizations in $\mathcal{V}$ and in $\mathcal{W}$ are not independent: they are related via G, which is lax symmetric monoidal. □

2.3. Commutative augmented monoids. Suppose F and G are normal. By Theorem 2.9, the geometric realization functor $|-| : s\mathcal{V} \to \mathcal{V}$ is normal lax symmetric monoidal, therefore it induces a functor

$$|-| : s\mathbf{CMon}(\mathcal{V})^{\mathrm{aug}} \to \mathbf{CMon}(\mathcal{V})^{\mathrm{aug}} \tag{2.13}$$

and similarly for $|-| : s\mathcal{W} \to \mathcal{W}$.

Since the natural transformation of Theorem 2.11 is monoidal, we can apply (A.4) and obtain a natural transformation

$$s\mathbf{CMon}(\mathcal{V})^{\mathrm{aug}} \underset{G|-|}{\overset{|G-|}{\rightrightarrows}} \mathbf{CMon}(\mathcal{W})^{\mathrm{aug}} \quad \Downarrow\tau \tag{2.14}$$

which is an isomorphism if G is strong.

3. The bar construction

Suppose F is normal. We can now define the *bar construction*

$$B : \mathbf{CMon}(\mathcal{V})^{\mathrm{aug}} \to \mathbf{CMon}(\mathcal{V})^{\mathrm{aug}} \tag{3.1}$$

as the composite $\mathbf{CMon}(\mathcal{V})^{\mathrm{aug}} \xrightarrow{B_\bullet} s(\mathbf{CMon}(\mathcal{V}))^{\mathrm{aug}} \xrightarrow{|-|} \mathbf{CMon}(\mathcal{V})^{\mathrm{aug}}$. Remark that the monoid structure on BA for $A \in \mathbf{CMon}(\mathcal{V})^{\mathrm{aug}}$ is induced by the simplicial map

$$A^{\otimes p} \otimes A^{\otimes p} \to A^{\otimes p} \tag{3.2}$$

which is the monoid structure on $A^{\otimes p}$.

If $\mathcal{V} = \mathcal{C}$ is cartesian and F is cartesian, we can take abelian group objects instead of augmented commutative monoids, and obtain a functor

$$B : \mathbf{Ab}(\mathcal{C}) \to \mathbf{Ab}(\mathcal{C}).$$

PROPOSITION 3.3. *Suppose F and G are normal. There is a natural transformation*

$$\begin{array}{ccc} \mathbf{CMon}(\mathcal{V})^{\text{aug}} & \xrightarrow{B} & \mathbf{CMon}(\mathcal{V})^{\text{aug}} \\ {\scriptstyle G}\downarrow & \Uparrow & \downarrow{\scriptstyle G} \\ \mathbf{CMon}(\mathcal{W})^{\text{aug}} & \xrightarrow[B]{} & \mathbf{CMon}(\mathcal{W})^{\text{aug}} \end{array}$$

which is an isomorphism if G is strong.

If $\mathcal{V}$, $\mathcal{W}$, F and G are cartesian, there is an analogous square for **Ab** *instead of* $\mathbf{CMon}^{\text{aug}}$.

PROOF. The natural transformation is the pasting of the following two.

$$\begin{array}{ccccc} \mathbf{CMon}(\mathcal{V})^{\text{aug}} & \xrightarrow{B_\bullet} & s\mathbf{CMon}(\mathcal{V})^{\text{aug}} & \xrightarrow{|-|} & \mathbf{CMon}(\mathcal{V})^{\text{aug}} \\ {\scriptstyle G}\downarrow & \Uparrow & \downarrow{\scriptstyle G} & \Uparrow & \downarrow{\scriptstyle G} \\ \mathbf{CMon}(\mathcal{W})^{\text{aug}} & \xrightarrow[B_\bullet]{} & s\mathbf{CMon}(\mathcal{W})^{\text{aug}} & \xrightarrow[|-|]{} & \mathbf{CMon}(\mathcal{W})^{\text{aug}} \end{array}$$

The left one comes from Corollary 1.6, and the right one is (2.14). For abelian groups it is entirely analogous. □

3.1. Remark: bicommutative Hopf monoids. By the results of A.2, the category $\mathbf{CoComon}(\mathcal{V})$ of cocommutative comonoids in $\mathcal{V}$ is cartesian, therefore admits a simplicial bar construction on its abelian group objects (Corollary 1.6). As remarked in A.3, $\mathbf{Ab}(\mathbf{CoComon}(\mathcal{V}))$ is equivalent to $\mathbf{BiHopf}(\mathcal{V})$, the category of bicommutative Hopf monoids in $\mathcal{V}$, so the simplicial bar construction in this case is actually a functor $B_\bullet : \mathbf{BiHopf}(\mathcal{V}) \to s(\mathbf{BiHopf}(\mathcal{V}))$. This functor coincides with the simplicial bar construction of Corollary 1.6 under the forgetful functors U from (simplicial) bicommutative Hopf monoids to (simplicial) commutative augmented monoids.

Suppose $F : s\mathbf{Set} \to \mathcal{V}$ is strong. As seen in Theorem 2.9, this implies that $|-| : s\mathcal{V} \to \mathcal{V}$ is strong symmetric monoidal. In the same fashion as above, we obtain an induced functor $|-| : s\mathbf{BiHopf}(\mathcal{V}) \to \mathbf{BiHopf}(\mathcal{V})$ which coincides with (2.13) under the forgetful functors U.

Composing the functors we get a functor

$$B : \mathbf{BiHopf}(\mathcal{V}) \to \mathbf{BiHopf}(\mathcal{V}) \tag{3.4}$$

which coincides with (3.1) under U.

3.2. Iterated bar constructions and ring structure. We can iterate the bar construction to obtain functors

$$B^n : \mathbf{CMon}(\mathcal{V})^{\text{aug}} \to \mathbf{CMon}(\mathcal{V})^{\text{aug}} \quad \text{for } n \geq 0,$$

where we define B^0 to be the identity functor.

If $\mathcal{V} = \mathcal{C}$ is cartesian, we similarly obtain functors

$$B^n : \mathbf{Ab}(\mathcal{C}) \to \mathbf{Ab}(\mathcal{C}) \quad \text{for } n \geq 0.$$

We will now put a graded ring structure on iterated bar constructions, provided we start with a ring object. To be able to carry this out, we make the following assumptions.

We let the monoidal structures in our categories $\mathcal{V}, \mathcal{W}$ be cartesian. We therefore change notation: we are given cartesian categories $\mathcal{C}, \mathcal{D}$, a cartesian functor $F : s\mathbf{Set} \to \mathcal{C}$ and a cartesian functor $G : \mathcal{C} \to \mathcal{D}$, both of which are left adjoints.

Thus we get induced geometric realizations on $s\mathcal{C}$ and on $s\mathcal{D}$ which are cartesian.

We can glue all the iterated bar constructions together into a single functor

$$B^* : \mathbf{Ab}(\mathcal{C}) \to \mathbf{GrAb}(\mathcal{C}), \qquad A \mapsto (B^n A)_{n \in \mathbb{N}}.$$

Here $\mathbf{GrAb}(\mathcal{C})$ stands for the category of $\mathbb{N}$-graded objects of $\mathbf{Ab}(\mathcal{C})$, i.e. the functor category $\operatorname{Fun}(\mathbb{N}, \mathbf{Ab}(\mathcal{C}))$.

THEOREM 3.5. a) *The functor B^* extends to a functor*

$$B^* : \mathbf{Ring}(\mathcal{C}) \to \mathbf{GrRing}(\mathcal{C}).$$

b) *There is a natural isomorphism*

$$\begin{array}{ccc} \mathbf{Ring}(\mathcal{C}) & \xrightarrow{B^*} & \mathbf{GrRing}(\mathcal{C}) \\ {\scriptstyle G}\downarrow & \Uparrow\!\!\!/ & \downarrow{\scriptstyle G} \\ \mathbf{Ring}(\mathcal{D}) & \xrightarrow[B^*]{} & \mathbf{GrRing}(\mathcal{D}). \end{array}$$

Here $\mathbf{GrRing}(\mathcal{C})$ stands for the category of graded ring objects in $\mathcal{C}$: its objects are sequences $(A_n)_{n \geq 0} \in \mathbf{GrAb}(\mathcal{C})$ together with a graded multiplication

$$A_n \times A_m \to A_{n+m}$$

which is associative, unital and distributive with respect to the abelian group operation of each A_i.

PROOF. a) Let $S \in \mathbf{Ring}(\mathcal{C})$. Denote by $\mu : S \times S \to S$ its multiplication. We define the graded multiplication $\smile_{n,m}: B^n S \times B^m S \to B^{n+m} S$ inductively.

For $n = m = 0$, $\smile_{0,0}: S \times S \to S$ is μ. Now let us define $\smile_{0,m+1}$ from $\smile_{0,m}$.

Consider, for $i = 1, \dots, p$, the composition

$$S \times (B^m S)^{\times p} \xrightarrow{\mathrm{id} \times \pi_i} S \times B^m S \xrightarrow{\smile_{0,m}} B^m S, \tag{3.6}$$

where π_i denotes the i-th projection map. By the universal property of the categorical product, these maps define a morphism

$$\varphi_m^p : S \times (B^m S)^{\times p} \to (B^m S)^{\times p}$$

in $\mathcal{C}$ which commutes with the faces and degeneracies of $B_\bullet B^m S$. Indeed, as an example, the commutativity with the differentials $d_1, \dots, d_{p-1}$ rests on the distributivity of $\smile_{0,m}$ with respect to the addition of $B^m S$, and this is obtained inductively: for $m = 0$ it is the mere distributivity of $\smile_{0,0} = \mu$ with respect to addition. We say more about the distributivity of the higher $\smile_{n,m}$ at the end of the proof.

We thus get a morphism

$$\varphi_m : S \times B_\bullet B^m S \to B_\bullet B^m S$$

in $s\mathcal{C}$, where S is seen as a constant simplicial object.

As the geometric realization of a constant simplicial object gives the original object (Remark 2.6) and as geometric realization is a cartesian functor, we obtain an induced map

$$S \times B^{m+1}S \xrightarrow{\cong} |S \times B_\bullet B^m S| \xrightarrow{|\varphi_m|} B^{m+1}S \tag{3.7}$$

which we call $\smile_{0,m+1}$.

The definition of $\smile_{n+1,m}$ from $\smile_{n,m}$ is symmetrical: replace (3.6) with

$$(B^n S)^{\times p} \times B^m S \xrightarrow{\pi_i \times \mathrm{id}} B^n S \times B^m S \xrightarrow{\smile_{n,m}} B^{n+m} S$$

and repeat the process.

The unit for this graded multiplication is the unit for the multiplication of S. Associativity and distributivity of $\smile$ follow from associativity and distributivity of μ; these are all straightforward verifications. As an example, here is the diagram expressing the distributivity of $\smile_{0,1}$ at the simplicial level,

$$\begin{array}{ccccc}
S \times S^{\times p} \times S^{\times p} & \xrightarrow{\Delta \times \mathrm{id} \times \mathrm{id}} & S \times S \times S^{\times p} \times S^{\times p} & \xrightarrow{\mathrm{id} \times \sigma \times \mathrm{id}} & S \times S^{\times p} \times S \times S^{\times p} \\
 & & & & \downarrow {\scriptstyle \varphi_0^p \times \varphi_0^p} \\
\downarrow {\scriptstyle \mathrm{id} \times +} & & & & S^{\times p} \times S^{\times p} \\
 & & & & \downarrow {\scriptstyle +} \\
S \times S^{\times p} & & \xrightarrow[\varphi_0^p]{} & & S^{\times p}
\end{array}$$

where $+ : S^{\times p} \times S^{\times p} \to S^{\times p}$ is the abelian group structure map of $S^{\times p}$ (3.2), $\Delta : S \to S \times S$ is the diagonal, and σ is the symmetry. Its commutativity follows from distributivity in S.

b) First of all, the vertical functors induced by G do exist since $G : \mathcal{C} \to \mathcal{D}$ is cartesian. Second, the commutativity at the level of each abelian group object follows by iterating Proposition 3.3.

For the compatibility of $\smile$-multiplications, first observe that when $n = m = 0$ this is just the definition of the multiplication in GS, for $S \in \mathbf{Ring}(\mathcal{C})$:

$$\begin{array}{ccc}
GS \times GS & \xrightarrow{\mu^{GS}} & GS \\
{\scriptstyle \cong}\downarrow & & \Vert \\
G(S \times S) & \xrightarrow[G\mu^S]{} & GS.
\end{array}$$

For general n and m, this amounts to the commutativity of the following diagram in $\mathcal{D}$,

$$\begin{CD}
B^n GS \times B^m GS @>{\smile^{GS}_{n,m}}>> B^{n+m}GS \\
@V{\cong}VV @. \\
GB^n S \times GB^m S @. @V{\cong}VV \\
@V{\cong}VV @. \\
G(B^n S \times B^m S) @>>{G(\smile^{S}_{n,m})}> G(B^{n+m}S)
\end{CD}$$

which holds since the definition of the $\smile$-multiplications only involves arrows which commute with G and B. □

REMARK 3.8. We really need that G be cartesian, since we need it to preserve ring objects, but we could ask that F be merely normal lax symmetric monoidal. This affects the proof only in that the isomorphism in (3.7) becomes just a morphism.

3.3. Cocommutative comonoids. If our categories are symmetric monoidal but not cartesian, we cannot a priori carry out the construction of the previous section. However, note that $s\mathbf{Set}$ *is* cartesian, therefore for any strong symmetric monoidal functor $F : s\mathbf{Set} \to \mathcal{V}$ we obtain a cartesian functor between cartesian categories

$$F : s\mathbf{Set} \to \mathbf{CoComon}(\mathcal{V})$$

as remarked in (A.2). Moreover, if $G : \mathcal{V} \to \mathcal{W}$ is strong symmetric monoidal, we also obtain a cartesian functor

$$G : \mathbf{CoComon}(\mathcal{V}) \to \mathbf{CoComon}(\mathcal{W}).$$

Thus, from Theorem 3.5 we obtain the following result.

COROLLARY 3.9. (1) *Let $\mathcal{V}$ be a cocomplete closed symmetric monoidal category and $F : s\mathbf{Set} \to \mathcal{V}$ be a strong symmetric monoidal functor which is a left adjoint. There is an iterated bar construction functor*

$$B^* : \mathbf{Ring}(\mathbf{CoComon}(\mathcal{V})) \to \mathbf{GrRing}(\mathbf{CoComon}(\mathcal{V})). \tag{3.10}$$

(2) *Let $\mathcal{W}$ be a cocomplete closed symmetric monoidal category and $G : \mathcal{V} \to \mathcal{W}$ be a strong symmetric monoidal functor which is a left adjoint. There is a natural isomorphism*

$$\begin{CD}
\mathbf{Ring}(\mathbf{CoComon}(\mathcal{V})) @>{B^*}>> \mathbf{GrRing}(\mathbf{CoComon}(\mathcal{V})) \\
@V{G}VV \Uparrow\!\!\!/ @VV{G}V \\
\mathbf{Ring}(\mathbf{CoComon}(\mathcal{W})) @>>{B^*}> \mathbf{GrRing}(\mathbf{CoComon}(\mathcal{W})).
\end{CD}$$

DEFINITION 3.11. A graded ring object in cocommutative comonoids in $\mathcal{V}$ is called a *coalgebraic* (or *Hopf*) *ring in $\mathcal{V}$*. We call a coalgebraic ring in **k-Mod** a *k-coalgebraic ring.*

The notion of k-coalgebraic ring was introduced in [**RW77**]: they called it "Hopf ring". We prefer this other term (which they also considered but didn't keep) since it is more explicit.

We can apply the second part of this Corollary to a strong symmetric monoidal, left adjoint functor $F : s\mathbf{Set} \to \mathcal{V}$, in which case the situation simplifies: there is a natural isomorphism

$$\begin{array}{ccc} s\mathbf{Ring} & \xrightarrow{B^*} & s\mathbf{GrRing} \\ {\scriptstyle F}\downarrow & \Uparrow & \downarrow{\scriptstyle F} \\ \mathbf{Ring}(\mathbf{CoComon}(\mathcal{V})) & \xrightarrow[B^*]{} & \mathbf{GrRing}(\mathbf{CoComon}(\mathcal{V})). \end{array} \tag{3.12}$$

REMARK 3.13. There is a forgetful functor

$$\mathbf{Ring}(\mathbf{CoComon}(\mathcal{V})) \to \mathbf{Ab}(\mathbf{CoComon}(\mathcal{V})) = \mathbf{BiHopf}(\mathcal{V}).$$

Thus, the functor (3.10) has each of its levels B^n forget down to the respective iteration of (3.4).

4. Examples

4.1. Simplicial sets. We start with the cartesian closed category $s\mathbf{Set}$ itself. Endowed with the Yoneda embedding $\Delta^\bullet$ as a cosimplicial simplicial set, the induced geometric realization

$$|-|_{\Delta^\bullet} : s^2\mathbf{Set} \to s\mathbf{Set}$$

is naturally isomorphic to the diagonal functor, as is well-known. In symbols,

$$\int^n X_{n,\bullet} \times \Delta^n \cong \operatorname{diag}(X).$$

A proof for this formula can be obtained by means of the density theorem (2.8), the coend formula for $\Delta^n \times \Delta^m$ (2.10) and Fubini's theorem for coends.

We obtain a functor

$$B : s\mathbf{Ab} \to s\mathbf{Ab}. \tag{4.1}$$

It is a classical result that this functor is weakly homotopy equivalent to the $\bar{W}$-construction of Eilenberg and Mac Lane [**EML53**]. Some discussion and references for this can be found in [**Ste12**] after Lemma 15. One way of obtaining this result is as follows. Duskin identified the functor $\bar{W} : s\mathbf{Ab} \to s\mathbf{Ab}$ with the functor $TB_\bullet :$ $s\mathbf{Ab} \to s\mathbf{Ab}$, where $T : s^2\mathbf{Ab} \to s\mathbf{Ab}$ is a functor going by several names, two of which are "Artin-Mazur diagonal" and "totalization". Then one needs to provide a natural weak homotopy equivalence $T \Rightarrow \operatorname{diag}$. This result has a complicated history: we refer to the aforementioned discussion by Stevenson, and to the more recent [**Zis15**]. If G is an abelian group and we view it as a constant simplicial abelian group, then Eilenberg and Mac Lane proved that $\bar{W}^n G$ gives a simplicial abelian group model for a $K(G, n)$; see [**GM74**, A.21].

Theorem 3.5 gives us a functor $B^* : \mathbf{Ring}(s\mathbf{Set}) \to \mathbf{GrRing}(s\mathbf{Set})$, i.e.

$$B^* : s\mathbf{Ring} \to s\mathbf{GrRing}. \tag{4.2}$$

When S is a constant simplicial ring, B^nS is a simplicial model for $K(S,n)$, and the graded multiplication is a simplicial model for the cup product in Eilenberg-Mac Lane simplicial sets, as we will see in the next section.

4.2. Topological spaces. Recall that **Top** denotes the cartesian closed category of compactly generated weakly Hausdorff spaces. Let

$$F = |-|^e : s\mathbf{Set} \to \mathbf{Top}$$

be the classical "extrinsic" geometric realization functor. It is cartesian: this is a well-known result of Milnor on the geometric realization of a product of simplicial sets. It is also a left adjoint: its right adjoint is the singular functor.

Note that this extrinsic geometric realization functor is defined following the pattern of Remark 2.5.2: the base (cartesian) monoidal category is **Set**, and we consider **Top** as a **Set**-category.

The cosimplicial object $|\Delta^\bullet|^e$ is the standard cosimplicial space, i.e. $|\Delta^n|^e$ is the topological n-simplex. Therefore the geometric realization $|-|_{|\Delta^\bullet|^e} : s\mathbf{Top} \to \mathbf{Top}$ is the standard geometric realization of a simplicial space, as considered e.g. in [**May72**]. It should be noted that our Theorem 2.9 gives a categorical proof that it preserves products (compare with [**May72**, 11.5]): the topology is contained entirely in Milnor's theorem that $|-|^e$ is cartesian.

The resulting functor

$$B : \mathbf{Ab}(\mathbf{Top}) \to \mathbf{Ab}(\mathbf{Top})$$

is Milgram's [**Mil67**] version of the bar construction of a topological abelian group, as observed by Mac Lane [**ML70**]. The space BG is an especially nice model for the classifying space of G, because it carries a strict topological abelian group structure. Thus if $G \in \mathbf{Ab}(\mathbf{Top})$ is discrete, then BG is a model for the Eilenberg-Mac Lane space $K(G,1)$, and B^nG is a model for a $K(G,n)$.

Theorem 3.5 applied to $F = \mathrm{id}_{s\mathbf{Set}}$ and $G = |-|^e : s\mathbf{Set} \to \mathbf{Top}$ gives a natural isomorphism

$$\begin{array}{ccc} s\mathbf{Ring} & \xrightarrow{B^*} & \mathbf{GrRing}(s\mathbf{Set}) \\ {\scriptstyle |-|^e}\downarrow & \Nearrow & \downarrow{\scriptstyle |-|^e} \\ \mathbf{Ring}(\mathbf{Top}) & \xrightarrow[B^*]{} & \mathbf{GrRing}(\mathbf{Top}). \end{array}$$

In other words, we have natural isomorphisms

$$B^n(|S|^e) \cong |B^nS|^e \tag{4.3}$$

compatible with the graded multiplications existing on each side, for $S \in s\mathbf{Ring}$.

When $S \in \mathbf{Ring}(\mathbf{Top})$ is a discrete topological ring, $B^*S = (K(S,n))_{n\geq 0}$ where $K(S,n)$ denotes an n-th Eilenberg-Mac Lane space of S, and the graded multiplication

$$\smile : K(S,n) \times K(S,m) \to K(S,n+m) \tag{4.4}$$

represents the cup product in ordinary cohomology with coefficients in S [**RW80**, 1.7]. Thus, viewing S as a constant simplicial ring, the graded multiplication in simplicial sets

$$B^nS \times B^mS \to B^{n+m}S \tag{4.5}$$

coincides with the cup product map (4.4) after geometric realization under the isomorphism (4.3), i.e. we have gotten a simplicial construction of the cup product map in Eilenberg-Mac Lane simplicial sets.

Let us pass to homology. Let E be a commutative ring spectrum. We denote by $E_* = \pi_*(E)$ its graded commutative ring of coefficients. Let $E_*(-) : \mathbf{Top} \to E_*\text{-}\mathbf{Mod}$ denote its associated unreduced homology theory on spaces taking values in E_*-graded modules.

The category $E_*\text{-}\mathbf{Mod}$ is symmetric monoidal with the tensor product $\otimes_{E_*}$. The functor $E_*(-)$ has a lax symmetric monoidal structure given by the homological cross product

$$E_*(X) \otimes_{E_*} E_*(Y) \to E_*(X \times Y). \tag{4.6}$$

Suppose E satisfies a Künneth isomorphism, i.e. (4.6) is an isomorphism for all spaces X and Y. In other words, $E_*(-)$ is a strong symmetric monoidal functor. As per (A.2), we get an induced cartesian functor $E_* : \mathbf{Top} \to E_*\text{-}\mathbf{CoCoalg}$, inducing a functor

$$E_* : \mathbf{GrRing}(\mathbf{Top}) \to \mathbf{GrRing}(E_*\text{-}\mathbf{CoCoalg}). \tag{4.7}$$

Thus for a topological ring S, $(E_*(B^n S))_{n\geq 0}$ is a graded E_*-coalgebraic ring (Definition 3.11). For S discrete this was discussed by Ravenel and Wilson [**RW80**].

As a particularly simple case, take $E = Hk$, the Eilenberg-Mac Lane spectrum of the commutative ring k. If k is a field, then Hk satisfies a Künneth isomorphism, but we will not need this in what follows.

Considering sets as discrete topological spaces and modules as graded modules concentrated in degree zero, the functor $(Hk)_* : \mathbf{Top} \to (Hk)_*\text{-}\mathbf{CoCoalg}$ restricts to the functor

$$k[-] : \mathbf{Set} \to k\text{-}\mathbf{CoCoalg} \tag{4.8}$$

which maps a set X to the free k-module $k[X]$ together with the comultiplication obtained by extending linearly the diagonal map on basis elements, $\Delta(x) = x \otimes x$ for $x \in X$. In other words, the cartesian functor (4.8) is obtained from the strong symmetric monoidal free functor $k[-] : \mathbf{Set} \to \mathbf{k\text{-}Mod}$ by passing to cocommutative comonoids.[1]

Thus, if S is a discrete ring, then $(Hk)_*(S)$ is the group ring[2] $k[S]$ as a k-coalgebraic ring concentrated in degree zero. This is the correct way of characterizing all the structure of the object $k[S]$: in particular, of characterizing the distributivity of the operations coming from the sum and multiplication of S.

It should be noted that there are not that many topological rings, since if G is an abelian topological group then G is homotopy equivalent to $\prod_{n\geq 0} K(\pi_n G, n)$.

[1]Remark that (4.8) cannot rightly be called a "free functor", since it is not the left adjoint to the "underlying set" functor: rather, it is the left adjoint to the "set of group-like elements" functor, which maps a coalgebra C to the set of elements $c \in C$ such that $\Delta(c) = c \otimes c$.

[2]More accurately, to be coherent with the naming convention for this kind of object, we should say "the ring k-coalgebraic ring $k[S]$".

4.3. Simplicial modules. Consider the closed symmetric monoidal category of k-modules $\mathcal{V} = \mathbf{k\text{-}Mod}$ with the tensor product $\otimes_k$. Then $\mathbf{CMon}(\mathbf{k\text{-}Mod})^{\mathrm{aug}}$ is the category $k\text{-}\mathbf{CAlg}^{\mathrm{aug}}$ of commutative augmented k-algebras, i.e. commutative k-algebras A with a k-algebra homomorphism $A \to k$. The functor

$$B_\bullet : k\text{-}\mathbf{CAlg}^{\mathrm{aug}} \to s(k\text{-}\mathbf{CAlg}^{\mathrm{aug}}) \tag{4.9}$$

coincides with the reduced (i.e. with coefficients in the trivial k-module) simplicial Hochschild functor $HH_\bullet(-, k)$, as is readily verified from the definitions. However, we cannot go any further with this example: we do not have a natural choice of a functor F from simplicial sets to k-modules. To put it differently, we have not been able to find a choice of a cosimplicial module which would yield an interesting realization of a simplicial module into a module.

We thus shift our attention to $\mathcal{V} = s(\mathbf{k\text{-}Mod})$, the closed symmetric monoidal category of simplicial k-modules with pointwise tensor product. It admits a natural strong symmetric monoidal functor from $s\mathbf{Set}$: the free simplicial k-module functor

$$F = k[-] : s\mathbf{Set} \to s(\mathbf{k\text{-}Mod}).$$

It has as right adjoint the functor that forgets the module structure at each level.

We claim that, as for simplicial sets (Example 4.1), the induced geometric realization functor

$$|-|_{k[\Delta^\bullet]} : s^2(\mathbf{k\text{-}Mod}) \to s(\mathbf{k\text{-}Mod})$$

is the diagonal functor. The proof is very similar, but uses instead the *enriched* density theorem [**Kel05**, 3.72].

As remarked in 1.3, the functor $B_\bullet : s(k\text{-}\mathbf{CAlg}^{\mathrm{aug}}) \to s^2(k\text{-}\mathbf{CAlg}^{\mathrm{aug}})$ is, degreewise, the simplicial bar construction (4.9). After geometric realizing (taking diagonals), we obtain a functor

$$B : s(k\text{-}\mathbf{CAlg}^{\mathrm{aug}}) \to s(k\text{-}\mathbf{CAlg}^{\mathrm{aug}})$$

which is the k-linear counterpart to the construction for simplicial abelian groups (4.1). It is weakly homotopy equivalent to the algebraic $\bar{W}$-construction of Eilenberg and Mac Lane (see [**GM74**], A.14 for the definition of $\bar{W}$ and A.20 for a proof of $BA \simeq \bar{W}A$).

Proposition 3.3 applied to $F = \mathrm{id}_{s\mathbf{Set}}$ and to $G = k[-]$ gives a natural isomorphism

$$\begin{array}{ccc} s\mathbf{CMon}^{\mathrm{aug}} & \xrightarrow{B} & s\mathbf{CMon}^{\mathrm{aug}} \\ {\scriptstyle k[-]}\downarrow & \Uparrow & \downarrow{\scriptstyle k[-]} \\ s(k\text{-}\mathbf{CAlg}^{\mathrm{aug}}) & \xrightarrow[B]{} & s(k\text{-}\mathbf{CAlg}^{\mathrm{aug}}). \end{array}$$

We cannot naively apply Theorem 3.5 to this situation because $\mathbf{k\text{-}Mod}$ is not cartesian. However, we can apply (3.12) to obtain a natural isomorphism

$$\begin{array}{ccc} s\mathbf{Ring} & \xrightarrow{B^*} & s\mathbf{GrRing} \\ {\scriptstyle k[-]}\downarrow & \Uparrow & \downarrow{\scriptstyle k[-]} \\ s\mathbf{Ring}(k\text{-}\mathbf{CoCoalg}) & \xrightarrow[B^*]{} & s\mathbf{GrRing}(k\text{-}\mathbf{CoCoalg}). \end{array}$$

The B^* in the upper line is (4.2). Thus, if S is a simplicial ring, then $k[B^*S] \cong B^*k[S]$ as simplicial graded coalgebraic rings. Let us see how the graded multiplication in $B^*k[S]$ passes to homotopy, just as we passed to homology in Example 4.2. For this we need k to be a field, so that the functor

$$\pi_* : s(\mathbf{k\text{-}Mod}) \to \mathbf{Gr}(\mathbf{k\text{-}Mod})$$

is strong symmetric monoidal. Indeed, we can decompose π_* as the normalized Moore functor N (which is lax symmetric monoidal with the shuffle product; see Example 4.4 for more details) followed by the homology functor. The Eilenberg-Zilber and algebraic Künneth theorems give the result. Thus, we get a functor

$$\pi_* : s\mathbf{GrRing}(k\text{-}\mathbf{CoCoalg}) \to \mathbf{GrRing}(k\text{-}\mathbf{CoCoalg}).$$

Let $A \in s\mathbf{Ring}(k\text{-}\mathbf{CoCoalg})$ be a constant simplicial object. By neglect of structure, A is an augmented commutative k-algebra, where the augmentation is given by the counit. Then the homotopy of B^nA gives $HH_*^{[n]}(A,k)$, Pirashvili's [**Pir00**] higher order reduced Hochschild homology of A, as noted in [**LR11**, Section 3.1]. The multiplication of A induces a graded multiplication

$$HH_*^{[n]}(A,k) \otimes HH_*^{[m]}(A,k) \to HH_*^{[n+m]}(A,k). \tag{4.10}$$

Now let S be a constant simplicial ring and consider $A = k[S]$. As seen in Example 4.2, B^*S is a model for the Eilenberg-Mac Lane graded simplicial ring $K(S,*)$. Thus, we have an isomorphism

$$HH_*^{[*]}(k[S],k) \cong k[K(S,*)]$$

and the graded multiplication (4.10) corresponds to the graded multiplication (4.5) corresponding to the cup product in cohomology.

The reader might want to jump to Section 4.5.3 where we analyze the analogous phenomena happening in *topological* Hochschild homology.

4.4. Differential graded modules. Let $\mathcal{W} = \mathbf{k\text{-}dgm}$ be the closed symmetric monoidal category of non-negatively graded chain complexes of k-modules with the tensor product $\otimes_k$. We would like the normalized Moore functor $N : s(\mathbf{k\text{-}Mod}) \to \mathbf{k\text{-}dgm}$ [**GJ99**, III.2] to play the role of our functor called G; however, it is not *strong* symmetric monoidal, it is merely colax (with an Alexander-Whitney map) and symmetric lax (with a shuffle product map, also called "Eilenberg-Zilber map")[3] which is why we do not take this example too far.

The functor N can be seen as an extrinsic geometric realization, following Remark 2.5.2. It can be considered a sort of linearization of $|-|^e : s\mathbf{Set} \to \mathbf{Top}$. Indeed, the category of chain complexes is enriched, tensored and cotensored over modules. The cosimplicial chain complex $Nk[\Delta^\bullet] : \mathbb{\Delta} \to \mathbf{k\text{-}dgm}$ begets a functor

$$|-|^e_{Nk[\Delta^\bullet]} : s(\mathbf{k\text{-}Mod}) \to \mathbf{k\text{-}dgm}.$$

The fact that $N \cong |-|^e_{Nk[\Delta^\bullet]}$ was observed in [**Kan58**, 6.3] by direct computation. Alternatively, if we know that the Dold-Kan correspondence is an *adjoint* equivalence [**Wei94**, 8.4.2], then just as in Lemma 2.3 we get that the right adjoint to $|-|^e_{Nk[\Delta^\bullet]}$ is the functor $A \mapsto \underline{\mathbf{k\text{-}dgm}}(Nk[\Delta^\bullet], A)$. This functor is known to be the right adjoint to N in the Dold-Kan correspondence, therefore $N \cong |-|^e_{Nk[\Delta^\bullet]}$ by uniqueness of left adjoints.

[3] It is even *bilax*: [**AM10**, Corollary 5.7].

The identification $N \cong |-|^e_{Nk[\Delta^\bullet]}$ is a priori a bit unsatisfactory, since to define $|-|^e_{Nk[\Delta^\bullet]}$ we use the cosimplicial chain complex $Nk[\Delta^\bullet]$ which depends on N. However, $Nk[\Delta^\bullet]$ can be defined independently: for example, it is the cellular chain complex on the cosimplicial space $|\Delta^\bullet|^e$.

At any rate, the cosimplicial chain complex $Nk[\Delta^\bullet] : \mathbb{\Delta} \to \mathbf{k\text{-}dgm}$ yields an interesting geometric realization

$$|-|_{Nk[\Delta^\bullet]} : s(\mathbf{k\text{-}dgm}) \to \mathbf{k\text{-}dgm}.$$

Indeed, $|-|_{Nk[\Delta^\bullet]}$ coincides with the functor $C : s(\mathbf{k\text{-}dgm}) \to \mathbf{k\text{-}dgm}$ which is the composition of the functors "normalized Moore in each internal degree", yielding a bicomplex, and the "totalization of a bicomplex" functor. One can prove that $C \cong |-|_{Nk[\Delta^\bullet]}$ by a computation similar to the one by Kan which proves that $N \cong |-|^e_{Nk[\Delta^\bullet]}$.

Gugenheim and May [**GM74**, A.2] call the functor C *condensation*, and they prove in Proposition A.3 that it is a normal lax symmetric monoidal functor (not strong), via a suitable totalized shuffle product, and colax, via a suitable Alexander-Whitney map.

Note that the induced bar construction

$$B : k\text{-}\mathbf{cdga}^{\mathrm{aug}} \to k\text{-}\mathbf{cdga}^{\mathrm{aug}}$$

is the usual bar construction of a commutative differential graded augmented k-algebra, as introduced in [**EML53**, Theorem 11.1]; see also [**GM74**, Page 69]. Its multiplication is the "shuffle product", since it is induced by the lax structure of N.

Proposition 3.3 applies, giving a natural transformation

$$\begin{array}{ccc} s(k\text{-}\mathbf{CAlg}^{\mathrm{aug}}) & \xrightarrow{B} & s(k\text{-}\mathbf{CAlg}^{\mathrm{aug}}) \\ {\scriptstyle N}\downarrow & \Uparrow & \downarrow{\scriptstyle N} \\ k\text{-}\mathbf{cdga}^{\mathrm{aug}} & \xrightarrow[B]{} & k\text{-}\mathbf{cdga}^{\mathrm{aug}}. \end{array}$$

It is not a natural isomorphism since N is not strong. Theorem 3.5 does not apply, since k-$\mathbf{cdga}^{\mathrm{aug}}$ is not cartesian, and Corollary 3.9 does not apply either, since N is not strong.

As a final remark, recall Dold and Puppe's version of the Eilenberg-Zilber theorem [**GJ99**, IV.2.4], which states that there is a natural quasi-isomorphism

$$N(\operatorname{diag} X_{\bullet,\bullet}) \to CN(X_{\bullet,\bullet})$$

for a bisimplicial module $X_{\bullet,\bullet}$. Rephrasing it in the terminology adopted here, there is a natural quasi-isomorphism $N|-| \Rightarrow |N-|$. It would be interesting to check whether it is a *monoidal* isomorphism, by comparison with Theorem 2.11.

4.5. Brave new algebra.

4.5.1. *Spectra.* Take $\mathcal{W}$ to be the closed symmetric monoidal category $\mathbb{S}$-**Mod** of S-modules [**EKMM97**], which we will refer to simply as "spectra". The monoidal structure of $\mathbb{S}$-**Mod** is given by the smash product $\wedge = \wedge_{\mathbb{S}}$ as the tensor product, and the sphere spectrum $\mathbb{S}$ as the unit.

We take as a functor G the functor

$$\Sigma_+^\infty : \mathbf{Top} \to \mathbb{S}\text{-}\mathbf{Mod} \tag{4.11}$$

which maps a topological space X to the suspension spectrum on the space X with an added disjoint basepoint. It is strong symmetric monoidal [**EKMM97**, II.1.2], and it is a left adjoint [**EKMM97**, Page 39]. We therefore consider $\mathbb{S}$-**Mod** endowed with the cosimplicial spectrum $\Sigma_+^\infty|\Delta^\bullet|^e : \mathbb{\Delta} \to \mathbb{S}\text{-}\mathbf{Mod}$ (recall from Example 4.2 that $|-|^e : s\mathbf{Set} \to \mathbf{Top}$ is the classical "extrinsic" geometric realization functor).

It should be noted that if $X \in \mathbf{Top}$ and $C \in \mathbb{S}\text{-}\mathbf{Mod}$, then $C \wedge \Sigma_+^\infty X$ defines the standard tensoring of $\mathbb{S}$-**Mod** over **Top** [**EKMM97**, III.1.1]. Therefore, the induced geometric realization

$$|-|_{\Sigma_+^\infty|\Delta^\bullet|^e} =: |-| : s(\mathbb{S}\text{-}\mathbf{Mod}) \to \mathbb{S}\text{-}\mathbf{Mod} \tag{4.12}$$

in our sense coincides with the one in [**EKMM97**, X.1.1].

By Theorem 2.9, $|-|$ is strong symmetric monoidal. This appears in [**EKMM97**, X.1.4]. Theorem 2.11 applies to prove that the natural isomorphism $\Sigma_+^\infty|X_\bullet| \cong |\Sigma_+^\infty X_\bullet|$, which appears in [**EKMM97**, X.1.3.i], is monoidal. This has not, to our knowledge, explicitly appeared in the literature.

We could apply Corollary 3.9 to the strong symmetric monoidal functor G of (4.11) right now. But instead, let us first delve into general R-modules and apply the machinery there (see (4.14)): it will give a more general result.

4.5.2. *R-modules and extension of scalars.* Let R be a commutative ring spectrum and take $\mathcal{W}$ to be the closed symmetric monoidal category R-**Mod** of R-modules, with smash product $\wedge = \wedge_R$ as tensor product, and as unit the R-module R.

We take as a functor G the strong symmetric monoidal functor of extension of scalars

$$R \wedge_{\mathbb{S}} - : \mathbb{S}\text{-}\mathbf{Mod} \to R\text{-}\mathbf{Mod},$$

whose right adjoint is the restriction of scalars functor. Thus R-**Mod** is endowed with the cosimplicial R-module $R \wedge_{\mathbb{S}} \Sigma_+^\infty|\Delta^\bullet|^e$.

Furthermore, if R' is another commutative ring spectrum and $f : R \to R'$ is a morphism, then in just the same fashion we obtain a strong symmetric monoidal functor G

$$R' \wedge_R - : R\text{-}\mathbf{Mod} \to R'\text{-}\mathbf{Mod}$$

endowing R'-**Mod** with the cosimplicial R'-module $R' \wedge_{\mathbb{S}} \Sigma_+^\infty|\Delta^\bullet|^e$.

Corollary 3.9 gives a natural isomorphism comparing the iterated bar construction together with its graded structure, whenever carried out in R-**Mod** or R'-**Mod**:

$$\begin{array}{ccc}
\mathbf{Ring}(R\text{-}\mathbf{CoCoalg}) & \xrightarrow{B^*} & \mathbf{GrRing}(R\text{-}\mathbf{CoCoalg}) \\
{\scriptstyle R'\wedge_R -}\downarrow & \Uparrow & \downarrow{\scriptstyle R'\wedge_R -} \\
\mathbf{Ring}(R'\text{-}\mathbf{CoCoalg}) & \xrightarrow[B^*]{} & \mathbf{GrRing}(R'\text{-}\mathbf{CoCoalg}).
\end{array}$$

We will now analyze what the bar construction in this context actually is.

4.5.3. *Topological Hochschild homology.* The simplicial bar construction in the category of R-modules,

$$B_\bullet : R\text{-}\mathbf{CAlg}^{\mathrm{aug}} \to s(R\text{-}\mathbf{CAlg}^{\mathrm{aug}}),$$

coincides with $THH^R_\bullet(-, R)$, the reduced simplicial topological Hochschild homology functor, by the same verification as in Section 4.3 for Hochschild homology.

Therefore,

$$B : R\text{-}\mathbf{CAlg}^{\mathrm{aug}} \to R\text{-}\mathbf{CAlg}^{\mathrm{aug}}$$

is an explicit model for the reduced topological Hochschild homology functor $THH^R(-, R)$, as noted in [**EKMM97**, IX.2].

Note that if one wants this $THH^R(A, R)$ to coincide with the derived smash product $R \wedge^L_{A^e} A$ using the model structure in R-**Mod** set in [**EKMM97**], it is necessary that R be a q-cofibrant commutative S-algebra and that A be a q-cofibrant commutative R-algebra. This is enough by a slight modification of Theorem 2.6 in [**EKMM97**].

The iterations of B,

$$B^n : R\text{-}\mathbf{CAlg}^{\mathrm{aug}} \to R\text{-}\mathbf{CAlg}^{\mathrm{aug}} \tag{4.13}$$

for $n \geq 0$ are an explicit model for higher reduced topological Hochschild homology $THH^{R,[n]}(-, R)$ as considered e.g. in [**BLP$^+$15**].

Indeed, $THH^R(A, R)$ can be expressed as $S^1 \odot A$, where $\odot$ denotes the tensoring of the category R-$\mathbf{CAlg}^{\mathrm{aug}}$ over pointed topological spaces [**Kuh04**, 7.1], and its higher version is

$$THH^{R,[n]}(A, R) = S^n \odot A.$$

We obtain natural isomorphisms

$$\begin{aligned} B^2(A) = THH^R(THH^R(A, R), R) = S^1 \odot (S^1 \odot A) \cong \\ \cong (S^1 \wedge S^1) \odot A \cong S^2 \odot A = THH^{R,[2]}(A, R) \end{aligned}$$

and similarly for higher powers.

We can now apply Corollary 3.9 to the functor

$$G = R[-] = R \wedge_{\mathbb{S}} \Sigma^\infty_+ : \mathbf{Top} \to R\text{-}\mathbf{Mod}, \tag{4.14}$$

and obtain a natural isomorphism

$$\begin{array}{ccc} \mathbf{Ring}(\mathbf{Top}) & \xrightarrow{\quad B^* \quad} & \mathbf{GrRing}(\mathbf{Top}) \\ {\scriptstyle R[-]}\big\downarrow & \Uparrow & \big\downarrow{\scriptstyle R[-]} \\ \mathbf{Ring}(R\text{-}\mathbf{CoCoalg}) & \xrightarrow[B^*]{} & \mathbf{GrRing}(R\text{-}\mathbf{CoCoalg}). \end{array}$$

We thus get, for $A \in \mathbf{Ring}(R\text{-}\mathbf{CoCoalg})$, a graded multiplication in higher THH:

$$THH^{R,[n]}(A, R) \wedge_R THH^{R,[m]}(A, R) \to THH^{R,[n+m]}(A, R) \tag{4.15}$$

and if $A = R[S]$ for $S \in \mathbf{Ring}(\mathbf{Top})$, then we get a natural isomorphism

$$THH^{R,[*]}(R[S], R) \cong R[B^* S] \tag{4.16}$$

of graded ring objects of R-**CoCoalg**, i.e. of coalgebraic rings in R-**Mod** (Definition 3.11). As noted in Example 4.2, when S is discrete the graded topological ring B^*S is a model for the Eilenberg-Mac Lane spaces $K(S,*)$.

Observe that by means of Remark 3.13, we get for every $n \geq 0$ an isomorphism

$$THH^{R,[n]}(R[G],R) \cong R[B^nG]$$

of bicommutative R-Hopf algebras, natural in the topological abelian group G. For $n=1$ and $R=\mathbb{S}$ there is a similar well-known unreduced result: $THH(\mathbb{S}[G]) \cong \mathbb{S}[B^{cy}G]$, where B^{cy} stands for the cyclic bar construction.

Let us pass to homology. Let $E \in R$-**CAlg** be a field, meaning that $E_* = \pi_*E$ is a graded field, i.e. every graded module over it is free. Then the E-homology functor $E_* : R\text{-}\mathbf{Mod} \to E_*\text{-}\mathbf{Mod}$ is strong symmetric monoidal, since the field hypothesis guarantees a Künneth isomorphism. It induces a functor

$$E_* : \mathbf{GrRing}(R\text{-}\mathbf{CoCoalg}) \to \mathbf{GrRing}(E_*\text{-}\mathbf{CoCoalg}). \tag{4.17}$$

Thus, $E_*(THH^{R,[*]}(R[S],R))$ is an E_*-coalgebraic ring. If in particular we take R to be $\mathbb{S}$, we can recover the coalgebraic ring of Ravenel-Wilson (see Example 4.2) in a different guise. Indeed, precomposing (4.17) with

$$\mathbb{S}[-] : \mathbf{GrRing}(\mathbf{Top}) \to \mathbf{GrRing}(R\text{-}\mathbf{CoCoalg})$$

gives the E-homology of topological rings (4.7). Then, if $S \in \mathbf{Ring}(\mathbf{Top})$, we get an isomorphism of E_*-coalgebraic rings

$$E_*(THH^{\mathbb{S},[*]}(\mathbb{S}[S],\mathbb{S})) \cong E_*(B^*S). \tag{4.18}$$

Appendix A. Monoidal category theory

We recall some results we will use from monoidal category theory. A modern, comprehensive reference is [**AM10**].

A.1. General facts.

- In monoidal categories $\mathcal{V}$ we can consider *(co)monoids* and morphisms of these. Thus we get categories $\mathbf{Mon}(\mathcal{V})$ and $\mathbf{Comon}(\mathcal{V})$ which are in turn symmetric monoidal provided $\mathcal{V}$ is symmetric. The symmetry is needed to have a natural multiplication on the tensor product of monoids. We will henceforth focus on symmetric monoidal categories, since this is the setting we need in the body of the paper.

  In symmetric monoidal categories, we can also consider (co)commutative (co)monoids, and thus get symmetric monoidal categories $\mathbf{CMon}(\mathcal{V})$ and $\mathbf{CoComon}(\mathcal{V})$.

  The unit for these categories of monoids and comonoids is the unit of $\mathcal{V}$, which has unique commutative monoid and cocommutative comonoid structures. It is an initial object for $\mathbf{Mon}(\mathcal{V})$ and $\mathbf{CMon}(\mathcal{V})$, and a final object for $\mathbf{Comon}(\mathcal{V})$ and for $\mathbf{CoComon}(\mathcal{V})$.
- A *lax monoidal functor* $F : \mathcal{V} \to \mathcal{W}$ between monoidal categories is the data of a functor F plus morphisms

  $$\nabla_{V,V'} : F(V) \otimes F(V') \to F(V \otimes V')$$

  natural in $V, V' \in \mathcal{V}$, and a unit map

  $$\nabla_0 : \mathbb{1}_{\mathcal{W}} \to F(\mathbb{1}_{\mathcal{V}})$$

satisfying associativity and unitality conditions. A lax monoidal functor is *normal* if ∇_0 is an isomorphism (see Section A.4.2 for an application of this concept). If the categories are symmetric monoidal, then a lax monoidal functor is *symmetric* if it commutes with the respective symmetries.

Dually, a *colax monoidal functor* $F : \mathcal{V} \to \mathcal{W}$ between monoidal categories is the data of a functor F plus natural morphisms

$$\Delta_{V,V'} : F(V \otimes V') \to F(V) \otimes F(V')$$

and a counit map

$$\Delta_0 : F(\mathbb{1}_{\mathcal{V}}) \to \mathbb{1}_{\mathcal{W}}$$

satisfying coassociativity and counitality conditions. If the categories are symmetric monoidal, then a colax monoidal functor is *symmetric* if it commutes with the respective symmetries.

- An important feature of (co)lax monoidal functors is that they send (co)monoids to (co)monoids. More precisely, if $F : \mathcal{V} \to \mathcal{W}$ is a (co)lax monoidal functor between symmetric monoidal categories, we get an induced monoidal functor $F : \mathbf{Mon}(\mathcal{V}) \to \mathbf{Mon}(\mathcal{W})$ (resp. $F : \mathbf{Comon}(\mathcal{V}) \to \mathbf{Comon}(\mathcal{W})$). For example, if $M \in \mathbf{Mon}(\mathcal{V})$ has multiplication μ and F is lax monoidal, then the multiplication in $F(M)$ is given by the composition

$$F(M) \otimes F(M) \xrightarrow{\nabla_{M,M}} F(M \otimes M) \xrightarrow{F(\mu)} F(M).$$

  If moreover F is symmetric (co)lax, we get an induced symmetric monoidal functor $F : \mathbf{CMon}(\mathcal{V}) \to \mathbf{CMon}(\mathcal{W})$ (resp. $F : \mathbf{CoComon}(\mathcal{V}) \to \mathbf{CoComon}(\mathcal{W})$).
- A *strong monoidal functor* $F : \mathcal{V} \to \mathcal{W}$ between monoidal categories is a lax monoidal functor such that $\nabla_{V,V'}$ and ∇_0 are isomorphisms (in particular, it is normal). Thus a strong monoidal functor is also a colax monoidal functor, by inverting the structure morphisms. Therefore, if the categories are symmetric then F induces monoidal functors $F : \mathbf{Mon}(\mathcal{V}) \to \mathbf{Mon}(\mathcal{W})$ and $F : \mathbf{Comon}(\mathcal{V}) \to \mathbf{Comon}(\mathcal{W})$. If F is symmetric, then it also induces symmetric monoidal functors between the categories of (co)commutative (co)monoids.
- *The Eckmann-Hilton argument.* Let $\mathcal{V}$ be a symmetric monoidal category. There is an equivalence of symmetric monoidal categories (i.e. a strong symmetric monoidal functor which is an equivalence of categories)

$$\mathbf{Mon}(\mathbf{Mon}(\mathcal{V})) \cong \mathbf{CMon}(\mathcal{V}).$$

  The symmetric monoidal category $\mathbf{CMon}(\mathcal{V})$ also coincides with the symmetric monoidal categories $\mathbf{Mon}(\mathbf{CMon}(\mathcal{V}))$, $\mathbf{CMon}(\mathbf{Mon}(\mathcal{V}))$, and $\mathbf{CMon}(\mathbf{CMon}(\mathcal{V}))$.

A.1.1. *2-categorical remarks.*

- One can define natural transformations of (symmetric) lax, colax, strong monoidal functors [**AM10**, 3.1.2]. For each of these alternatives there is a corresponding 2-category [**AM10**, 3.3.3, C.1.1]. The natural transformations of (co)lax monoidal functors are called *monoidal transformations*. The natural transformations of (symmetric) strong monoidal functors and normal (co)lax monoidal functors are the natural transformations of the underlying (co)lax monoidal functors.

Let us make the lax case explicit. A monoidal transformation

$$\mathcal{V} \underset{G}{\overset{F}{\Downarrow \tau}} \mathcal{W}$$

of lax monoidal functors is a natural transformation such that the following diagrams commute for every $V, V' \in \mathcal{V}$.

$$\begin{array}{ccc} FV \otimes FV' & \xrightarrow{\nabla^F_{V,V'}} & F(V \otimes V') \\ \downarrow{\scriptstyle \tau_V \otimes \tau_{V'}} & & \downarrow{\scriptstyle \tau_{V\otimes V'}} \\ GV \otimes GV' & \xrightarrow[\nabla^G_{V,V'}]{} & G(V \otimes V'). \end{array} \qquad \begin{array}{ccc} & & F\mathbb{1}_{\mathcal{V}} \\ & \nearrow^{\nabla^F_0} & \downarrow{\scriptstyle \tau_{\mathbb{1}_{\mathcal{V}}}} \\ \mathbb{1}_{\mathcal{W}} & & \\ & \searrow_{\nabla^G_0} & \\ & & G\mathbb{1}_{\mathcal{V}} \end{array}$$

- Let $\mathcal{V} \underset{G}{\overset{F}{\Downarrow \tau}} \mathcal{W}$ be a 2-cell in the 2-category of symmetric monoidal categories, lax symmetric monoidal functors and monoidal transformations. There is an induced 2-cell

(A.1) $$\mathbf{CMon}(\mathcal{V}) \underset{G}{\overset{F}{\Downarrow \tau}} \mathbf{CMon}(\mathcal{W})$$

in the same 2-category. The fact that τ is a monoidal transformation and not merely a natural transformation guarantees that if $A \in \mathbf{CMon}(\mathcal{V})$, then $\tau_A : FA \to GA$ is a morphism of commutative monoids in $\mathcal{W}$.

A.2. Cartesian monoidal categories.

- A symmetric monoidal category $\mathcal{C}$ is *cartesian* if the monoidal structure defines category-theoretical binary products: the tensor product is a categorical product and the unit is a final object.

  Any category where finite products exist is cartesian monoidal, after a definite, functorial choice of finite products.

  In a cartesian category, more structure is available than in a mere symmetric monoidal category: we have projections and diagonal maps.
- A strong symmetric monoidal functor between cartesian categories is called *cartesian*. They are also called "finite product-preserving functors", because they send the chosen product diagram $A \leftarrow A \times B \to B$ to a product diagram.
- If $\mathcal{V}$ is a symmetric monoidal category, then $\mathbf{CoComon}(\mathcal{V})$ is cartesian.
- Conversely, if $\mathcal{C}$ is a cartesian category, then every object of $\mathcal{C}$ is uniquely a cocommutative comonoid for the product: $\mathcal{C} = \mathbf{CoComon}(\mathcal{C})$, thanks to the existence of diagonal maps $C \to C \times C$.

In particular, if we are given a colax symmetric monoidal functor $\mathcal{C} \to \mathcal{V}$ from a cartesian category to a symmetric monoidal category, then we get an induced colax symmetric monoidal functor

$$\mathcal{C} \to \mathbf{CoComon}(\mathcal{V}) \tag{A.2}$$

which is cartesian if $\mathcal{C} \to \mathcal{V}$ is strong.

- In a cartesian category $\mathcal{C}$, one can internalize algebras for a Lawvere theory (also called "algebraic theories" or "finite product theories"; see [**Bor94**, Chapter 3]), and these categories of algebras are in turn cartesian. A cartesian functor between cartesian categories induces cartesian functors between the categories of algebras.

  We will be considering abelian groups objects $\mathbf{Ab}(\mathcal{C})$, ring objects $\mathbf{Ring}(\mathcal{C})$, and $\mathbb{N}$-graded ring objects $\mathbf{GrRing}(\mathcal{C})$.

A.3. Hopf monoids.

- A *Hopf monoid* [**AM10**, 1.2.5] in a symmetric monoidal category $\mathcal{V}$ is an object of $\mathcal{V}$ together with compatible monoid and comonoid structures, plus an antipode morphism: the prototypical example is a Hopf algebra, which is a Hopf monoid in a category of modules over a commutative ring. A Hopf monoid is *bicommutative* if it is commutative and cocommutative; these form a symmetric monoidal category denoted by $\mathbf{BiHopf}(\mathcal{V})$. It turns out that in this case,

$$\mathbf{BiHopf}(\mathcal{V}) \cong \mathbf{Ab}(\mathbf{CoComon}(\mathcal{V})), \tag{A.3}$$

  the category of abelian group objects in the cartesian category of cocommutative comonoids in $\mathcal{V}$ ([**MP12**, 20.3.4], see also [**AM10**, 1.22.iii]). In particular, $\mathbf{BiHopf}(\mathcal{V})$ is cartesian.
- A strong symmetric monoidal functor $F : \mathcal{V} \to \mathcal{W}$ between symmetric monoidal categories induces a cartesian functor $F : \mathbf{BiHopf}(\mathcal{V}) \to \mathbf{BiHopf}(\mathcal{W})$. Indeed, we can break it up in two parts: first, F induces a cartesian functor between the categories of cocommutative monoids. Then, this functor induces a functor in the categories of abelian group objects. We are done by the characterization (A.3).
- There is a forgetful functor $\mathbf{BiHopf}(\mathcal{V}) \to \mathbf{CMon}(\mathcal{V})^{\mathrm{aug}}$, where the augmentation to be considered is the counit of the given Hopf monoid. This forgetting procedure is compatible with strong symmetric monoidal functors $\mathcal{V} \to \mathcal{W}$.

A.4. Overcategories.

A.4.1. *Augmented objects.*

- Let $\mathcal{C}$ be a category with an initial object I. Define the category $\mathcal{C}^{\mathrm{aug}}$ of *augmented objects* of $\mathcal{C}$ to be the slice category $\mathcal{C}^{/I}$, i.e. the category of objects of $\mathcal{C}$ with a chosen morphism to I, with the obvious morphisms.
- In $\mathcal{C}^{\mathrm{aug}}$, the object $I \xrightarrow{\mathrm{id}} I$ is a zero object (i.e. both initial and final). The forgetful functor $\mathcal{C}^{\mathrm{aug}} \to \mathcal{C}$ is an equivalence of categories if and only if $\mathcal{C}$ has a zero object. In particular, $(\mathcal{C}^{\mathrm{aug}})^{\mathrm{aug}} \cong \mathcal{C}^{\mathrm{aug}}$.

A.4.2. *Augmented objects in categories of monoids and abelian groups.*

- As we have already observed, $\mathbf{Mon}(\mathcal{V})$ and $\mathbf{CMon}(\mathcal{V})$ have $\mathbb{1}$ as their initial object. Thus, we can consider $\mathbf{Mon}(\mathcal{V})^{\mathrm{aug}}$ and $\mathbf{CMon}(\mathcal{V})^{\mathrm{aug}}$, the categories of *augmented (commutative) monoids*. They have induced symmetric monoidal structures.
- Let $F : \mathcal{V} \to \mathcal{W}$ be a normal lax symmetric monoidal functor. There is an induced normal lax symmetric monoidal functor $F : \mathbf{CMon}(\mathcal{V})^{\mathrm{aug}} \to \mathbf{CMon}(\mathcal{W})^{\mathrm{aug}}$. Normality is needed so that an augmentation of a monoid $A \to \mathbb{1}$ in $\mathcal{V}$ gets mapped to an augmentation $FA \to F\mathbb{1} \stackrel{\cong}{\leftarrow} \mathbb{1}$ of FA in $\mathcal{W}$.

  If τ is a monoidal transformation between normal lax symmetric monoidal functors, there is an induced 2-cell like (A.1) but on augmented commutative monoids:

$$\mathbf{CMon}(\mathcal{V})^{\mathrm{aug}} \underset{G}{\overset{F}{\rightrightarrows}} \mathbf{CMon}(\mathcal{W})^{\mathrm{aug}} \quad \Downarrow\tau \tag{A.4}$$

  Similar remarks hold for augmented monoids, normal lax monoidal functors and monoidal transformations.
- If $\mathcal{C}$ is a cartesian category then $\mathbb{1}$ is a final object, and $\mathbf{Ab}(\mathcal{C})$ has $\mathbb{1}$ as a zero object. Thus, $\mathbf{Ab}(\mathcal{C})^{\mathrm{aug}} \cong \mathbf{Ab}(\mathcal{C})$.

A.5. Simplicial objects.

- If $\mathcal{C}$ is any category, we denote by $s\mathcal{C}$ its category of simplicial objects, i.e. the functor category $\mathrm{Fun}(\mathbb{\Delta}^{\mathrm{op}}, \mathcal{C})$ where $\mathbb{\Delta}$ is the category of non-empty finite ordinals and order-preserving maps. If $\mathcal{C}$ is (symmetric) monoidal or cartesian, then so is $s\mathcal{C}$, with pointwise tensor product:

$$(X_\bullet \otimes Y_\bullet)_n = X_n \otimes Y_n.$$

  Its unit is the constant simplicial object $\mathbb{1}$. Any functor $F : \mathcal{C} \to \mathcal{D}$ induces a functor $F : s\mathcal{C} \to s\mathcal{D}$, which is cartesian or (symmetric, normal) lax, colax, strong monoidal if the original one is.

  Let $s^2\mathcal{C} = \mathrm{Fun}(\mathbb{\Delta}^{\mathrm{op}} \times \mathbb{\Delta}^{\mathrm{op}}, \mathcal{C})$ be the category of bisimplicial objects in $\mathcal{C}$. By adjunction, we can identify $s^2\mathcal{C}$ with $s(s\mathcal{C})$. Thus, we will think of a bisimplicial object $X_{\bullet,\bullet} \in s^2\mathcal{C}$ as a simplicial object $([n] \mapsto X_{n,\bullet})$ in $s\mathcal{C}$.
- For $\mathcal{V}$ a symmetric monoidal category, there is an equivalence of categories $s(\mathcal{V}^{/\mathbb{1}}) \cong (s\mathcal{V})^{/\mathbb{1}}$ (the slice category notation was explained in A.4.1). Similarly, for $\mathcal{C}$ a category with an initial object I, we have $s(\mathcal{C}^{\mathrm{aug}}) \cong (s\mathcal{C})^{\mathrm{aug}}$.
- Let T be the cartesian category associated to a Lawvere theory (cf. the last bullet point of A.2) and $\mathcal{C}$ be a cartesian category. Then the category of $\mathcal{C}$-valued T-algebras is the category of cartesian functors $T \to \mathcal{C}$.

  By adjunction, the category of $s\mathcal{C}$-valued T-algebras coincides with the category of functors from $\mathbb{\Delta}^{\mathrm{op}}$ to $\mathcal{C}$-valued T-algebras. In symbols,

$$\mathrm{CFun}(T, s\mathcal{C}) = \mathrm{Fun}(\mathbb{\Delta}^{\mathrm{op}}, \mathrm{CFun}(T, \mathcal{C}))$$

  where CFun denotes cartesian functors.

In particular, $s(\mathbf{Ab}(\mathcal{C})) = \mathbf{Ab}(s\mathcal{C})$, and similarly for rings and graded rings.

- This also works for ((co)commutative) (co)monoids in a (symmetric) monoidal category $\mathcal{V}$. Thus, $s(\mathbf{Mon}(\mathcal{V})) = \mathbf{Mon}(s\mathcal{V})$ and similarly for commutative monoids and (co)commutative comonoids.

References

[AM10] Marcelo Aguiar and Swapneel Mahajan, *Monoidal functors, species and Hopf algebras*, CRM Monograph Series, vol. 29, American Mathematical Society, Providence, RI, 2010. With forewords by Kenneth Brown and Stephen Chase and André Joyal. MR2724388

[BLP+15] Irina Bobkova, Ayelet Lindenstrauss, Kate Poirier, Birgit Richter, and Inna Zakharevich, *On the higher topological Hochschild homology of $\mathbb{F}_p$ and commutative $\mathbb{F}_p$-group algebras*, Women in topology: collaborations in homotopy theory, Contemp. Math., vol. 641, Amer. Math. Soc., Providence, RI, 2015, pp. 97–122. MR3380071

[Bor94] Francis Borceux, *Handbook of categorical algebra. 2*, Encyclopedia of Mathematics and its Applications, vol. 51, Cambridge University Press, Cambridge, 1994. Categories and structures. MR1313497

[EKMM97] A. D. Elmendorf, I. Kriz, M. A. Mandell, and J. P. May, *Rings, modules, and algebras in stable homotopy theory*, Mathematical Surveys and Monographs, vol. 47, American Mathematical Society, Providence, RI, 1997. With an appendix by M. Cole. MR1417719

[EML53] Samuel Eilenberg and Saunders Mac Lane, *On the groups $H(\Pi, n)$. I*, Ann. of Math. (2) **58** (1953), 55–106. MR0056295

[GJ99] Paul G. Goerss and John F. Jardine, *Simplicial homotopy theory*, Progress in Mathematics, vol. 174, Birkhäuser Verlag, Basel, 1999. MR1711612

[GM74] V. K. A. M. Gugenheim and J. Peter May, *On the theory and applications of differential torsion products*, American Mathematical Society, Providence, R.I., 1974. Memoirs of the American Mathematical Society, No. 142. MR0394720

[Kan58] Daniel M. Kan, *Functors involving c.s.s. complexes*, Trans. Amer. Math. Soc. **87** (1958), 330–346. MR0131873

[Kel05] G. M. Kelly, *Basic concepts of enriched category theory*, Repr. Theory Appl. Categ. **10** (2005), vi+137. Reprint of the 1982 original [Cambridge Univ. Press, Cambridge; MR0651714]. MR2177301

[Kuh04] Nicholas J. Kuhn, *The McCord model for the tensor product of a space and a commutative ring spectrum*, Categorical decomposition techniques in algebraic topology (Isle of Skye, 2001), Progr. Math., vol. 215, Birkhäuser, Basel, 2004, pp. 213–236. MR2039768

[Lac00] Stephen Lack, *A coherent approach to pseudomonads*, Adv. Math. **152** (2000), no. 2, 179–202. MR1764104

[LR11] Muriel Livernet and Birgit Richter, *An interpretation of E_n-homology as functor homology*, Math. Z. **269** (2011), no. 1-2, 193–219. MR2836066

[May72] J. P. May, *The geometry of iterated loop spaces*, Springer-Verlag, Berlin-New York, 1972. Lectures Notes in Mathematics, Vol. 271. MR0420610

[Mil67] R. James Milgram, *The bar construction and abelian H-spaces*, Illinois J. Math. **11** (1967), 242–250. MR0208595

[ML70] Saunders Mac Lane, *The Milgram bar construction as a tensor product of functors*, The Steenrod Algebra and its Applications (Proc. Conf. to Celebrate N. E. Steenrod's Sixtieth Birthday, Battelle Memorial Inst., Columbus, Ohio,1970), Lecture Notes in Mathematics, Vol. 168, Springer, Berlin, 1970, pp. 135–152. MR0273618

[ML98] Saunders Mac Lane, *Categories for the working mathematician*, 2nd ed., Graduate Texts in Mathematics, vol. 5, Springer-Verlag, New York, 1998. MR1712872

[MP12] J. P. May and K. Ponto, *More concise algebraic topology*, Chicago Lectures in Mathematics, University of Chicago Press, Chicago, IL, 2012. Localization, completion, and model categories. MR2884233

[Pir00] Teimuraz Pirashvili, *Hodge decomposition for higher order Hochschild homology* (English, with English and French summaries), Ann. Sci. École Norm. Sup. (4) **33** (2000), no. 2, 151–179. MR1755114

[Rie14] Emily Riehl, *Categorical homotopy theory*, New Mathematical Monographs, vol. 24, Cambridge University Press, Cambridge, 2014. MR3221774

[RW80] Douglas C. Ravenel and W. Stephen Wilson, *The Morava K-theories of Eilenberg-Mac Lane spaces and the Conner-Floyd conjecture*, Amer. J. Math. **102** (1980), no. 4, 691–748. MR584466

[RW77] Douglas C. Ravenel and W. Stephen Wilson, *The Hopf ring for complex cobordism*, J. Pure Appl. Algebra **9** (1976/77), no. 3, 241–280. MR0448337

[Shu] Mike Shulman, *Coherence theorem for symmetric lax monoidal functors*, MathOverflow, URL:http://mathoverflow.net/q/243442 (version: 2016-07-12).

[Ste12] Danny Stevenson, *Décalage and Kan's simplicial loop group functor*, Theory Appl. Categ. **26** (2012), No. 28, 768–787. MR3065943

[Wei94] Charles A. Weibel, *An introduction to homological algebra*, Cambridge Studies in Advanced Mathematics, vol. 38, Cambridge University Press, Cambridge, 1994. MR1269324

[Zis15] Michel Zisman, *Comparaison de deux diagonales pour un ensemble bisimplicial* (French, with English summary), J. Homotopy Relat. Struct. **10** (2015), no. 4, 1013–1021. MR3423083

Université Paris 13, Sorbonne Paris Cité, LAGA, CNRS, UMR 7539, 93430 Villetaneuse, France

Email address: `stonek@math.univ-paris13.fr`

Contemporary Mathematics
Volume **708**, 2018
http://dx.doi.org/10.1090/conm/708/14268

Double homotopy (co)limits for relative categories

Kay Werndli

Abstract. We answer the question to what extent homotopy (co)limits in categories with weak equivalences allow for a Fubini-type interchange law. The main obstacle is that we do not assume our categories with weak equivalences to come equipped with a calculus for homotopy (co)limits, such as a derivator.

0. Introduction

In modern, categorical homotopy theory, there are a few different frameworks that formalise what exactly a "homotopy theory" should be. Maybe the two most widely used ones nowadays are Quillen's model categories (see e.g. [**8**] or [**13**]) and $(\infty, 1)$-categories. The latter again comes in different guises, the most popular one being quasicategories (see e.g. [**16**] or [**17**]), originally due to Boardmann and Vogt [**3**]. Both of these contexts provide enough structure to perform homotopy invariant analogues of the usual categorical limit and colimit constructions and more generally Kan extensions.

An even more stripped-down notion of a "homotopy theory" (that arguably lies at the heart of model categories) is given by just working with categories that come equipped with a class of weak equivalences. These are sometimes called *relative categories* [**2**], though depending on the context, this might imply some restrictions on the class of weak equivalences. Even in this context, we can still say what a homotopy (co)limit is and do have tools to construct them, such as model approximations due to Chachólski and Scherer [**5**] or more generally, left and right deformation retracts as introduced in [**9**] and generalised in section 3 below.

To actually obtain a calculus of homotopy (co)limits, maybe the most general structure is that of a *derivator* due to Franke, Heller [**12**] and Grothendieck (see also [**11**]). Every model category has an underlying derivator [**5**, **6**] and therefore a good calculus of homotopy (co)limits. The same is true for a bicomplete $(\infty, 1)$-category, as shown by Riehl and Verity in [**20**].

Now, for a relative category, we might be able to construct homotopy (co)limits but not a derivator structure. The goal of this paper is to answer the question to what extent such relative categories still allow for a Fubini-type homotopy (co)limit interchange law, even in the absence of a derivator structure.

2010 *Mathematics Subject Classification.* Primary 55U35, 18A30, 18A40, 18G10, 18G55.
The author was partially supported by Swiss NSF grant 200020 149118.

For us, the main motivation to pursue such a result was that a proof based on universal properties, rather than explicit constructions, is much more conceptually satisfying. Even in nice contexts (e.g. when working with model categories), explicit constructions of homotopy colimits can become quite involved when no additional hypotheses (such as cofibrant generation) are imposed (c.f. [**5**]). For example, checking a seemingly simple claim such as 3.15 with explicit constructions seems rather tedious. For completeness, we also provide a few examples 9.8, 9.9, 9.10 immediately after our main theorem that are not nice model categories but where our theorem is applicable.

In section 1, we first derive a universal property of the (strict) localisation of a category that we are going to need and which is usually only proven for explicit constructions of the localisation (e.g. for the homotopy category of a model category). In section 2, we restate the usual definition of a derived functor and, more importantly, of an absolute derived functor. We then give an alternative, external characterisation of these, which don't seem to have been previously known but turn out to be very useful to us later on.

In section 3, we slightly generalise the approach of [**9**] for the construction of derived functors and see that the usual construction of derived functors via (co)fibrant replacements is not confined to model categories. In sections 4 and 5, for lack of references, we quickly recall the definition, coherence and Beck-Chevalley interchange condition for the calculus of mates. These are standard tools in the world of derivators but less well-known outside of it.

In the section that follows and motivated by [**19**], we investigate the concept of derived adjunctions in our general context and show the relation between absolute derived functors and adjoints thereof. It is here that we obtain our first homotopically substantial result.

COROLLARY 6.4. *Let $F \dashv G$ be an adjunction of functors between categories with weak equivalences such that an absolute total right derived functor $\mathbb{R}G$ of G exists. Then $F \dashv G$ is derivable if and only if $\mathbb{R}G$ has a left adjoint.* □

Here, an adjunction between two categories with weak equivalences is called *derivable* if the left and right adjoint respectively have absolute total left and right derived functors and these form an adjunction between the corresponding homotopy categories (this generalises the *deformable adjunctions* of [**9**]). The example to keep in mind here are Quillen adjunctions between model categories.

As we learned later, the same result was already obtained in the yet unpublished [**10**] but using different techniques. In *op. cit.*, the author used internal, diagrammatic methods, whereas we use our external characterisation of absolute Kan extensions obtained in section 2.

In section 7, we apply this result to the notion of homotopy (co)limits and obtain two equivalent definitions. Finally, sections 8 and 9 contain our main results about the pointwiseness of certain homotopy colimits. What we mean by this is best exemplified by the often used case of a double homotopy pushout. Given a 3×3-diagram of the form

$$\begin{array}{ccccc} \bullet & \longleftarrow & \bullet & \longrightarrow & \bullet \\ \uparrow & & \uparrow & & \uparrow \\ \bullet & \longleftarrow & \bullet & \longrightarrow & \bullet \\ \downarrow & & \downarrow & & \downarrow \\ \bullet & \longleftarrow & \bullet & \longrightarrow & \bullet \end{array},$$

most homotopy theorists take it for granted that the homotopy colimit of this diagram can be calculated as a double homotopy pushout in two ways: rows first or columns first. Under suitably nice hypotheses (e.g. working in a cofibrantly generated model category) this is certainly true, though there doesn't seem to be a formal proof of this relying on universal properties. Rather, explicit formulae for the homotopy colimits or descriptions by cofibrant replacements are used.

It is important to note that the pointwiseness of homotopy colimits (say for double homotopy pushouts as above) consists of two parts; namely that the objects in the span obtained by the row- or columnwise homotopy colimit are what one would expect and that the same is true for the morphisms between them. Our first result concerning the pointwiseness of homotopy colimits is now the following.

THEOREM 9.5. *Let $\mathcal{C}$ be a category with weak equivalences, $\mathcal{I}$, $\mathcal{J}$ two index categories and $J \in \mathcal{J}$ such that all derived adjunctions in the left-hand square*

$$\begin{array}{ccc} \mathrm{Ho}\big((\mathcal{C}^{\mathcal{I}})^{\mathcal{J}}\big) \cong \mathrm{Ho}\big((\mathcal{C}^{\mathcal{J}})^{\mathcal{I}}\big) & \underset{\mathrm{Ho}\,\Delta}{\overset{\mathrm{hocolim}_{\mathcal{I}}}{\rightleftarrows}}\ (\perp) & \mathrm{Ho}(\mathcal{C}^{\mathcal{J}}) \\ \mathrm{Ho}\,\mathrm{ev}_J \downarrow \dashv \uparrow \mathbb{R}J_* & & \mathrm{Ho}\,\mathrm{ev}_J \downarrow \dashv \uparrow \mathbb{R}J_* \\ \mathrm{Ho}(\mathcal{C}^{\mathcal{I}}) & \underset{\mathrm{Ho}\,\Delta}{\overset{\mathrm{hocolim}_{\mathcal{I}}}{\rightleftarrows}}\ (\perp) & \mathrm{Ho}\,\mathcal{C} \end{array} \qquad \begin{array}{ccc} \mathrm{Ho}\big((\mathcal{C}^{\mathcal{I}})^{\mathcal{J}}\big) \cong \mathrm{Ho}\big((\mathcal{C}^{\mathcal{J}})^{\mathcal{I}}\big) & \xleftarrow{\mathrm{Ho}\,\Delta} & \mathrm{Ho}(\mathcal{C}^{\mathcal{J}}) \\ \downarrow \mathrm{Ho}\,\mathrm{ev}_J & & \downarrow \mathrm{Ho}\,\mathrm{ev}_J \\ \mathrm{Ho}(\mathcal{C}^{\mathcal{I}}) & \xleftarrow[\mathrm{Ho}\,\Delta]{} & \mathrm{Ho}\,\mathcal{C} \end{array}$$

exist and $\mathrm{Ho}\,\Delta$ *composes with* $\mathbb{R}J_*$. *Then the right-hand square (filled with the identity) satisfies the (horizontal) Beck-Chevalley condition. In particular, there is an isomorphism*

$$\sigma\colon \underset{\mathcal{I}}{\mathrm{hocolim}} \circ \mathrm{Ho}\,\mathrm{ev}_J \cong \mathrm{Ho}\,\mathrm{ev}_J \circ \underset{\mathcal{I}}{\mathrm{hocolim}}, \qquad \textit{natural in } J \in \mathcal{J}.$$

Here, the stated hypothesis that $\mathrm{Ho}\,\Delta$ *compose* with $\mathbb{R}J_*$ means that the composite of right-derived functors $\mathbb{R}J_* \circ \mathrm{Ho}\,\Delta$ is the right-derived functor of the composite $J_* \circ \Delta$; something which is automatic when only considering Quillen pairs but not in our general context.

It is this theorem 9.5, which allows us to conclude that the most common homotopy colimits, such as double homotopy pushouts, coproducts and telescopes can always be calculated pointwise, assuming they exist and, curiously, assuming the base category has a terminal object.

In the above theorem, applied to the double homotopy pushout example, the existence of an isomorphism $\mathrm{hocolim}_{\mathcal{I}} \circ \mathrm{Ho}\,\mathrm{ev}_J \cong \mathrm{Ho}\,\mathrm{ev}_J \circ \mathrm{hocolim}_{\mathcal{I}}$ ensures that the objects of the row- or columnwise homotopy colimit are the ones one would expect, while the naturality in J ensures the same for the morphisms between these objects.

It is this essential naturality condition at the end, which is hard to prove. Since the mixing of homotopy colimits and right derived functors can be inconvenient (as one might be able to construct left derived functors such as homotopy colimits but not right derived ones), we then go on to investigate how we can use the left adjoints to evaluation functors to obtain the sought for naturality condition. To that extent, we have the following result.

THEOREM 9.16. *Let $\mathcal{C}$ be a category with weak equivalences and $\mathcal{I}$, $\mathcal{J}$ two index categories such that for all $J \in \mathcal{J}$, the two evaluation functors ev_J indicated below* *have fully faithful left adjoints $J_!$ and the derived adjunctions in the square*

$$\begin{array}{ccc}
\operatorname{Ho}\big((\mathcal{C}^{J})^{\mathcal{J}}\big) \cong \operatorname{Ho}\big((\mathcal{C}^{\mathcal{J}})^{J}\big) & \underset{\operatorname{Ho}\Delta}{\overset{\operatorname{hocolim}_{\mathcal{J}}}{\rightleftarrows}} \ (\bot) & \operatorname{Ho}(\mathcal{C}^{J}) \\
\mathbb{L}J_! \uparrow \dashv \downarrow \operatorname{Ho}\operatorname{ev}_J & & \mathbb{L}J_! \uparrow \dashv \downarrow \operatorname{Ho}\operatorname{ev}_J \\
\operatorname{Ho}(\mathcal{C}^{\mathcal{J}}) & \underset{\operatorname{Ho}\Delta}{\overset{\operatorname{hocolim}_{\mathcal{J}}}{\rightleftarrows}} \ (\bot) & \operatorname{Ho}\mathcal{C}
\end{array}$$

exist and there is an isomorphism $\operatorname{hocolim}_{\mathcal{J}} \circ \operatorname{Ho}\operatorname{ev}_J \cong \operatorname{Ho}\operatorname{ev}_J \circ \operatorname{hocolim}_{\mathcal{J}}$. *Then there is a family of isomorphisms*

$$\beta_J\colon \operatorname*{hocolim}_{\mathcal{J}} \circ \operatorname{Ho}\operatorname{ev}_J \cong \operatorname{Ho}\operatorname{ev}_J \circ \operatorname*{hocolim}_{\mathcal{J}} \qquad \textit{natural in } J.$$

1. Localisations reviewed

By the localisation of a category $\mathcal{C}$ with respect to a class of morphisms $\mathcal{W}$ is meant the (comparison functor with) the category (possibly not locally small) obtained by formally inverting the morphisms in $\mathcal{W}$.

DEFINITION 1.1. Let $\mathcal{C}$ be a category and $\mathcal{W}$ a class of arrows in $\mathcal{C}$ (which we usually call the *weak equivalences* of $\mathcal{C}$). A *weak localisation* of $\mathcal{C}$ is a functor $H\colon \mathcal{C} \to \mathcal{C}[\mathcal{W}^{-1}]$ sending all arrows in $\mathcal{W}$ to isomorphisms and having the following universal property:

(1) Whenever we have a functor $F\colon \mathcal{C} \to \mathcal{D}$ sending all arrows in $\mathcal{W}$ to isomorphisms, then there is some $\bar{F}\colon \mathcal{C}[\mathcal{W}^{-1}] \to \mathcal{D}$ together with a natural isomorphism $F \cong \bar{F} \circ H$;

(2) For every category $\mathcal{D}$, the precomposition $H^*\colon \mathcal{D}^{\mathcal{C}[\mathcal{W}^{-1}]} \to \mathcal{D}^{\mathcal{C}}$ is fully faithful. That is to say, for any two $F, G\colon \mathcal{C}[\mathcal{W}^{-1}] \to \mathcal{D}$ and any transformation $\tau\colon F \circ H \Rightarrow G \circ H$, there is a unique $\bar{\tau}\colon F \Rightarrow G$ such that $\tau = \bar{\tau}_H$.

OBSERVATION 1.2. One easily sees that $\mathcal{C}[\mathcal{W}^{-1}]$ is unique up to equivalence; that the extension $\bar{F}$ in property (1) is unique up to unique isomorphism and the natural isomorphism $F \cong \bar{F} \circ H$ is unique up to unique automorphism of $\bar{F}$.

Most authors will define localisations differently. Namely, $\bar{F}$ in (1) needs to be unique and the natural isomorphism $F \cong \bar{F} \circ H$ is required to be an identity, while property (2) is left out entirely. We call this a *strict localisation*.

DEFINITION 1.3. A *(strict) localisation* of a category $\mathcal{C}$ at a class of arrows $\mathcal{W}$ (again called the *weak equivalences*) is a functor $H\colon \mathcal{C} \to \mathcal{C}[\mathcal{W}^{-1}]$ that sends arrows in $\mathcal{W}$ to isomorphisms and such that every functor $F\colon \mathcal{C} \to \mathcal{D}$ that does so factors uniquely through H as $F = \bar{F} \circ H$. It follows that $\mathcal{C}[\mathcal{W}^{-1}]$ is unique up to isomorphism. We will seldomly have the situation where we are given two different classes of weak equivalences in $\mathcal{C}$ and will thus just write $\operatorname{Ho}\mathcal{C} := \mathcal{C}[\mathcal{W}^{-1}]$, leaving the class $\mathcal{W}$ implicit.

EXAMPLE 1.4. If every arrow in $\mathcal{W}$ is already an isomorphism, then $\mathrm{id}_{\mathcal{C}}\colon \mathcal{C} \to \mathcal{C}$ is a localisation of $\mathcal{C}$ at $\mathcal{W}$. In particular, if $\mathcal{C}$ is a groupoid, then $\mathrm{id}_{\mathcal{C}}$ is a localisation at any class of weak equivalences.

EXAMPLE 1.5. If $\mathcal{C}$ is a model category, then the canonical functor $\mathcal{C} \to \operatorname{Ho}\mathcal{C}$ is a localisation of $\mathcal{C}$ with respect to its weak equivalences. In fact, having well-behaved localisations is the reason why the theory of model categories was developed in the first place.

CONVENTION 1.6. If $\mathcal{W}$ is a class of weak equivalences in a category $\mathcal{C}$, one can always add the isomorphisms of $\mathcal{C}$ to $\mathcal{W}$ without changing the localisation and so we shall always assume that isomorphisms are weak equivalences.

It seems curious that part (2) from the definition of a localisation is omitted entirely in the strict version. After all, it has some consequences, e.g. for derived functors, which seem to be important enough to be proven for an explicit construction of a localisation (e.g. [**8**, 5.9] for the homotopy category of a model category). But using an explicit construction is not necessary as the following proof shows.

PROPOSITION 1.7. *If $H\colon \mathcal{C} \to \operatorname{Ho}\mathcal{C}$ is a strict localisation of a category $\mathcal{C}$ at a class $\mathcal{W}$ of morphisms, then, for every category $\mathcal{D}$, the precomposition functor $H^*\colon \mathcal{D}^{\operatorname{Ho}\mathcal{C}} \to \mathcal{D}^{\mathcal{C}}$ is fully faithful.*

PROOF. Let $[1]$ be the *interval category* with two objects 0, 1 and exactly one non-identity morphism $i\colon 0 \to 1$. There is the canonical twist isomorphism $(\mathcal{D}^{\mathcal{C}})^{[1]} \cong (\mathcal{D}^{[1]})^{\mathcal{C}}$ so that pairs of functors $F, G\colon \mathcal{C} \to \mathcal{D}$ together with $\tau\colon F \Rightarrow G$ correspond to functors $\mathcal{C} \to \mathcal{D}^{[1]}$. Now given $F, G\colon \operatorname{Ho}\mathcal{C} \to \mathcal{D}$ together with $\tau\colon F \circ H \Rightarrow G \circ H$, these determine

$$T\colon \mathcal{C} \to \mathcal{D}^{[1]} \text{ with } (TC)i = \tau_C \text{ and } (Tf)_0 = FHf,\ (Tf)_1 = GHf \text{ for } f\colon C \to C'.$$

Under T, arrows $f \in \mathcal{W}$ are sent to isomorphisms because HFf and GFf are invertible and so we get a unique $\overline{T}\colon \operatorname{Ho}\mathcal{C} \to \mathcal{D}^{[1]}$ such that $T = \overline{T} \circ H$. Under the above twist isomorphism, this corresponds to a unique pair of functors

$$F',\ G'\colon \operatorname{Ho}\mathcal{C} \to \mathcal{D} \quad \text{defined by} \quad F' = \mathrm{ev}_0 \circ \overline{T} \text{ and } G' = \mathrm{ev}_1 \circ \overline{T}$$

together with $\bar{\tau}\colon F' \Rightarrow G'$ defined by $\bar{\tau}_C = (\overline{T}C)i$ that satisfies $\tau = \bar{\tau}_H$. □

NOMENCLATURE 1.8. For brevity reasons, if $\mathcal{C}$ and $\mathcal{D}$ are categories with weak equivalences, we call a functor $F\colon \mathcal{C} \to \mathcal{D}$ *homotopical* iff it preserves weak equivalences. Moreover, by a *natural weak equivalence*, we mean a natural transformation $\tau\colon F \Rightarrow G$, all of whose components are weak equivalences.

Assuming we have an explicit localisation construction (e.g. using zig-zags [**4**]), we note that Ho is strictly 2-functorial. In particular, any adjunction $F \dashv G$ of homotopical functors yields an adjunction $\operatorname{Ho}F \dashv \operatorname{Ho}G$ between the correspondig homotopy categories.

OBSERVATION 1.9. Let $\mathcal{C}$ be a category with weak equivalences and $H\colon \mathcal{C} \to \operatorname{Ho}\mathcal{C}$ a localisation. If an object 0 is initial in $\mathcal{C}$, then $H0$ is initial in $\operatorname{Ho}\mathcal{C}$.

PROOF. Both adjoints in $0\colon \{*\} \rightleftarrows \mathcal{C} :!$ are homotopical. □

2. Derived functors

Consider a category $\mathcal{C}$ with weak equivalences. Time and again, one finds oneself in the situation where one wants to study some $F\colon \mathcal{C} \to \mathcal{D}$ that doesn't map weak equivalences to isomorphisms (so that it cannot be extended along the localisation). Sometimes this is conceived as a defect of F (e.g. for F a [co]limit functor) and at other times as an interesting peculiarity that can be exploited to construct invariants (e.g. using Hom-functors, tensoring or global sections). In either case, the approach is usually to approximate F as well as possible by a functor on $\operatorname{Ho}\mathcal{C}$.

DEFINITION 2.1. Let $H\colon \mathcal{C} \to \operatorname{Ho}\mathcal{C}$ be a localisation of a category with weak equivalences and $F\colon \mathcal{C} \to \mathcal{D}$. Recall that a *left derived functor* $(LF, \lambda\colon LF \circ H \Rightarrow F)$ of F is a right Kan extension of F along H. That is, for every $L\colon \operatorname{Ho}\mathcal{C} \to \mathcal{D}$ together with $\tau\colon L \circ H \Rightarrow F$, there is a unique transformation $\bar{\tau}\colon L \Rightarrow LF$ such that $\lambda \circ \bar{\tau}_H = \tau$. The universal 2-arrow λ is called the *counit* of the left derived functor. Dually (more precisely "*co*-dually", i.e. just reversing 2-cells), one defines a *right derived functor* (RF, ρ) of F, whose universal 2-arrow is called its *unit*.

If $\mathcal{D}$ also comes with a class of weak equivalences and a localisation functor $H'\colon \mathcal{D} \to \operatorname{Ho}\mathcal{D}$, we will usually be more interested in the *total left derived functor*, which is the left derived functor $\mathbb{L}F := L(H' \circ F)$ of $H' \circ F$. Dually for the *total right derived functor* $\mathbb{R}F := R(H' \circ F)$.

REMARK 2.2. Being a terminal object in the category $H^* \downarrow F$, a left derived functor (LF, λ) is unique up to unique isomorphism. In particular, for a fixed LF, its counit λ is unique up to precomposition with α_H for some unique automorphism α of LF.

As a special case of this universal property, we obtain that the operations L (resp. $\mathbb{L}$) and R (resp. $\mathbb{R}$) are functorial.

DEFINITION 2.3. Let F, $F'\colon \mathcal{C} \to \mathcal{D}$ have right Kan extensions (LF, λ) and (LF', λ') along some $H\colon \mathcal{C} \to \mathcal{H}$ and let $\sigma\colon F \Rightarrow F'$ be a natural transformation. Then there is a unique $L\sigma := \overline{\sigma \circ \lambda}\colon LF \Rightarrow LF'$ such that $\sigma \circ \lambda = \lambda' \circ (L\sigma)_H$. Dually, if G, $G'\colon \mathcal{C} \to \mathcal{D}$ have left Kan extensions (RG, ρ) and (RG', ρ') along H and $\tau\colon G \Rightarrow G'$, then there is a unique $R\tau := \overline{\rho' \circ \sigma}\colon RG \Rightarrow RG'$ such that $(R\tau)_H \circ \rho = \rho' \circ \tau$. The unicity of these shows that L and R strictly preserve vertical compositions of natural transformations as well as identities. Similarly for the totally derived versions $\mathbb{L}\sigma := L(H'\sigma)$ and $\mathbb{R}\tau := R(H'\tau)$ for $H'\colon \mathcal{D} \to \mathcal{H}'$.

There is the following convenient external characterisation of derived functors. Unfortunately, there are some class-theoretic difficulties involved (cf. the remark below).

PROPOSITION 2.4. *If (LF, λ) is a right Kan extension of $F\colon \mathcal{C} \to \mathcal{D}$ along some functor $H\colon \mathcal{C} \to \mathcal{H}$, then*

$$\varphi_X\colon \operatorname{Nat}(X, LF) \to \operatorname{Nat}(X \circ H, F),\ \tau \mapsto \lambda \circ \tau_H$$

is a bijection, natural in $X\colon \mathcal{H} \to \mathcal{D}$. Conversely, if there is a functor $LF\colon \mathcal{H} \to \mathcal{D}$ together with a natural family of bijections

$$\varphi_X\colon \operatorname{Nat}(X, LF) \to \operatorname{Nat}(X \circ H, F)$$

indexed by all functors $X\colon \mathcal{H} \to \mathcal{D}$, then $(LF, \varphi_{LF}\operatorname{id}_{LF})$ is a left derived functor of F. Moreover, these two constructions are inverse to each other.

PROOF. The map $\tau \mapsto \lambda \circ (H\tau)$ being a bijection is just the universal property of a right Kan extension and naturality is easy. Conversely, if there is a natural family $(\varphi_X)_X$ of bijections as in the proposition, we put $\lambda := \varphi_{LF}\,\mathrm{id}_{LF}$. Now for any $X\colon \mathcal{H} \to \mathcal{D}$ together with $\tau\colon X \circ H \Rightarrow F$ there is a unique $\bar{\tau}\colon X \Rightarrow LF$ such that $\varphi_X \bar{\tau} = \tau$. Chasing id_{LF} around the naturality square

$$\begin{array}{ccc} \mathrm{Nat}(LF, LF) & \xrightarrow{\varphi_{LF}} & \mathrm{Nat}(LF \circ H, F) \\ {\scriptstyle \bar{\tau}^*}\big\downarrow & & \big\downarrow{\scriptstyle \bar{\tau}^*_H} \\ \mathrm{Nat}(X, LF) & \xrightarrow[\varphi_X]{} & \mathrm{Nat}(X \circ H, F) \end{array} \tag{2.1}$$

we see that $\bar{\tau}$ is the unique transformation $X \Rightarrow LF$ such that $\lambda \circ \bar{\tau}_H = \tau$. □

REMARK 2.5. In ordinary NBG class theory, one cannot properly formalise the above proposition because in general, there is no class of all functors $X\colon \mathcal{H} \to \mathcal{D}$. The obvious remedies for this (apart from never using the external characterisation) is to assume the existence of a universe or switch to a higher class theory that has 2-classes, whose relation to ordinary classes is the same as that of classes to sets (such as in [**1**]).

Of particular importance for homotopy theory are the so-called *absolute* Kan extensions, i.e. those preserved by any morphism, as was observed by Maltsiniotis [**19**]. Their importance comes from the fact that all derived functors that arise from Quillen adjunctions are absolute and that there is a nice interplay between adjunctions and absolute derived functors as we shall see in section 6.

DEFINITION 2.6. A right Kan extension (LF, λ) of a functor $F\colon \mathcal{C} \to \mathcal{D}$ along $H\colon \mathcal{C} \to \mathcal{H}$ (and in particular a [total] left derived functor) is called *absolute* iff for every $Y\colon \mathcal{D} \to \mathcal{E}$, the composite $(Y \circ LF, Y\lambda)$ is a right Kan extension of $Y \circ F$ along H. Dually for left Kan extensions (and in particular [total] right derived functors).

EXAMPLE 2.7. Given two functors $F\colon \mathcal{C} \rightleftarrows \mathcal{D} : G$, then $F \dashv G$ with unit η iff (G, η) is an absolute left Kan extension of $\mathrm{id}_{\mathcal{C}}$ along F (cf. 4.1). Dually, $F \dashv G$ with counit ε iff (F, ε) is an absolute right Kan extension of $\mathrm{id}_{\mathcal{D}}$ along G.

EXAMPLE 2.8. If $\mathcal{C}$ and $\mathcal{D}$ are two categories, each equipped with a class of weak equivalences and $F\colon \mathcal{C} \to \mathcal{D}$ is homotopical, then $\mathrm{Ho}\, F\colon \mathrm{Ho}\,\mathcal{C} \to \mathrm{Ho}\,\mathcal{D}$ together with the identity transformation is both an absolute total left and an absolute total right derived functor of F.

REMARK 2.9. We have already seen that taking Kan extensions preserves vertical compositions of natural transformations. In the absolute case, this is also true for horizontal compositions (which doesn't even make sense in the non-absolute case). To wit, consider two pairs of parallel functors F, $F'\colon \mathcal{C} \to \mathcal{D}$, Y, $Y'\colon \mathcal{D} \to \mathcal{E}$ together with $\sigma\colon F \Rightarrow F'$, $\tau\colon Y \Rightarrow Y'$ such that F and F' have absolute right Kan extensions (LF, λ) and (LF', λ') along an $H\colon \mathcal{C} \to \mathcal{H}$. Then $L(\tau \star \sigma) = \tau \star L\sigma$ (where $\star$ is horizontal composition of natural transformations).

PROOF. For an arbitrary object $C \in \mathcal{C}$, we easily calculate

$$\begin{aligned} Y'\lambda'_C \circ (\tau \star L\sigma)_{HC} &= Y'\lambda'_C \circ Y'(L\sigma)_{HC} \circ \tau_{(LF)HC} = Y'\sigma_C \circ Y'\lambda_C \circ \tau_{(LF)HC} \\ &\overset{\tau\ \mathrm{nat}}{=} Y'\sigma_C \circ \tau_{FC} \circ Y\lambda_C = (\tau \star \sigma)_C \circ G\lambda_C. \end{aligned}$$

□

As an obvious next step, we can adapt the external characterisation 2.4 to the case of absolute derived functors (again stated in larger generality).

PROPOSITION 2.10. *If (LF, λ) is an absolute right Kan extension of $F\colon \mathcal{C} \to \mathcal{D}$ along some functor $H\colon \mathcal{C} \to \mathcal{H}$, then*

$$\varphi_{\mathcal{E},X,Y}\colon \operatorname{Nat}(X, Y \circ LF) \to \operatorname{Nat}(X \circ H, Y \circ F),\ \tau \mapsto Y\lambda \circ \tau_H$$

is a bijection natural in $\mathcal{E}$, $X\colon \mathcal{H} \to \mathcal{E}$ and $Y\colon \mathcal{D} \to \mathcal{E}$. Conversely, if there is $LF\colon \mathcal{H} \to \mathcal{D}$ together with a family of bijections

$$\varphi_{\mathcal{E},X,Y}\colon \operatorname{Nat}(X, Y \circ LF) \to \operatorname{Nat}(X \circ H, Y \circ F)$$

natural in the category $\mathcal{E}$ and the functor $X\colon \mathcal{H} \to \mathcal{E}$ (naturality in $Y\colon \mathcal{D} \to \mathcal{E}$ is automatic), then $(LF, \varphi_{\mathcal{D},LF,\mathrm{id}_\mathcal{D}} \mathrm{id}_{LF})$ is a total left derived functor of F. Moreover, these two constructions are inverse to each other.

PROOF. The bijectivity of $\tau \mapsto Y\lambda \circ \tau_H$ is just the universal property of the right Kan extension $(Y \circ LF, Y\lambda)$ and naturality is easy. For the converse claim, we put $\lambda := \varphi_{\mathcal{D},LF,\mathrm{id}_\mathcal{D}} \mathrm{id}_{LF}$. For $Y\colon \mathcal{D} \to \mathcal{E}$, $X\colon \mathcal{H} \to \mathcal{E}$ and $\tau\colon X \circ H \Rightarrow Y \circ F$ there is a unique $\bar{\tau}\colon X \Rightarrow Y \circ LF$ such that $\varphi_{\mathcal{E},X,Y}\bar{\tau} = \tau$ and chasing id_{LF} around the commutative diagram

$$\begin{array}{ccc}
\operatorname{Nat}(LF, LF) & \xrightarrow{\varphi_{\mathcal{D},LF,\mathrm{id}_\mathcal{D}}} & \operatorname{Nat}(LF \circ H, F) \\
{\scriptstyle Y_*}\big\downarrow & & \big\downarrow{\scriptstyle Y_*} \\
\operatorname{Nat}(Y \circ LF, Y \circ LF) & \xrightarrow{\varphi_{\mathcal{E},Y\circ LF,Y}} & \operatorname{Nat}(Y \circ LF \circ H, Y \circ F) \\
{\scriptstyle \bar{\tau}^*}\big\downarrow & & \big\downarrow{\scriptstyle \bar{\tau}_H^*} \\
\operatorname{Nat}(X, Y \circ LF) & \xrightarrow[\varphi_{\mathcal{E},X,Y}]{} & \operatorname{Nat}(X \circ H, Y \circ F)
\end{array}$$

we obtain that indeed, $\bar{\tau}$ is the unique transformation satisfying $\tau = Y\lambda \circ \bar{\tau}_H$. □

Because we shall usually work with total derived functors, let us quickly restate this theorem for the total derived case.

COROLLARY 2.11. *Let $\mathcal{C}$, $\mathcal{D}$ be categories with weak equivalences and $H_\mathcal{C}$, $H_\mathcal{D}$ the corresponding localisations. Moreover, let $F\colon \mathcal{C} \to \mathcal{D}$ and $\mathbb{L}F\colon \operatorname{Ho}\mathcal{C} \to \operatorname{Ho}\mathcal{D}$. If $\mathbb{L}F$ is an absolute total left derived functor of F with counit λ, then $\tau \mapsto Y\lambda \circ \tau_{H_\mathcal{C}}$ defines a bijection*

$$\varphi_{\mathcal{E},X,Y}\colon \operatorname{Nat}(X, Y \circ \mathbb{L}F) \cong \operatorname{Nat}(X \circ H_\mathcal{C}, Y \circ H_\mathcal{D} \circ F)$$

natural in $\mathcal{E}$, $X\colon \operatorname{Ho}\mathcal{C} \to \mathcal{E}$ and $Y\colon \operatorname{Ho}\mathcal{D} \to \mathcal{E}$. Conversely, if there is such a natural family of bijections, then $\mathbb{L}F$ is an absolute total left derived functor of F with counit $\varphi_{\operatorname{Ho}\mathcal{D},\mathbb{L}F,\mathrm{id}_{\operatorname{Ho}\mathcal{D}}} \mathrm{id}_{\mathbb{L}F}$. These assignments are mutually inverse. Dually, if we have $G\colon \mathcal{D} \to \mathcal{C}$ and $\mathbb{R}G\colon \operatorname{Ho}\mathcal{D} \to \operatorname{Ho}\mathcal{C}$ such that $\mathbb{R}G$ is an absolute total right derived functor of G with unit ρ, then $\tau \mapsto \tau_{H_\mathcal{D}} \circ E'\rho$ defines a bijection

$$\psi_{\mathcal{E}',E',Y}\colon \operatorname{Nat}(E' \circ \mathbb{R}G, Y) \cong \operatorname{Nat}(E' \circ H_\mathcal{C} \circ G, Y \circ H_\mathcal{D})$$

natural in $\mathcal{E}'$, $E'\colon \operatorname{Ho}\mathcal{C} \to \mathcal{E}'$, $Y\colon \operatorname{Ho}\mathcal{D} \to \mathcal{E}'$ and conversely given such a natural family of bijections, $\mathbb{R}G$ is an absolute total right derived functor of G with unit $\psi_{\operatorname{Ho}\mathcal{C},\mathrm{id}_{\operatorname{Ho}\mathcal{C}},\mathbb{R}G} \mathrm{id}_{\mathbb{R}G}$. Again, these assignments are mutually inverse. □

3. Construction of derived functors

Virtually all derived functors that occur "in nature" are constructed by means of resolutions (a.k.a. replacements, a.k.a. approximations) and we shall quickly abstract these constructions to our context. Our treatment has been greatly inspired by **[9]** and our explicit goal is to remove a functoriality condition, which is present in *op. cit.* For this, we fix some category $\mathcal{C}$ equipped with a class of weak equivalences and write $H\colon \mathcal{C} \to \operatorname{Ho}\mathcal{C}$ for its localisation.

CONVENTION 3.1. If not stated otherwise, we will always equip a subcategory $\mathcal{C}_0 \subseteq \mathcal{C}$ with the weak equivalences coming from $\mathcal{C}$. I.e. an arrow in $\mathcal{C}_0$ is a weak equivalence iff it is one in $\mathcal{C}$.

DEFINITION 3.2. Following the nomenclature in **[9]** (and weakening their notion) a *left deformation retract* of $\mathcal{C}$ is a full subcategory $\mathcal{C}_0 \hookrightarrow \mathcal{C}$ with localisation $H_0\colon \mathcal{C}_0 \to \operatorname{Ho}\mathcal{C}_0$ such that there exist

(1) a map $Q\colon \operatorname{Ob}\mathcal{C} \to \operatorname{Ob}\mathcal{C}_0$ and for every $f\colon C \to C'$ in $\mathcal{C}$ a $Qf\colon QC \to QC'$ in $\mathcal{C}_0$ such that H_0Q defines a functor $H_0Q\colon \mathcal{C} \to \operatorname{Ho}\mathcal{C}_0$ that sends weak equivalences to isomorphisms and thus induces a unique $\widetilde{Q}\colon \operatorname{Ho}\mathcal{C} \to \operatorname{Ho}\mathcal{C}_0$ such that $\widetilde{Q} \circ H = H_0Q$;
(2) for each $C \in \mathcal{C}$ a weak equivalence $q_C\colon QC \to C$ such that for every $f\colon C \to C'$

$$\begin{array}{ccc} QC & \xrightarrow{Qf} & QC' \\ {\scriptstyle q_C}\downarrow & & \downarrow{\scriptstyle q_{C'}} \\ C & \xrightarrow[f]{} & C' \end{array} \qquad \text{commutes in } \mathcal{C}.$$

The triple $(\mathcal{C}_0, Q, q)$ is then called a *left deformation retraction* (of $\mathcal{C}$ to $\mathcal{C}_0$). Dually, one defines a *right deformation retract*.

OBSERVATION 3.3. If the weak equivalences in $\mathcal{C}$ satisfy 2-out-of-3, then the requirement that H_0Q send weak equivalences to isomorphisms is superfluous. Indeed, for $f\colon C \to C'$ in $\mathcal{C}$ we have $q'_C \circ Qf = f \circ q_C$ and so f is a weak equivalence iff Qf is.

As already mentioned, our notion of a deformation retract is weaker than the one in **[9]**. There, a left deformation retract is defined as a full subcategory $I\colon \mathcal{C}_0 \hookrightarrow \mathcal{C}$ together with a homotopical functor $Q\colon \mathcal{C} \to \mathcal{C}_0$ and a natural weak equivalence $q\colon I \circ Q \Rightarrow \mathrm{id}_{\mathcal{C}}$. One reason why one would like to have this stronger condition is the following.

REMARK 3.4. If Q in the above definition is a functor $\mathcal{C} \to \mathcal{C}_0$, we can lift the deformation retract to diagram categories. To wit, if $\mathcal{C}_0 \subseteq \mathcal{C}$ is a left deformation retract as above, with $Q\colon \mathcal{C} \to \mathcal{C}_0$ a functor and $\mathcal{I}$ is a small index category, then $\mathcal{C}_0^{\mathcal{I}} \subseteq \mathcal{C}^{\mathcal{I}}$ is a left deformation retract by applying Q pointwise (though this is not enough to obtain homotopy colimits). This is not possible if Q only becomes a functor when passing to the homotopy category, since $\operatorname{Ho}(\mathcal{C}^{\mathcal{I}}) \not\cong \operatorname{Ho}(\mathcal{C})^{\mathcal{I}}$ in general.

EXAMPLE 3.5. For $\mathcal{M}$ a model category (not necessarily with functorial factorisations), the category of cofibrant objects $\mathcal{M}_c$ together with some chosen cofibrant replacements $q_C\colon QC \to C$ (i.e. QC is cofibrant and q_C an acyclic fibration) and chosen lifts $Qf\colon QC \to QC'$ for $f\colon C \to C'$ forms a left deformation retraction. Indeed, as shown in [**8**, Lemma 5.1], if $f\colon C \to C'$ is any morphism in $\mathcal{M}$ and $q_C\colon QC \to C$, $q_{C'}\colon QC' \to C'$ are fixed cofibrant replacements, a dotted morphism making the square

$$\begin{array}{ccc} QC & \dashrightarrow & QC' \\ {\scriptstyle q_C}\big\downarrow{\scriptstyle \sim} & & {\scriptstyle \sim}\big\downarrow{\scriptstyle q_{C'}} \\ C & \xrightarrow[f]{} & C' \end{array}$$

commute in $\mathcal{M}$ always exists (i.e. we can choose a Qf as required) and is unique up to (left) homotopy. In particular, all such lifts are mapped to the same morphism under $H_0\colon \mathcal{M}_c \to \operatorname{Ho}\mathcal{M}_c$ (cf. [**8**, Lemma 5.10], whose proof goes through even though $\mathcal{M}_c$ need not be a model category) and thus H_0Q is a functor.

If $\mathcal{M}$ has functorial factorisations, then we even obtain the stronger version of a left deformation retraction, where $Q\colon \mathcal{M} \to \mathcal{M}_c$ is actually a functor.

Example 3.6. More generally, if $L\colon \mathcal{M} \rightleftarrows \mathcal{C} : R$ is a *left model approximation* in the sense of [**5**] (left adjoint on the left and again with $\mathcal{M}$ not necessarily having functorial factorisations), we let $\mathcal{C}_0 \subseteq \mathcal{C}$ be the full subcategory comprising all (objects isomorphic to) images of cofibrant objects in $\mathcal{M}$ under L. Choosing a cofibrant replacement Q for $\mathcal{M}$ as in the last example

$$(C \xrightarrow{f} C') \quad \mapsto \quad (LQRC \xrightarrow{LQRf} LQRC') \quad \text{together with the} \quad LQRC \xrightarrow{q^\flat_{RC}} C$$

defines a left deformation retraction of $\mathcal{C}$ to $\mathcal{C}_0$. Indeed, the $q^\flat_{RC}$ are weak equivalences because their adjuncts are and by definition of a left model approximation. Moreover, our cofibrant replacement is functorial when passing to the homotopy category because H_0LQR is just

$$\mathcal{C} \xrightarrow{R} \mathcal{M} \xrightarrow{H_{\mathcal{M}}} \operatorname{Ho}\mathcal{M} \xrightarrow{\widetilde{Q}} \operatorname{Ho}\mathcal{M}_c \xrightarrow{\operatorname{Ho}L} \operatorname{Ho}\mathcal{C}_0$$

(where for the last functor, we used that L sends weak equivalences between cofibrant objects to weak equivalences). Finally, the "naturality squares" for the $q^\flat_{RC}\colon LQRC \to C$ commute because their adjunct squares do.

Although it is usually convenient to require that Q itself be a functor, this has the undesirable consequence of needing to have functorial factorisations on a model category for the theory to apply. For the construction of derived functors, it is much more important that Q become functorial when passing to the homotopy category.

Proposition 3.7. *Let $(\mathcal{C}_0, Q, q)$ be a left deformation retraction of $\mathcal{C}$ and $I\colon \mathcal{C}_0 \hookrightarrow \mathcal{C}$ the inclusion. Then the families $(Hq_C)_{C\in\mathcal{C}}$ and $(H_0q_{C_0})_{C_0\in\mathcal{C}_0}$ define natural isomorphisms*

$$\operatorname{Ho}I \circ \widetilde{Q} \cong \mathrm{id}_{\operatorname{Ho}\mathcal{C}} \qquad \textit{and} \qquad \widetilde{Q} \circ \operatorname{Ho}I \cong \mathrm{id}_{\operatorname{Ho}\mathcal{C}_0}.$$

Proof. By 1.7, it suffices to check that the two families define natural isomorphisms

$$\operatorname{Ho}I \circ \widetilde{Q} \circ H = \operatorname{Ho}I \circ H_0Q \overset{\cong}{\Rightarrow} H \qquad \text{and} \qquad \widetilde{Q} \circ \operatorname{Ho}I \circ H_0 = H_0Q \circ I \overset{\cong}{\Rightarrow} H_0,$$

which is simple because we already have "naturality squares" in $\mathcal{C}$. For example, the naturality of the first family corresponds to the commutativity of

$$\begin{array}{cccccccccc} C & (\operatorname{Ho} I)\widetilde{Q}HC & = & (\operatorname{Ho} I)H_0QC & = & HIQC & = & HQC & \xrightarrow{Hq_C} & HC \\ {\scriptstyle f}\downarrow & \downarrow{\scriptstyle (\operatorname{Ho} I)\widetilde{Q}Hf} & & \downarrow{\scriptstyle (\operatorname{Ho} I)H_0Qf} & & \downarrow{\scriptstyle HIQf} & & \downarrow{\scriptstyle HQf} & & \downarrow{\scriptstyle Hf} \\ C' & (\operatorname{Ho} I)\widetilde{Q}HC' & = & (\operatorname{Ho} I)H_0QC' & = & HIQC' & = & HQC' & \xrightarrow[Hq_{C'}]{} & HC' \end{array}$$

for all $f\colon C \to C'$ in $\mathcal{C}$. □

REMARK 3.8. Because $\operatorname{Ho} I$ has now been shown to be fully faithful (in fact, it is even an equivalence), we will usually view $\operatorname{Ho}\mathcal{C}_0$ as a full subcategory of $\operatorname{Ho}\mathcal{C}$ and under this identification $H_0 = H|_{\mathcal{C}_0}$.

As already mentioned at the beginning of this section, the importance of deformation retracts comes from the fact that we can use them to "deform" functors in order to obtain derived ones.

DEFINITION 3.9. Let $H_{\mathcal{C}}\colon \mathcal{C} \to \operatorname{Ho}\mathcal{C}$ and $H_{\mathcal{D}}\colon \mathcal{D} \to \operatorname{Ho}\mathcal{D}$ be two localisation functors and $F\colon \mathcal{C} \to \mathcal{D}$. A left deformation retract $\mathcal{C}_0 \subseteq \mathcal{C}$ is called a *left F-deformation retract* iff the restriction $F|_{\mathcal{C}_0}\colon \mathcal{C}_0 \to \mathcal{D}$ is homotopical. Consequently, a left deformation retraction $(\mathcal{C}_0, Q, q)$ is called a *left F-deformation retraction* iff $\mathcal{C}_0$ is a left F-deformation retract. Dually for right deformation retracts.

REMARK 3.10. Given $F\colon \mathcal{C} \to \mathcal{D}$ together with a small category $\mathcal{I}$, we always get an induced functor $F^{\mathcal{I}}\colon \mathcal{C}^{\mathcal{I}} \to \mathcal{D}^{\mathcal{I}}$ by applying F pointwise. Just as in 3.4, any left F-deformation retraction $(\mathcal{C}_0, Q, q)$ where Q is actually a functor $\mathcal{C} \to \mathcal{C}_0$ can be lifted to a left $F^{\mathcal{I}}$-deformation retraction by applying Q pointwise.

THEOREM 3.11. *Let $(\mathcal{C}_0, Q, q)$ be a left deformation retraction of $\mathcal{C}$ and consider a functor $F\colon \mathcal{C} \to \mathcal{D}$ that maps weak equivalences in $\mathcal{C}_0$ to isomorphisms. Then*

$$LF\colon \operatorname{Ho}\mathcal{C} \xrightarrow{\widetilde{Q}} \operatorname{Ho}\mathcal{C}_0 \xrightarrow{\overline{F|_{\mathcal{C}_0}}} \mathcal{D} \quad \textit{together with} \quad (Fq_C\colon FQC \to FC)_{C\in\mathcal{C}}$$

is an absolute left derived functor of F. In particular, if $\mathcal{D}$ is also equipped with a class of weak equivalences and $(\mathcal{C}_0, Q, q)$ is a left F-deformation retraction of $\mathcal{C}$, then

$$\mathbb{L}F\colon \operatorname{Ho}\mathcal{C} \xrightarrow{\widetilde{Q}} \operatorname{Ho}\mathcal{C}_0 \xrightarrow{\operatorname{Ho}(F|_{\mathcal{C}_0})} \operatorname{Ho}\mathcal{D} \quad \textit{together with} \quad (H_{\mathcal{D}}Fq_C\colon FQC \to FC)_{C\in\mathcal{C}}$$

is an absolute total left derived functor of F.

PROOF. Let us write $\bar{F} := \overline{F|_{\mathcal{C}_0}}\colon \operatorname{Ho}\mathcal{C}_0 \to \mathcal{D}$ for the functor induced by $F|_{\mathcal{C}_0}$. We need to check that the Fq_C are natural in $C \in \mathcal{C}$; i.e. that the square

$$\begin{array}{ccc} FQC & \xrightarrow{\bar{F}\widetilde{Q}Hf} & FQC' \\ {\scriptstyle Fq_C}\downarrow & & \downarrow{\scriptstyle Fq_{C'}} \\ FC & \xrightarrow[Ff]{} & FC' \end{array} \tag{3.1}$$

in $\mathcal{D}$ commutes for all morphisms $f\colon C \to C'$ in $\mathcal{C}$. This is clear because we have $\bar{F}\widetilde{Q}Hf = \bar{F}HQf = FQf$. To see that this transformation is universal, let $G\colon \operatorname{Ho}\mathcal{C} \to \mathcal{D}$ and $\tau\colon G \circ H \Rightarrow F$. If there is $\bar{\tau}\colon G \Rightarrow LF$ such that $Fq \circ \bar{\tau}_H = \tau$, then

$$GHC \xrightarrow{\bar{\tau}_{HC}} LFHC = FQC \xrightarrow{Fq_C} FC \quad = \quad GHC \xrightarrow{\tau_C} FC \qquad \text{for all } C \in \mathcal{C}.$$

For $C \in \mathcal{C}_0$, the arrow Fq_C is invertible and so $\bar{\tau}_{HC} = (Fq_C)^{-1}\tau_C$. For the general case, the naturality of $\bar{\tau}$ gives a commutative diagram in $\mathcal{D}$ as follows:

$$\begin{array}{ccccc} GHQC & \xrightarrow{\bar{\tau}_{HQC}} & LFHQC & = & FQQC \\ \downarrow{\scriptstyle GHq_C} & & \downarrow{\scriptstyle LFHq_C} & & \downarrow{\scriptstyle FQq_C} \\ GHC & \xrightarrow[\bar{\tau}_{HC}]{} & LFHC & = & FQC \end{array}.$$

But GHq_C is invertible, so that

$$\bar{\tau}_{HC} = FQq_C \circ \bar{\tau}_{HQC} \circ (GHq_C)^{-1} = FQq_C \circ (Fq_{QC})^{-1} \circ \tau_{QC} \circ (GHq_C)^{-1}.$$

We check that this does indeed define a transformation $\bar{\tau}_H\colon G \circ H \Rightarrow LF \circ H$ (thus determining $\bar{\tau}$) by considering the following commutative diagram for $f\colon C \to C'$ in $\mathcal{C}$ (where for the last square, we note that while Q is not necessarily a functor, $FQ = \bar{F}H_0Q = \bar{F}\widetilde{Q}H$ certainly is):

$$\begin{array}{ccccccccc} GHC & \xrightarrow{(GHq_C)^{-1}} & GHQC & \xrightarrow{\tau_{QC}} & FQC & \xrightarrow{(Fq_{QC})^{-1}} & FQQC & \xrightarrow{FQq_C} & FQC \\ \downarrow{\scriptstyle GHf} & & \downarrow{\scriptstyle GHQf} & & \downarrow{\scriptstyle FQf} & & \downarrow{\scriptstyle FQQf} & & \downarrow{\scriptstyle FQf} \\ GHC' & \xrightarrow[(GHq_{C'})^{-1}]{} & GHQC' & \xrightarrow[\tau_{QC'}]{} & FQC' & \xrightarrow[(Fq_{QC'})^{-1}]{} & FQQC' & \xrightarrow[FQq_{C'}]{} & FQC' \end{array}.$$

Moreover, $\bar{\tau}$ does indeed satisfy $Fq \circ \bar{\tau}_H = \tau$ because if $C \in \mathcal{C}$, then

$$\begin{aligned} Fq_C \circ \bar{\tau}_{HC} &= Fq_C \circ FQq_C \circ (Fq_{QC})^{-1} \circ \tau_{QC} \circ (GHq_C)^{-1} \\ &\overset{Fq \text{ nat}}{=} Fq_C \circ Fq_{QC} \circ (Fq_{QC})^{-1} \circ \tau_{QC} \circ (GHq_C)^{-1} \\ &= Fq_C \circ \tau_{QC} \circ (GHq_C)^{-1} \\ &\overset{\tau \text{ nat}}{=} \tau_C \circ GHq_C \circ (GHq_C)^{-1} \end{aligned}$$

(for the second equality, put $C \rightsquigarrow QC$, $C' \rightsquigarrow C$, $f \rightsquigarrow q_C$ in 3.1 above). Finally, for the absoluteness claim, observe that if $F'\colon \mathcal{D} \to \mathcal{E}$ is another functor, then $F' \circ F$ again maps weak equivalences in $\mathcal{C}_0$ to isomorphisms. □

EXAMPLE 3.12. As already mentioned in 3.5, the full subcategory of cofibrant objects in a model category $\mathcal{M}$, together with some chosen cofibrant replacements, is a left deformation retract of $\mathcal{M}$. Consequently, a functor $F\colon \mathcal{M} \to \mathcal{N}$ that sends weak equivalences between cofibrant objects to isomorphisms has an absolute left derived functor in the above manner. By Ken Brown's lemma, a left Quillen functor $F\colon \mathcal{M} \to \mathcal{N}$ between two model categories (which preserves cofibrations and acyclic cofibrants) has an absolute total left derived functor.

EXAMPLE 3.13. More generally, in [**5**], a left model approximation $L\colon \mathcal{M} \rightleftarrows \mathcal{C} : R$ is called *good* for a functor $F\colon \mathcal{C} \to \mathcal{D}$ between two categories with weak equivalences iff $F \circ L$ sends weak equivalences between cofibrant objects to weak equivalences. This immediately implies that $\mathcal{C}_0 \subseteq \mathcal{C}$ as in 3.6 is a left F-deformation retract and so we obtain $\mathbb{L}F$.

An important instance of this is that if $\mathcal{M}$ is a model category, $F\colon \mathcal{I} \to \mathcal{J}$ a functor between small categories and $\mathcal{N}(\mathcal{I}) := \int_{\mathbf{\Delta}} N(\mathcal{I})$ the category of simplices of the nerve of $\mathcal{I}$, there is the so-called Bousfield-Kan left model approximation

$$\mathrm{Fun}^b(\mathcal{N}(\mathcal{I}), \mathcal{M}) \rightleftarrows \mathrm{Fun}(\mathcal{I}, \mathcal{M}),$$

which is good for the functor $F_!$ given by taking the left Kan extension along F. In particular, there is a homotopy left Kan extension functor $\mathbb{L}F_!\colon \mathrm{Ho}(\mathcal{M}^{\mathcal{I}}) \to \mathrm{Ho}(\mathcal{M}^{\mathcal{J}})$.

EXAMPLE 3.14. Consider a functor $F\colon \mathcal{C} \to \mathcal{D}$, a small indexing category $\mathcal{I}$ and $(\mathcal{C}_0, Q, q)$ a left F-deformation retraction such that $Q\colon \mathcal{C} \to \mathcal{C}_0$ is even a functor. As mentioned in 3.10, if $F^{\mathcal{I}}\colon \mathcal{C}^{\mathcal{I}} \to \mathcal{D}^{\mathcal{I}}$ is postcomposition with F, then $(\mathcal{C}_0^{\mathcal{I}}, Q^{\mathcal{I}}, q)$ is a left $F^{\mathcal{I}}$-deformation retraction (where $q_X\colon Q^{\mathcal{I}}X = Q \circ X \Rightarrow X$ has components q_{XI}). In this case $\mathbb{L}F$ is given by $\mathrm{Ho}(F \circ Q)$ and $\mathbb{L}F^{\mathcal{I}}$ is $\mathrm{Ho}(F^{\mathcal{I}} \circ Q^{\mathcal{I}}) = \mathrm{Ho}\big((F \circ Q)^{\mathcal{I}}\big)$. More colloquially, the derived functor of $F^{\mathcal{J}}$ is just the derived functor of F, applied pointwise.

To finish off this section, let us consider one instance where it is useful to have functorial factorisations in a model category. In that case, there is the following conceptual proof that the derived functors of left Quillen functors preserve homotopy colimits (even if the existence of projective model structures is not assumed). Here, we assume known that homotopy colimits in model categories can always be constructed, as is shown e.g. in [**5**].

PROPOSITION 3.15. *If $F\colon \mathcal{M} \rightleftarrows \mathcal{N} :G$ is a Quillen adjunction between two model categories with functorial factorisations and $\mathcal{I}$ is any small indexing category, there is a natural isomorphism*

$$\mathbb{L}F \circ \operatorname*{hocolim}_{\mathcal{I}} \cong \operatorname*{hocolim}_{\mathcal{I}} \circ \mathbb{L}F^{\mathcal{I}}.$$

PROOF. First off, even though $\mathcal{M}^{\mathcal{I}}$ and $\mathcal{N}^{\mathcal{I}}$ do not come with model structures, the derived functors $\mathbb{L}F^{\mathcal{I}}$ and $\mathbb{R}G^{\mathcal{I}}$ of $F^{\mathcal{I}}\colon \mathcal{M}^{\mathcal{I}} \to \mathcal{N}^{\mathcal{I}}$ and $G^{\mathcal{I}}\colon \mathcal{N}^{\mathcal{I}} \to \mathcal{M}^{\mathcal{I}}$ can still be constructed, as seen in the previous example. Writing Δ for the constant diagram inclusion, to prove the claim, we can switch to adjoints and it suffices to prove that $\mathrm{Ho}\,\Delta \circ \mathbb{R}G \cong \mathbb{R}G^{\mathcal{I}} \circ \mathrm{Ho}\,\Delta$. This is not hard because if $R\colon \mathcal{N} \to \mathcal{N}_f$ is a fibrant replacement functor, $\mathbb{R}G \cong \mathrm{Ho}(G \circ R)$ and $\mathbb{R}G^{\mathcal{I}} \cong \mathrm{Ho}(G^{\mathcal{I}} \circ R^{\mathcal{I}})$. $\square$

4. The yoga of mates

As is well-known, adjunctions are just a special instance of Kan extensions (then again, what isn't?) and the external characterisation makes it really obvious.

PROPOSITION 4.1. *Let $F\colon \mathcal{C} \rightleftarrows \mathcal{D} :G$ be two functors. Then $F \dashv G$ with counit ε iff (F, ε) is an absolute right Kan extension of $\mathrm{id}_{\mathcal{D}}$ along G and dually, $F \dashv G$ with unit η iff (G, η) is an absolute left Kan extension of $\mathrm{id}_{\mathcal{C}}$ along F. Indeed, if $F \dashv G$ with unit η and counit ε, then $\tau \mapsto Y\varepsilon \circ \tau_G$ defines a bijection*

$$\varphi_{\mathcal{E},X,Y}\colon \mathrm{Nat}(X, Y \circ F) \cong \mathrm{Nat}(X \circ G, Y) \qquad \textit{with inverse} \qquad \sigma_F \circ X\eta \leftarrow\!\shortmid \sigma$$

and this is natural in $\mathcal{E}$, $X\colon \mathcal{C} \to \mathcal{E}$ and $Y\colon \mathcal{D} \to \mathcal{E}$. Conversely, given such a natural family of bijections, then $F \dashv G$ with unit $\varphi^{-1}_{\mathcal{C},\mathrm{id}_{\mathcal{C}},G}(\mathrm{id}_G)$ and counit $\varphi_{\mathcal{D},F,\mathrm{id}_{\mathcal{D}}}(\mathrm{id}_F)$. These two constructions are inverse to each other.

PROOF. The whole proof is simply about finding the correct naturality conditions to apply but let's do it anyway for the sake of completeness. Starting with an adjunction $(F \dashv G, \eta, \varepsilon)$ and defining $\varphi_{\mathcal{E},X,Y}$ as in the claim, the naturality of φ in $\mathcal{E}$ and X is immediate. For the naturality in Y, we assume that we have $\alpha\colon Y \Rightarrow Y'$ and need to show that

$$\begin{array}{ccc} \mathrm{Nat}(X, Y \circ F) & \xrightarrow{\varphi_{\mathcal{E},X,Y}} & \mathrm{Nat}(X \circ G, Y) \\ {\scriptstyle (\alpha_F)_*}\downarrow & & \downarrow{\scriptstyle \alpha_*} \\ \mathrm{Nat}(X, Y' \circ F) & \xrightarrow[\varphi_{\mathcal{E},X,Y'}]{} & \mathrm{Nat}(X \circ G, Y') \end{array}$$

commutes. Chasing some τ through the square, this means that

$$\alpha \circ Y\varepsilon \circ \tau_G = Y'\varepsilon \circ \alpha_{FG} \circ \tau_G,$$

which follows from naturality of α. Moreover, the assignments

$$\tau \mapsto Y\varepsilon \circ \tau_G \qquad \text{and} \qquad \sigma_F \circ X\eta \mapsfrom \sigma$$

are indeed inverse to each other, because, starting with τ, we have

$$\tau \longmapsto Y\varepsilon \circ \tau_G \longmapsto (Y\varepsilon \circ \tau_G)_F \circ X\eta \overset{\tau \text{ nat}}{=} Y\varepsilon_{FC} \circ YF\eta_C \circ \tau_C \overset{\Delta\text{-id}}{=} \tau_C$$

and similarly the other way around.

Conversely, starting with a natural family of $\varphi_{\mathcal{E},X,Y}$, we put $\eta := \varphi^{-1}_{\mathcal{C},\mathrm{id}_{\mathcal{C}},G}(\mathrm{id}_G)$ and $\varepsilon := \varphi_{\mathcal{D},F,\mathrm{id}_{\mathcal{D}}}(\mathrm{id}_F)$ as in the claim and need to verify the triangle identities. For example, to show $\varepsilon_F \circ F\eta = \mathrm{id}_F$ (the other one being similar), we just consider the commutative diagram

$$\begin{array}{ccccc} \mathrm{Nat}(\mathrm{id}_{\mathcal{C}}, G \circ F) & \xrightarrow{F_*} & \mathrm{Nat}(F, F \circ G \circ F) & \xrightarrow{(\varepsilon_F)_*} & \mathrm{Nat}(F, F) \\ {\scriptstyle \varphi_{\mathcal{C},\mathrm{id}_{\mathcal{C}},G}}\downarrow & & \downarrow{\scriptstyle \varphi_{\mathcal{D},F,FG}} & & \downarrow{\scriptstyle \varphi_{\mathcal{D},F,\mathrm{id}_{\mathcal{D}}}} \\ \mathrm{Nat}(G, G) & \xrightarrow[F_*]{} & \mathrm{Nat}(F \circ G, F \circ G) & \xrightarrow[\varepsilon_*]{} & \mathrm{Nat}(F \circ G, \mathrm{id}_{\mathcal{D}}), \end{array}$$

where the left square commutes by naturality of $\varphi_{\mathcal{E},X,Y}$ in the variable $\mathcal{E}$, while the right square commutes by naturality in Y. Chasing $\eta = \varphi^{-1}_{\mathcal{C},\mathrm{id}_{\mathcal{C}},G}(\mathrm{id}_G)$ around the diagram, we find

$$\varphi_{\mathcal{D},F,\mathrm{id}_{\mathcal{D}}}(\varepsilon_F \circ F\eta) = \varepsilon = \varphi_{\mathcal{D},F,\mathrm{id}_{\mathcal{D}}}(\mathrm{id}_F)$$

and the claim follows.

Finally, the two assignments $(\eta, \varepsilon) \mapsto \varphi$ and $(\eta, \varepsilon) \mapsfrom \varphi$ as in the claim are mutually inverse, because, starting from (η, ε), constructing $\varphi_{\mathcal{E},X,Y}\colon \tau \mapsto Y\varepsilon \circ \tau_G$ and taking the associated unit and counit, the new counit is

$$\varphi_{\mathcal{D},F,\mathrm{id}_{\mathcal{D}}}(\mathrm{id}_F) = \mathrm{id}_{\mathcal{D}}\,\varepsilon \circ (\mathrm{id}_F)_G = \varepsilon,$$

which also shows that the new unit is again η because the counit determines the unit and vice versa. Conversely, starting with φ, and taking the associated unit $\eta := \varphi^{-1}_{\mathcal{C},\mathrm{id}_{\mathcal{C}},G}(\mathrm{id}_G)$ and counit $\varepsilon := \varphi_{\mathcal{D},F,\mathrm{id}_{\mathcal{D}}}(\mathrm{id}_F)$, the associated natural family φ' is defined as

$$\varphi'_{\mathcal{E},X,Y}\colon \tau \mapsto Y\varepsilon \circ \tau_G = Y\varphi_{\mathcal{D},F,\mathrm{id}_{\mathcal{D}}}(\mathrm{id}_F) \circ \tau_G.$$

Now, τ is natural transformation $X \Rightarrow Y \circ F$ and by naturality of $\varphi_{\mathcal{E},X,Y}$ in the variable X, the square

$$\begin{array}{ccc} \mathrm{Nat}(Y \circ F, Y \circ F) & \xrightarrow{\varphi_{\mathcal{E},Y\circ F,Y}} & \mathrm{Nat}(Y \circ F \circ G, Y) \\ {\scriptstyle \tau^*}\big\downarrow & & \big\downarrow{\scriptstyle (\tau_G)^*} \\ \mathrm{Nat}(X, Y \circ F) & \xrightarrow[\varphi_{\mathcal{E},X,Y}]{} & \mathrm{Nat}(X \circ G, Y) \end{array}$$

commutes. Chasing $\mathrm{id}_{Y\circ F}$ through the square, we find that

$$\varphi'_{\mathcal{E},X,Y}(\tau) = \varphi_{\mathcal{E},X,Y}(\tau \circ \mathrm{id}_{Y\circ F}) = \varphi_{\mathcal{E},X,Y}(\tau).$$

□

As always, by dualising (really *op*-dualising, i.e. inverting 1-arrows), we get an alternative external characterisation of adjunctions, which exhibits adjoints as *Kan lifts* and also fixes the counter-intuitive aspect that in the above formula, the left adjoint appears on the right.

PROPOSITION 4.2. *Given two functors* $F\colon \mathcal{C} \rightleftarrows \mathcal{D} : G$ *we have* $F \dashv G$ *with unit* η *and counit* ε *iff* $\sigma \mapsto G\sigma \circ \eta_X = \sigma^\sharp$ *defines a bijection*

$$\mathrm{Nat}(F \circ X, Y) \cong \mathrm{Nat}(X, G \circ Y) \qquad \textit{with inverse} \qquad \tau^\flat = \varepsilon_Y \circ F\tau \leftarrow\!\shortmid \tau,$$

which is natural in $\mathcal{E}$, $X\colon \mathcal{E} \to \mathcal{C}$ *and* $Y\colon \mathcal{E} \to \mathcal{D}$. *Conversely, given such a natural family of bijections, then* $F \dashv G$ *with unit* $\psi_{\mathcal{C},\mathrm{id}_\mathcal{C},F}(\mathrm{id}_F)$ *and counit* $\psi^{-1}_{\mathcal{D},G,\mathrm{id}_\mathcal{D}}(\mathrm{id}_G)$. *These two constructions are inverse to each other.*

Combining the two bijections from these propositions leads to the well-known *mating-bijection*

$$\begin{array}{ccc} \sigma & \mapsto & \sigma^{\infty} \\ \mathrm{Nat}(F' \circ X, Y \circ F) & \cong & \mathrm{Nat}(X \circ G, G' \circ Y) \\ {}^{\infty}\tau & \leftarrow\!\shortmid & \tau \end{array} \tag{4.1}$$

for categories and functors
$$\begin{array}{ccc} \mathcal{C} & \underset{G}{\overset{F}{\rightleftarrows}}\,\bot & \mathcal{D} \\ {\scriptstyle X}\big\downarrow & & \big\downarrow{\scriptstyle Y} \\ \mathcal{C}' & \underset{G'}{\overset{F'}{\rightleftarrows}}\,\bot & \mathcal{D}' \end{array}.$$

This bijection is natural in X and Y and is explicitly given by

$$\sigma \mapsto G'Y\varepsilon \circ G'\sigma_G \circ \eta'_{XG} = (Y\varepsilon \circ \sigma_G)^\sharp, \qquad (\tau_F \circ X\eta)^\flat = \varepsilon'_{YF} \circ F'\tau_F \circ F'X\eta \leftarrow\!\shortmid \tau.$$

DEFINITION 4.3. We write $\sigma \infty \tau$ and say $\sigma\colon F'X \Rightarrow YF$ and $\tau\colon XG \Rightarrow G'Y$ are *mates* (we should really say that $\sigma = {}^{\infty}\tau$ is the *left mate* and $\tau = \sigma^{\infty}$ is the *right mate*) iff they correspond to each other under this mating bijection. For the special case where X and Y are identities, we will occasionally use Mac Lane's nomenclature from [**18**] and speak of *conjugate* transformations.

REMARK 4.4. A useful consequence of the mating bijection's naturality in X and Y is that for $\alpha\colon X \Rightarrow X'$, $\beta\colon Y \Rightarrow Y'$ and two squares of natural transformations

$$\begin{array}{ccc} F'\circ X & \xrightarrow{\sigma} & Y\circ F \\ {\scriptstyle F\alpha}\downarrow & & \downarrow{\scriptstyle \beta_F} \\ F'\circ X' & \xrightarrow[\sigma']{} & Y'\circ F \end{array} \qquad \begin{array}{ccc} X\circ G & \xrightarrow{\tau} & G'\circ Y \\ {\scriptstyle \alpha_G}\downarrow & & \downarrow{\scriptstyle G'\beta} \\ X'\circ G & \xrightarrow[\tau']{} & G'\circ Y' \end{array}$$

where the two horizontal pairs are mates, $\beta_F \circ \sigma$ and $G'\beta \circ \tau$ as well as $\sigma' \circ F\alpha$ and $\tau' \circ \alpha_G$ are again mates. Consequently, the left-hand square commutes iff the right-hand square commutes.

Due to a lack of references, we shall briefly establish some standard results about mates, which one expects to be true and whose proofs are purely formal (whence the term *yoga*).

EXAMPLE 4.5. If X and Y are identities and the two adjunctions $F \dashv G$, $F' \dashv G'$ are the same, then $\mathrm{id}_F \mathrel{\circ\!\!\circ} \mathrm{id}_G$, which is just a complicated way to state the triangle identities. More generally, if the two adjunctions are the same, while X and Y are endofunctors equipped with $\alpha\colon X \Rightarrow \mathrm{id}_{\mathcal{C}}$ and $\beta\colon \mathrm{id}_{\mathcal{D}} \Rightarrow Y$, then by naturality of the mating bijection, $\beta_F \circ F\alpha \mathrel{\circ\!\!\circ} G\beta \circ \alpha_G$. Dually for the directions of α and β reversed.

PROPOSITION 4.6. *In the same situation as in the definition, for two natural transformations $\sigma\colon F'X \Rightarrow YF$ and $\tau\colon XG \Rightarrow G'Y$, the following are equivalent:*

(1) $\sigma \mathrel{\circ\!\!\circ} \tau$, *i.e.* $\sigma = (\tau_F \circ X\eta)^\flat$ *or equivalently* $\tau = (Y\varepsilon \circ \sigma_G)^\sharp$;
(2) $\sigma^\sharp = G'\sigma \circ \eta'_X = \tau_F \circ X\eta$ *or equivalently* $\tau^\flat = \varepsilon'_Y \circ F'\tau = Y\varepsilon \circ \sigma_G$;
(3) *for all $C \in \mathcal{C}$ and all $D \in \mathcal{D}$ the rectangle*

$$\begin{array}{ccc} \mathcal{D}(FC, D) & \xrightarrow{\cong} & \mathcal{C}(C, GD) \\ {\scriptstyle Y}\downarrow & & \downarrow{\scriptstyle X} \\ \mathcal{D}'(YFC, YD) & & \mathcal{C}'(XC, XGD) \\ {\scriptstyle \sigma_C^*}\downarrow & & \downarrow{\scriptstyle (\tau_D)_*} \\ \mathcal{D}'(F'XC, YD) & \xrightarrow[\cong]{} & \mathcal{C}'(XC, G'YD) \end{array}$$

commutes, where the horizontal arrows are the tuning bijections of the two adjunctions.

PROOF. "(1) $\Leftrightarrow$ (2)": Trivial.
"(2) $\Leftrightarrow$ (3)": Condition (2) is just the commutativity of the diagram for $D = FC$ and $C = GD$ respectively. Conversely, this implies the commutativity for all C and D by naturality of the tuning bijection for the adjunction. $\square$

PROPOSITION 4.7. *Taking mates is compatible with vertical and horizontal pasting of squares in the following sense:*

(1) *Given categories and functors*

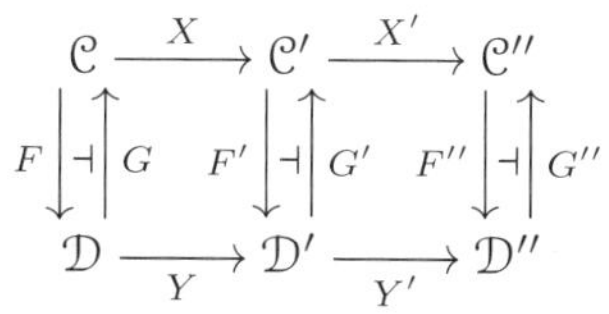

together with transformations $\sigma\colon F'X \Rightarrow YF$ *and* $\sigma'\colon F''X' \Rightarrow Y'F'$ *having mates* $\tau = \sigma^{\infty}\colon XG \Rightarrow G'Y$ *and* $\tau' = \sigma'^{\infty}\colon X'G' \Rightarrow G''Y'$ *respectively, then* $Y'\sigma \circ \sigma'_X$ *and* $\tau'_Y \circ X'\tau$ *are again mates. In particular, if* X, X', Y *and* Y' *are identities, then* $\sigma \circ \sigma'$ *and* $\tau' \circ \tau$ *are conjugate.*

(2) *Given categories and functors*

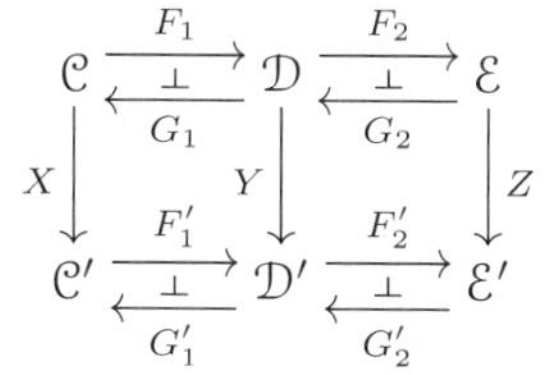

together with transformations $\sigma_1\colon F'_1X \Rightarrow YF_1$ *and* $\sigma_2\colon F'_2Y \Rightarrow ZF_2$ *having mates* $\tau_1\colon XG_1 \Rightarrow G'_1Y$ *and* $\tau_2\colon YG_2 \Rightarrow G'_2Z$*, then* $\sigma_{2F_1} \circ F'_2\sigma_1$ *and* $G'_1\tau_2 \circ \tau_{1G_2}$ *are again mates. In particular, if* X, Y *and* Z *are identities, then the horizontal composites* $\sigma' \star \sigma$ *and* $\tau \star \tau'$ *are conjugate. Still more specially, if in addition* $F_2 = F'_2$ *and* σ_2 *(whence* τ_2*) is the identity, we obtain that* $F_2\sigma_1$ *and* τ_{1G_2} *are conjugate.*

Proof. *Ad* (1): Writing (η,ε), (η',ε') and (η'',ε'') for the unit-counit pairs of the three adjunctions we easily calculate

$$\begin{aligned}(Y'\sigma \circ \sigma'_X)^\sharp &= G''Y'\sigma \circ G''\sigma'_X \circ \eta''_{X'X} = G''Y'\sigma \circ \sigma'^\sharp_X \\ &= G''Y'\sigma \circ \tau'_{F'X} \circ X'\eta'_X \overset{\tau'\text{ nat}}{=} \tau'_{YF} \circ X'G'\sigma \circ X'\eta_X \\ &= \tau'_{YF} \circ X'\sigma^\sharp = \tau'_{YF} \circ X'\tau_F \circ X'X\eta.\end{aligned}$$

Ad (2): Again, writing (η_1,ε_1), (η_2,ε_2) (η'_1,ε'_1) and (η'_2,ε'_2) for the unit-counit pairs this is just a routine calculation:

$$\begin{aligned}(\sigma_{2F_1} \circ F'_2\sigma_1)^\sharp &= G'_1G'_2\sigma_{2F_1} \circ G'_1G'_2F'_2\sigma_1 \circ (G'_1\eta'_{2F'_1} \circ \eta'_1)_X \\ &\overset{\eta'_2\text{ nat}}{=} G'_1G'_2\sigma_{2F_1} \circ G'_1\eta'_{2YF_1} \circ G'_1\sigma_1 \circ \eta'_{1X} \\ &= G'_1\sigma^\sharp_{2F_1} \circ \sigma^\sharp_1 = G'_1\tau_{2F_2F_1} \circ G'_1Y\eta_{2F_1} \circ \tau_{1F_1} \circ X\eta_1 \\ &\overset{\tau_1\text{ nat}}{=} G'_1\tau_{2F_2F_1} \circ \tau_{1G_2F_2F_1} \circ XG_1\eta_{2F_1} \circ X\eta_1 \\ &= \big(G'_1\tau_2 \circ \tau_{1G_1}\big)_{F_2F_1} \circ X\big(G_1\eta_{2F_1} \circ \eta_1\big).\end{aligned}$$

□

Corollary 4.8. *Given two adjunctions* $(F \dashv G, \eta, \varepsilon)$, $(F' \dashv G', \eta', \varepsilon')\colon \mathcal{C} \rightleftarrows \mathcal{D}$ *and conjugate transformations* $\sigma\colon F \Rightarrow F'$, $\tau\colon G' \Rightarrow G$*, then* σ *is an isomorphism iff* τ *is one.*

Proof. Let $\tau' \colon G' \Rightarrow G$ be the mate of σ^{-1}. Then by point (1) in the proposition (with $F'' = F$, $G'' = G$ and X, X', Y, Y' all identities) $\sigma \circ \sigma^{-1} = \mathrm{id}_F$ and $\tau' \circ \tau$ are mates, so that $\tau' \circ \tau = \mathrm{id}_{G'}$. Similarly, $\sigma^{-1} \circ \sigma = \mathrm{id}_{F'}$ and $\tau \circ \tau'$ are mates, which proves our claim. $\square$

5. Beck-Chevalley condition

Later on, we will find ourselves in the situation where we have a mating square as in 4.1 with a natural isomorphism $\tau \colon XG \cong G'Y$ but where we would really like its mate $\sigma \colon F'X \Rightarrow YF$ to be an isomorphism.

Definition 5.1. Consider a square of categories and functors together with a natural transformation

$$\begin{array}{ccc} \mathcal{C} & \xleftarrow{\;G\;} & \mathcal{D} \\ {\scriptstyle X}\downarrow & \overset{\tau}{\Searrow} & \downarrow{\scriptstyle Y} \\ \mathcal{C}' & \xleftarrow[\;G'\;]{} & \mathcal{D}' \end{array}$$

where G and G' have left adjoints

$$(F \dashv G, \eta, \varepsilon),\ (F' \dashv G', \eta', \varepsilon').$$

We then say that the square satisfies the *Beck-Chevalley condition* (or that it's a *Beck-Chevalley square*) iff the mate ${}^{\infty}\tau$ is an isomorphism $F' \circ X \cong Y \circ F$. Dually, there is the *dual Beck-Chevalley condition*, where we start with F, F' and σ and then require the mate σ^{∞} to be an isomorphism. Usually, τ is some sort of canonical isomorphism and is then often not explicitly mentioned. However, possible confusion can arise if the functors X and Y themselves have left adjoints. In that case, we shall speak of the *horizontal* and *vertical Beck-Chevalley condition* according to whether one considers the horizontal or vertical pairs of adjunctions; the case in the above definition being the horizontal one.

Example 5.2. By 4.8, if X and Y are identities, then the square from the definition satisfies the Beck-Chevalley condition iff τ is an isomorphism.

Example 5.3. The mate of τ is explicitly given by $\varepsilon'_{YF} \circ F'\tau_F \circ F'X\eta$; so if τ is an isomorphism and F, G' are fully faithful (i.e. η and ε' are isomorphisms), then the square satisfies the Beck-Chevalley condition.

Example 5.4. More importantly for us, if $\mathcal{I}$, $\mathcal{J}$ are index categories and $\mathcal{C}$ is a category with $\mathcal{I}$-colimits, then $\mathcal{C}^{\mathcal{J}}$ has $\mathcal{I}$-colimits, too. A colimit functor is given by

$$\operatorname{colim} \colon \mathcal{C}^{\mathcal{I} \times \mathcal{J}} \cong (\mathcal{C}^{\mathcal{J}})^{\mathcal{I}} \to \mathcal{C}^{\mathcal{J}},\ X \mapsto \Big(J \mapsto \operatorname{colim} X(-, J)\Big)$$

(i.e. colimits are calulcated pointwise). If we write η' for the unit of the adjunction $\operatorname{colim} \colon \mathcal{C}^{\mathcal{I}} \rightleftarrows \mathcal{C} : \Delta$, whose components are just the universal cocones, then a unit of $\operatorname{colim} \colon \mathcal{C}^{\mathcal{I} \times \mathcal{J}} \rightleftarrows \mathcal{C}^{\mathcal{J}} : \Delta$ is given by $\eta_{X,I,J} := \eta'_{X(-,J),I}$. All in all, the square

$$\begin{array}{ccc} \mathcal{C}^{\mathcal{I} \times \mathcal{J}} & \xleftarrow{\;\Delta\;} & \mathcal{C}^{\mathcal{J}} \\ {\scriptstyle \mathrm{ev}_J}\downarrow & \overset{\mathrm{id}}{\Searrow} & \downarrow{\scriptstyle \mathrm{ev}_J} \\ \mathcal{C}^{\mathcal{I}} & \xleftarrow[\;\Delta\;]{} & \mathcal{C} \end{array}$$

satisfies the Beck-Chevalley condition for all $J \in \mathcal{J}$. Indeed, the mate of the identity is again the identity and $\mathrm{ev}_J\,\eta = \eta'_{\mathrm{ev}_J}$.

OBSERVATION 5.5. According to 4.7, horizontal and vertical composites of Beck-Chevalley squares (defined in the obvious manner) are again Beck-Chevalley squares.

The situation gets really interesting when X and Y do have right adjoints. In that case, there is the following interchange law for mates as stated e.g. in [**11**, Lemma 1.20].

THEOREM 5.6 (Beck-Chevalley Interchange). *Consider a square of categories and functors together with a natural transformation* $\tau\colon XG \Rightarrow G'Y$ *as in the above definition and where all four functors have adjoints*

$$(F \dashv G, \eta, \varepsilon), \quad (F' \dashv G', \eta', \varepsilon'), \quad (X \dashv S, \theta, \zeta) \quad \text{and} \quad (Y \dashv T, \theta', \zeta'),$$

so that τ *has a horizontal left mate* $\sigma\colon F'X \Rightarrow YF$ *as well as a vertical right mate* $\rho\colon GT \Rightarrow SG'$*. Then* σ *and* ρ *are conjugate natural transformations. In particular,* σ *is an isomorphism iff* ρ *is one and thus the square satisfies the horizontal Beck-Chevalley condition iff it satisfies the vertical dual Beck-Chevalley condition.*

PROOF. Recall that the two mates σ and ρ are defined by

$$G'\sigma \circ \eta'_X = \tau_F \circ X\eta \qquad \text{and} \qquad S\tau \circ \theta_G = \rho_Y \circ G\theta'$$

and we need to check that σ and ρ are conjugate with respect to the composite adjunctions

$$\begin{array}{ccccc} \mathcal{C} & \underset{G}{\overset{F}{\underset{\perp}{\rightleftarrows}}} & \mathcal{D} & \underset{T}{\overset{Y}{\underset{\perp}{\rightleftarrows}}} & \mathcal{D}' \\ \| & & & & \| \\ \mathcal{C} & \underset{S}{\overset{X}{\underset{\perp}{\rightleftarrows}}} & \mathcal{C}' & \underset{G'}{\overset{F'}{\underset{\perp}{\rightleftarrows}}} & \mathcal{D}' \end{array} .$$

For this, we just need to take the adjunct of σ, which is

$$\begin{aligned} \sigma^\sharp &= S(G'\sigma \circ \eta'_X) \circ \theta = S(\tau_F \circ X\eta) \circ \theta = S\tau_F \circ SX\eta \circ \theta \\ &= S\tau_F \circ \theta_{GF} \circ \eta = (S\tau \circ \theta_G)_F \circ \eta = (\rho_Y \circ G\theta')_F \circ \eta = \rho_{YF} \circ G\theta'_F \circ \eta \end{aligned}$$

and $G\theta'_F \circ \eta$ is the unit of the upper composite adjunction. □

COROLLARY 5.7. *Consider a square as in* 5.1 *with* $\tau\colon XG \cong G'Y$ *an isomorphism. If* Y *has a fully faithful right adjoint and* X *is itself fully faithful and has a right adjoint, then the square satisfies the Beck-Chevalley condition.*

PROOF. The square satisfies the (horizontal) Beck-Chevalley condition iff it satisfies the vertical dual Beck-Chevalley condition. Taking adjoints

$$(X \dashv S, \theta, \zeta)\colon \mathcal{C} \underset{\perp}{\rightleftarrows} \mathcal{C}' \qquad \text{and} \qquad (Y \dashv T, \theta', \zeta')\colon \mathcal{D} \underset{\perp}{\rightleftarrows} \mathcal{D}'$$

the mate of τ for the vertical adjoints is $SG'\zeta' \circ S\tau_T \circ \theta_{GT}$. But τ is an isomorphism, same as ζ' (since T is fully faithful) and θ (since X is fully faithful). □

COROLLARY 5.8. *Given a mating square with* X *and* Y *equivalences*

$$\begin{array}{ccc} \mathcal{C} & \underset{G}{\overset{F}{\underset{\perp}{\rightleftarrows}}} & \mathcal{D} \\ {\scriptstyle X}\downarrow & & \downarrow{\scriptstyle Y} \\ \mathcal{C}' & \underset{G'}{\overset{F'}{\underset{\perp}{\rightleftarrows}}} & \mathcal{D}' \end{array} ,$$

then $\sigma\colon F'X \Rightarrow YF$ *is an isomorphism iff its mate* $\tau\colon XG \Rightarrow G'Y$ *is so.*

PROOF. Every equivalence is fully faithful and has a fully faithful left and right adjoint. □

Unfortunately, if the two vertical arrows X and Y in 5.1 have left adjoints rather than right ones, there is no nice interchange law. However, we can still use such adjoints to our advantage.

THEOREM 5.9. *Again consider a square as in* 5.1 *with* $\tau\colon XG \cong G'Y$ *a natural isomorphism and where all four functors have left adjoints*

$$(F \dashv G, \eta, \varepsilon), \quad (F' \dashv G', \eta', \varepsilon'), \quad (M \dashv X, \theta, \zeta) \quad \text{and} \quad (N \dashv Y, \theta', \zeta').$$

(1) *If* X *and* N *are fully faithful, then the square satisfies the Beck-Chevalley condition.*
(2) *If both* M *and* N *or both* X *and* Y *are fully faithful, then the square satisfies the Beck-Chevalley condition iff there is an isomorphism* $F'X \cong YF$.

PROOF. The mate σ of τ is the unique natural transformation that satisfies $G'\sigma \circ \eta'_X = \tau_F \circ X\eta$ and so, it suffices to construct a natural isomorphism σ subject to this equation. For this, we consider the conjugate of τ with respect to the two composite adjunctions, which is the unique isomorphism

$$\rho\colon N \circ F' \cong F \circ M \qquad \text{satisfying} \qquad G'Y\rho \circ G'\theta'_{F'} \circ \eta' = \tau_{FM} \circ X\eta_M \circ \theta.$$

Evaluating this equation at X and postcomposing with $G'YF\zeta$ yields

$$\begin{aligned} G'(YF\zeta \circ Y\rho_X \circ \theta'_{F'X}) \circ \eta'_X &= G'YF\zeta \circ \tau_{FMX} \circ X\eta_{MX} \circ \theta_X \\ &= \tau_F \circ XGF\zeta \circ X\eta_{MX} \circ \theta_X = \tau_F \circ X\eta \circ X\zeta \circ \theta_X = \tau_F \circ X\eta, \end{aligned}$$

so that the mate of τ is $\sigma = YF\zeta \circ Y\rho_X \circ \theta'_{F'X}$. Under the hypotheses of (1), this is an isomorphism since ζ and θ' are invertible, thus proving the first point. For point (2), assume that M and N are fully faithful and there is an isomorphism $F'X \cong YF$. We need to check that $YF\zeta$ is invertible, which is obvious since $YF \cong F'X$ and $X\zeta = \theta_X^{-1}$ is invertible. Similarly if X and Y are fully faithful. □

6. Derived adjunctions

As already mentioned, there is a beauti- and useful interplay between adjunctions and absolute derived functors; the most well-known instance of it perhaps being the famous Quillen adjoint functor theorem. Most proofs of it rely heavily on the explicit construction of a derived functor by means of (co)fibrant replacements whereas our approach really gets down to its bare bones. For this, let us fix two categories $\mathcal{C}$, $\mathcal{D}$ each equipped with a class of weak equivalences and let us write $H_{\mathcal{C}}$ and $H_{\mathcal{D}}$ for the corresponding localisations.

The results proven in this section have been obtained independently by González [**10**] though using very different, more diagrammatic methods.

THEOREM 6.1. *Let* $(F \dashv G, \eta, \varepsilon)\colon \mathcal{C} \rightleftarrows \mathcal{D}$ *be an adjunction such that* G *has an absolute total right derived functor* $(\mathbb{R}G, \rho)$. *If this in turn has a left adjoint* $(\dot{F} \dashv \mathbb{R}G, \dot{\eta}, \dot{\varepsilon})$, *then* $(\dot{F}, \lambda)$ *is an absolute total left derived functor of* F, *where* $\lambda := \dot{\varepsilon}_{H_{\mathcal{D}}F} \circ \dot{F}\rho_F \circ \dot{F}H_{\mathcal{C}}\eta$.

Proof. We need a bijection $\mathrm{Nat}(L, E \circ F') \cong \mathrm{Nat}(L \circ H_{\mathcal{C}}, E \circ H_{\mathcal{D}} \circ F)$, natural in $\mathcal{E}$, $E\colon \operatorname{Ho}\mathcal{D} \to \mathcal{E}$ and $L\colon \operatorname{Ho}\mathcal{C} \to \mathcal{E}$ and can construct one such by

$$\begin{array}{rcll} \mathrm{Nat}(L, E \circ \dot{F}) & \cong & \mathrm{Nat}(L \circ \mathbb{R}G, E) & \tau \mapsto E\dot{\varepsilon} \circ \tau_{\mathbb{R}G} \\ & \cong & \mathrm{Nat}(L \circ H_{\mathcal{C}} \circ G, E \circ H_{\mathcal{D}}) & \tau \mapsto \tau_{H_{\mathcal{D}}} \circ L\rho \\ & \cong & \mathrm{Nat}(L \circ H_{\mathcal{C}}, E \circ H_{\mathcal{D}} \circ F) & \tau \mapsto \tau_F \circ (L \circ H_{\mathcal{C}})\eta. \end{array}$$

Upon putting $L := \dot{F}$, $E := \mathrm{id}_{\operatorname{Ho}\mathcal{D}}$ and chasing $\mathrm{id}_{\dot{F}}$ through the bijections, the counit λ has indeed the claimed form. □

Theorem 6.2. *Let $(F \dashv G, \eta, \varepsilon)\colon \mathcal{C} \rightleftarrows \mathcal{D}$. If F has an absolute total left derived functor $(\mathbb{L}F, \lambda)$ and G has an absolute total right derived functor $(\mathbb{R}G, \rho)$, then we get an adjunction $(\mathbb{L}F \dashv \mathbb{R}G, \dot{\eta}, \dot{\varepsilon})$ where $\dot{\eta}$ is the unique $\mathrm{id}_{\operatorname{Ho}\mathcal{C}} \Rightarrow \mathbb{R}G \circ \mathbb{L}F$ such that $\mathbb{R}G\lambda \circ \dot{\eta}_{H_{\mathcal{C}}} = \rho_F \circ H_{\mathcal{C}}\eta$ and $\dot{\varepsilon}$ is the unique $\mathbb{L}F \circ \mathbb{R}G \Rightarrow \mathrm{id}_{\operatorname{Ho}\mathcal{D}}$ such that $H_{\mathcal{D}}\varepsilon \circ \lambda_G = \dot{\varepsilon}_{H_{\mathcal{D}}} \circ \mathbb{L}F\rho$.*

Proof. We need to construct a bijection $\mathrm{Nat}(E' \circ \mathbb{R}G, E) \cong \mathrm{Nat}(E', E \circ \mathbb{L}F)$ natural in $\mathcal{E}$, $E'\colon \operatorname{Ho}\mathcal{C} \to \mathcal{E}$ and $E\colon \operatorname{Ho}\mathcal{D} \to \mathcal{E}$. We can do so by

$$\begin{array}{rcll} \mathrm{Nat}(E' \circ \mathbb{R}G, E) & \cong & \mathrm{Nat}(E' \circ H_{\mathcal{C}} \circ G, E \circ H_{\mathcal{D}}) & \sigma \mapsto \sigma_{H_{\mathcal{D}}} \circ E'\rho \\ & \cong & \mathrm{Nat}(E' \circ H_{\mathcal{C}}, E \circ H_{\mathcal{D}} \circ F) & \sigma \mapsto \sigma_F \circ E'H_{\mathcal{C}}\eta, \\ & & & EH_{\mathcal{D}}\varepsilon \circ \tau_G \leftarrow\!\shortmid \tau \\ & \cong & \mathrm{Nat}(E', E \circ \mathbb{L}F) & E\lambda \circ \tau_{H_{\mathcal{C}}} \leftarrow\!\shortmid \tau \end{array}$$

and it plainly follows that $\dot{\eta}$ and $\dot{\varepsilon}$ are of the required form. □

The explicit descriptions of the unit and counit in the last theorem is not very enlightening and it might be clearer (although the author is not convinced) to draw the corresponding diagrams:

$$\begin{array}{ccc} H_{\mathcal{C}} & \xrightarrow{H_{\mathcal{C}}\eta} & H_{\mathcal{C}} \circ G \circ F \\ {\scriptstyle \dot{\eta}_{H_{\mathcal{C}}}}\big\downarrow & & \big\downarrow{\scriptstyle \rho_F} \\ \mathbb{R}G \circ \mathbb{L}F \circ H_{\mathcal{C}} & \xrightarrow[\mathbb{R}G\lambda]{} & \mathbb{R}G \circ H_{\mathcal{D}} \circ F \end{array} \qquad \begin{array}{ccc} \mathbb{L}F \circ H_{\mathcal{C}} \circ G & \xrightarrow{\mathbb{L}F\rho} & \mathbb{L}F \circ \mathbb{R}G \circ H_{\mathcal{D}} \\ {\scriptstyle \lambda_G}\big\downarrow & & \big\downarrow{\scriptstyle \dot{\varepsilon}_{H_{\mathcal{D}}}} \\ H_{\mathcal{D}} \circ F \circ G & \xrightarrow[H_{\mathcal{D}}\varepsilon]{} & H_{\mathcal{D}} \end{array} \tag{6.1}$$

Even better, in the situation of the first theorem 6.1, these two diagrams again commute (by naturality of all arrows involved and the triangle identities).

Definition 6.3. By a *derived adjunction* of $(F \dashv G, \eta, \varepsilon)\colon \mathcal{C} \rightleftarrows \mathcal{D}$ we mean an adjunction $(\mathbb{L}F \dashv \mathbb{R}G, \dot{\eta}, \dot{\varepsilon})\colon \operatorname{Ho}\mathcal{C} \rightleftarrows \operatorname{Ho}\mathcal{D}$ together with transformations λ and ρ satisfying

(1) $(\mathbb{L}F, \lambda)$ is an absolute total left derived functor of F;
(2) $(\mathbb{R}G, \rho)$ is an absolute total right derived functor of G;
(3) $\dot{\eta}$ and $\dot{\varepsilon}$ are the unique natural transformations making the squares 6.1 commute.

If such a derived adjunction exists, we say that $F \dashv G$ is *derivable*.

With this definition, we can summarise the results of the two theorems above by the following (less precise) corollary.

Corollary 6.4. *Let $F \dashv G$ be an adjunction of functors between categories with weak equivalences such that an absolute total right derived functor $\mathbb{R}G$ of G exists. Then $F \dashv G$ is derivable if and only if $\mathbb{R}G$ has a left adjoint.* □

Also, using the theorems, we can easily study the question of when derived functors compose. Unfortunately, they do not in general but at least we can check it on adjoints.

COROLLARY 6.5. *Let $\mathcal{C}$, $\mathcal{D}$ and $\mathcal{E}$ be three categories with weak equivalences and*

$$(F \dashv G, \eta, \varepsilon)\colon \mathcal{C} \rightleftarrows \mathcal{D}, \qquad (F' \dashv G', \eta', \varepsilon')\colon \mathcal{D} \rightleftarrows \mathcal{E}$$

with derived adjunctions $(\mathbb{L}F \dashv \mathbb{R}G, \dot\eta, \dot\varepsilon)$ and $(\mathbb{L}F' \dashv \mathbb{R}G', \dot\eta', \dot\varepsilon')$. If $(\mathbb{R}G \circ \mathbb{R}G', \rho'')$ is an absolute total right derived functor of $G \circ G'$, then $\mathbb{L}F' \circ \mathbb{L}F$ is an absolute total left derived functor of $F' \circ F$ with counit

$$(\dot\varepsilon' \circ \mathbb{L}F'\dot\varepsilon_{\mathbb{R}G'})_{H_\mathcal{E} F'F} \circ \mathbb{L}F'\mathbb{L}F\rho''_{F'F} \circ \mathbb{L}F'\mathbb{L}FH_\mathcal{C}(G\dot\eta_F \circ \eta).$$

PROOF. Composing the two adjunctions as well as their derived adjunctions yields

$$(F \circ F \dashv G \circ G', G\eta'_F \circ \eta, \varepsilon' \circ F'\varepsilon_{G'})\colon \mathcal{C} \rightleftarrows \mathcal{E} \qquad \text{and}$$

$$(\mathbb{L}F' \circ \mathbb{L}F \dashv \mathbb{R}G \circ \mathbb{R}G', \mathbb{R}G\dot\eta'_{\mathbb{L}F} \circ \dot\eta, \dot\varepsilon' \circ \mathbb{L}F'\dot\varepsilon_{\mathbb{R}G'})\colon \operatorname{Ho}\mathcal{C} \rightleftarrows \operatorname{Ho}\mathcal{E}.$$

Now if $(\mathbb{R}G \circ \mathbb{R}G', \rho')$ is an absolute total right derived functor of $G \circ G'$, then by 6.1 $\mathbb{L}F' \circ \mathbb{L}F$ is indeed an absolute total left derived functor of $F' \circ F$ with counit

$$(\dot\varepsilon' \circ \mathbb{L}F'\dot\varepsilon_{\mathbb{R}G'})_{H_\mathcal{E} F'F} \circ \mathbb{L}F'\mathbb{L}F\rho''_{F'F} \circ \mathbb{L}F'\mathbb{L}FH_\mathcal{C}(G\eta'_F \circ \eta).$$

□

As a special instance of this corollary, we can consider the case where ρ'' is the canonical candidate $\rho'' = \mathbb{R}G\rho' \circ \rho_{G'}$ for a unit of $\mathbb{R}G \circ \mathbb{R}G'$. It then follows that a counit of the total left derived functor $\mathbb{L}F' \circ \mathbb{L}F$ is again given by the canonical candidate and vice versa. Note that this result is neither stronger nor weaker than the previous one.

COROLLARY 6.6. *In the same situation as in the last corollary, let us write*

$$(\mathbb{L}F, \lambda), \quad (\mathbb{R}G, \rho), \quad (\mathbb{L}F', \lambda') \quad \text{and} \quad (\mathbb{R}G', \rho')$$

for the absolute total derived functors. Then the composite $(\mathbb{L}F' \circ \mathbb{L}F, \lambda'_F \circ \mathbb{L}F'\lambda)$ is an absolute total left derived functor of $F' \circ F$ if and only if $(\mathbb{R}G \circ \mathbb{R}G', \mathbb{R}G\rho' \circ \rho_{G'})$ is an absolute total right derived functor of $G \circ G'$.

PROOF. If $(\mathbb{R}G \circ \mathbb{R}G', \mathbb{R}G\rho' \circ \rho_{G'})$ is an absolute total right derived functor, we apply the last corollary to $\rho'' := \mathbb{R}G\rho' \circ \rho_{G'}$ and the claim follows by a routine (albeit tedious) calculation:

$$\begin{aligned}
&\dot\varepsilon'_{H_\mathcal{E} F'F} \circ \mathbb{L}F'\dot\varepsilon_{\mathbb{R}G' H_\mathcal{E} F'F} \circ \mathbb{L}F'\mathbb{L}F\mathbb{R}G\rho'_{F'F} \circ \mathbb{L}F'\mathbb{L}F\rho_{G'F'F} \circ \mathbb{L}F'\mathbb{L}FH_\mathcal{C}G\eta'_F \\
&\qquad \circ \mathbb{L}F'\mathbb{L}FH_\mathcal{C}\eta \\
&\overset{\varepsilon\text{ nat}}{=} \dot\varepsilon'_{H_\mathcal{E} F'F} \circ \mathbb{L}F'\rho_{F'F} \circ \mathbb{L}F'\dot\varepsilon_{H_\mathcal{D} G'F'F} \circ \mathbb{L}F'\mathbb{L}F\rho_{G'F'F} \circ \mathbb{L}F'\mathbb{L}FH_\mathcal{C}G\eta'_F \\
&\qquad \circ \mathbb{L}F'\mathbb{L}FH_\mathcal{C}\eta \\
&\overset{6.1}{=} H_\mathcal{E}\varepsilon'_{F'F} \circ \lambda'_{G'F'F} \circ \mathbb{L}F'H_\mathcal{D}\varepsilon_{G'F'F} \circ \mathbb{L}F'\lambda_{GG'F'F} \circ \mathbb{L}F'\mathbb{L}FH_\mathcal{C}G\eta'_F \\
&\qquad \circ \mathbb{L}F'\mathbb{L}FH_\mathcal{C}\eta \\
&\overset{\lambda\text{ nat}}{=} H_\mathcal{E}\varepsilon'_{F'F} \circ \lambda'_{G'F'F} \circ \mathbb{L}F'H_\mathcal{D}\varepsilon_{G'F'F} \circ \mathbb{L}F'H_\mathcal{D}FG\eta'_F \circ \mathbb{L}F'H_\mathcal{D}F\eta \circ \mathbb{L}F'\lambda \\
&\overset{\lambda'\text{ nat}}{=} H_\mathcal{E}\varepsilon'_{F'F} \circ H_\mathcal{E}F'\varepsilon_{G'F'F} \circ H_\mathcal{E}F'FG\eta'_F \circ H_\mathcal{E}F'F\eta \circ \lambda'_F \circ \mathbb{L}F'\lambda \\
&\overset{\varepsilon\text{ nat}}{=} H_\mathcal{E}\varepsilon'_{F'F} \circ H_\mathcal{E}F'\bar\eta_F \circ H_\mathcal{E}F'\varepsilon_F \circ H_\mathcal{E}F'F\eta \circ \lambda'_F \circ \mathbb{L}F'\lambda \overset{\Delta\text{-id}}{=} \lambda'_F \circ \mathbb{L}F'\lambda.
\end{aligned}$$

The other direction is dual. □

DEFINITION 6.7. Let $\mathcal{C}$, $\mathcal{D}$ and $\mathcal{E}$ be categories equipped with weak equivalences and let $F\colon \mathcal{C} \to \mathcal{D}$, $F'\colon \mathcal{D} \to \mathcal{E}$ have absolute total left derived functors $(\mathbb{L}F, \lambda)$ and $(\mathbb{L}F', \lambda')$. By abuse of notation, we say that $\mathbb{L}F$ *composes* with $\mathbb{L}F'$ iff $(\mathbb{L}F' \circ \mathbb{L}F, \lambda'_F \circ \mathbb{L}F'\lambda)$ is an absolute total left derived functor of $F' \circ F$. Dually for right derived functors.

REMARK 6.8. Clearly, this definition is equivalent to requiring that the composite $F' \circ F$ have a total left derived functor $\big(\mathbb{L}(F' \circ F), \lambda''\big)$ and that the natural transformation

$$\mathbb{L}F' \circ \mathbb{L}F \Rightarrow \mathbb{L}(F' \circ F) \qquad \text{induced by } \lambda'_F \circ \mathbb{L}F'\lambda \text{ be an isomorphism.}$$

REMARK 6.9. With the above definition, the last corollary 6.6 can be restated by saying that for two derived adjunctions, the left adjoints compose iff the right adjoints compose.

EXAMPLE 6.10. Obviously, if F' is homotopical, then $\mathbb{L}F' \cong \operatorname{Ho} F'$, so that $\mathbb{L}F$ and $\mathbb{L}F'$ compose. Also, total left derived functors of left Quillen functors between model categories compose. Dually for right Quillen functors.

EXAMPLE 6.11. More generally, let $\mathcal{C}$, $\mathcal{D}$ and $\mathcal{E}$ be equipped with weak equivalences and $H_\mathcal{C}$, $H_\mathcal{D}$, $H_\mathcal{E}$ their respective localisations. If we are given

$$\mathcal{C} \xrightarrow{F} \mathcal{D} \xrightarrow{F'} \mathcal{E}$$

together with a left F-deformation retract $\mathcal{C}_0 \subseteq \mathcal{C}$ and a left F'-deformation retract $\mathcal{D}_0 \subseteq \mathcal{D}$ such that $F\mathcal{C}_0 \subseteq \mathcal{D}_0$, then $\mathbb{L}F'$ and $\mathbb{L}F$ compose.

PROOF. Let us write $\bar{F} := \overline{F|_{\mathcal{C}_0}}$, $\bar{F}' := \overline{F'|_{\mathcal{D}_0}}$ and fix some left deformation retractions $(\mathcal{C}_0, Q, q)$, $(\mathcal{D}_0, Q', q')$, so that we may choose

$$\mathbb{L}F = \bar{F} \circ \widetilde{Q},\ \mathbb{L}F' = \bar{F}' \circ \widetilde{Q}' \qquad \text{with counits} \qquad \lambda := H_\mathcal{D} F q,\ \lambda' := H_\mathcal{E} F' q'.$$

By hypothesis $(\mathcal{C}_0, Q, q)$ is a left $(F'F)$-deformation retraction and $\overline{F'F} := \overline{F'F|_{\mathcal{C}_0}}$ comes with a natural isomorphism $\omega\colon \mathbb{L}F' \circ \mathbb{L}F \cong \overline{F'F} \circ \widetilde{Q}$ determined by the components

$$\omega_{H_\mathcal{C} C}\colon \mathbb{L}F'\mathbb{L}F H_\mathcal{C} C = H_\mathcal{E} F'Q'FQC \xrightarrow{H_\mathcal{E} F' q'_{FQC}} H_\mathcal{E} F'FQC \qquad \text{for } C \in \mathcal{C}.$$

Moreover, ω is compatible with the claimed counit of $\mathbb{L}F' \circ \mathbb{L}F$ in the sense that

$$H_\mathcal{E} F'Fq_C \circ \omega_{H_\mathcal{C} C} = H_\mathcal{E} F'(Fq_C \circ q'_{FQC}) = H_\mathcal{E} F'(q'_{FC} \circ Q'Fq_C) = \lambda'_{FC} \circ \mathbb{L}F'\lambda_C.$$

□

EXAMPLE 6.12. As a special instance of the previous example, that is relevant to our main topic of the pointwiseness of homotopy colimits is the following. Let $\mathcal{C}$ is a category with weak equivalences, $\mathcal{I}$, $\mathcal{J}$ index categories and $\mathcal{D} \subseteq \mathcal{C}^\mathcal{I}$ is a left colim-deformation retract with a functorial replacement $Q\colon \mathcal{C}^\mathcal{I} \to \mathcal{D}$ (i.e. a left deformation retract in the sense of [**9**]). Then $\mathbb{L}\operatorname{colim}_\mathcal{I} \cong \operatorname{Ho}(\operatorname{colim}_\mathcal{I} \circ Q)$ and for every $J \in \mathcal{J}$, the evaluation functor $\operatorname{Ho}\operatorname{ev}_J\colon \mathcal{C}^{\mathcal{I}\times\mathcal{J}} \to \mathcal{C}^\mathcal{I}$ composes with $\mathbb{L}\operatorname{colim}_\mathcal{I}\colon \mathcal{C}^\mathcal{I} \to \mathcal{C}$. To wit, as observed in 3.4, we can just apply Q pointwise, yielding a replacement functor $Q^\mathcal{J}\colon \mathcal{C}^{\mathcal{I}\times\mathcal{J}} \to \mathcal{D}^\mathcal{J}$. With this

$$\mathbb{L}\operatorname*{colim}_\mathcal{I} \circ \operatorname{Ho}\operatorname{ev}_J \cong \operatorname{Ho}(\operatorname*{colim}_\mathcal{I} \circ Q \circ \operatorname{ev}_J) = \operatorname{Ho}(\operatorname*{colim}_\mathcal{I} \circ \operatorname{ev}_J \circ Q^\mathcal{J}) \cong \mathbb{L}(\operatorname*{colim}_\mathcal{I} \circ \operatorname{ev}_J).$$

Finally, we will quickly go one dimension higher and study the connection between mates and derivable adjunctions. For this, we consider two derivable adjunctions $F \dashv G$, $F' \dashv G'$ (as always writing λ, λ', ρ and ρ' for the universal morphisms of $\mathbb{L}F$, $\mathbb{L}F'$, $\mathbb{R}G$ and $\mathbb{R}G'$) together with two homotopical functors X and Y as in the following diagram:

$$\begin{array}{ccc} \mathcal{C} & \underset{G}{\overset{F}{\rightleftarrows}} & \mathcal{D} \\ {\scriptstyle X}\downarrow & & \downarrow{\scriptstyle Y} \\ \mathcal{C}' & \underset{G'}{\overset{F'}{\rightleftarrows}} & \mathcal{D}' \end{array} \qquad \text{yielding} \qquad \begin{array}{ccc} \operatorname{Ho}\mathcal{C} & \underset{\mathbb{R}G}{\overset{\mathbb{L}F}{\rightleftarrows}} & \operatorname{Ho}\mathcal{D} \\ {\scriptstyle \operatorname{Ho}X}\downarrow & & \downarrow{\scriptstyle \operatorname{Ho}Y} \\ \operatorname{Ho}\mathcal{C}' & \underset{\mathbb{R}G'}{\overset{\mathbb{L}F'}{\rightleftarrows}} & \operatorname{Ho}\mathcal{D}' \end{array}.$$

In addition, we require $\operatorname{Ho}X$ to compose with $\mathbb{L}F'$ and $\operatorname{Ho}Y$ to compose with $\mathbb{R}G'$ (all other reasonable combinations automatically compose as remarked in 6.10).

THEOREM 6.13. *In the above situation, if $\sigma\colon F'X \Rightarrow YF$ and $\tau\colon XG \Rightarrow G'Y$ are mates, then the induced transformations between the derived functors*

$$\mathbb{L}\sigma\colon \mathbb{L}F' \circ \operatorname{Ho}X \Rightarrow \operatorname{Ho}Y \circ \mathbb{L}F \qquad \textit{and} \qquad \mathbb{R}\tau\colon \operatorname{Ho}X \circ \mathbb{R}G \Rightarrow \mathbb{R}G' \circ \operatorname{Ho}Y$$

are mates, too (note that for this to even make sense, we need that $\operatorname{Ho}X$ *composes with $\mathbb{L}F'$ and that* $\operatorname{Ho}Y$ *composes with $\mathbb{R}G'$). In particular, for $\mathcal{C} = \mathcal{C}'$, $\mathcal{D} = \mathcal{D}'$ and X, Y identities, if σ and τ are conjugate, so are $\mathbb{L}\sigma$ and $\mathbb{R}\tau$.*

PROOF. First recall that the composite counits of $\mathbb{L}F' \circ \operatorname{Ho}X$ and $\operatorname{Ho}Y \circ \mathbb{L}F$ are respectively λ'_X and $(\operatorname{Ho}Y)\lambda$ while the composite units of $\operatorname{Ho}X \circ \mathbb{R}G$ and $\mathbb{R}G' \circ \operatorname{Ho}Y$ are $(\operatorname{Ho}X)\rho$ and ρ'_Y. It follows that $\mathbb{L}\sigma$ and $\mathbb{R}\tau$ are defined by the equations

$$(\operatorname{Ho}Y)\lambda \circ (\mathbb{L}\sigma)_{H_{\mathcal{C}}} = H_{\mathcal{D}'}\sigma \circ \lambda'_X \qquad \text{and} \qquad (\mathbb{R}\tau)_{H_{\mathcal{D}}} \circ (\operatorname{Ho}X)\rho = \rho'_Y \circ H_{\mathcal{C}'}\tau.$$

We now need to check that $(\mathbb{L}\sigma)^\sharp = \mathbb{R}G'\mathbb{L}\sigma \circ \dot{\eta}'_{\operatorname{Ho}X} = (\mathbb{R}\tau)_{\mathbb{L}F} \circ X\dot{\eta}$, which are natural transformations $\operatorname{Ho}X \Rightarrow \mathbb{R}G' \circ \operatorname{Ho}Y \circ \mathbb{L}F$. By the universal property of a localisation and $\mathbb{L}F$ being an absolute total left derived functor, it suffices to check this after precomposition with $H_{\mathcal{C}}$ and postcomposition with $(\mathbb{R}G' \circ \operatorname{Ho}Y)\lambda$, where then

$$\begin{aligned} (\mathbb{R}G' \circ \operatorname{Ho}Y)\lambda \circ (\mathbb{R}G'\mathbb{L}\sigma \circ \dot{\eta}'_{\operatorname{Ho}X})_{H_{\mathcal{C}}} &= \mathbb{R}G'\big((\operatorname{Ho}Y)\lambda \circ \mathbb{L}\sigma_{H_{\mathcal{C}}}\big) \circ \dot{\eta}'_{H_{\mathcal{C}'}X} \\ &= \mathbb{R}G'(H_{\mathcal{D}'}\sigma \circ \lambda'_X) \circ \dot{\eta}'_{H_{\mathcal{C}'}X} = \mathbb{R}G'H_{\mathcal{D}'}\sigma \circ (\mathbb{R}G'\lambda' \circ \dot{\eta}'_{H_{\mathcal{C}'}})_X \\ &= \mathbb{R}G'H_{\mathcal{D}'}\sigma \circ (\rho'_{F'} \circ H_{\mathcal{C}'}\eta')_X = \mathbb{R}G'H_{\mathcal{D}'}\sigma \circ \rho'_{F'X} \circ H_{\mathcal{C}'}\eta'_X \\ &= \rho'_{YF} \circ H_{\mathcal{C}'}G'\sigma \circ H_{\mathcal{C}'}\eta'_X = \rho'_{YF} \circ H_{\mathcal{C}'}(G'\sigma \circ \eta'_X) \\ &= \rho'_{YF} \circ H_{\mathcal{C}'}(\tau_F \circ X\eta) = (\rho'_Y \circ H_{\mathcal{C}'}\tau)_F \circ H_{\mathcal{C}'}X\eta \\ &= (\mathbb{R}\tau)_{H_{\mathcal{D}}F} \circ (\operatorname{Ho}X)\rho_F \circ (\operatorname{Ho}X)H_{\mathcal{C}}\eta = (\mathbb{R}\tau)_{H_{\mathcal{D}}F} \circ (\operatorname{Ho}X)(\rho_F \circ H_{\mathcal{C}}\eta) \\ &= (\mathbb{R}\tau)_{H_{\mathcal{D}}F} \circ (\operatorname{Ho}X)(\mathbb{R}G\lambda \circ \dot{\eta}_{H_{\mathcal{C}}}) = (\mathbb{R}\tau)_{H_{\mathcal{D}}F} \circ (\operatorname{Ho}X \circ \mathbb{R}G)\lambda \circ (\operatorname{Ho}X)\dot{\eta}_{H_{\mathcal{C}}} \\ &= (\mathbb{R}G' \circ \operatorname{Ho}Y)\lambda \circ (\mathbb{R}\tau)_{\mathbb{L}FH_{\mathcal{C}}} \circ (\operatorname{Ho}X)\dot{\eta}_{H_{\mathcal{C}}} \\ &= \mathbb{R}G'(\operatorname{Ho}Y)\lambda \circ \big((\mathbb{R}\tau)_{\mathbb{L}F} \circ (\operatorname{Ho}X)\dot{\eta}\big)_{H_{\mathcal{C}}} \end{aligned}$$

□

7. Two notions of homotopy colimits

As an application of our theorems, we are going to show that homotopy colimits in the sense of derivator theory are the same as homotopy colimits in the Quillen model category sense. To make this precise, let us fix the following convention for convenience.

CONVENTION 7.1. Whenever we have a category $\mathcal{C}$ equipped with a class of weak equivalences as well as some index category $\mathcal{I}$ we always equip the diagram category $\mathcal{C}^{\mathcal{I}}$ with the class of pointwise weak equivalences. That is to say, a natural transformation $\tau\colon X \Rightarrow Y$ of diagrams $X, Y\colon \mathcal{I} \to \mathcal{C}$ is a weak equivalence iff $\tau_I\colon X_I \to Y_I$ is a weak equivalence in $\mathcal{C}$ for all $I \in \mathcal{I}$.

With this natural definition of weak equivalences in a diagram category, one has the problem that in general, colimit and limit functors do not preserve them. The classical example is given by the commutative diagram in **Top**

$$\begin{array}{ccccc} D^n & \hookleftarrow & S^{n-1} & \hookrightarrow & D^n \\ \downarrow & & \| & & \downarrow \\ * & \hookleftarrow & S^{n-1} & \hookrightarrow & * \end{array},$$

where all vertical arrows are (weak) homotopy equivalences but the pushout of the top row is S^n, while the one of the bottom row is $*$ and these spaces are not (weakly) homotopy equivalent. So, in general, we cannot form Ho(colim) and the best we can do is trying to take a derived functor.

OBSERVATION 7.2. The constant diagram functor $\Delta\colon \mathcal{C} \to \mathcal{C}^{\mathcal{I}}$ is homotopical and thus induces a unique $\operatorname{Ho}\Delta\colon \operatorname{Ho}\mathcal{C} \to \operatorname{Ho}(\mathcal{C}^{\mathcal{I}})$ such that $\operatorname{Ho}\Delta \circ H_{\mathcal{C}} = H_{\mathcal{C}^{\mathcal{I}}} \circ \Delta$ (where $H_{\mathcal{C}}$ and $H_{\mathcal{C}^{\mathcal{I}}}$ are the corresponding localisations).

DEFINITION 7.3. For $\mathcal{C}$ a category equipped with weak equivalences and $\mathcal{I}$ an index category, a *homotopy colimit functor* is a functor $\operatorname{hocolim}_{\mathcal{I}}\colon \operatorname{Ho}(\mathcal{C}^{\mathcal{I}}) \to \operatorname{Ho}\mathcal{C}$ left adjoint to $\operatorname{Ho}\Delta$. If $\mathcal{C}$ has $\mathcal{I}$-colimits (so that $\Delta\colon \mathcal{C} \to \mathcal{C}^{\mathcal{I}}$ has a left adjoint colim) a *derived colimit functor* is an absolute total left derived functor $\mathbb{L}\operatorname{colim}_{\mathcal{I}}$ of $\operatorname{colim}_{\mathcal{I}}$.

PROPOSITION 7.4. *If $\mathcal{C}$ has $\mathcal{I}$-colimits, then a homotopy colimit functor exists iff a derived colimit functor exists and they are the same.*

PROOF. The functor $\Delta\colon \mathcal{C} \to \mathcal{C}^{\mathcal{I}}$ is homotopical and so $\mathbb{R}\Delta = \operatorname{Ho}\Delta$ exists. $\square$

Although, by this result the existence of homotopy colimits is weaker than the existence of derived colimits (one does not need the existence of strict colimits), we prefer the more restrictive notion for what follows.

DEFINITION 7.5. We say that a category $\mathcal{C}$ with weak equivalences *has homotopy colimits of type* $\mathcal{I}$ (or simply that it has $\mathcal{I}$-*homotopy colimits*) iff the adjunction $\operatorname{colim}_{\mathcal{I}}\colon \mathcal{C}^{\mathcal{I}} \rightleftarrows \mathcal{C} :\Delta$ exists and is derivable. Also, when writing "$\operatorname{hocolim}_{\mathcal{I}}$", we always mean "$\mathbb{L}\operatorname{colim}_{\mathcal{I}}$".

8. Evaluation and endomorphisms

To say that colimits in a diagram category $\mathcal{C}^{\mathcal{J}}$ are calculated pointwise means that

$\operatorname{colim} \circ \operatorname{ev}_J = \operatorname{ev}_J \circ \operatorname{colim}$ (or rather that there is a natural isomorphism)

for all $J \in \mathcal{J}$, where ev_J are evaluation functors. So in order to prove a similar statement for homotopy colimits, we should try to better understand these.

PROPOSITION 8.1. *Let $\mathcal{C}$ be a category and $\mathcal{J}$ an index category. For $J \in \mathcal{J}$ with inclusion $\mathrm{In}_J \colon \{J\} \hookrightarrow \mathcal{J}$, the evaluation functor $\mathrm{ev}_J \colon \mathcal{C}^{\mathcal{J}} \to \mathcal{C}$ has a right adjoint*

$$J_* \colon \mathcal{C} \to \mathcal{C}^{\mathcal{J}} \qquad \text{given by} \qquad J_*C = C^{\mathcal{J}(-,J)}$$

(with the obvious arrow map), granted all these powers in $\mathcal{C}$ exist. Moreover, if $\mathrm{End}_{\mathcal{J}}(J) = \{\mathrm{id}_J\}$, then J_* is fully faithful or equivalently, a counit ε of the adjunction is invertible. Dually, ev_J has a left adjoint

$$J_! \colon \mathcal{C} \to \mathcal{C}^{\mathcal{J}} \qquad \text{given by} \qquad J_!C = \mathcal{J}(J,-) \cdot C,$$

granted all these copowers exist and if $\mathrm{End}_{\mathcal{J}}(J) = \{\mathrm{id}_J\}$, then $J_!$ is fully faithful.

PROOF. Note that $\mathrm{ev}_J = J^*$ is given by precomposition with $\mathrm{In}_J \colon \{J\} \hookrightarrow \mathcal{J}$ and thus has a right adjoint given by taking the pointwise right Kan extension along J (granted that all these exist)

$$C \mapsto J_*C = \lim(- \downarrow \mathrm{In}_J \to \{C\} \hookrightarrow \mathcal{C}),$$

so that $(J_*C)_{J'} = C^{\pi_0(J' \downarrow \mathrm{In}_J)}$ is the limit of a constant diagram for $J' \in \mathcal{J}$. But $J' \downarrow \mathrm{In}_J$ is a discrete category with objects $\mathcal{J}(J', J)$. The counit's component at $C \in \mathcal{C}$ is $\mathrm{pr}_{\mathrm{id}_J} \colon C^{\mathrm{End}_{\mathcal{J}}(J)} \to C$, which is invertible if $\mathrm{End}_{\mathcal{J}}(J) = \{\mathrm{id}_J\}$. □

EXAMPLE 8.2. For $\mathcal{J}$ any direct or inverse category (in the sense of the theory of Reedy categories), then $J_!$ and J_* (assuming they exist) are fully faithful for all $J \in \mathcal{J}$. In particular, this is true for the indexing categories of the most common homotopy colimits, such as homotopy pushouts, coproducts, telescopes and coequalisers.

OBSERVATION 8.3. Given a category $\mathcal{C}$, index categories $\mathcal{I}$, $\mathcal{J}$ and an object $J \in \mathcal{J}$ such that $J_* \colon \mathcal{C} \to \mathcal{C}^{\mathcal{J}}$ as above exists, then so does $J_* \colon \mathcal{C}^{\mathcal{I}} \to (\mathcal{C}^{\mathcal{I}})^{\mathcal{J}} \cong \mathcal{C}^{\mathcal{I} \times \mathcal{J}}$ (all powers are calculated pointwise) and an easy calculation shows that

$$\begin{array}{ccc} \mathcal{C}^{\mathcal{J}} & \xrightarrow{\mathrm{ev}_J} & \mathcal{C} \\ {\scriptstyle\Delta}\downarrow & & \downarrow{\scriptstyle\Delta} \\ \mathcal{C}^{\mathcal{I} \times \mathcal{J}} & \xrightarrow[\mathrm{ev}_J]{} & \mathcal{C}^{\mathcal{I}} \end{array}$$

(filled with the identity) satisfies the dual Beck-Chevalley condition. Indeed, the identity's mate is again the identity, for which we only need to check that $\Delta\varepsilon = \varepsilon'_{\Delta}$, where ε is the counit of the top adjunction and ε' the counit of the bottom one. So let $C \in \mathcal{C}$, $I \in \mathcal{I}$ and then

$$(\varepsilon'_{\Delta C})_I = (\mathrm{pr}_{\mathrm{id}_J})_I \colon \left((\Delta C)^{\mathcal{J}(J,J)}\right)_I = \left((\Delta C)I\right)^{\mathcal{J}(J,J)} = C^{\mathcal{J}(J,J)} \xrightarrow{\mathrm{pr}_{\mathrm{id}_J}} C = (\Delta C)_I$$

while $(\Delta\varepsilon_C)_I = \varepsilon_C = \mathrm{pr}_{\mathrm{id}_J}$. If a right adjoint J_* to ev_J exists but is not calculated pointwise as in the proposition, then the square still satisfies the dual Beck-Chevalley condition, granted that $\mathcal{C}$ has $\mathcal{J}$-colimits. This follows from Beck-Chevalley interchange and 5.4.

COROLLARY 8.4. *If $\mathcal{C}$ is a category with weak equivalences that has a terminal object, $\mathcal{J}$ a preorder and $J \in \mathcal{J}$, then $J_* \colon \mathcal{C} \to \mathcal{C}^{\mathcal{J}}$ exists and is homotopical. For a general indexing category $\mathcal{J}$ (not necessarily a preorder), if $\mathcal{C}$ has all powers and* *weak equivalences are stable under them, then all functors $J_*\colon \mathcal{C} \to \mathcal{C}^{\mathcal{J}}$ exist and are homotopical.*

PROOF. The second part follows directly from the formula for J_* given in the above proposition. For the first part, note that in a preorder, every $\mathcal{J}(J, J')$ is either empty or a singleton, so that all components of a $J_*w\colon J_*C \to J_*D$ induced by a weak equivalence $w\colon C \to D$ in $\mathcal{C}$ are either w itself or the identity for the terminal object of $\mathcal{C}$. □

In case $\mathcal{C}$ is a model category (in particular bicomplete) and $\mathcal{C}^{\mathcal{J}}$ carries the projective model structure, we even obtain two Quillen adjunctions. The less trivial part of this result can also be found e.g. in [**7**, Lemma 3.1.12].

COROLLARY 8.5. *If $\mathcal{C}$ is a model category and $\mathcal{J}$ an index category such that the projective model structure on $\mathcal{C}^{\mathcal{J}}$ exists, then $J_! \dashv \mathrm{ev}_J \dashv J_*$ are both Quillen adjunctions for all $J \in \mathcal{J}$.*

PROOF. The adjunction $J_! \dashv \mathrm{ev}_J$ is easy because ev_J preserves fibrations and acyclic fibrations. For the other adjunction, let $f\colon C \to D$ be an (acyclic) fibration in $\mathcal{C}$ and $J' \in \mathcal{J}$. Then

$$J_*(C \xrightarrow{f} D)_{J'} = C^{\mathcal{J}(J',J)} \xrightarrow{f^{\mathcal{J}(J',J)}} D^{\mathcal{J}(J',J)}$$

and a product of (acyclic) fibrations is again an (acyclic) fibration. □

EXAMPLE 8.6. Let $\mathcal{C}$ be a category with weak equivalences that has a terminal object. If $\mathcal{J}$ is a small preorder, then ev_J has a fully faithful right adjoint J_* and the adjunction

$$\mathcal{C}^{\mathcal{J}} \underset{J_*}{\overset{\mathrm{ev}_J}{\underset{\longleftarrow}{\overset{\longrightarrow}{\bot}}}} \mathcal{C} \qquad \text{lifts to} \qquad \mathrm{Ho}(\mathcal{C}^{\mathcal{J}}) \underset{\mathrm{Ho}\,J_*}{\overset{\mathrm{Ho}\,\mathrm{ev}_J}{\underset{\longleftarrow}{\overset{\longrightarrow}{\bot}}}} \mathrm{Ho}\,\mathcal{C}$$

for all $J \in \mathcal{J}$. Dually, if $\mathcal{C}$ has an initial object and $\mathcal{J}$ is a preorder, then ev_J has a fully faithful left adjoint $J_!$ and the adjunction

$$\mathcal{C}^{\mathcal{J}} \underset{\mathrm{ev}_J}{\overset{J_!}{\underset{\longrightarrow}{\overset{\longleftarrow}{\bot}}}} \mathcal{C} \qquad \text{lifts to} \qquad \mathrm{Ho}(\mathcal{C}^{\mathcal{J}}) \underset{\mathrm{Ho}\,\mathrm{ev}_J}{\overset{\mathrm{Ho}\,J_!}{\underset{\longrightarrow}{\overset{\longleftarrow}{\bot}}}} \mathrm{Ho}\,\mathcal{C}\,.$$

9. Derived functors on diagram categories

A common trick in diagrammatic homotopy theory is to first draw a 3×3-diagram of the form

$$\begin{array}{ccccc} \bullet & \longleftarrow & \bullet & \longrightarrow & \bullet \\ \uparrow & & \uparrow & & \uparrow \\ \bullet & \longleftarrow & \bullet & \longrightarrow & \bullet \\ \downarrow & & \downarrow & & \downarrow \\ \bullet & \longleftarrow & \bullet & \longrightarrow & \bullet \end{array}$$

and then use that the homotopy colimit of this diagram can be calculated as a double homotopy pushout in two ways: rows first or columns first. This is usually taken for granted and under suitably nice hypotheses (e.g. working in a cofibrantly generated model category) very easy to derive from the corresponding result for strict colimits.

When analysing this technique in order to weaken the hypotheses as far as possible, one sees that the whole trick is a combination of two results. First one uses a Fubini type result, which says that a homotopy colimit indexed by $\mathcal{I} \times \mathcal{J}$ can be calculated as a double homotopy colimit and second, one uses that homotopy colimits in a diagram category $\mathcal{C}^{\mathcal{J}}$ can be calculated pointwise (cf. 5.4). More specifically, the *Fubini theorem* is the following.

THEOREM 9.1. *Let $\mathcal{C}$ be a category with weak equivalences and $\mathcal{I}$, $\mathcal{J}$ index categories such that $\mathcal{C}^{\mathcal{I}}$ has $\mathcal{J}$-homotopy colimits while $\mathcal{C}$ has $\mathcal{I}$-homotopy colimits. Then $\mathcal{C}$ has $\mathcal{I} \times \mathcal{J}$-homotopy colimits and they are given by*

$$\operatorname{Ho}(\mathcal{C}^{\mathcal{I}\times\mathcal{J}}) \cong \operatorname{Ho}\big((\mathcal{C}^{\mathcal{I}})^{\mathcal{J}}\big) \xrightarrow{\operatorname{hocolim}_{\mathcal{J}}} \operatorname{Ho}(\mathcal{C}^{\mathcal{I}}) \xrightarrow{\operatorname{hocolim}_{\mathcal{I}}} \operatorname{Ho}\mathcal{C}.$$

PROOF. Look at the right adjoints. □

For the second result, recall that if $\mathcal{C}$ is a cocomplete category, then every diagram category $\mathcal{C}^{\mathcal{J}}$ is again cocomplete and the colimits are calculated pointwise. Unfortunately, for homotopy colimits, the matter is not that simple. For instance, there is no reason that the existence of homotopy colimits in a category $\mathcal{C}$ (equipped with a class of weak equivalences) should imply the existence of homotopy colimits in a diagram category $\mathcal{C}^{\mathcal{J}}$ (with pointwise weak equivalences). However if we assume their existence, we can study interactions between homotopy colimits in different categories.

PROPOSITION 9.2. *Let $\mathcal{C}$ be a category with weak equivalences, $\mathcal{I}$, $\mathcal{J}$ two index categories and $J \in \mathcal{J}$ such that $J_! \dashv \operatorname{ev}_J \dashv J_*$ exist and are both derivable while $\operatorname{End}_{\mathcal{J}}(J) = \{\operatorname{id}_J\}$. If $\mathcal{C}^{\mathcal{J}}$ has $\mathcal{I}$-homotopy colimits, then $\mathcal{C}$ has $\mathcal{I}$-homotopy colimits, too.*

PROOF. We obtain a left adjoint to $\operatorname{Ho}\Delta\colon \operatorname{Ho}\mathcal{C} \to \operatorname{Ho}(\mathcal{C}^{\mathcal{I}})$ by

$$\operatorname{Ho}(\mathcal{C}^{\mathcal{I}}) \underset{\operatorname{Ho}\operatorname{ev}_J}{\overset{\mathbb{L}J_!}{\rightleftarrows}} \operatorname{Ho}\big((\mathcal{C}^{\mathcal{I}})^{\mathcal{J}}\big) \cong \operatorname{Ho}(\mathcal{C}^{\mathcal{I}\times\mathcal{J}}) \cong \operatorname{Ho}\big((\mathcal{C}^{\mathcal{J}})^{\mathcal{I}}\big) \underset{\operatorname{Ho}\Delta}{\overset{\operatorname{hocolim}_{\mathcal{I}}}{\rightleftarrows}} \operatorname{Ho}(\mathcal{C}^{\mathcal{J}}) \underset{\mathbb{R}J_*}{\overset{\operatorname{Ho}\operatorname{ev}_J}{\rightleftarrows}} \operatorname{Ho}\mathcal{C}$$

because $\mathbb{R}J_*$ and $\operatorname{Ho}(\operatorname{ev}_J \circ \Delta)$ compose (cf. 6.10) and $\operatorname{ev}_J \circ \Delta \circ J_* \cong \Delta$, where we need that $\operatorname{End}_{\mathcal{J}}(J) = \{\operatorname{id}_J\}$ for the last isomorphism. □

EXAMPLE 9.3. The conditions of this proposition are satisfied if $\mathcal{C}$ has an initial and a terminal object while J is such that $\mathcal{J}(J', J)$ and $\mathcal{J}(J, J')$ contain at most one element for all $J' \in \mathcal{J}$ (e.g. $\mathcal{J}$ a preorder).

REMARK 9.4. If $\mathcal{M}$ is a model category and $\mathcal{I}$ an arbitrary indexing category, then $\mathcal{M}^{\mathcal{I}}$ need not be a model category. However, it has a left model approximation in the sense of [**5**] and we can therefore still construct homotopy colimits in $\mathcal{M}^{\mathcal{I}}$.

As this proof illustrates, a basic tactic for establishing results about $\operatorname{hocolim}_{\mathcal{I}}$ is to switch to adjoints and work with $\operatorname{Ho}\Delta$ instead. With this in mind, a direct attack on the "pointwiseness" of homotopy colimits in a diagram category might go as follows: If $\operatorname{Ho}\Delta$ composes with $\mathbb{R}J_*$, then two composite right adjoints

$$\mathbb{R}J_* \circ \operatorname{Ho}\Delta = \mathbb{R}(J_* \circ \Delta) = \mathbb{R}(\Delta \circ J_*) = \operatorname{Ho}\Delta \circ \mathbb{R}J_*$$

are the same and so its conjugate is an isomorphism between composite left adjoints

$$\operatorname{Ho}\operatorname{ev}_J \circ \operatorname*{hocolim}_{\mathcal{I}} \cong \operatorname*{hocolim}_{\mathcal{I}} \circ \operatorname{Ho}\operatorname{ev}_J .$$

Although already a first step, this line of thought only tells us that we can calculate homotopy colimits pointwise on the object level and we cannot conclude anything on the level of arrows. More specifically, if $j\colon J \to J'$ is an arrow, we obtain $\mathrm{ev}_j\colon \mathcal{C}^{\mathfrak{J}} \to \mathcal{C}^{[1]}$, which induces $\mathrm{Ho}\,\mathrm{ev}_j\colon \mathrm{Ho}(\mathcal{C}^{\mathfrak{J}}) \to \mathrm{Ho}(\mathcal{C}^{[1]})$, so that for $X \in \mathrm{Ho}(\mathcal{C}^{\mathcal{J}\times\mathfrak{J}})$,

$$(\mathrm{Ho}\,\mathrm{ev}_j)\operatorname*{hocolim}_{\mathcal{J}} X \in \mathrm{Ho}(\mathcal{C}^{[1]}).$$

On the other hand, if we view $X(-,j)\colon X(-,J) \to X(-,J')$ as a morphism in $\mathrm{Ho}(\mathcal{C}^{\mathcal{J}})$ (i.e. an object in $\mathrm{Ho}(\mathcal{C}^{\mathcal{J}})^{[1]}$), then applying $\mathrm{hocolim}_{\mathcal{J}}$ gives a morphism in $\mathrm{Ho}\,\mathcal{C}$ (i.e. an object in $(\mathrm{Ho}\,\mathcal{C})^{[1]}$) and we need to show that this is the same as

$$\overline{(H_{\mathcal{C}})_*}(\mathrm{Ho}\,\mathrm{ev}_j)\operatorname*{hocolim}_{\mathcal{J}} X, \quad \text{where } \overline{(H_{\mathcal{C}})_*} \text{ is defined by} \quad \begin{array}{ccc} \mathcal{C}^{[1]} & \xrightarrow{H_{\mathcal{C}^{[1]}}} & \mathrm{Ho}(\mathcal{C}^{[1]}) \\ & {\scriptstyle (H_{\mathcal{C}})_*}\searrow & \downarrow{\scriptstyle \overline{(H_{\mathcal{C}})_*}} \\ & & (\mathrm{Ho}\,\mathcal{C})^{[1]} \end{array}$$

being commutative. In fact, it is not even clear why $\mathrm{hocolim}_{\mathcal{J}} X(-,j)$ should lie in the image of $H_{\mathcal{C}}$ (up to conjugation with a natural isomorphism). A more informal way to summarise this is that we would like to have an isomorphism $\mathrm{hocolim}_{\mathcal{J}} \circ \mathrm{Ho}\,\mathrm{ev}_J \cong \mathrm{Ho}\,\mathrm{ev}_J \circ \mathrm{hocolim}_{\mathcal{J}}$ that is natural in $J \in \mathfrak{J}$; i.e. such that for all $j\colon J \to J'$ in $\mathfrak{J}$ the following square in $\mathrm{Ho}\,\mathcal{C}$ is commutative:

$$\begin{array}{ccc} \mathrm{hocolim}_{\mathcal{J}} X(-,J) & \cong & (\mathrm{hocolim}_{\mathcal{J}} X)_J \\ {\scriptstyle \mathrm{hocolim}_{\mathcal{J}} X(-,j)}\big\downarrow & & \big\downarrow{\scriptstyle (\mathrm{hocolim}_{\mathcal{J}} X)_j} \\ \mathrm{hocolim}_{\mathcal{J}} X(-,J') & \cong & (\mathrm{hocolim}_{\mathcal{J}} X)_{J'} \end{array}.$$

Again inspired by the corresponding strict result in the form of 5.4, the (horizontal) Beck-Chevalley condition comes to our rescue.

THEOREM 9.5. *Let $\mathcal{C}$ be a category with weak equivalences, $\mathcal{J}$, $\mathfrak{J}$ two index categories and $J \in \mathfrak{J}$ such that all derived adjunctions in the left-hand square*

$$\begin{array}{ccc} \mathrm{Ho}\big((\mathcal{C}^{\mathcal{J}})^{\mathfrak{J}}\big) \cong \mathrm{Ho}\big((\mathcal{C}^{\mathfrak{J}})^{\mathcal{J}}\big) & \underset{\mathrm{Ho}\,\Delta}{\overset{\mathrm{hocolim}_{\mathcal{J}}}{\rightleftarrows}} & \mathrm{Ho}(\mathcal{C}^{\mathfrak{J}}) \\ {\scriptstyle \mathrm{Ho}\,\mathrm{ev}_J}\downarrow \dashv \uparrow{\scriptstyle \mathbb{R}J_*} & & {\scriptstyle \mathrm{Ho}\,\mathrm{ev}_J}\downarrow \dashv \uparrow{\scriptstyle \mathbb{R}J_*} \\ \mathrm{Ho}(\mathcal{C}^{\mathcal{J}}) & \underset{\mathrm{Ho}\,\Delta}{\overset{\mathrm{hocolim}_{\mathcal{J}}}{\rightleftarrows}} & \mathrm{Ho}\,\mathcal{C} \end{array} \qquad \begin{array}{ccc} \mathrm{Ho}\big((\mathcal{C}^{\mathcal{J}})^{\mathfrak{J}}\big) \cong \mathrm{Ho}\big((\mathcal{C}^{\mathfrak{J}})^{\mathcal{J}}\big) & \xleftarrow{\mathrm{Ho}\,\Delta} & \mathrm{Ho}(\mathcal{C}^{\mathfrak{J}}) \\ {\scriptstyle \mathrm{Ho}\,\mathrm{ev}_J}\downarrow & & \downarrow{\scriptstyle \mathrm{Ho}\,\mathrm{ev}_J} \\ \mathrm{Ho}(\mathcal{C}^{\mathcal{J}}) & \xleftarrow[\mathrm{Ho}\,\Delta]{} & \mathrm{Ho}\,\mathcal{C} \end{array}$$

exist and $\mathrm{Ho}\,\Delta$ *composes with* $\mathbb{R}J_*$. *Then the right-hand square (filled with the identity) satisfies the (horizontal) Beck-Chevalley condition. In particular, there is an isomorphism*

$$\sigma\colon \operatorname*{hocolim}_{\mathcal{J}} \circ \mathrm{Ho}\,\mathrm{ev}_J \cong \mathrm{Ho}\,\mathrm{ev}_J \circ \operatorname*{hocolim}_{\mathcal{J}}, \qquad \textit{natural in } J \in \mathfrak{J}.$$

PROOF. By Beck-Chevalley interchange, it suffices to check that the right-hand square satisfies the vertical dual Beck-Chevalley condition. For this, we observe that $\mathrm{Ho}\,\mathrm{ev}_J$ composes with $\mathrm{hocolim}_{\mathcal{J}}$ since $\mathrm{Ho}\,\Delta$ composes with $\mathbb{R}J_*$ (cf. 6.9) and that on the strict level, the right-hand square from the proposition satisfies the Beck-Chevalley condition (cf. 8.3). Now use 6.13. Finally, the "in particular"-part is just an instance of 4.4. To wit, for $j\colon J \to J'$ in $\mathcal{J}$, the left-hand square below commutes iff the right-hand square does, where the two horizontal pairs are mates:

$$\begin{array}{ccccccc} \operatorname{hocolim}_{\mathcal{I}} \circ \operatorname{Ho} \operatorname{ev}_J & \xrightarrow{\sigma} & \operatorname{Ho} \operatorname{ev}_J \circ \operatorname{hocolim}_{\mathcal{I}} & & \operatorname{Ho} \operatorname{ev}_J \circ \operatorname{Ho} \Delta & \xrightarrow{\operatorname{id}} & \operatorname{Ho} \Delta \circ \operatorname{Ho} \operatorname{ev}_j \\ \downarrow {\scriptstyle \operatorname{hocolim}_{\mathcal{I}} \operatorname{Ho} \operatorname{ev}_j} & & \downarrow {\scriptstyle (\operatorname{Ho} \operatorname{ev}_j)_{\operatorname{hocolim}_{\mathcal{I}}}} & & \downarrow {\scriptstyle (\operatorname{Ho} \operatorname{ev}_j)_{\operatorname{Ho} \Delta}} & & \downarrow {\scriptstyle (\operatorname{Ho} \Delta) \operatorname{Ho} \operatorname{ev}_j} \\ \operatorname{hocolim}_{\mathcal{I}} \circ \operatorname{Ho} \operatorname{ev}_{J'} & \xrightarrow[\sigma']{} & \operatorname{Ho} \operatorname{ev}_{J'} \circ \operatorname{hocolim}_{\mathcal{I}} & & \operatorname{Ho} \operatorname{ev}_{J'} \circ \operatorname{Ho} \Delta & \xrightarrow[\operatorname{id}]{} & \operatorname{Ho} \Delta \circ \operatorname{Ho} \operatorname{ev}_{J'}. \end{array}$$

□

COROLLARY 9.6. *Let $\mathcal{C}$, $\mathcal{I}$, $\mathcal{J}$ be as in the theorem and assume that $\mathcal{C}^{\mathcal{J}}$ has homotopy $\mathcal{I}$-colimits. If $\mathcal{C}$ has a terminal object and $\mathcal{J}$ is a preorder, there are isomorphisms*

$$\sigma\colon \operatorname*{hocolim}_{\mathcal{I}} \circ \operatorname{Ho} \operatorname{ev}_J \cong \operatorname{Ho} \operatorname{ev}_J \circ \operatorname*{hocolim}_{\mathcal{I}}, \qquad \textit{natural in } J \in \mathcal{J}.$$

In particular, we always have such natural isomorphisms for double homotopy pushouts, coproducts and telescopes.

PROOF. If $\mathcal{J} = \emptyset$ the statement is void, so we assume we can pick some $J \in \mathcal{J}$ to construct homotopy $\mathcal{I}$-colimits in $\mathcal{C}$ as shown in 9.2. By our hypotheses, all functors J_* exist and are homotopical, as observed in 8.4. But then $\operatorname{Ho} \Delta$ and $\mathbb{R}J_* = \operatorname{Ho} J_*$ clearly compose and we can invoke the theorem. □

COROLLARY 9.7. *Let $\mathcal{C}$, $\mathcal{I}$, $\mathcal{J}$ be as in the theorem and assume that both $\mathcal{C}$ and $\mathcal{C}^{\mathcal{J}}$ have homotopy $\mathcal{I}$-colimits. If $\mathcal{C}$ has powers and its weak equivalences are stable under them, there are isomorphisms*

$$\sigma\colon \operatorname*{hocolim}_{\mathcal{I}} \circ \operatorname{Ho} \operatorname{ev}_J \cong \operatorname{Ho} \operatorname{ev}_J \circ \operatorname*{hocolim}_{\mathcal{I}}, \qquad \textit{natural in } J \in \mathcal{J}.$$

PROOF. Again by 8.4, all functors J_* exist and are homotopical, so that $\operatorname{Ho} \Delta$ and $\mathbb{R}J_* = \operatorname{Ho} J_*$ compose. □

EXAMPLE 9.8. The category $\operatorname{Ch}_{\geq 0}(\mathbf{Ab}_{\mathrm{fg}})$ of non-negatively graded chain complexes of *finitely generated* abelian groups inherits a model structure from the usual model category of non-negatively graded chain complexes, where the weak equivalences are the quasi-isomorphisms, the fibrations are the levelwise surjections and the cofibrations the levelwise injections with projective (i.e. free) cokernel. There are no obvious functorial cofibrant replacements within $\operatorname{Ch}_{\geq 0}(\mathbf{Ab}_{\mathrm{fg}})$ and all the usual techniques fail since our base category is only finitely cocomplete, though the author was unable to find a proof in the literature that $\operatorname{Ch}_{\geq 0}(\mathbf{Ab}_{\mathrm{fg}})$ is not cofibrantly generated.

Still, $\operatorname{Ch}_{\geq 0}(\mathbf{Ab}_{\mathrm{fg}})$ is a model category (albeit only a finitely bicomplete one) and we can construct finite homotopy colimits in it, as well as in all diagram categories $\operatorname{Ch}_{\geq 0}(\mathbf{Ab}_{\mathrm{fg}})^{\mathcal{I}}$ with $\mathcal{I}$ finite (see 9.4). Since quasi-isomorphisms are stable under (finite) direct sums, the above corollary applies.

EXAMPLE 9.9. As a more involved example, in [**14**] and [**15**], Isaksen constructed model structures on the category of pro-simplicial sets and more general pro-objects in model categories. Being model categories, we can again construct homotopy colimits in them as well as all diagram categories. However, as shown in *op. cit.*, the model structures in question are not cofibrantly generated and so, there are no projective model structures at our disposal. Still, our theorem applies and we at least always have 9.6 (since the stability of weak equivalences under powers depends on the model category we take pro-objects in).

Finally, let us mention an easy example that is not accessible to any of the previously mentioned techniques (model structures or left model approximations).

EXAMPLE 9.10. Let $\mathbf{\Delta}$ be the usual indexing category for simplicial sets and $\mathbf{\Delta}_+ \subset \mathbf{\Delta}$ the wide subcategory of injections with inclusion $I\colon \mathbf{\Delta}_+ \hookrightarrow \mathbf{\Delta}$. The category of *δ-sets* (or *semi-simplicial sets*) is the presheaf category $\delta\text{-}\mathbf{Sets} := \mathbf{Sets}^{\mathbf{\Delta}_+}$. The left Kan extension $I_!$ of a δ-set along I corresponds to freely adjoining degeneracies and we define a map of δ-sets to be a weak equivalence iff its image under $I_!$ is one. Equivalently, a map of δ-sets is a weak equivalence iff it is one in **Top** after geometric realisation.

It was remarked by Peter May (who himself learned this from his student Matthew Thibault) that there cannot be a model structure with these weak equivalences. Still, we can construct homotopy colimits in δ-**Sets** (or any category of diagrams in it) by passing to **sSets** via $I_!$, forming them there and then passing back to δ-**Sets** via I^*.

In some situations (e.g. when using left model approximations for the construction of homotopy colimits as in [**5**]), it might not be possible to verify the hypotheses of the above theorem. Indeed, it might not even be possible to construct the right adjoint $\mathbb{R}J_*$. However, we can still use the left adjoint $\mathbb{L}J_!$ of $\operatorname{Ho}\operatorname{ev}_J$ rather than $\mathbb{R}J_*$ to force naturality.

PROPOSITION 9.11. *Let $\mathcal{C}$ be a category with weak equivalences, $\mathcal{I}$, $\mathcal{J}$ two index categories and $J \in \mathcal{J}$ such that all derived adjunctions in the square*

$$\begin{array}{ccc}
\operatorname{Ho}\big((\mathcal{C}^{\mathcal{I}})^{\mathcal{J}}\big) \cong \operatorname{Ho}\big((\mathcal{C}^{\mathcal{J}})^{\mathcal{I}}\big) & \underset{\operatorname{Ho}\Delta}{\overset{\operatorname{hocolim}_{\mathcal{I}}}{\rightleftarrows}} \ (\perp) & \operatorname{Ho}(\mathcal{C}^{\mathcal{J}}) \\
\mathbb{L}J_! \uparrow \dashv \downarrow \operatorname{Ho}\operatorname{ev}_J & & \mathbb{L}J_! \uparrow \dashv \downarrow \operatorname{Ho}\operatorname{ev}_J \\
\operatorname{Ho}(\mathcal{C}^{\mathcal{I}}) & \underset{\operatorname{Ho}\Delta}{\overset{\operatorname{hocolim}_{\mathcal{I}}}{\rightleftarrows}} \ (\perp) & \operatorname{Ho}\mathcal{C}
\end{array}$$

exist and whose units we denote by η, θ, θ' and η' (top to bottom, left to right). Then there is

$$\alpha\colon \operatorname*{hocolim}_{\mathcal{I}} \circ \mathbb{L}J_! \cong \mathbb{L}J_! \circ \operatorname*{hocolim}_{\mathcal{I}}$$

compatible with the composite units (and counits for that matter) in the sense that

$$\operatorname{Ho}(\operatorname{ev}_J \circ \Delta)\alpha \circ (\operatorname{Ho}\operatorname{ev}_J)\eta_{\mathbb{L}J_!} \circ \theta = (\operatorname{Ho}\Delta)\theta'_{\operatorname{hocolim}_{\mathcal{I}}} \circ \eta'.$$

PROOF. Obviously $\operatorname{ev}_J \circ \Delta = \Delta \circ \operatorname{ev}_J$, whence the square of right adjoints commutes. □

While the compatibility formula in this proposition might seem useless at first, note that if the $\mathbb{L}J_!$ are fully faithful (or equivalently, if θ and θ' are isomorphisms), we get a good starting point for comparing η and η'.

COROLLARY 9.12. *Under the hypotheses of* 9.11, *if the two $\mathbb{L}J_!$ are fully faithful and there is some isomorphism* $\operatorname{hocolim}_{\mathcal{I}} \circ \operatorname{Ho}\operatorname{ev}_J \cong \operatorname{Ho}\operatorname{ev}_J \circ \operatorname{hocolim}_{\mathcal{I}}$ *(e.g. if* $\operatorname{hocolim}_{\mathcal{I}}$ *and* $\operatorname{Ho}\operatorname{ev}_J$ *compose), then there is even an isomorphism*

$\beta_J\colon \operatorname*{hocolim}_{\mathcal{I}} \circ \operatorname{Ho}\operatorname{ev}_J \cong \operatorname{Ho}\operatorname{ev}_J \circ \operatorname*{hocolim}_{\mathcal{I}}$ *such that* $(\operatorname{Ho}\operatorname{ev}_J)\eta = (\operatorname{Ho}\Delta)\beta_J \circ \eta'_{\operatorname{Ho}\operatorname{ev}_J}$.

PROOF. The compatibility condition of α at its $(\operatorname{Ho}\operatorname{ev}_J)X$-component for a diagram $X \in \operatorname{Ho}((\mathcal{C}^{\mathcal{I}})^{\mathcal{J}})$ reads

$$(\operatorname{Ho}\Delta)\Big(\theta'^{-1}_{\operatorname{hocolim}_{\mathcal{J}}} \circ (\operatorname{Ho}\operatorname{ev}_J)\alpha\Big)_{(\operatorname{Ho}\operatorname{ev}_J)X} \circ (\operatorname{Ho}\operatorname{ev}_J)\eta_{\mathbb{L}J_!(\operatorname{Ho}\operatorname{ev}_J)X} \circ \theta_{(\operatorname{Ho}\operatorname{ev}_J)X} = \eta'_{(\operatorname{Ho}\operatorname{ev}_J)X}$$

and we need to show that $(\operatorname{Ho}\operatorname{ev}_J)\eta_{\mathbb{L}J_!\operatorname{ev}_J X} = (\operatorname{Ho}\operatorname{ev}_J)\eta_X$ up to composition with suitable natural isomorphisms. For this, let us write ζ for the counit of $\mathbb{L}J_! \dashv \operatorname{Ho}\operatorname{ev}_J$ and note that $(\operatorname{Ho}\operatorname{ev}_J)\zeta$ is invertible with inverse $\theta_{\operatorname{Ho}\operatorname{ev}_J}$. The naturality of η gives a commutative square

$$\begin{array}{ccc} (\operatorname{Ho}\operatorname{ev}_J)\mathbb{L}J_!(\operatorname{Ho}\operatorname{ev}_J)X & \xrightarrow{(\operatorname{Ho}\operatorname{ev}_J)\eta_{\mathbb{L}J_!(\operatorname{Ho}\operatorname{ev}_J)X}} & (\operatorname{Ho}\operatorname{ev}_J)(\operatorname{Ho}\Delta)\operatorname{hocolim}_{\mathcal{J}}\mathbb{L}J_!(\operatorname{Ho}\operatorname{ev}_J)X \\ {\scriptstyle (\operatorname{Ho}\operatorname{ev}_J)\zeta_X}\downarrow\cong & & {\scriptstyle (\operatorname{Ho}\operatorname{ev}_J)(\operatorname{Ho}\Delta)\operatorname{hocolim}_{\mathcal{J}}\zeta_X}\downarrow\cong \\ (\operatorname{Ho}\operatorname{ev}_J)X & \xrightarrow[(\operatorname{Ho}\operatorname{ev}_J)\eta_X]{} & (\operatorname{Ho}\operatorname{ev}_J)(\operatorname{Ho}\Delta)\operatorname{hocolim}_{\mathcal{J}} X \end{array},$$

where the right-hand arrow $(\operatorname{Ho}\Delta)(\operatorname{Ho}\operatorname{ev}_J)\operatorname{hocolim}_{\mathcal{J}}\zeta_X$ is invertible because $\operatorname{Ho}\operatorname{ev}_J$ commutes with $\operatorname{hocolim}_{\mathcal{J}}$, while the left-hand arrow cancels with $\theta_{(\operatorname{Ho}\operatorname{ev}_J)X}$ and so $\eta'_{(\operatorname{Ho}\operatorname{ev}_J)X}$ equals

$$(\operatorname{Ho}\Delta)\Big(\theta'^{-1}_{\operatorname{hocolim}_{\mathcal{J}}(\operatorname{Ho}\operatorname{ev}_J)} \circ (\operatorname{Ho}\operatorname{ev}_J)\alpha_{\operatorname{Ho}\operatorname{ev}_J} \circ \big((\operatorname{Ho}\operatorname{ev}_J)\operatorname{hocolim}_{\mathcal{J}}\zeta\big)^{-1}\Big)_X \circ (\operatorname{Ho}\operatorname{ev}_J)\eta_X.$$

□

The last obstacle to overcome before assembling this all is that we need some criterion to decide when the $\mathbb{L}J_!$ are fully faithful. This turns out to be rather simple and it suffices if $J_!$ is so (e.g. as in 8.1).

PROPOSITION 9.13. *Let $\mathcal{C}$ be a category with weak equivalences, $\mathcal{J}$ some index category and $J \in \mathcal{J}$ such that a fully faithful left adjoint $J_!$ to ev_J exists and $J_! \dashv \operatorname{ev}_J$ is derivable. Then $\mathbb{L}J_!$ is fully faithful.*

PROOF. Choosing compatible units and counits of the derived adjunction, 6.1 gives us

$$\rho_F \circ H_{\mathcal{C}}\eta = (\operatorname{Ho}\operatorname{ev}_J)\lambda \circ \dot{\eta}_{H_{\mathcal{C}}},$$

which is an isomorphism because $\mathbb{R}\operatorname{ev}_J = \operatorname{Ho}\operatorname{ev}_J$ (i.e. ρ is invertible) and $J_!$ is fully faithful (i.e. η is invertible). Now note that $\mathbb{L}J_!$ and $\operatorname{Ho}\operatorname{ev}_J$ compose, whence $\operatorname{Ho}\operatorname{ev}_J \circ \mathbb{L}J_! \cong \mathbb{L}(\operatorname{ev}_J \circ J_!)$. But $\operatorname{ev}_J \circ J_! \cong \operatorname{id}_{\mathcal{C}}$, so that $\mathbb{L}(\operatorname{ev}_J \circ J_!) \cong \operatorname{id}_{\operatorname{Ho}\mathcal{C}}$ and λ must be invertible. It follows that $\dot{\eta}_{H_{\mathcal{C}}}$ (or equivalently $\dot{\eta}$) is invertible, thus proving our claim. □

REMARK 9.14. Obviously, this proposition generalises mutatis mutandis to an arbitrary derivable adjunction with a fully faithful left and a homotopical right adjoint.

EXAMPLE 9.15. The proposition in particular applies in both of the following cases:

(1) $\mathcal{C}$ has an initial object and $\mathcal{J}$ is a preorder;
(2) $\mathcal{C}$ is a model category such that the projective model structure on $\mathcal{C}^{\mathcal{J}}$ exists and $\operatorname{End}_{\mathcal{J}}(J) = \{\operatorname{id}_J\}$ (e.g. $\mathcal{C}$ any model category and $\mathcal{J}$ direct).

THEOREM 9.16. *Let $\mathcal{C}$ be a category with weak equivalences and $\mathcal{I}$, $\mathcal{J}$ two index categories such that for all $J \in \mathcal{J}$, the two evaluation functors ev_J indicated below have fully faithful left adjoints $J_!$ and the derived adjunctions in the square*

$$\begin{array}{ccc}
\mathrm{Ho}\big((\mathcal{C}^{\mathcal{I}})^{\mathcal{J}}\big) \cong \mathrm{Ho}\big((\mathcal{C}^{\mathcal{J}})^{\mathcal{I}}\big) & \underset{\mathrm{Ho}\,\Delta}{\overset{\mathrm{hocolim}_{\mathcal{I}}}{\rightleftarrows}} \ (\perp) & \mathrm{Ho}(\mathcal{C}^{\mathcal{J}}) \\
\mathbb{L}J_! \uparrow \dashv \downarrow \mathrm{Ho}\,\mathrm{ev}_J & & \mathbb{L}J_! \uparrow \dashv \downarrow \mathrm{Ho}\,\mathrm{ev}_J \\
\mathrm{Ho}(\mathcal{C}^{\mathcal{I}}) & \underset{\mathrm{Ho}\,\Delta}{\overset{\mathrm{hocolim}_{\mathcal{I}}}{\rightleftarrows}} \ (\perp) & \mathrm{Ho}\,\mathcal{C}
\end{array}$$

exist and there is an isomorphism $\mathrm{hocolim}_{\mathcal{I}} \circ \mathrm{Ho}\,\mathrm{ev}_J \cong \mathrm{Ho}\,\mathrm{ev}_J \circ \mathrm{hocolim}_{\mathcal{I}}$. *Then there is a family of isomorphisms*

$$\beta_J \colon \underset{\mathcal{I}}{\mathrm{hocolim}} \circ \mathrm{Ho}\,\mathrm{ev}_J \cong \mathrm{Ho}\,\mathrm{ev}_J \circ \underset{\mathcal{I}}{\mathrm{hocolim}} \qquad \textit{natural in } J.$$

PROOF. For $J \in \mathcal{J}$ let us choose β_J as in 9.12 and note that for $j \colon J \to J'$ in $\mathcal{J}$ the arrow $\mathrm{hocolim}_{\mathcal{I}} X(-,j)$ is the unique $f \colon \mathrm{hocolim}_{\mathcal{I}} X(-,J) \to \mathrm{hocolim}_{\mathcal{I}} X(-,J')$ in $\mathrm{Ho}\,\mathcal{C}$ making

$$\begin{array}{ccc}
X(-,J) & \xrightarrow{\eta'_{X(-,J)}} & (\mathrm{Ho}\,\Delta)\,\mathrm{hocolim}_{\mathcal{I}} X(-,J) \\
{\scriptstyle X(-,j)}\downarrow & & \vdots\, {\scriptstyle (\mathrm{Ho}\,\Delta) f} \\
X(-,J') & \xrightarrow[\eta'_{X(-,J')}]{} & (\mathrm{Ho}\,\Delta)\,\mathrm{hocolim}_{\mathcal{I}} X(-,J')
\end{array}$$

in $\mathrm{Ho}(\mathcal{C}^{\mathcal{I}})$ commute. Extending this square to the right using $\beta_{J,X}$ and $\beta_{J',X}$, the naturality of $\mathrm{Ho}\,\mathrm{ev}_j$ yields a commutative rectangle of solid arrows in $\mathrm{Ho}(\mathcal{C}^{\mathcal{I}})$

$$\begin{array}{ccccc}
X(-,J) & \xrightarrow{\eta'_{X(-,J)}} & (\mathrm{Ho}\,\Delta)\,\mathrm{hocolim}_{\mathcal{I}} X(-,J) & \xrightarrow[\cong]{(\mathrm{Ho}\,\Delta)\beta_{J,X}} & (\mathrm{Ho}\,\Delta)(\mathrm{hocolim}_{\mathcal{I}} X)J \\
{\scriptstyle X(-,j)}\downarrow & & \vdots & & \downarrow {\scriptstyle (\mathrm{Ho}\,\Delta)(\mathrm{hocolim}_{\mathcal{I}} X)j} \\
X(-,J') & \xrightarrow[\eta'_{X(-,J')}]{} & (\mathrm{Ho}\,\Delta)\,\mathrm{hocolim}_{\mathcal{I}} X(-,J) & \xrightarrow[(\mathrm{Ho}\,\Delta)\beta_{J',X}]{\cong} & (\mathrm{Ho}\,\Delta)\,\mathrm{hocolim}_{\mathcal{I}} X(-,J')
\end{array}$$

with the top and bottom composites being $(\mathrm{Ho}\,\mathrm{ev}_J)\eta_X$ and $(\mathrm{Ho}\,\mathrm{ev}_{J'})\eta_X$ respectively. Upon defining $f := \beta_{J',X}^{-1} \circ (\mathrm{hocolim}_{\mathcal{I}} X)j \circ \beta_{J,X}$ the dotted arrow $(\mathrm{Ho}\,\Delta)f$ renders the entire diagram commutative and the claim follows. □

References

[1] Jiří Adámek, Horst Herrlich, and George E. Strecker, *Abstract and concrete categories*, Pure and Applied Mathematics (New York), John Wiley & Sons, Inc., New York, 1990. The joy of cats; A Wiley-Interscience Publication. MR1051419

[2] C. Barwick and D. M. Kan, *Relative categories: another model for the homotopy theory of homotopy theories*, Indag. Math. (N.S.) **23** (2012), no. 1-2, 42–68. MR2877401

[3] J. M. Boardman and R. M. Vogt, *Homotopy invariant algebraic structures on topological spaces*, Lecture Notes in Mathematics, Vol. 347, Springer-Verlag, Berlin-New York, 1973. MR0420609

[4] Francis Borceux, *Handbook of categorical algebra. 1*, Encyclopedia of Mathematics and its Applications, vol. 50, Cambridge University Press, Cambridge, 1994. Basic category theory. MR1291599

[5] Wojciech Chachólski and Jérôme Scherer, *Homotopy theory of diagrams*, Mem. Amer. Math. Soc. **155** (2002), no. 736, x+90. MR1879153

[6] Denis-Charles Cisinski, *Images directes cohomologiques dans les catégories de modèles* (French, with French summary), Ann. Math. Blaise Pascal **10** (2003), no. 2, 195–244. MR2031269

[7] Denis-Charles Cisinski, *Les préfaisceaux comme modèles des types d'homotopie*, Astérisque (2006), no. 308.

[8] W. G. Dwyer and J. Spaliński, *Homotopy theories and model categories*, Handbook of algebraic topology, North-Holland, Amsterdam, 1995, pp. 73–126. MR1361887

[9] William G. Dwyer, Philip S. Hirschhorn, Daniel M. Kan, and Jeffrey H. Smith, *Homotopy limit functors on model categories and homotopical categories*, Mathematical Surveys and Monographs, vol. 113, American Mathematical Society, Providence, RI, 2004. MR2102294

[10] Beatriz Rodríguez González, *A derivability criterion based on the existence of adjunctions*, 2 2012.

[11] Moritz Groth, *Derivators, pointed derivators and stable derivators*, Algebr. Geom. Topol. **13** (2013), no. 1, 313–374. MR3031644

[12] Alex Heller, *Homotopy theories*, Mem. Amer. Math. Soc. **71** (1988), no. 383, vi+78. MR920963

[13] Mark Hovey, *Model categories*, Mathematical Surveys and Monographs, vol. 63, American Mathematical Society, Providence, RI, 1999. MR1650134

[14] Daniel C. Isaksen, *A model structure on the category of pro-simplicial sets*, Trans. Amer. Math. Soc. **353** (2001), no. 7, 2805–2841. MR1828474

[15] Daniel C. Isaksen, *Strict model structures for pro-categories*, Categorical decomposition techniques in algebraic topology (Isle of Skye, 2001), Progr. Math., vol. 215, Birkhäuser, Basel, 2004, pp. 179–198. MR2039766

[16] And Joyal, *Notes on quasi-categories*, 6 2008.

[17] Jacob Lurie, *Higher topos theory*, Annals of Mathematics Studies, vol. 170, Princeton University Press, Princeton, NJ, 2009. MR2522659

[18] Saunders Mac Lane, *Categories for the working mathematician*, 2nd ed., Graduate Texts in Mathematics, vol. 5, Springer-Verlag, New York, 1998. MR1712872

[19] Georges Maltsiniotis, *Le théorème de Quillen, d'adjonction des foncteurs dérivés, revisité* (French, with English and French summaries), C. R. Math. Acad. Sci. Paris **344** (2007), no. 9, 549–552. MR2323740

[20] Emily Riehl and Dominic Verity, *Kan extensions and the calculus of modules for ∞-categories*, Algebr. Geom. Topol. **17** (2017), no. 1, 189–271. MR3604377

Mathematisch Instituut, Universiteit Utrecht, Utrecht, Netherlands
Email address: k.r.werndli@uu.nl